Select Problem Collection

to accompany

FUNDAMENTALS OF PHYSICS

Select Problem Collection

to accompany

FUNDAMENTALS OF PHYSICS, 5/E

by HALLIDAY, RESNICK, AND WALKER

by Jearl Walker

Cleveland State University

with contributions from

Harry Dulaney

Georgia Institute of Technology

Laurent Hodges

Iowa State University

Fred F. Tomblin

New Jersey Institute of Technology

JOHN WILEY & SONS, INC.

New York • Chichester • Weinheim • Brisbane • Singapore • Toronto

ACQUISITIONS EDITOR Stuart Johnson
MARKETING MANAGER Catherine Beckham
SENIOR PRODUCTION EDITOR Jeanie Berke
DESIGNER Dawn L. Stanley
ILLUSTRATION EDITOR Edward Starr

This book was set in 10/12 New Baskerville by Progressive Information Technologies and printed and bound by Port City Press. The cover was printed by Phoenix Color Corp.

This book is printed on acid-free paper. ∞

The paper in this book was manufactured by a mill whose forest management programs include sustained yield harvesting of its timberlands. Sustained yield harvesting principles ensure that the numbers of trees cut each year does not exceed the amount of new growth.

ISBN 0-471-19777-7 (pbk)

Printed in the United States of America

10 9 8 7 6 5 4 3 2

Preface

This collection, a supplement to Halliday, Resnick, and Walker, *Fundamentals of Physics,* fifth edition, provides 565 new Exercises & Problems, 408 new checkpoint-type Questions, and 26 new problems devoted to graphing calculators. It also includes three new features: **Clustered Problems, Tutorial Problems,** and detailed instructions for **Graphing Calculators.**

Clustered Problems are problems grouped around a common theme or technique of solution to advance the reader from easier situations to more difficult situations. There are 145 Clustered Problems here.

Tutorial Problems provide detailed answers just after the problem statements to show the reader how physics problems are thought through and solved. There are 38 Tutorial Problems here.

The Graphing Calculator sections give 26 new problems for that type of calculator to supplement the Electronic Computation problems in the textbook. The new problems range widely in subject matter, from fractal dimensions in Chapter 1, to fly-fishing in Chapter 8, to Poe's pendulum in Chapter 16, to huddling emperor penguins in Chapter 19. As a new feature, the Graphing Calculator sections also explain how to use the TI-85 and TI-86 calculators, giving step-by-step instructions on how to solve not only problems written for graphing calculators but also problems written in the traditional fashion. Here there are 41 Sample Problems and Problem Solving Tactics devoted to these calculators.

The Questions continue those of the textbook, requiring reasoning, decisions, and mental calculations of the reader rather than use of a calculator. They are written with the implicit question, "If you really understand the physics, can you do this?" Some have the quality of a game in which the solution is plain to view but at first hard to see.

In addition, the Questions sections include a new feature: **Organizing Questions.** These special questions focus on either a general organization of the physics in a chapter or a type of problem common to the chapter. Many of them help bridge the gap between the conceptional Questions and the data-filled Exercises & Problems.

The Clustered Problems, written by Fred F. Tomblin of the New Jersey Institute of Technology, provide a treasure chest of problems that can be assigned either individually or in the designed clusters. The Tutorial Problems, written by Laurent Hodges of Iowa State University, offer a unique opportunity for students to follow the steps taken by a skilled physicist in unraveling a problem. The traditional Exercises & Problems were written by Harry Dulaney of the Georgia Institute of Technology and Walker. Many of Dulaney's problems actually made Walker smile with curiosity as he worked them, and occasionally, as a surprising answer appeared, Walker even laughed out loud (a rarity for him).

Walker wrote the Questions and the Graphing Calculator sections. He also edited all the material in this collection; so any error in it is his. Please send any correction or any other comment about this collection or the HRW textbook to Jearl Walker, Physics Department, Cleveland State University, Cleveland OH 44115 USA, or fax him at 216-687-2424.

ACKNOWLEDGMENTS

Walker extends a belated but extremely grateful thanks to Frank G. Jacobs of Evanston, Illinois, for his thorough checking of the Questions and the Exercises & Problems in the fifth edition of the textbook.

Walker is also very grateful to Glen Terrell of the University of Texas at Arlington for introducing him to Texas Instruments graphing calculators. He also thanks Will Roberts, an engineering student at Cleveland State University, for introducing him to the technique of storing notes on a TI calculator by using graphing text commands.

Contents

CHAPTER 21

CHAPTER 22

CHAPTER 23

CHAPTER 24

CHAPTER 25

CHAPTER 26

CHAPTER 27

CHAPTER 28

CHAPTER 29

CHAPTER 30

Graphing Calculator Guidance

Chapter 1

Chapter 2

Chapter 3

Chapter 4

Chapter 5

Chapter 6

Chapter 7

Select Problem Collection

to accompany

FUNDAMENTALS OF PHYSICS

Chapter One
Measurement

EXERCISES & PROBLEMS

40. During the summers at high latitudes, ghostly, silver-blue clouds occasionally appear after sunset when common clouds are in Earth's shadow and are no longer visible. The ghostly clouds have been called *noctilucent clouds* (NLC), which means "luminous night clouds," but now are often called *mesospheric clouds,* after the *mesosphere,* the name of the atmosphere at the altitude of the clouds.

These clouds were first seen in June 1885, after dust and water from the massive 1883 volcanic explosion of Krakatoa Island (near Java in the Southeast Pacific) reached the high altitudes in the Northern Hemisphere. In the low temperatures of the mesosphere, the water collected and froze on the volcanic dust (and perhaps on comet and meteor dust already present there) to form the particles that made up the first clouds. Since then, mesospheric clouds have generally increased in occurrence and brightness, probably because of the increased production of methane by industries, rice paddies, landfills, and livestock flatulence. The methane works its way into the upper atmosphere, undergoes changes, and results in an increase of water molecules there, and thus also in bits of ice for the mesospheric clouds.

If mesospheric clouds are spotted 38 min after sunset and then quickly dim, what is their altitude? (*Hint:* See Sample Problem 1-4.)

41. An old English cookbook carries a recipe for cream of nettle soup: Boil stock of the following amount: 1 breakfastcup plus 1 teacup plus 6 tablespoons plus 1 dessertspoon. Using gloves, separate the nettle tops until you have 0.5 quart; add the tops to the boiling stock. Add 1 tablespoon of cooked rice and 1 saltspoon of salt. Simmer for 15 min. The following table gives some of the conversions among old (premetric) British measures and common (still premetric) U.S. measures. (These measures scream for metrication.) In converting liquid measures between the two systems, 1 British teaspoon = 1 U.S. teaspoon. For dry measures, 1 British teaspoon = 2 U.S. teaspoons and 1 British quart = 1 U.S. quart. Translate the old recipe into U.S. measures.

OLD BRITISH MEASURES	U.S. MEASURES
saltspoon	
teaspoon = 2 saltspoons	teaspoon
dessertspoon = 2 teaspoons	tablespoon = 3 teaspoons
tablespoon = 2 dessertspoons	half cup = 8 tablespoons
teacup = 8 tablespoons	cup = 2 half cups
breakfastcup = 2 teacups	

42. The cubit is an ancient unit of length based on the distance between the elbow and the tip of the middle finger of the measurer, usually 43 to 53 cm. If ancient drawings indicate that a cylindrical pillar in a tomb was to have the length 9 cubits, what would have been the length in (a) meters and (b) millimeters? (c) If the diameter of the pillar is 2 cubits, what would have been the volume of the pillar in cubic meters?

43. If a ball of string were as big as Earth, how many meters of string (to the nearest order of magnitude) would it contain?

44. To the nearest order of magnitude, how many standard toilet paper sheets would be needed to "TP" the shortest path between Cleveland and Los Angeles?

45. To the nearest order of magnitude, how many times per year do you take a breath?

Graphing Calculators

SAMPLE PROBLEM 1-5*

Linear Regression and Fractal Dimension. A flat sheet of paper is two-dimensional (that is, its dimension is $d = 2.0$), and a solid cube of paper is three-dimensional ($d = 3.0$). However, if we wad up the sheet into a ball, its two-dimensional surface is then embedded in three dimensions, and the sheet has a *fractal* dimension d that can lie between 2.0 and 3.0. A value that is close to 2.0 implies that the sheet tends to avoid itself; one that is close to 3.0 implies the opposite.

The mass m of the paper and the diameter D of the wadded-up ball are related by dimension d according to

$$m = kD^d,$$

where k is an unknown constant. If we measure m for a range

*Adapted from "Fractal Geometry in Crumpled Paper Balls," by M. A. F. Gomes, *American Journal of Physics,* 1987, Vol. 55, pp. 649–650, and "A Simple Experiment That Demonstrates Fractal Behavior," by R. H. Ko and C. P. Bean, *The Physics Teacher,* Feb. 1991, Vol. 29, pp. 78–79.

of D, using the same type of paper but wadding up different-sized balls, we can determine d from a plot of the data.

To collect the data, we begin with a large sheet of paper, wad the sheet into a ball, measure its mass m on a balance, and then determine an average diameter D of the ball by averaging any two widths. After smoothing out the sheet, we tear it in half and repeat the process with each half. Then we tear one of those halves in half and repeat the process with each smaller half. We continue this process until we reach the limit of our ability to measure the mass or the diameter.

In such an experiment, using relatively stiff paper (with an initial area of about 0.80 m^2), the masses m were 112, 56.6, 55.5, 25.9, 30.0, 15.2, 14.8, 7.57, 7.71, 3.85, 3.89, 2.05, and 1.85 grams. The corresponding diameters D were 27.5, 20.0, 19.0, 14.5, 15.5, 10.0, 9.0, 7.8, 6.5, 6.0, 4.8, 4.9, and 4.8 cm.

(a) *Creating Data Lists.* What is the fractal dimension d of the wadded-up paper?

SOLUTION: If we take the natural logarithm of the given equation, we have

$$\ln m = \ln kD^d = \ln k + \ln D^d = \ln k + d \ln D.$$

The result is in the form of the generic linear equation $y = a + bx$. Here, the variable y is $\ln m$, the y intercept a is $\ln k$, the slope b is d, and the variable x is $\ln D$. So, we can find the fractal dimension d if we do a **linear regression** of the ln m values versus the ln D values and then take the slope. That is, we have the calculator find the straight line that best fits through a plot of the data and then tell us the slope of that best-fit line.

Here is our procedure on a TI-85/86: Press ON. If the display shows anything but the cursor, press CLEAR and EXIT until the screen is empty. To set the style in which calculated numbers will appear on the screen, press 2nd MODE (that is, press the key marked 2nd, release it, and then press the key with MODE marked above it). If Normal on the first line lacks a highlighting background, press ENTER to choose this style. Then, using the cursor keys, move the cursor to 2 on the second line and again press ENTER. This step sets the number of decimal places of a calculated number to 2. If Dec (for decimal) on the sixth line lacks a highlighting background, move to it and then press ENTER. Finally, press EXIT to leave the menu and return to the *home screen.*

We need to put the m and D values into lists, so press 2nd LIST (the key labeled 2nd and then the key with LIST written above it). The list menu appears at the screen's bottom. To begin a list, press the F1 key to choose the symbol { from the menu. Then begin pressing in the values for m in order, separated by a comma (use the comma key).

If you make a mistake, use the cursor keys to back up to the error. If you want to delete what is there, press the DEL key one or more times. If you want to overwrite what is there, just begin pressing in what you want. If you want to insert a correction without overwriting, first press 2nd and then the key with INS printed above; you can then insert as much as you want. Insertion stops when you next press one of the cursor keys.

When the last value is pressed in, press the F2 key for the } symbol to end the list. So that we can use this list in a calculation, store it under the name M by pressing the STO→ key and then the key with M printed above it. (As soon as the STO→ is pressed, the calculator is in *alphabetic mode*—note the symbol A on the cursor; this means that if we press any keys with letters printed above them, the letters appear on the screen.) After M is pressed in, press ENTER to complete the storage. The calculator then displays the list as

{112.00 56.60 55.50 . . .

where . . . indicates that the list extends rightward off the screen. To see the rest of the list, press the rightward cursor key to scroll the list across the screen.

Next, repeat the procedure to create a list of the D values, storing the list under the name D.

We want to do a linear regression of ln m values (not m values) versus ln D values (not D values). To make the switch we press the LN key (for natural logarithm), then ALPHA (to change to alphabetic mode), then M (that is, the key with M printed above it), then STO→, then L, then M, and then ENTER. This process takes the natural logarithm of each value in the M list and then stores the results, in order, in a list called LM. The calculator shows the first portion of this new list as

{4.72 4.04 4.02 3.25 . . .

where again the list extends off screen. Now repeat the process to store ln D values in a list called LD. The calculator shows

{3.31 3.00 2.94 2.67 . . .

as the list.

The procedure for performing a linear regression on a TI-85 differs from that on a TI-86. On the former, press STAT (for statistical) and then F1 (for the CALC option, that is, for calculations). The screen asks for the names of the lists to be used. For the x list name, press in the letters L and D. (The calculator is automatically in alphabetic mode; if you accidentally press ALPHA to turn off that mode, press that key once again to turn on the mode.) Then press ENTER. For the y list name, press the letters L and M, and then press ENTER. Then press F2 for the LINR (linear regression) option of the menu at the bottom of the screen. The calculator then finds the linear regression of LM versus LD.

On a TI-86, press 2nd STAT and then F1 for the CALC option. Then press F3 for the LinR option. Next, using the ALPHA key, finish the line that the calculator begins so that it reads

LinR LD,LM

which requests the linear regression of LM versus LD. Then press ENTER.

On both the TI-85 and the TI-86, the results are that the y-intercept a is -2.37 and the slope b is 2.15 (for the $n = 13$ pairs of data). Recall that the slope gives the fractal dimension d of the wadded-up paper. Thus

$$d = 2.15, \qquad \text{(Answer)}$$

which is only slightly more than the initial dimension of 2.0. The result implies that when the stiff paper is wadded up, it tends to avoid itself rather than form a solid ball.

(To later delete a list from memory, press 2nd MEM [for memory], then F2 for the DELETE option in the menu, and then F4 for the LIST option. Move the cursor to the list to be deleted and then press ENTER. Press EXIT to return to the home screen.)

(b) *Statistical Plotting.* Plot ln m versus ln D and then include the result of the linear regression.

SOLUTION: After finding the linear regression, check the graphing mode of the calculator: Press 2nd MODE to access the mode menu. If Func (for functional graphing) on the fifth line lacks a highlighting background, move the cursor to it and then press ENTER. Then exit the menu with EXIT.

To set the range of the axes on the graph of ln m versus ln D, we need to know the maximum values of those two variables. Using the ALPHA key, press in LD and then press ENTER. Scroll the list to find the maximum value; it is less than 3.5. Next, press in LM and then press ENTER. The maximum value in this list is less than 5.0. So, on our graph, the horizontal axis should extend from 0 to, say, 3.5, and the vertical axis should extend from 0 to, say, 5.0.

To enter this data, press GRAPH and then F2 to gain what is called the RANGE menu on a TI-85 and the WIND (for window) menu on a TI-86. Moving the cursor line by line, press in 0 for xMin, 3.5 for xMax, 1 xScl (the spacing between tick marks on the horizontal axis), 0 for yMin, 5 for yMax, and 1 for yScl (the spacing between tick marks on the vertical axis).

Press F1 for the y= option on the screen menu. If an equation appears with a highlighted equals mark, deactivate the equation (or it too will be plotted) by moving to the equation line and pressing F5 for SELECT. The dark background disappears. (Instead, you could erase the equation by moving to it and pressing CLEAR.)

Next press STAT on a TI-85 or 2nd STAT on a TI-86. Then choose the DRAW option from the menu and then press F2 for the SCAT option (scatter plot of the data). A graph of the data appears, with ln m plotted vertically and ln D plotted horizontally. To see the whole graph, remove the menu by pressing CLEAR. To make the menu reappear, press STAT on a TI-85 or 2nd STAT on a TI-86. To include the result of the linear regression on the graph, we need the DRREG (draw regression) option in a screen menu. On a TI-85, that option is F4 under the DRAW menu. On a TI-86, it is also under the DRAW menu, but you must press MORE to get it on screen. Once the straight line is added to the graph, press CLEAR to see the whole graph.

(c) *Editing a List.* Change the second diameter in the D list from 20.0 cm to 21.0 cm.

SOLUTION: If the home screen is not present, press EXIT one or more times until it is. Then press 2nd LIST and choose F4 for the EDIT option.

On a TI-85: Press in D as the name of the list to be changed and then press ENTER. The data of the D list are then displayed. Move the cursor to the second data line (which shows e2 = 20 as the second *element* of the list), overwrite the data with 21, and then press ENTER. Then press EXIT to return to the home screen.

On a TI-86: The data of a list are displayed in columns. If you have just gone through the graphing of part (b), column 1 has the data of list LD (called xStat here) and column 2 has the data of list LM (called yStat here). With the cursor at the top of the screen, move rightward to the first empty column. Press in D as the name of the list you want displayed in that column and then press ENTER. Move down to the second element in the column and then press in 21 as the new value. Press ENTER to make the change in the list. If you then want to remove the list from the editor so that it is not there when you next return to the editor, move to the name of the list at the top of the column and press DEL. (The list still exists in memory.) To return to the home screen, press EXIT.

46. Repeat Sample Problem 1-5 to find dimension d with the second diameter value of 21.0 cm instead of 20.0 cm. Does that change in the data imply an increase or a decrease in (a) the tendency of the paper to avoid itself and (b) the value of d? (c) What is the new value of d?

47. Repeat the procedure of Sample Problem 1-5 but substitute plastic food wrap and then a large, flat square of either Silly Putty or American cheese. What dimensions do the wadded-up balls have? Interpret the results in terms of the tendency of the material to avoid itself when wadded up.

Chapter Two
Motion Along a Straight Line

QUESTIONS

14. Figure 2-32 gives the acceleration $a(t)$ of a Chihuahua as it chases a German shepherd along an axis. In which of the time periods indicated, if any, does the Chihuahua move at constant speed?

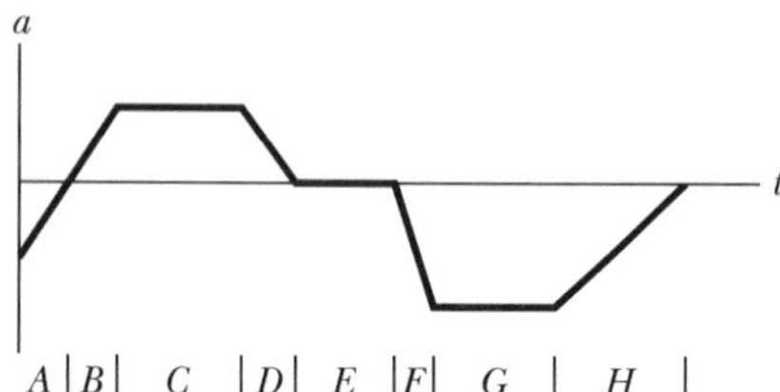

Figure 2-32 Question 14.

15. Figure 2-33 shows four paths along which objects move from a starting point to a final point, all in the same time. The paths pass over a grid of equally spaced straight lines. Rank the paths according to (a) the average velocity of the objects and (b) the average speed of the objects, greatest first.

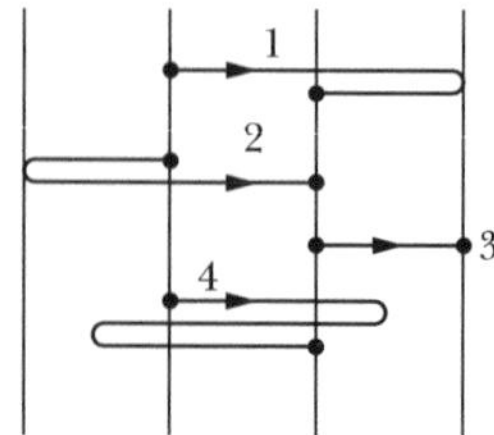

Figure 2-33 Question 15.

16. In Fig. 2-34, a cream tangerine is thrown directly upward past three evenly spaced windows of equal heights. Rank the windows according to (a) the average speed of the cream tangerine while passing them, (b) the time the cream tangerine takes to pass them, (c) the magnitude of the acceleration of the cream tangerine while passing them, and (d) the change Δv in the speed of the cream tangerine during the passage, greatest first.

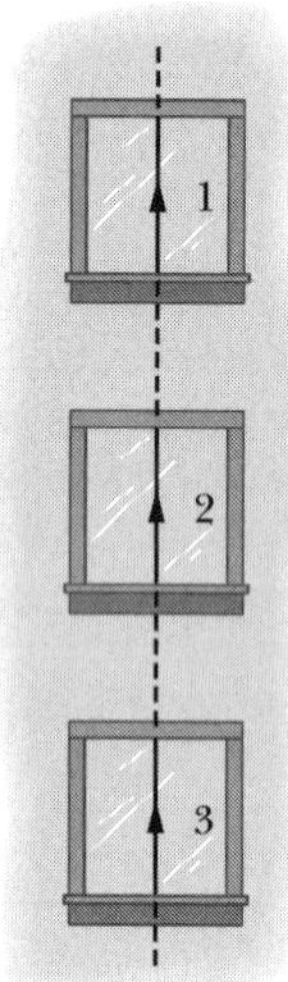

Figure 2-34 Question 16.

17. In Fig. 2-35, assume that the vertical axis pertains to the velocity v of an object moving along an x axis. Then determine which of the 10 plots of v versus time t best describes the motion for the following four situations. (Plots 2, 3, 7, and 9 are straight; the others are curved.)

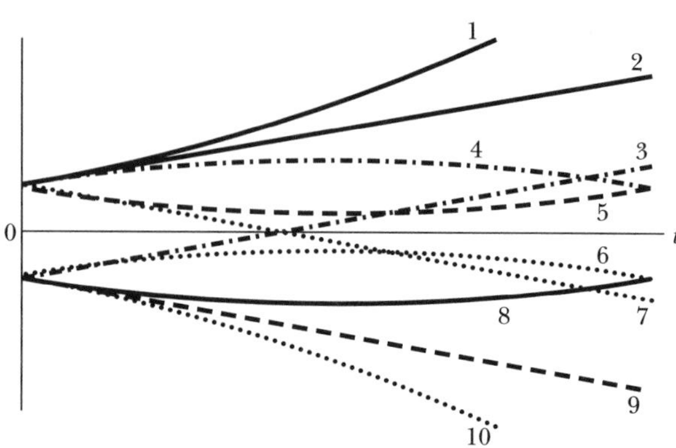

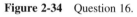

Figure 2-35 Questions 17 and 18.

SITUATION	a	b	c	d
Initial x (m)	+10	−10	+10	−10
Initial v (m/s)	+5	−5	−5	+5
Constant a (m/s²)	+2	−2	+2	−2

18. Question 17 continued: If, instead, the vertical axis pertains to the position x of the object, then which of the 10 plots best describes the motion for the given four sets of values?

19. Figure 2-36a shows a radio-controlled model pickup truck moving along an x axis. For four situations, here are data about the truck's motion, for which the acceleration has a constant value. Match the situations with the $x(t)$ curves given in Fig. 2-36b and tell whether the acceleration value is positive, negative, or zero. (Curves 1 and 4 are straight.)

SITUATION	a	b	c	d
Initial position (m)	3	−3	3	−3
Initial velocity (m/s)	4	−5	−4	5

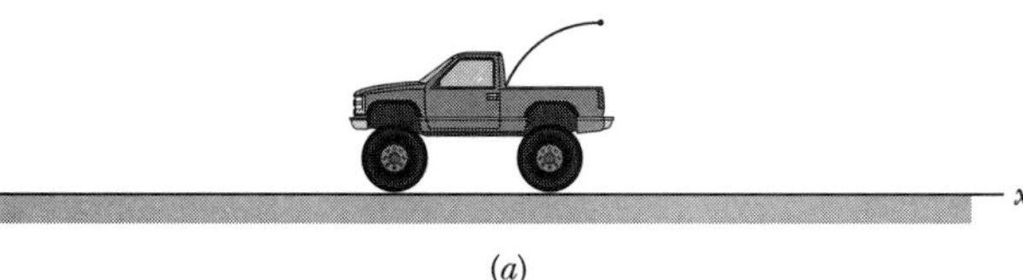

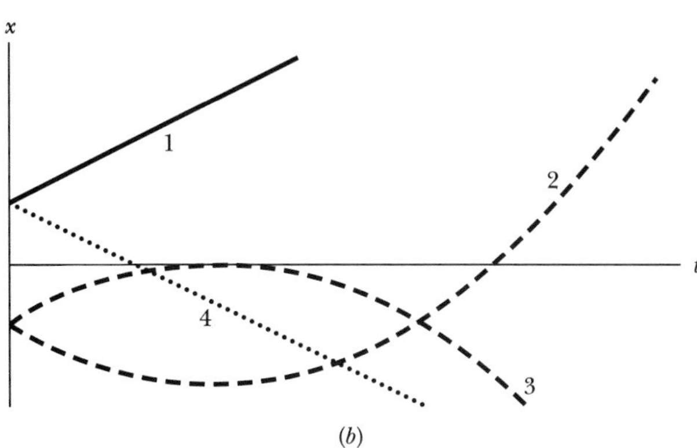

Figure 2-36 Question 19.

20. Suppose that the position of a particle is given by the expression $x = (1.0)t^n$, where n is an integer, x is in meters, and t is in seconds. That expression is graphed in Fig. 2-37a for $n =$ 1, 2, 3, and 4. (a) Which graphed curve corresponds to which value of n and what are the corresponding units of the coefficient 1.0 in the expression for $x(t)$? Rank the values of n according to the particle's displacement during (b) the first 0.5 s of motion ($t = 0$ to $t = 0.5$ s) and (c) the first 1.5 s of motion, greatest first.

(d) The acceleration $a(t)$ of the particle is graphed in Fig. 2-37b for the four values of n. (Curve H lies along the horizontal axis.) Which $a(t)$ curve corresponds to which $x(t)$ curve? (e) Rank the $a(t)$ curves according to the rate at which the velocity is changing at time $t = 0.20$ s, greatest first.

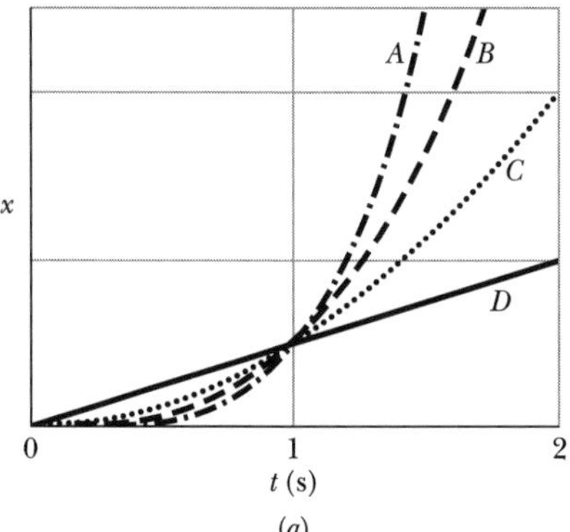

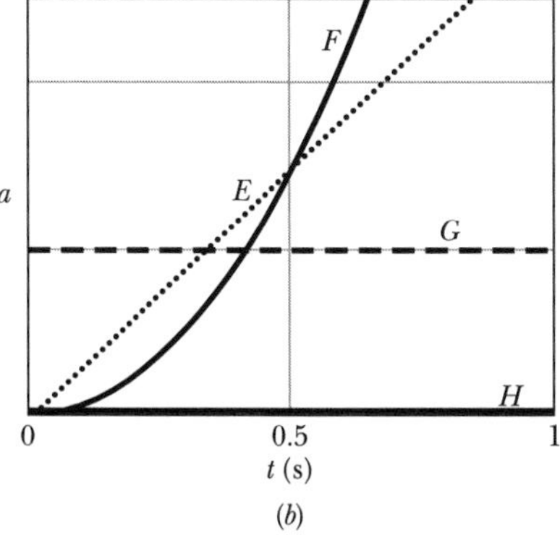

Figure 2-37 Question 20.

21. *Organizing question:* At the National Enchilada Proving Grounds, enchiladas are launched either directly upward or directly downward from a high overhang with an initial speed of 10 m/s. Set up equations, complete with known data, to find the time t an upward-launched enchilada takes to reach a point (a) 2 m above the launch point and (b) 2 m below the launch point. (c) Rewrite the equation for (b) for a downward launch. Next, set up equations, complete with known data, to find the enchilada's displacement $y - y_0$ from the launch point when it is traveling at a speed of 12 m/s after being launched (d) directly upward and (e) directly downward.

22. *Math Tool Time.* What are the results of differentiating with respect to time t the position functions (a) $x = 3t^2 + 4t + 5$ and (b) $x = 3t^{-2}$? What are the results of the integrals (c) $\int 3t^2\,dt$ and (d) $\int (2t + 5)\,dt$?

EXERCISES & PROBLEMS

95. An electric vehicle starts from rest and accelerates at a rate of 2.0 m/s^2 in a straight line until it reaches a speed of 20 m/s. The vehicle then slows at a constant rate of 1.0 m/s^2 until it stops. (a) How much time elapses from start to stop? (b) How far does the vehicle travel from start to stop?

96. The position of a particle as it moves along the x axis is given by $x = 15e^{-t}$ m, where t is in seconds. (a) What is the position of the particle at $t = 0$, 0.50, and 1.0 s? (b) What is the average velocity of the particle between $t = 0$ and $t = 1.0$ s? (c) What is the instantaneous velocity of the particle at $t =$ 0, 0.50, and 1.0 s? (d) Plot x versus t for the particle for $0 \leq t \leq 1.0$ s, and estimate the instantaneous velocity at $t = 0.50$ s from the graph.

97. A ball is thrown vertically downward from the top of a 36.6 m tall building. The ball passes the top of a window that is 12.2 m above the ground 2.00 s after being thrown. What is the speed of the ball as it passes the top of the window?

98. Figure 2-38 depicts the motion of a particle moving along an x axis with a constant acceleration. What is the value of the acceleration?

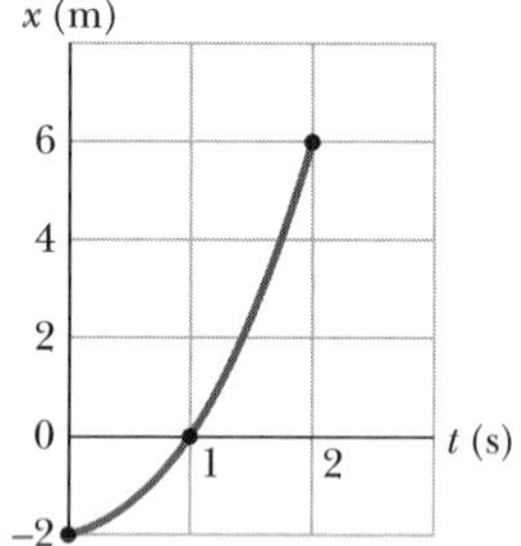

Figure 2-38 Problem 98.

99. A graph of x versus t for a particle in straight-line motion is shown in Fig. 2-39. (a) What is the average velocity of the particle between $t = 0.50$ s and $t = 4.5$ s? (b) What is the instantaneous velocity of the particle at $t = 4.5$ s? (c) What is the average acceleration of the particle between $t = 0.50$ s and $t = 4.5$ s? (d) What is the instantaneous acceleration of the particle at $t =$ 4.5 s?

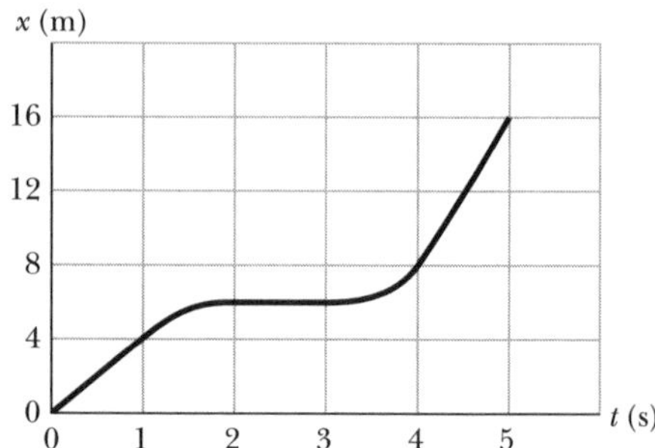

Figure 2-39 Problem 99.

100. A rock is projected vertically upward from the edge of the top of a tall building. The rock reaches its maximum height above the top of the building 1.60 s after being launched. Then, after barely missing the edge of the building as it falls downward, the rock strikes the ground 6.00 s after it was launched. In SI units: (a) with what upward velocity was the rock initially projected, (b) what maximum height above the top of the building is reached by the rock, (c) how tall is the building?

101. The position of a particle as it moves along the y axis is given by

$$y = 2.0 \sin\left(\frac{\pi}{4}t\right),$$

where t is in seconds and y is in centimeters. (a) What is the average velocity of the particle between $t = 0$ and $t = 2.0$ s? (b) What is the instantaneous velocity of the particle at $t = 0$, 1.0, and 2.0 s? (c) What is the average acceleration of the particle between $t = 0$ and $t = 2.0$ s? (d) What is the instantaneous acceleration of the particle at $t = 0$, 1.0, and 2.0 s? (e) Plot v versus t for $0 \leq t \leq 2.0$ s, and estimate the instantaneous acceleration at $t = 1.0$ s from the graph.

102. The speed of a bullet is measured to be 640 m/s as the bullet emerges from a barrel of length 1.2 m. Assuming constant acceleration, find the time that the bullet spends in the barrel after it is fired.

103. A particle moving along the x axis has a position given by $x = 16te^{-t}$ m, where t is in seconds. How far is the particle from the origin when it momentarily stops? (Do not consider its stop at $t = \infty$.)

104. A motorcyclist who is moving along an x axis directed toward the east has an acceleration given by $a_x(t) =$ $(6.1 - 1.2t)$ m/s^2 for $0 \leq t \leq 1.0$ s. At $t = 0$, the velocity and position of the cyclist are 2.7 m/s and 7.3 m. (a) What is the maximum speed achieved by the cyclist? (b) What total distance does the cyclist travel during the time interval between $t = 0$ and 6.0 s?

105. The position of a particle moving along an x axis is given by $x = 12t^2 - 2t^3$, where x is in meters and t is in seconds. (a) Determine the position, velocity, and acceleration of the particle at $t = 3.0$ s. (b) What is the maximum positive coordinate reached by the particle and at what time is it reached? (c) What is the maximum positive velocity reached by the particle and at what time is it reached? (d) What is the acceleration of the particle at the instant the particle is not moving (other than at $t = 0$)? (e) Determine the average velocity of the particle between $t = 0$ and $t = 3$ s.

106. A particle starts from the origin at $t = 0$ and moves along the positive x axis. A graph of the velocity of the particle as a function of the time is shown in Fig. 2-40. (a) What is the coordinate of the particle at $t = 5.0$ s? (b) What is the velocity of the particle at $t = 5.0$ s? (c) What is the acceleration of the particle at $t = 5.0$ s? (d) What is the average velocity of the particle between $t = 1.0$ s and $t = 5.0$ s? (e) What is the average acceleration of the particle between $t = 1.0$ s and $t = 5.0$ s?

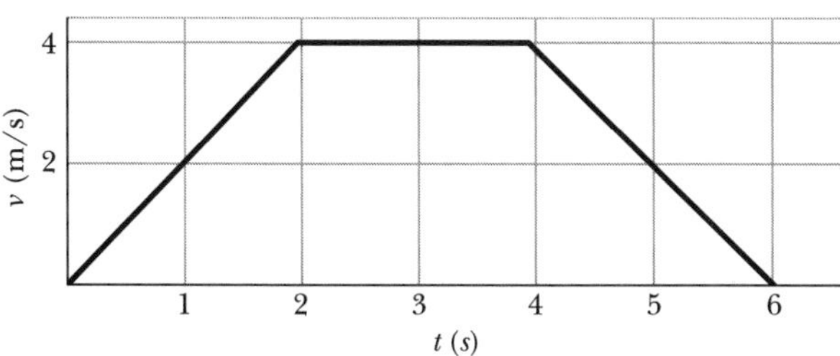

Figure 2-40 Problem 106.

107. A graph of acceleration versus time for a particle as it moves along an x axis is shown in Fig. 2-41. At $t = 0$ the coordinate of the particle is 4.0 m and the velocity is 2.0 m/s. (a) What is the velocity of the particle at $t = 2.0$ s? (b) Write an expression for the velocity as a function of the time that is valid for the interval $2.0 \text{ s} \leq t \leq 4.0$ s.

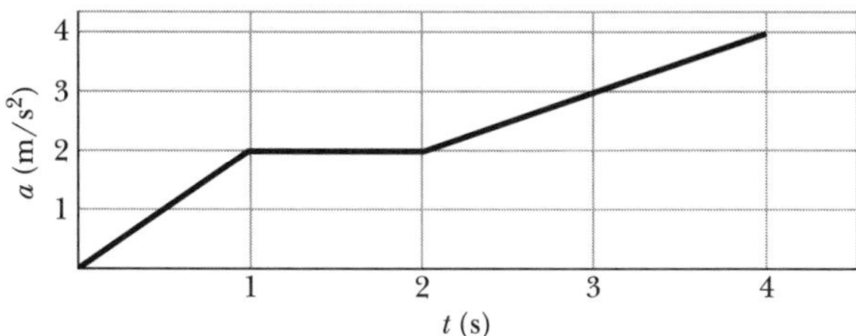

Figure 2-41 Problem 107.

108. A shuffleboard disk is accelerated at a constant rate from rest to a speed of 6.0 m/s over a 1.8 m distance by a player using a cue. At this point the disk loses contact with the cue and slows at a constant rate of 2.5 m/s^2 until it stops. (a) How much time elapses from when the disk begins to accelerate until it stops? (b) What total distance does the disk travel?

109. A certain sprinter has a top speed of 11.0 m/s. If the sprinter starts from rest and accelerates at a constant rate, he is able to reach his top speed in a distance of 12.0 m. He is then able to maintain this top speed for the remainder of a 100 m race. (a) What is his time for the 100 m race? (b) In order to improve his time, the sprinter tries to decrease the distance required for him to reach his top speed. What must this distance be if he is to achieve a time of 10.0 s for the race?

110. An iceboat has a constant velocity toward the east when a sudden gust of wind causes the iceboat to have a constant acceleration toward the east for a period of 3.0 s. A plot of x versus t is shown in Fig. 2-42, where $t = 0$ is taken to be the instant the wind starts to blow and the positive x axis is toward the east. (a) What is the acceleration of the iceboat during the 3.0 s interval? (b) What is the velocity of the iceboat at the end of the 3.0 s interval? (c) If the acceleration remains constant for an additional 3.0 s, how far will the iceboat travel during this second 3.0 s interval?

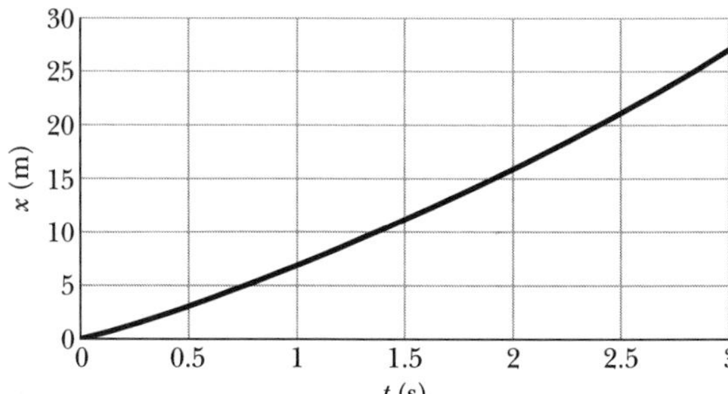

Figure 2-42 Problem 110.

111. A rock is dropped (from rest) from the top of a 60 m tall building. How far above the ground is the rock 1.2 s before it reaches the ground?

112. A rock is thrown vertically upward from ground level. The rock passes the top of a tall tower 1.5 s after being thrown and then reaches its maximum height above ground level 1.0 s after passing the top of the tower. What is the height of the tower?

113. A ball is shot vertically upward from the surface of a planet in a distant solar system. A plot of y versus t for the ball is shown in Fig. 2-43, where y is the height of the ball above its starting point and $t = 0$ at the instant the ball is shot. (a) What is the magnitude of the acceleration due to gravity on the planet? (b) What is the magnitude of the initial velocity of the ball?

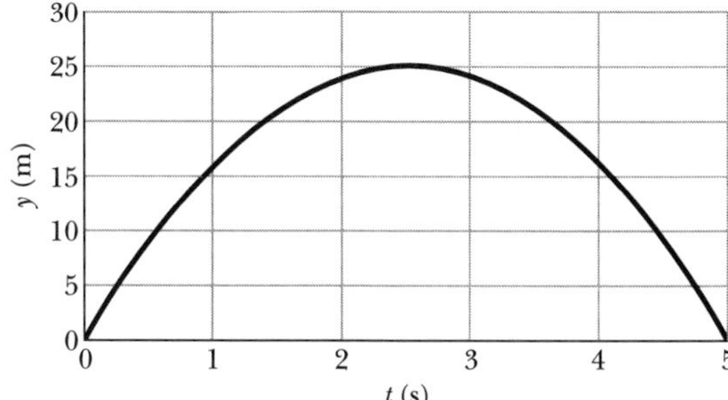

Figure 2-43 Problem 113.

Clustered Problems

Cluster 1

In the problems of this cluster, points A, B, and C lie, in that order, along a straight line.

114. A car speeds up at a constant rate from rest beginning at point A. It achieves a speed of 10.0 m/s at point B, which is 40.0 m from point A. From point B the car travels at constant speed, arriving at point C 10.0 s after leaving point B. (a) Find the acceleration of the car from point A to point B. (b) Find the time required to travel from point A to point B. (c) Find the distance from point B to point C. (d) Find the average velocity of the car as it moves from point A to point C.

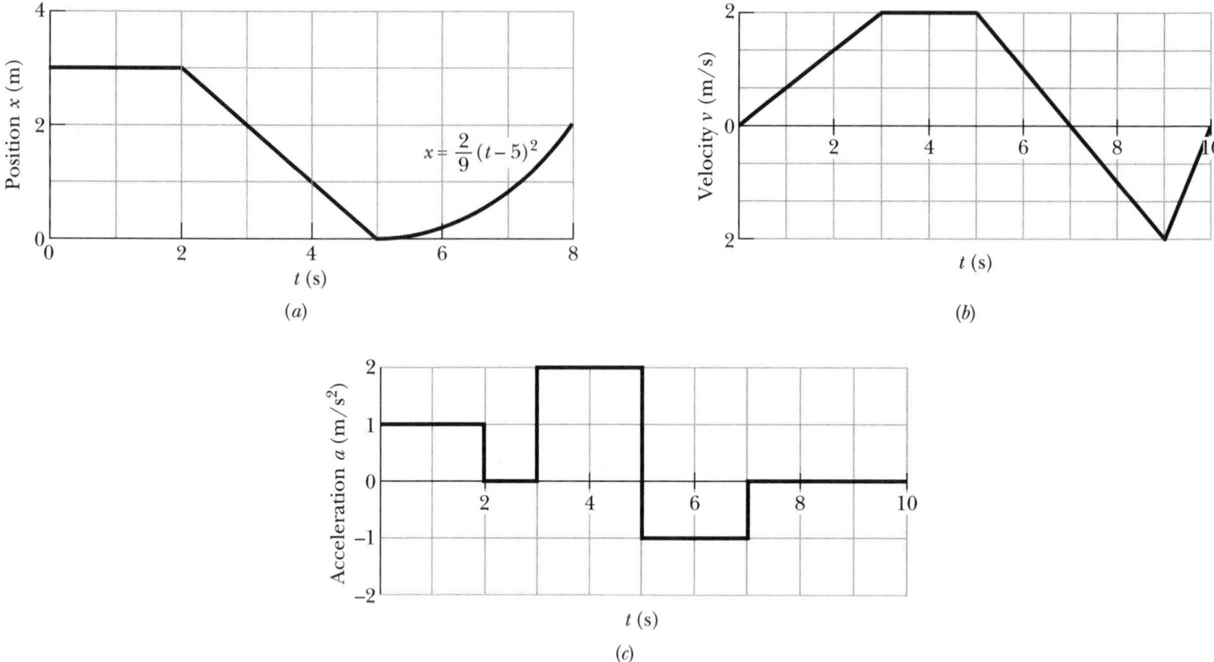

Figure 2-44 Problems 120 through 122.

115. A car passes point A moving at 20.0 m/s. Accelerating uniformly for 10.0 s, it passes point B moving at 30.0 m/s. After passing point B it slows uniformly to a speed of 15.0 m/s at point C, which is 150 m from point B. (a) Find the distance from point A to point B. (b) Find the time required to travel from point B to point C. (c) Find the average velocity of the car as it moves from point A to point C. (d) Find the average acceleration of the car between points A and C.

116. A car passes point A moving at 20 m/s. It travels for 5.0 s at constant velocity to point B and then slows to a stop at point C in 10 s. (a) How far did the car travel from point A to point C? (b) What was the deceleration of the car as it moved from point B to point C?

117. A car travels from point A to point B in 5.00 s at constant velocity and then decelerates to point C in 20.0 s, where it is traveling 10.0 m/s. Point C is 300 m from point A. (a) What is the velocity of the car at point A? (b) What is the rate of change of the velocity from point B to point C?

118. A car travels from point A to point B in 5.00 s at constant velocity and then slows at a uniform rate of -0.500 m/s^2, stopping at point C. Point C is 250 m from point A. (a) What is the velocity of the car at point A? (b) How long does it take to go from point A to point C?

119. A car starts from rest at point A, accelerates at a constant rate to reach point B in 20 s, then accelerates at a different rate to reach point C in 40 s. The car is going 50 m/s at point C, and point A is 1300 m from point C. (a) Find the velocity of the car at point B. (b) Find the distance from point A to point B. Find the acceleration (c) from point A to point B and (d) from point B to point C.

Cluster 2

120. Figure 2-44*a* gives the position x versus time t of an object moving along an x axis. Sketch corresponding graphs of (a) the object's velocity versus time and (b) its acceleration versus time.

121. Figure 2-44*b* gives the velocity v versus time t of an object moving along an x axis; the object is at $x = 0$ at $t = 0$. Sketch corresponding graphs of (a) the object's position versus time and (b) its acceleration versus time.

122. Figure 2-44*c* gives the acceleration a versus time t of an object moving along an x axis; the object is at rest at $t = 0$ and $x = 0$. Sketch corresponding graphs of (a) the object's velocity versus time and (b) its position versus time.

Graphing Calculators

PROBLEM SOLVING TACTICS

TACTICS 10 THROUGH 20: *Basic Procedures*

A caution first: Using a graphing calculator can require many steps of both thought and procedure, and the requirements can appear daunting to someone new to such a calculator. However, if the requirements are practiced, a graphing calculator can speed you through homework and allow you to avoid many of the errors frequently made by students who do not use such a calculator. If you want that homework speed and error reduction, learn how to use your graphing calculator by solving many homework problems with it. Only with practice will the required steps become easy; hopefully, the physics will also.

Because of space limitations, this booklet considers only the procedures required of a TI-85 and a TI-86 graphing calculator. If you have another type of graphing calculator, you will need to study its manual.

TACTIC 10: *Mode Setting*

The mode setting of a TI-85/86 determines how calculated values appear on the screen. To access the mode menu, press 2nd MODE (that is, press the 2nd key, release it, and then press the key with MODE printed above it). For problems in Chapter 2, you will probably want calculated values to appear in *scientific notation,* such as 3.134E3, which is equivalent to 3134 in *normal notation.* So, if the first line in the menu is not *highlighted* on SCI, use the cursor keys to move to SCI and then press ENTER.

The second line of the menu is used to set the number of decimal places shown in a calculated value. You will probably want three decimal places. (The last one will be rounded off in any displayed result of a calculation.) So, use the cursor keys to move to 3 in the second line of the menu and then press ENTER. To exit the mode menu, press EXIT.

If you do a calculation and find that the displayed result is not in the style you wish, press 2nd MODE to access the mode menu, change the mode to what you want (pressing ENTER for each change), press EXIT to exit the menu, and then press ENTER to reevaluate your calculation (you do not need to again press in the steps of the calculation).

TACTIC 11: *Subtraction and Negation Signs*

Because a TI-85/86 distinguishes between a subtraction sign and a negation (or negative) sign, you must be alert about which key to press. For example, what do you get for the subtraction of 3 from 5 if you press 5, then (erroneously) the negation key—marked as (−)—, then 3, and finally ENTER? In general, you need to press the negation key for a negative value that starts a line of calculation or a calculation just after a parenthesis.

TACTIC 12: *Dividing by a Group of Numbers or Terms*

In a textbook or in handwritten notes, the expression $A/3B$ means that A is divided by the term $3B$. Generally, whatever numbers or terms are shown to the right of a division sign (the symbol "/") are taken as being the divisor. However, on a TI-85/86 the divisor is usually only the *first* number or term after the divide sign. For example, if you substitute 3 for A and 4 for B, a calculator's evaluation of $A/3B$ as written yields 4, not 1/4. Instead, you must enter 3/3/4 or 3/(3*4). Obviously this error would be very difficult to spot during the pressure of an exam; your best defense against it is to practice using your calculator on the course homework to the point at which entering data on it is automatic.

TACTIC 13: *Reevaluating an Expression*

If you press in an expression of numbers and operations on a TI-85/86 and then press ENTER to evaluate the expression, you can change the numbers or the operations for another evaluation without repressing everything. Press 2nd ENTRY (that is, press the 2nd key, release it, and then press the key with ENTRY printed above it). The expression reappears on the screen; use the cursor keys to move to where you want a change. If you next press in new numbers or operations, those overwrite what is on the screen, starting at the place of the cursor. If you need to insert numbers or operations without overwriting what is there, first press 2nd INS (that is, press the 2nd key, release it, and then press the key with INS printed above it). Then press in the insertion, pressing ENTER when you finish.

On a TI-86, you can make several previous expressions of numbers and operations reappear on the screen. Repeat pressing 2nd ENTRY to march back through the expressions; how far you can go depends on how much memory the expressions require.

TACTIC 14: *Using an Answer in a Next Calculation*

If you have just calculated a value on a TI-85/86, you can use that answer in another calculation without pressing it in again. (Nicely, the answer is used complete with all its significant figures, not just the rounded-off version that appears on the screen.) If the answer is the first item in that next calculation, just start pressing in the other steps of the calculation. For example, if you want to square the answer, just press the x^2 key, release it, and then press ENTER. Or if you want to take a square root of the answer, just press the ^ key, then press in .5 for a power of $\frac{1}{2}$, and finally press ENTER.

If the answer is not the first item in the next calculation, when you reach where it is needed, press the 2nd key, release it, press the ANS key, and then continue pressing in any remaining steps in that next calculation. Finally, press ENTER to perform the calculation.

TACTIC 15: *Conversion of Units*

Press the 2nd key, release it, and then press the key with CONV (for conversion) printed above it. As an example, let's convert 10 mi/h to meters per second. Press in 10, press the MORE key twice to find the SPEED option in the menu, then press F1 to choose that option. In the new menu, press F3 for mi/hr, then F2 for m/s, and finally ENTER to make the conversion. The calculator quickly indicates that 4.5 m/s is the equivalent of 10 mi/h.

TACTIC 16: *Entering Alphabetic Symbols*

You can make uppercase and lowercase letters appear on the screen by using the ALPHA key, which acts as an on–off switch for the alphabetic mode. When the mode is turned on, either the uppercase symbol A or the lowercase symbol a appears on the cursor to indicate the case. Here are some examples of how to enter letters:

H	press ALPHA, then H
HEY	press ALPHA twice (the double press locks in the uppercase alphabetic mode), then H, then E, then Y
h	press 2nd, then alpha, then H
hey	press 2nd, alpha, ALPHA (to lock in the lowercase alphabetic mode), then H, then E, then Y
H2	press ALPHA, then H, then 2
HEY2	press ALPHA twice, then H, then E, then Y, then ALPHA (to turn off the uppercase alphabetic mode), then 2
hey2	press 2nd, alpha, ALPHA, then H, then E, then Y, then ALPHA twice (to turn off the lowercase alphabetic mode), then 2

Locking in the uppercase alphabetic mode is described as ALPHA-lock; turning off that mode requires one press of the ALPHA key. Locking in the lowercase alphabetic mode is described as alpha-lock; turning off that mode requires a double press of the ALPHA key.

Warning: Some symbols, such as g and h, are reserved by the calculator to represent certain constants. For example, the number 9.80665 is stored as the symbol g. So, if you use the symbol g in a calculation, the calculator inserts that number for the symbol when it performs the calculation.

TACTIC 17: *Stored Constants*

The value of π is available on the keyboard. Other constants commonly used in physics can be accessed by pressing 2nd CONS and then F1 for the BLTIN (built-in) option. The only constant of use before Chapter 14 is the value for the free-fall acceleration g (press the MORE key once to see the symbol). If you are pressing in a calculation involving g, you can enter the symbol either by choosing g from this menu (press F2) or by using the alphabetic mode as explained in Tactic 16. Neither method is worth the effort; just pressing in 9.8 is easier.

TACTIC 18: *Storing and Recalling a Value*

If you want to store a number for later use, press the STO→ key and then a key with a letter printed above it. Then press ENTER. (The STO→ key automatically makes the keyboard ALPHA-locked; see Tactic 16.) For example, if the calculator shows 500, pressing in STO→U will store the 500 as the symbol U. You can then use the symbol U in a calculation. Press, say, 2U and then press ENTER. Depending on the mode settings, the calculator shows 1000 or its equivalent. If you want to see the stored number, press 2nd RCL U ENTER or press ALPHA U ENTER.

If the calculator signals an error when you attempt to store a value as a symbol, the trouble might be that the symbol is reserved for another use. For example, you cannot store a number as the reserved symbol g (but you can store it as the unreserved symbol G).

TACTIC 19: *Storing a Few Equations and Notes*

You can store equations and notes on a TI-85/86 as a program that can function as a notebook. Here is the procedure to store the equations of Table 2-1. (However, if you have a TI-86 and wish to store extensive notes and create your own menu to call them up, skip this Tactic and see instead Tactic 21 among the Sample Problems here.)

Press the PRGM key. Choose EDIT from the screen menu so that you can create the program. The calculator is then in ALPHA-lock (see Tactic 16) and asks for a name for the program. Let's call the program C2 for Chapter 2. Press the key with C printed above it, then the ALPHA key (to turn off the ALPHA-lock), then 2, and then ENTER. Each line of a program begins with a colon, which is automatically provided when you press ENTER.

Choose I/O (for input/output) from the menu and then Disp (for display) from the secondary menu. You will not be entering a computing program but one that simply displays some equations and notes. To do this, you need to enclose the equations and notes with quotation marks (symbol "). To find the symbol, press the MORE key once on a TI-85 or twice on a TI-86. To choose the symbol, press F5 on a TI-85 or F1 on a TI-86.

To enter the first equation of Table 2-1, you must use letters, numbers, and the equation symbols = and +. Using the ALPHA key to turn on and off the alphabetic mode, press

ALPHA ALPHA V =
V ALPHA 0 + ALPHA ALPHA A T

and then choose symbol " from the menu once more to close the expression. Then press ENTER to jump to the next line in the program.

To enter the next equation of Table 2-1, press the MORE key to find the Disp option, choose that option, press the MORE key as before to find the " symbol, and then press

ALPHA X − ALPHA X 0 ALPHA ALPHA
= V ALPHA 0 ALPHA T
+ 0 . 5 ALPHA ALPHA A T ALPHA ^ 2

and then choose symbol " again. Next press ENTER.

You could now continue this procedure with the rest of the equations in Table 2-1. However, let's stop to see how the equations so far will appear on the screen. Press EXIT once to erase the bottom menu on the screen and then again to return to the home screen. To call up the equations, press the PRGM key and then choose NAMES and also C2 (the name of your program now appears in the menu). Then press ENTER. The calculator displays

```
C2
V=V0+AT
X-X0=V0T+0.5AT^2
```

as your version of the equations so far.

To add more of the equations of Table 2-1, press PRGM, choose EDIT and also C2, and then press ENTER. Move down past the equations. Then, starting with the I/O choice above, continue the procedure for the rest of the equations.

You can enter notes with the same procedure. For example, on the C2 program you might enter the helpful note "DOWNWARD DISPLACEMENTS ARE NEGATIVE" because that fact is frequently forgotten. To insert a blank to separate words, press ALPHA and then the negation (−) key. If you want to mimic the lowercase style of the symbols in Table 2-1, use the lowercase alphabetic procedure explained in Tactic 16. You can also insert blanks to make an equation easier to read.

To delete a program from the calculator, press 2nd MEM and choose DELET. Press MORE to find the PRGM option. Choose that option. Move the pointer to the name of the program to delete. Press ENTER to delete it. Then press EXIT to return to the home screen.

TACTIC 20: *Putting Greek Letters on the Screen*

To find the Greek letters (not all are available), press 2nd CHAR, choose GREEK, and then press MORE. The only symbol of use in this and the next three chapters is the Greek uppercase delta Δ.

SAMPLE PROBLEM 2-12

Quadratic Equation Solutions. A hot-air balloon is rising at the rate of 2.0 m/s when, at a height of 35 m, a banana is accidentally dropped overboard. How long does the banana take to reach the ground?

SOLUTION: From Eq. 2-15 (or Eq. 2-22), we know that the displacement Δy of the banana is given by

$$\Delta y = v_0 t + \tfrac{1}{2}(-g)t^2.$$

Here the displacement from the balloon to the ground is $\Delta y = -35$ m, the initial velocity is $v_0 = 2.0$ m/s, and t is the time of the fall. Substituting the data and rearranging give us

$$-4.9t^2 + 2.0t + 35 = 0.$$

On a TI-85/86, press 2nd and then POLY (for polynomial). Then press 2 and ENTER to indicate that we have a second-order polynomial equation (a quadratic equation). The calculator provides, as a generic form of such an equation,

$$\mathrm{a_2x\hat{}2 + a_1x + a_0 = 0}$$

and then requires the coefficients a_2, a_1, and a_0, in that order. Press in −4.9. (remember to use the negation key) and then ENTER, 2 and then ENTER, and finally 35. (The data entered can be numbers, such as these, or expressions, such as −.5*9.8 for a_2 here.) Next press F5 for the SOLVE option in the displayed menu. The calculator then shows the two solutions to the quadratic equation. For the data here, depending on the mode setting for decimal places, the solutions are 2.88 and −2.48. The positive answer is the one we seek. Thus the time of fall is

$$t = 2.88 \text{ s} \approx 2.9 \text{ s}, \qquad \text{(Answer)}$$

where we have rounded off the answer to match the number of significant figures of the given data.

If you wish to change the coefficients and solve a new version of the equation, choose the COEFS option in the displayed menu. Then use the cursor keys to move to any of the displayed coefficients, and change them by entering new numbers or by pressing the DEL key to delete a character, the CLEAR key to clear a line, or the F1 key to clear all the lines.

123. In Sample Problem 2-12, what is the banana's time of fall if it is thrown downward from the balloon with an initial speed of 8.0 m/s relative to the balloon? (Answer = 2.1 s.)

SAMPLE PROBLEM 2-13

Using a List in a Calculation. See Sample Problem 2-11(b), in which a pitcher tosses a baseball straight up. Consider the magnitudes 9.0, 10, 11, 12, and 13 m/s for the launch velocity. Using a list of those magnitudes, generate a list of the corresponding maximum heights reached by the ball.

SOLUTION: To reach the list menu on a TI-85/86, press 2nd LIST (that is, press the 2nd key, release it, and then press the key with LIST printed above it). To start a list, press F1 to choose the symbol { from the menu. Then press in the velocity magnitudes, value after value, separated by commas. End the list by pressing F2 to choose } from the menu. The screen shows

{9.0,10,11,12,13}

where you could enter 9 instead of 9.0 without any effect.

To store the list under the name of, say, V, press STO→ and then the key with V printed above it, and then ENTER. (When you press STO→, the calculator is automatically in ALPHA-lock; see Problem Solving Tactic 16 in these pages.) The screen then displays the list with the style of calculated values set in the mode-setting menu. For example, with settings of Sci and 3 decimal places, the screen shows 9.000E0 1.000E1 1.1... . The rest of the list extends rightward off the screen. To see it, use the rightward cursor key to scroll the numbers across the display.

From Sample Problem 2-11(b), we know that the maximum height of the ball is given by

$$y = \frac{(v_0^2 - v^2)}{2g} = \frac{(v_0^2 - 0)}{2 \times 9.8}.$$

To evaluate this function for the v_0 values in list V, press ALPHA (to enter a letter) and then press in V²/(2*9.8). (To get the square, either press the x² key or press the ^ key and then the 2 key.) Note we could press in V²/2/9.8 or V²/19.6, but we should not press in V²/2*9.8 because the calculator will first divide V² by 2 and then multiply the result by 9.8.

Press ENTER to evaluate the function. Each v_0 value in

list V is squared and then divided by (2*9.8), and the calculator returns the results as the list

{4.133E0 5.102E0 6.173E0 7.347E0 8.622E0} (Answer)

which we must scroll across the screen to see fully. You can save this list under the name of, say, H by pressing STO→ H and then ENTER.

To delete a list from memory, press 2nd MEM (for memory), then F2 for the DELET option in the menu, and then F4 for the LIST option. Move the cursor to the list to be deleted and then press ENTER. Press EXIT to return to the home screen.

124. In Sample Problem 2-8(a), produce a list of the accelerations required (in kilometers per hour-squared) if the initial and final speeds are, respectively, (a) 85 km/h and 65 km/h, (b) 80 km/h and 60 km/h, and (c) 50 km/h and 40 km/h. The displacement is still 88 m.

SAMPLE PROBLEM 2-14

Introduction to Programming. First, a caution. Setting up a homework solution as a program on a calculator so that it can be called up and run may be worthwhile if later you must use that same solution again with different data. However, if the solution varies in other ways, then the program probably will not help. As an example, we shall set up a program here for a certain situation, see how it can be applied to the situation of Sample Problem 2-11 in the textbook, and then see how it can be useless if the situation is changed somewhat.

A penguin slides across frictionless ice and along an x axis that the mathematically gifted penguin has scratched in the ice. The penguin passes initial point $X_0 = 0$ with initial velocity V_0 while an opposing wind gives it a constant acceleration A in the opposite direction. This situation means that the penguin slows and eventually stops momentarily before it begins to move in the direction of the wind.

(a) Assume that we are given data for V_0 and A. Set up a program that gives the time T the penguin takes to come to the momentary stop and the displacement X from the initial point $X_0 = 0$ to that stopping point.

SOLUTION: For the time T, we use Eq. 2-11 ($v = v_0 + at$) to write

$$0 = V_0 + AT,$$

where we have substituted the given symbols and also 0 for the velocity v at the point of stopping. Solving for time T, we have

$$T = -\frac{V_0}{A}.$$

For the displacement X, we use Eq. 2-15 ($x - x_0 = v_0t + \frac{1}{2}at^2$) to write

$$X = V_0T + \tfrac{1}{2}AT^2.$$

We shall set up a program called HALT (see Table 2-3) to find T and X with the last two equations. Press PRGM and choose the EDIT option from the menu. The calculator, in ALPHA-lock, requests a name for the program. Press in the name HALT using the keys with the corresponding letters printed above them. Then press ENTER. Each line in a program begins with a colon, which is automatically provided when you press ENTER.

When we run the program, we want to be able to enter the given data for V_0 and A. To do this, we shall want the calculator to *prompt* (ask) us for the data. To set up the prompt for V_0, choose I/O (for input/output) from the primary menu and then choose Promp from the I/O menu. Then press the ALPHA key, then the key with V printed above it, and then 0 (zero). To end the line, press ENTER.

Now repeat the procedure to set up the prompt for A. Next press the sequence

ALPHA T ALPHA = −ALPHA V 0 ÷ ALPHA A

to enter the equation for T. Be sure to use the negation key (−) and not the subtraction key and also to use the zero key 0 and not the letter O (the subscripts in our equations are zeros and not letters). The ÷ key produces a "slash." Press ENTER to end the line. Then press the sequence

ALPHA X ALPHA = ALPHA V 0 × ALPHA T + .5 × ALPHA A × ALPHA T ^ 2

to put in the equation for x. The × key produces a "star." Again press ENTER.

When we run the program, the prompted values for V_0 and A will be used in these two equations. We next must set up the steps that will *display* (show) the results. To do this, choose Disp from the I/O menu. Next press MORE to find the quote symbol " in the menu, and then choose it. Then press

ALPHA T ALPHA

and then again choose the quote symbol. Next press the comma key, then ALPHA, then the key with T printed above it, and then ENTER. Repeat this display procedure for X. The program is now complete. Press EXIT twice to return to the home screen.

(b) Run the program with the values $V_0 = 4.0$ m/s and $A = -2.0$ m/s² for the penguin. Run the program again for the situation of Sample Problem 2-11 in the textbook.

TABLE 2-3 SAMPLE PROBLEM 2-14

```
PRGM:HALT
:Prompt V0
:Prompt A
:T=-V0/A
:X=V0*T+.5*A*T^2
:Disp"T",T
:Disp"X",X
```

SOLUTION: For the penguin, press PRGM, choose NAMES to find the program, choose HALT to choose the program, and then press ENTER to run that program. When the first prompt appears, press in 4 and then press ENTER. When the second prompt appears press in -2 (using the negation key) and then press ENTER. The calculator quickly displays that T = 2.0 and X = 4.0 (the style of the results, such as number of decimal places, depends on the mode settings you have chosen for the calculator). These results mean that the time to the stopping point is

$$t = 2.0 \text{ s} \qquad \text{(Answer)}$$

and the displacement to the stopping point is

$$x = 4.0 \text{ m.} \qquad \text{(Answer)}$$

For Sample Problem 2-11, repeat the process by pressing ENTER to rerun the program, this time putting in 12 at the first prompt and -9.8 at the second prompt. The calculator then displays the results as T = 1.2 and X = 7.3, which means that

$$t = 1.2 \text{ s} \qquad \text{(Answer)}$$

and

$$y = 7.3 \text{ m,} \qquad \text{(Answer)}$$

just as we found in Sample Problem 2-11.

The notation there differs from here, because there we used the symbol y and took the initial point y_0 as being zero. But the physical situation there matches the penguin situation here. Moreover, the given data and the requested data in the two situations are parallel, allowing us to use the same program. However, the program would not help, for example, with a problem where the acceleration a and the time t to the stopping point are known, and you are to find the initial speed v_0.

PROBLEM SOLVING TACTICS

TACTIC 21: *Storing Many Lines of Equations and Notes*

Tactic 19 shows one way to store notes on either a TI-85 or a TI-86. But if you have a TI-86 and want to store extensive notes, you probably want the smaller character size that is available with the graphing menu (the characters are more difficult to read but you can put many more on the screen simultaneously). You probably also want to sort your notes by chapters and be able to call them up to the screen via your own menu. As an example, we shall here create a program called PHY1 that sorts notes for Chapter 1 under the name CH 1 and notes for Chapter 2 under the name CH 2. You can store notes for up to 15 chapters under PHY1. To store notes for more chapters, you can create PHY2, and so on. Writing PHY1 is tedious, but running it is easy: pressing in PHY1 on the home screen and then pressing ENTER puts a menu of the chapter names on the screen. To see your notes on, say, Chapter 1, you simply choose the CH 1 from the menu.

TABLE 2-4 PROBLEM SOLVING TACTIC 21

`PROGRAM: TRNOFF`	Turns off normal graphing
`:AxesOff`	
`:CoordOff`	
`:FnOff`	
`:GridOff`	
`:Return`	

TRNOFF. We first write a program called TRNOFF (see Table 2-4) that turns off the usual features of a graph, such as axes. Press PRGM, choose EDIT, and then press in TRNOFF (the calculator is already ALPHA-locked; see Problem Solving Tactic 16). Press ENTER. (Each command on a calculator begins with a colon; pressing ENTER automatically provides one.) To turn off the axes, press 2nd CATLG-VARS (for catalog and variables), and then choose CATLG. Move down the list of the calculator's catalog of commands and operations until you reach AxesOff. Press ENTER to choose it and then once again to end the line in the program.

Repeat this process to put in CoordOff (which turns off the coordinates), FnOff (which turns off the graphing functions), and GridOff (which turns off the grid). Each time press ENTER once to choose the command and then again to end the line. From the screen menu, choose CTL (for control), press MORE until Retur appears, and then choose it to end this program. Press EXIT twice to leave the programming options.

PHY1. We next write the PHY1 program (see Table 2-5). Press PRGM to return to the programming options. Choose EDIT. Press in PHY, then turn off the ALPHA-lock by pressing ALPHA. Then press 1 and ENTER.

First, turn on ALPHA-lock and press in TRNOFF and then press ENTER to end the line. When PHY1 is run, that turn-off program will be executed first. To create a menu for your notes, choose CTL, press MORE to find Menu and then choose it. Then press in

1, "CH 1",CH1,2,"CH 2", CH2

to finish the line. To get the quotation symbol, first choose the I/O option by pressing 2nd M3. Then press MORE until the option for the symbol appears in the menu. Choose it as needed. To get the blanks, press the negation key (−) with the calculator in ALPHA mode. Press ENTER to end the line.

The elements of this line of programming mean the following: box 1 of your menu will have the name "CH 1" and will contain the notes you store in this program under the label CH1. And box 2 will have the name "CH 2" and will contain the notes you store in this program under the label CH2.

CH1 Label. To get the label command, choose CTL by pressing 2nd and M4. Then press MORE until Lbl appears and then choose it. Then, using the ALPHA key, press in CH1 as the label for the Chapter 1 notes and then press ENTER to end the line.

When PHY1 is run, you want to erase the screen at this

TABLE 2-5 PROBLEM SOLVING TACTIC 21

PROGRAM: PHY1	
:TRNOFF	Runs TRNOFF
:Menu(1,"CH 1",CH1,2,"CH 2",CH2	Sets up a menu
:Lbl CH1	Label for Chapter 1
:ZStd	Clears screen
:Text(0,0,"10^24 yotta Y"	
:Text(6,0,"10^21 zetta Z"	
:Text(12,0,"10^18 exa E"	
:Text(18,0,"10^15 peta P"	First screen of notes for Chapter 1
:Text(24,0,"10^12 tera T"	
:Text(30,0,"10^9 giga G"	
:Text(36,0,"10^6 mega M"	
:Text(42,0,"10^3 kilo k"	
:Text(48,0,"10^2 hecto h"	
:Text(54,0,"10^1 deka da"	
:Pause	
:ZStd	Second screen of notes for Chapter 1
:Text(0,0,"10^−1 deci d"	
:Text(6,0,"10^−2 centi c"	
:Text(12,0,"10^−3 milli m"	
:Text(18,0,"10^−6 micro μ"	
:Pause	Label for Chapter 2
:Lbl CH2	
:ZStd	
:Text(0,0,"v=v0+at"	
:Text(6,0,"x−x0=v0t+0.5at²"	First screen of notes for Chapter 2
:Text(12,0,"v²=v0²+2a(x−x0)"	
:Text(18,0,"x−x0=0.5(v0+v)t"	
:Text(24,0,"x−x0=vt−0.5at²"	
:Text(30,0,"g = 9.8 m/s²"	
:Text(36,0,"DOWN IS NEGATIVE"	
:Pause	Runs TRNON
:TRNON	

point before the notes appear. To set this up, go back to the catalog, find ZStd, press ENTER to choose the command and then once again to end the line.

To press in your notes, you need the quotation mark (which is in one screen menu) and the command called Text (which is in another screen menu). To put both those menus on the screen simultaneously, first choose I/O by pressing 2nd M3. Then press MORE until the option for the quotation symbol appears in the menu. Then press GRAPH and then MORE until the DRAW option appears. Choose it and then press MORE until the Text option appears.

Using the Text command you can determine what information appears where on the screen when the program runs. We shall have notes begin at the upper left of the screen, at position 0,0 (for row 0 and column 0). The next line will begin at position 6,0 (for row 6 and column 0). (The character height will be five pixels, or elementary dots on the screen. One blank pixel will separate lines. Thus we start the second line in row 6 of the pixels.)

To get started, let's store the first line of Table 1-2 in the textbook. Choose the Text option and then complete the line by pressing in

0,0,"10^24 yotta Y"

by choosing the quotation option with 2nd M1 and using the ALPHA key to get the letters and the blanks (again see Problem Solving Tactic 16). Press ENTER to end the line.

Now, keep in mind that we are just exploring how to write a program that stores notes. Having the meaning of the yotta prefix at your fingertips may not be worth your time in writing this part of the program and it may not be worth the memory space on the calculator. If you do feel more secure about life with scientific prefixes at your fingertips, you could enter more of Table 1-2 with the same procedure, advancing the row number by 6 each time, as shown here in Table 2-5. Stop when you reach row 54, which is the last row you can use in a full screen of information.

At this point you want the program to pause so that the full screen of information can be read instead of scrolled upward out of view as more information scrolls into view. To insert a pause here, press EXIT and then choose CTL by pressing 2nd M4. Then press MORE until the Pause option appears, choose it, then press ENTER to end the line.

To begin the next screen of notes, find Zstd from the catalog as explained previously, press ENTER to choose it, and then press ENTER to end the line. This command will erase the first screen of notes if you advance to the second screen of notes when running the program. Then repeat the procedure given to fill out more Text lines for the second screen. And repeat it again for a third screen, and so on. There is a way to avoid tediously reconstructing the Text instructions over and over (see program TEX later).

CH2 label. When you have finished your notes for Chapter 1, putting in the last Pause, press in the label CH2 for Chapter 2 notes and then put in another Zstd. Then fill out the Text lines for this chapter, screen by screen. (See, for example, Table 2-5.)

TRNON. To close the program and return to the home screen, run a TRNON program that turns back on the graphing parameters turned off by TRNOFF. To write TRNON, press EXIT once or twice to reach the home screen. Then press PRGM to return to programming. Choose EDIT and then write the program of Table 2-6 as we wrote TRNOFF. CILCD, which clears the screen, is found in the I/O menu.

Editing. To edit your notes later, press PRGM and choose EDIT. Then choose PHY1 from the menu. Press ENTER. Move to where you want a change. Press DEL to delete characters, press 2nd INS to insert characters, or overwrite characters. To insert a blank line, choose INSc from the menu. To delete a line, choose DELc. If you delete a line by error, press UNDEL to bring it back.

TABLE 2-6 PROBLEM SOLVING TACTIC 21

`PROGRAM:TRNON`	Turns on normal graphing
`:AxesOn`	
`:CoordOn`	
`:FnOn`	
`:CILCD`	Clears home screen
`:Stop`	

To delete an entire program from memory, press 2nd MEM from the home screen. Choose DELET, press MORE to find PRGM, choose it, move down to the name of the program you wish to delete, and press ENTER. (Be careful. You cannot bring a program back with an undelete.) Press EXIT.

Running PHY1. You can run the program PHY1 from the home screen by pressing in PHY1 (using the ALPHA key) and then pressing ENTER. Or you can press PRGM, choose NAMES, and then choose PHY1 from the menu. If you then choose, say, CH 1 from your menu, you see the first screen of notes for Chapter 1. Pressing ENTER erases that screen of notes and brings up the second screen of notes. Continuing in this way you move screen by screen through your notes for Chapter 1 and then Chapter 2. If, instead, you choose CH 2 from the menu, you jump directly to the notes for Chapter 2.

The more notes you store under PHY1, the slower it will run. So, if your purpose with this program is to have speedy notes at your fingertips, you should store only the barest of notes for each chapter, or store only notes for a few chapters with PHY1 and store other notes with, say, PHY2.

TEX. To avoid reconstructing Text instructions over and over, write the program called TEX shown in Table 2-7. Then whenever you want to add another screen of notes to PHY1, first go to the line after the Pause or Lbl line where you want the additional notes. Insert a blank line by using the INSc option in the primary programming menu. Next press 2nd RCL, then press in TEX, and then press ENTER. The contents of TEX are recalled at that point in PHY1. To press in the new notes, go to the first quotation mark of a line and press 2nd INS to insert. If you don't have enough notes for a full screen, you might leave some of the lines incomplete in case you return with more notes later.

TABLE 2-7 PROBLEM SOLVING TACTIC 21

```
PROGRAM:TEX
:ZStd
:Text(0,0,""
:Text(6,0,""
:Text(12,0,""
:Text(18,0,""
:Text(24,0,""
:Pause
```

SAMPLE PROBLEM 2-15

Functional Graphing and Graphical Solutions. Two particle-like rockets A and B are shot along a horizontal y axis. The position of rocket A, moving at a constant speed of 5.00 m/s and located at $y = 100$ m at $t = 0$, is given by

$$y_A = 5.00t + 100.$$

Rocket B, launched at $t = 0$, accelerates for about a minute, with its position given by

$$y_B = t^{1.5} + 0.100t^{2.5} - 0.0100t^3$$

during the acceleration. In both functions, t is in seconds and y is in meters.

(a) At what time and where on the y axis does rocket B reach rocket A?

SOLUTION: *Intersection of Two Graphed Curves.* We can answer the question by graphing both functions and then finding the intersection of the two curves. Rocket B reaches rocket A when they have the same position at the same time. Here is the procedure for a TI-85/86 calculator.

Press 2nd MODE. If FUNC (for functional graphing) is not highlighted, move down to it and then press ENTER to choose such graphing. Next press GRAPH and then choose the y(x)= option from the menu.

On a TI-85/86, we must let the generic variable x represent the variable t in our rocket functions. So, at y1= press in 5x + 100 for the function y_A, pressing in the x with either the F1 key or the x-VAR key. Pressing ENTER causes the equals mark to be highlighted, indicating that this equation will be graphed. Next, at y2= press in x^1.5 + .1x^2.5 − .01x^3 for the function y_B and then press ENTER to highlight the equals mark. If any other function is on the screen and has a highlighted equals mark, move to its line and then either erase it by pressing CLEAR or deactivate it by choosing the SELECT option from the menu. Then it will not be graphed.

To prepare the range of the x and y values on the graph, press 2nd M2 to choose from the higher menu. On a TI-85, this choice is for RANGE; on a TI-86, it is for WIND (for window); the names differ but the purposes are identical. Moving the cursor down line by line, press in the following parameters: 0 for xMin, 35 for xMax, 5 for xScl (the tick-mark spacing on the x axis), 0 for yMin, 280 for yMax, and 100 for yScl (the tick-mark spacing on the y axis).

Next press MORE to see more of the menu at the bottom of the screen. From the new options, choose the format option. Moving the cursor and pressing ENTER to make any changes, set up the format as

RectGC for rectangular coordinate system,

CoordOn so that a cursor location on a graphed curve is indicated,

DrawLine to draw curves with lines instead of dots,

SimulG to draw the curves simultaneously instead of one after the other,

GridOff if grid lines are not to be shown,

AxesOn if coordinate axes are to be shown, and

LabelOff if the x and y labels are not to be shown on the graph.

Now to graph the functions y1 and y2, choose GRAPH from the screen menu. Wait until the moving vertical dash at the upper right disappears (if you want to stop the process, press ON). The curves of the two functions appear on the screen, along with their intersection. To make the screen menu disappear so that the whole graph is visible, press CLEAR; to make it reappear, press GRAPH. (Generally, if nothing shows up on a graph, the first items to check are the values entered under RANGE or WIND—are they appropriate?)

To compute values of x and y (that is, t and y) for the intersection, press MORE to change the menu, choose MATH, press MORE, and then choose ISECT (intersection). The cursor appears on one of the curves; press ENTER to choose this curve for analysis. The cursor then appears on the other curve. Press ENTER again to choose this curve. (At this point, a TI-86 asks about a guess; just press ENTER once more.) Soon the x (that is, t) and y values for the intersection appear: x = 20.94358 and y = 204.7179. Thus, we find that rocket B reaches rocket A at

$$t = 20.9 \text{ s} \quad \text{and} \quad y = 205 \text{ m}. \qquad \text{(Answer)}$$

(b) Where is rocket B 3.0 s before it reaches rocket A?

SOLUTION: *Evaluating a Point on a Graphed Curve.* Press GRAPH to put the first graphing menu on the screen, press MORE twice, and then choose EVAL (for evaluate). We want to evaluate our graph for rocket B (the y2 curve) for time $t = 17.9$ s. So press in 17.9 for the requested x value and then press ENTER. The calculator first evaluates y1 for that time (because y1 is the first in the list of functions), marking the point on the graphed curve. Use the up or down cursor key to go to the other curve, the one for y2. We find that for x = 17.9, y ≈ 154. Thus, at $t = 17.9$, rocket B is at

$$y \approx 154 \text{ m}. \qquad \text{(Answer)}$$

(c) What is the speed of rocket B when it reaches rocket A?

SOLUTION: *Finding a Slope on a Graphed Curve.* To find the speed of rocket B, press GRAPH to see the first graphing menu. Then press MORE, choose MATH, and then choose dy/dx. The cursor appears on the straight line corresponding to y1 and rocket A. Press the up or down cursor key to move to the curved line corresponding to y2 and rocket B. While checking the cursor location that appears at the bottom of the screen, press the rightward cursor key to move to the intersection. When the cursor location is approximately at the intersection (x is about 20.9 and y is about 205), press ENTER. The slope dy/dx of y_B is soon given: about 17.7. This slope is the speed dy/dt of rocket B at the intersection. Thus we find

$$\text{speed} \approx 17.7 \text{ m/s}. \qquad \text{(Answer)}$$

(d) Store the data concerning the graph so that the data can be redrawn later.

SOLUTION: *Storing Databases for Graphs.* If you go to the trouble of drawing a graph and anticipate that you shall need to draw the same graph again or one like it, store the associated equations and the RANGE or WIND parameters. Then later you can quickly call up all that data and redraw the graph as before or with modifications to the equations or parameters.

For example, if you think you shall need to draw a graph of $x = x_0 + v_0t + \frac{1}{2}at^2$, draw such a graph with assumed values for the equation's variables and for the RANGE or WIND parameters. Then store the database. If you do need to graph that equation later, you can call up the database, modify the values of the variables and parameters, and then draw the graph; this procedure might be faster than starting fresh.

To store a graph database, go to the primary GRAPH menu, press MORE, choose STGDB (store graph database), and then, as requested on the screen, press in a name under which to store the data. The keyboard is automatically ALPHA-locked (see Problem Solving Tactic 16). Finally, press ENTER. To retrieve the data later, choose RCGDB (receive graph database) from the primary GRAPH menu. The name you pressed in earlier appears in a secondary menu; choosing it causes the graph to be redrawn. If you want to modify the equations or the RANGE or WIND parameters, press MORE to find the corresponding options in the menu.

125. Two rockets travel along a y axis as in Sample Problem 2-15, except that the positions are now given by

$$y_A = 10 + t^2e^{-t/10}$$

and

$$y_B = 1.5t$$

for $t \geq 0$, with t in seconds and y in meters. (a) Where and when do the rockets meet? (b) What is the velocity of rocket A just then?

126. *Minimum Trailing Distance.* Your car trails a highway patrol car at separation d, with both cars at speed v_0, when suddenly the patrol car begins to decelerate to a stop at the constant negative rate a_1. After a time t_R to react to the patrol car's change (to perceive the change and then to begin applying your brake), you begin to decelerate at the constant negative rate a_2.

What is the stopping distance of (a) the patrol car and (b) your car? (c) What is the difference in those two stopping distances? This difference is the least separation d_{min} you should have originally if you are to avoid rear-ending the patrol car. (d) Make a single graph of d_{min} versus v_0 for v_0 ranging from 0 to 50 m/s and for the following values of a_1 and a_2, used in lists:

(1) −5.00 and −5.00 m/s²,

(2) −5.00 and −4.50 m/s²,

(3) −4.50 and −4.00 m/s²,

(4) −2.00 and −1.50 m/s².

Assume a response time $t_R = 1.00$ s. Set the separation of tick-marks on the vertical axis to be the typical car length of 4.5 m.

Using the evaluation option for the graph, evaluate d_{min} at $v_0 =$ 30 m/s for the (e) second pair and (f) third pair of acceleration values.

(g) According to a conventional rule in the United States, d_{min} should be one car length for every 10 mi/h (= 4.5 m/s) in speed. Plot this dependence on the existing graph. (h) If the conventional rule is to be valid, what must be the relation between a_1 and a_2? (i) If this condition is not met, what is the error between the actual d_{min} and that predicted by the conventional rule? (j) Plot this error versus the given range of v_0 and for the given four pairs of accelerations. (k) Generally, does the error increase or decrease for weaker braking on the two cars? Evaluate the error at $v_0 = 30$ m/s for the (l) second pair and (m) third pair of acceleration values.

Chapter Three
Vectors

QUESTIONS

18. The two vectors shown in Fig. 3-39 lie in an xy plane. What are the signs of the x and y components, respectively, of (a) $\mathbf{d}_1 + \mathbf{d}_2$, (b) $\mathbf{d}_1 - \mathbf{d}_2$, and (c) $\mathbf{d}_2 - \mathbf{d}_1$?

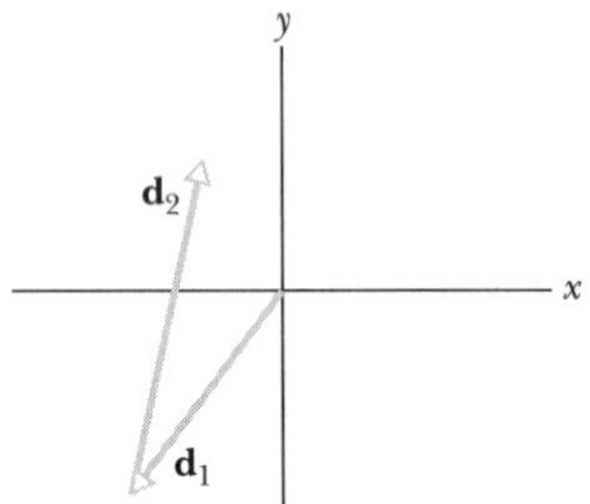

FIGURE 3-39 Question 18.

19. Figure 3-40 shows a vector **R** and two coordinate systems (the primed system $x'y'$ and the double-primed system $x''y''$, with a common origin). (a) Is component $R_{x'}$ greater than, less than, or the same as component $R_{x''}$? (b) Is component $R_{y'}$ greater than, less than, or the same as component $R_{y''}$? (c) Is the angle that **R** makes with the x' axis greater than, less than, or the same as the angle it makes with the x'' axis? (d) Is the value of $\sqrt{(R_{x'})^2 + (R_{y'})^2}$ greater than, less than, or the same as the value of $\sqrt{(R_{x''})^2 + (R_{y''})^2}$?

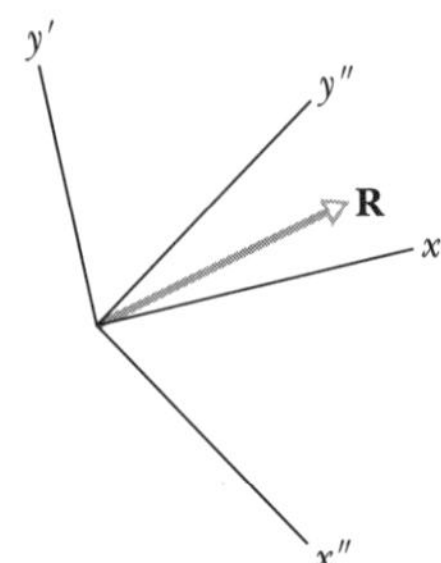

FIGURE 3-40 Question 19.

20. Figure 3-41 shows vectors **r** and **F**, both in the xy plane. What is the direction of (a) $\mathbf{r} \times \mathbf{F}$ and (b) $\mathbf{r} \times (-\mathbf{F})$?

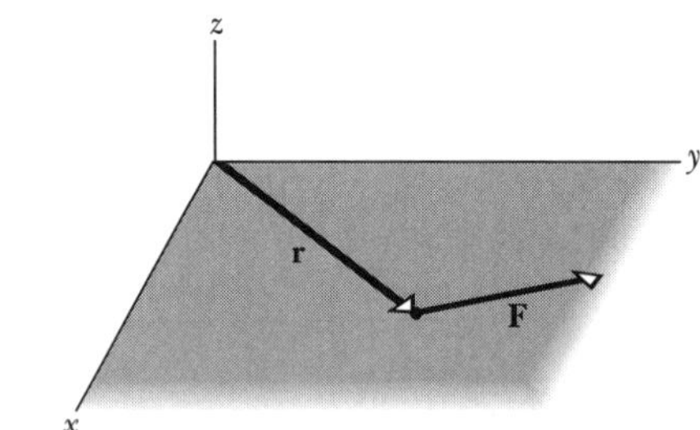

FIGURE 3-41 Question 20.

21. Suppose that a vector **F** is given by

$$\mathbf{F} = q(\mathbf{v} \times \mathbf{B}),$$

where q is a constant and **v** and **B** are vectors. (a) What are the directions of the cross product $\mathbf{v} \times \mathbf{B}$ for the three situations in Fig. 3-42? What are the directions of **F** for those situations if q is (b) a positive quantity and (c) a negative quantity?

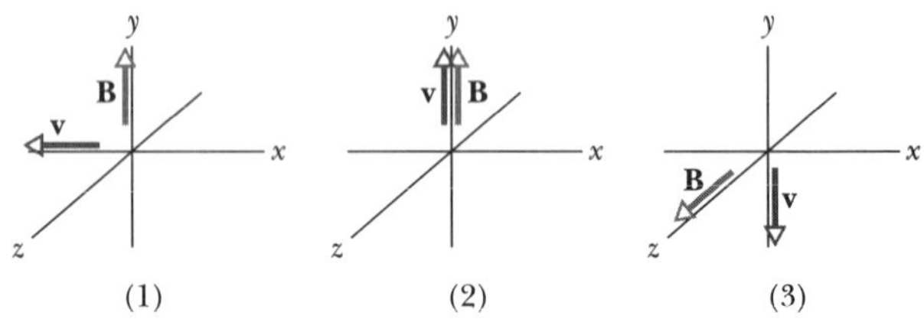

FIGURE 3-42 Question 21.

22. Using the expression for **F** given in Question 21, determine the direction of **B** for the three situations of Fig. 3-43 if q is (a) a positive quantity and (b) a negative quantity. If the situation is impossible, state that fact.

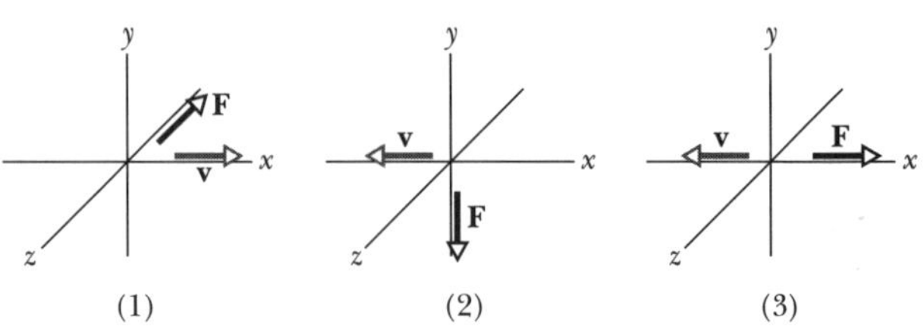

FIGURE 3-43 Question 22.

EXERCISES & PROBLEMS

59. Oasis B is 25 km due east of oasis A. Starting from oasis A, a camel walks 24 km in a direction 15° south of east and then walks 8.0 km due north. How far is the camel then from oasis B?

60. A vector **B**, which has a magnitude of 8.0, is added to a vector **A**, which lies along an x axis. The sum of these two vectors is a third vector, which lies along the y axis and has a magnitude that is twice the magnitude of **A**. What is the magnitude of **A**?

61. In the product $\mathbf{F} = q\mathbf{v} \times \mathbf{B}$, take $q = 2$,

$$\mathbf{v} = 2.0\mathbf{i} + 4.0\mathbf{j} + 6.0\mathbf{k}$$

and $$\mathbf{F} = 4.0\mathbf{i} - 20\mathbf{j} + 12\mathbf{k}.$$

What then is **B** in unit-vector notation if $B_x = B_y$?

62. An ant, crazed by the Sun on a hot Texas afternoon, darts over an xy plane scratched in the dirt. The x and y components of four consecutive darts are the following, all in centimeters: (30.0, 40.0), (b_x, −70.0), (−20.0, c_y), (−80.0, −70.0). The overall displacement of the four darts has the xy components (−140, −20.0). What are (a) component b_x, (b) component c_y, and (c) the overall displacement in magnitude-angle notation?

63. What is the sum of the following four vectors in (a) unit-vector notation and (b) magnitude-angle notation? Positive angles are counterclockwise from the positive direction of the x axis; negative angles are clockwise.

$\mathbf{A} = 2.00\mathbf{i} + 3.00\mathbf{j}$ **B**: 4.00, at +65.0°

$\mathbf{C} = -4.00\mathbf{i} - 6.00\mathbf{j}$ **D**: 5.00, at −235°

64. What is the sum of the following four vectors in (a) unit-vector notation and (b) magnitude-angle notation? For the latter, give the angle in both degrees and radians. Positive angles are counterclockwise from the positive direction of the x axis; negative angles are clockwise.

E: 6.00, at +0.900 rad **F**: 5.00, at −75.0°

G: 4.00, at +1.20 rad **H**: 6.00, at −210°

65. Find the sum of the following four vectors in (a) unit-vector notation and (b) magnitude-angle notation.

P: 10.0, at 25.0° counterclockwise from $+x$

Q: 12.0, at 10.0° counterclockwise from $+y$

R: 8.00, at 20.0° clockwise from $-y$

S: 9.00, at 40.0° counterclockwise from $-y$

66. In the sum $\mathbf{A} + \mathbf{B} = \mathbf{C}$, vector **A** has magnitude 12.0 and is angled 40.0° counterclockwise from the $+x$ direction, and vector **C** has magnitude 15.0 and is angled 20.0° counterclockwise from the $-x$ direction. What is **B** in magnitude-angle notation?

67. **A** has magnitude 12.0 and is angled 60.0° counterclockwise from the positive direction of an x axis of an xy coordinate system. Also, $\mathbf{B} = 12.0\mathbf{i} + 8.00\mathbf{j}$ on that coordinate system. We now rotate the system counterclockwise about the origin by 20.0° to form an $x'y'$ system. On this new system, what are (a) **A** and (b) **B**, both in unit-vector notation?

68. For the following three vectors, what is the result of $3\mathbf{C} \cdot (2\mathbf{A} \times \mathbf{B})$?

$$\mathbf{A} = 2.00\mathbf{i} + 3.00\mathbf{j} - 4.00\mathbf{k}$$

$$\mathbf{B} = -3.00\mathbf{i} + 4.00\mathbf{j} + 2.00\mathbf{k}$$

$$\mathbf{C} = 7.00\mathbf{i} - 8.00\mathbf{j}$$

Graphing Calculators

SAMPLE PROBLEM 3-9

Expressing and Adding Vectors. In a memory experiment, a trained mouse runs through a horizontal maze on which the experimenter has superimposed an xy coordinate system. The mouse takes three straight-line runs: 0.25 m at 36°, 0.38 m at 120°, and 0.15 m at 210°, with each angle measured counterclockwise from the positive direction of the x axis. At the end of the three runs, what is the net displacement of the mouse from the starting point? Express the answer in both magnitude-angle notation and unit-vector notation.

SOLUTION: On a TI-85/86, we first adjust the mode settings. Press 2nd MODE (that is, press the 2nd key, release it, and then press the key with MODE printed above it). Because the angles are given in degrees, we want the calculator to be in the degree mode. If DEGREE in the menu is not highlighted, move to it and then press ENTER.

To get the answer in magnitude-angle notation, we want the calculator to be in spherical coordinates. If SphereV (for spherical values) is not highlighted, move to it and then press ENTER. Also, just for practice, set the first line of the menu as Sci (for scientific notation) and the second line as 2 (to display two decimal places). Then exit the menu by pressing EXIT.

On a TI-85/86, we can represent a vector in the generic form of

[magnitude ∠ angle in degrees or radians]

where the brackets indicate we have a vector. We get the brackets and the angle symbol ∠ by first pressing the 2nd key and then the keys above which the symbols are printed. Since we have set the mode to DEG, the data we enter for any angle will be interpreted by the calculator to be in degrees rather than in radians.

To sum the displacement vectors for the mouse's three runs, we press

2nd [.25 2nd ∠ 36 2nd] + 2nd [.38 2nd ∠ 120 2nd]
+ 2nd [.15 2nd ∠ 210 2nd]

The screen should then show

[.25∠36]+[.38∠120]+[.15∠210]

as our calculation. Pressing ENTER, we find the result

[4.18E−1∠1.06E2]

which is in magnitude-angle notation because of our choice of SphereV in the mode menu. The result means that

$$\text{magnitude} \approx 0.42 \text{ m}$$

and angle counterclockwise from $+x \approx 110°$, (Answer)

where we round off to match the significant figures of the given data.

To express the answer in terms of unit vectors, we change the calculator mode. Press 2nd MODE, move down to RectV (for rectangular values), and then press ENTER, EXIT, and finally ENTER again. That last ENTER causes the calculator to repeat the last calculation made, which here is the vector addition. This time, with our new choice of RectV, the display is

[−1.18E−1 4.01E−1]

which means the net displacement has the (rounded-off) components of

x component ≈ -0.12 m and y component ≈ 0.40 m.

Thus in unit vector notation, the net displacement is

$$(-0.12 \text{ m})\mathbf{i} + (0.40 \text{ m})\mathbf{j}. \qquad \text{(Answer)}$$

PROBLEM SOLVING TACTICS

TACTIC 6: *Mode Settings for Vectors*

Look back over Sample Problem 3-9. What adjustments to the mode settings must be made before you begin a calculation? The selection of SphereV for magnitude-angle notation or RectV for unit-vector notation affects how the answer appears, not how you enter the data for a vector. So, that adjustment can wait until you are about to get an answer or, as we saw in the Sample Problem, even after you have already gotten an answer. Similarly, decisions about the use of normal notation or scientific notation and about the number of decimal places can also wait or can even be changed later.

However, if you want to enter data for angles, then you should first choose either degree mode or radian mode. If you enter data in, say, degrees, when the calculator is in radian mode, the data will be interpreted as being in radians and any calculations made with the data will be quite wrong. *This error is very common, especially during exams.* If you happen to notice that the calculation is wrong (that is usually difficult to do), you do not necessarily have to re-enter all the data. If that calculation is the last one performed, just go to the mode menu, correct the mode, exit the menu, and then press ENTER to repeat the calculation. (Tactic 8 explains how to override the mode setting.)

In Sample Problem 3-9 we entered vectors in the generic form [magnitude ∠ angle]. In Sample Problem 3-10 we shall enter vectors in unit-vector notation with the generic form [x component, y component, z component]. With either notation, a vector sometimes does not fit onto the screen. Although calculations wrap around to stay on a screen, long vectors extend rightward off the screen. To see the rest of such a vector, you can use the rightward cursor key to scroll the vector across the screen.

TACTIC 7: *Switching between Magnitude-Angle Notation and Unit-Vector Notation*

You can quickly switch a vector from one type of an expression to the other type. For example, press in a vector in magnitude-angle notation, set the mode to RectV, exit the mode menu, and then press ENTER to see the vector in unit-vector notation. Or, conversely, press in a vector in unit-vector notation, set the mode to SphereV, exit the menu, and then press ENTER to see the vector in magnitude-angle notation.

Warning: If you press 2nd VECTR and then choose the MATH option, you find a choice called unitV (for unit vectors). That choice will translate magnitude-angle notation into a type of unit-vector notation, but not the type used in this book. (The vector components will be divided by the magnitude of the vector.)

TACTIC 8: *Overriding the Mode Setting for Angles*

You can enter angles in degrees, radians, or both degrees and radians regardless of the mode setting by explicitly indicating degrees with the symbol ° and radians with the symbol ^{r}. To find these symbols, press 2nd MATH and then choose the ANGLE option. For example,

$$\cos 0.7854^r + \cos 60°$$

yields 1.21 regardless of the mode setting for angles.

SAMPLE PROBLEM 3-10

(a) Given two vectors, $\mathbf{A} = 3.00\mathbf{i} + 4.00\mathbf{j}$ and $\mathbf{B} = -2.00\mathbf{i} - 4.00\mathbf{j} + 3.00\mathbf{k}$, find the combination $\mathbf{B} - 5.00\mathbf{A}$ in unit-vector notation.

SOLUTION: *Vector Addition.* On a TI-85/86, because the given vectors are in unit-vector notation, we enter them in the generic form

[x component, y component, z component]

where the components are separated by commas (made with the comma key). Here, press

2nd [(−) 2 , (−) 4 , 3 2nd] − 5 2nd [3 , 4 , 0]

where the symbol (−) means the negation key and the symbol − means the subtraction key. (We must distinguish between a negative, or negation, sign and a subtraction sign.) The screen should now show

$$[-2,-4,3]-5[3,4,0]$$

where each component of the vector [3,4,0] will be multiplied by the scalar 5.

Note that we can safely drop trailing zeros in the data (we pressed in −2 instead of the given −2.00). However, here we cannot drop a zero component (we must press in 0 for the z component of **A**) or there will be a mismatch between the forms of the vectors we are combining, something that the calculator signals as an error.

To have the answer appear in unit-vector notion, we call up the mode menu and highlight RectV (rectangular values). Just for practice, also set the first line to NORMAL (for a normal expression of numbers) and the second line to 2 (for two decimal places). After exiting the mode menu, press ENTER to make the calculation. The screen then shows

$$[-17.00 \quad -24.00 \quad 3.00]$$

which means that

$$\mathbf{B} - 5.00\mathbf{A} = -17.0\mathbf{i} - 24.0\mathbf{j} + 3.00\mathbf{k}. \quad \text{(Answer)}$$

(We have rounded off to match the significant figures of the given data.)

(b) *Finding Magnitude and Angle.* Find the magnitude and angle of vectors **A** and 5**A**.

SOLUTION: Press in vector **A** as [3,4], where we neglect the zero z component. (This is what we usually do when we work with vectors that are only on the xy plane.) Change the mode setting to SphereV. After exiting the mode menu, press ENTER to express the vector in spherical coordinates. We find [5.00∠53.13], which means

$$\text{magnitude of } \mathbf{A} = 5.00$$

and

$$\text{angle of } \mathbf{A} = 53.1°$$

counterclockwise from the positive x direction. (Answer)

If, instead, we press in **A** as [3,4,0], we find

$$[5.00\angle 53.13\angle 90.00]$$

which means

$$\text{magnitude of } \mathbf{A} = 5.00,$$

angle of **A** in the xy plane

$= 53.1°$ counterclockwise from the positive x direction,

and angle of **A** from the positive z direction $= 90.0°$. (Answer)

The last feature indicates that **A** lies in the xy plane.

To find the vector 6**A**, let's use the result in the SphereV mode. Press in 6[5.00∠53.13] and then press ENTER. The calculator returns [30.00∠53.13]. So

$$\text{magnitude of } 6\mathbf{A} = 30.0$$

and

$$\text{angle of } 6\mathbf{A} = 53.1°$$

counterclockwise from the positive x direction. (Answer)

Note that only the magnitude of **A** (and not the angle) is multiplied by the scalar 6.

(c) *Scalar (or Dot) Product.* Find the dot product 6.00**A** · **B**.

SOLUTION: Access the first vector menu by pressing 2nd VECTR. Next choose the MATH option and then the dot option. The screen then starts a line with the symbols dot(which we finish so that the line shows

$$\text{dot}(6[3,4,0],[-2,-4,3])$$

as the calculation to be made. Note that the two vectors must be of the same form. (We can take the product [2] dot [4] and [2,0] dot [4,5] but not [2] dot [4,5].)

Pressing ENTER, we get the result

$$6.00\mathbf{A} \cdot \mathbf{B} = -132. \quad \text{(Answer)}$$

(d) *Vector (or Cross) Product.* Find the product 6.00**A** × **B** in both unit-vector notation and magnitude-angle notation.

SOLUTION: Access the first vector menu by pressing 2nd VECTR. Then press F3 for the MATH option. Then press F1 for the cross option in the new menu. The screen then starts a line with the symbols cross(which we finish so that the line shows

$$\text{cross}(6[3,4,0],[-2,-4,3])$$

as the calculation to be made. Note that, here too, the two vectors must be of the same form.

Go to the mode menu, choose a setting of RectV, exit the menu, and then press ENTER. We find

$$[72.00 \quad -54.00 \quad -24.00]$$

which means that, in rounded-off numbers,

$$6.00\mathbf{A} \times \mathbf{B} = 72.0\mathbf{i} - 54.0\mathbf{j} - 24.0\mathbf{k}. \quad \text{(Answer)}$$

Resetting the mode menu to SphereV and pressing ENTER to repeat the calculation, we find

$$[93.15\angle -36.87\angle 104.93]$$

which means that 6**A** × **B** produces a vector with

$$\text{magnitude} = 93.2,$$

angle in the xy plane

$= 36.9°$ clockwise from the positive x direction,

and angle from the positive z direction $= 105°$. (Answer)

(e) *Storing Vectors.* Store vectors **A** and **B** on the calculator and then repeat parts (a) through (d).

SOLUTION: On a TI-85/86, we can store **A** under a name (say, A) with the following steps. Press in the vector as [3,4,0] and then press the STO→ key. The calculator is automatically ALPHA-locked. So, press the key with A printed above it and then ENTER. The calculator displays the vector.

Similarly, to store vector **B** under the name B, press in [−2,−4,3] and then press STO→ B ENTER. (Remember to use the negation key.)

Then, to find **B** − 5.00**A** in part (a), we press

ALPHA B − 5 ALPHA A

where we use the ALPHA key to enter a letter. Pressing ENTER gives the answer to part (a).

To find the magnitude and angle of **A** in part (b), we set the mode to SphereV, then press ALPHA, then A, and finally ENTER.

To find 6.00**A** · **B** in part (c), we access the dot option as we did previously, but in finishing the line, we now press

6 ALPHA A , ALPHA B)

so that the line reads DOT(6A,B) as the calculation to be made. Pressing ENTER gives us the result. Similarly, to find 6.00**A** × **B**, we access the cross option and then finish the line so that it reads CROSS(6A,B) as the calculation to be made. Pressing ENTER gives us the result.

Chapter Four
Motion in Two and Three Dimensions

QUESTIONS

18. Figure 4-44 shows the initial position i and the final position f of a particle. What are the particle's (a) initial position vector $\mathbf{r}_i$ and (b) final position vector $\mathbf{r}_f$, both in unit-vector notation? (c) What is the x component of the particle's displacement $\Delta\mathbf{r}$?

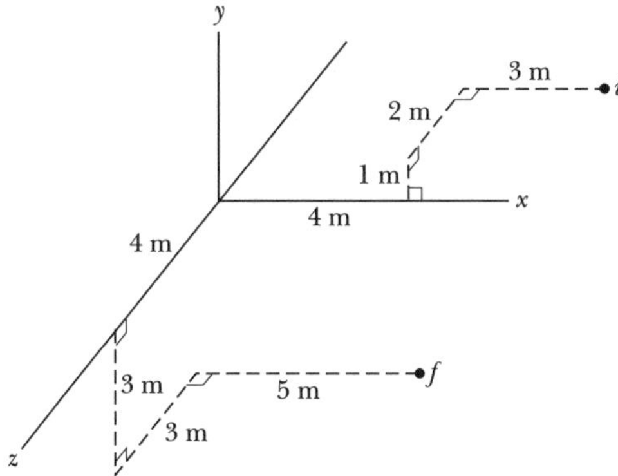

FIGURE 4-44 Question 18.

19. The position vectors of four infant wombats are given in the following table in terms of a coordinate system. Without using a calculator, rank the wombats according to their distances from the origin of the coordinate system, greatest first.

WOMBAT	POSITION VECTOR
a	$(-2\text{ m})\mathbf{i} + (3\text{ m})\mathbf{j} + (4\text{ m})\mathbf{k}$
b	$(2\text{ m})\mathbf{i} + (-3\text{ m})\mathbf{j} + (4\text{ m})\mathbf{k}$
c	$(3\text{ m})\mathbf{i} + (2\text{ m})\mathbf{j} + (4\text{ m})\mathbf{k}$
d	$(4\text{ m})\mathbf{i} + (-1\text{ m})\mathbf{j} + (2\text{ m})\mathbf{k}$

20. Figure 4-45 shows an overhead view of the path taken by a water strider as it darts over the surface of a pond. What is the direction of its velocity vector at (a) point A and (b) point B?

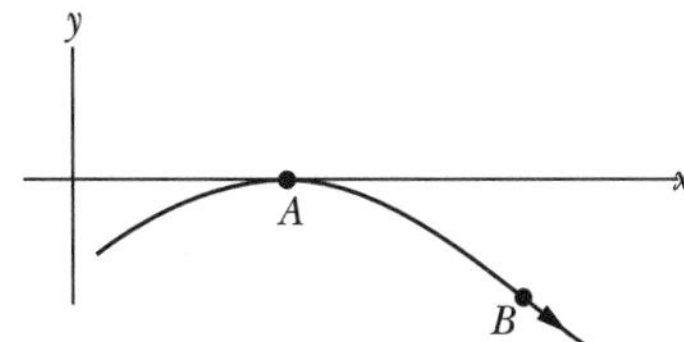

FIGURE 4-45 Question 20.

21. Figure 4-46*a*, *b*, and *c* gives, for three situations, the coordinates $x(t)$ and $y(t)$ of a particle moving in the xy plane. In each situation, the acceleration of the particle is constant and the graphs are drawn to the same scale. For each situation, which of the vectors shown in Fig. 4-46*d* best represents the acceleration of the particle?

22. In Fig. 4-47, a cream tangerine is thrown up past windows 1, 2, and 3, which are identical in size and regularly spaced ver-

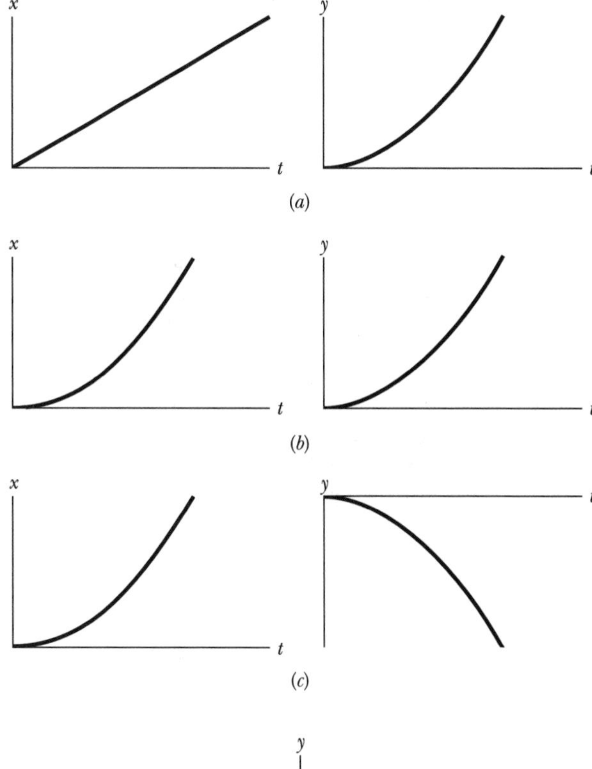

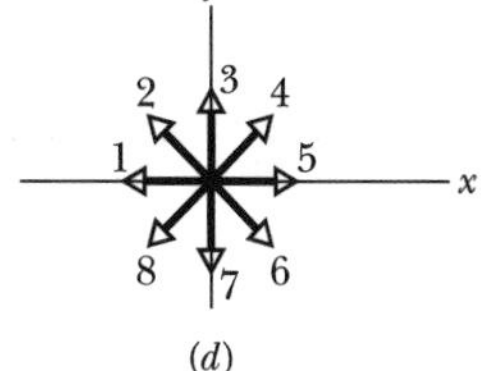

FIGURE 4-46 Question 21.

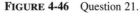

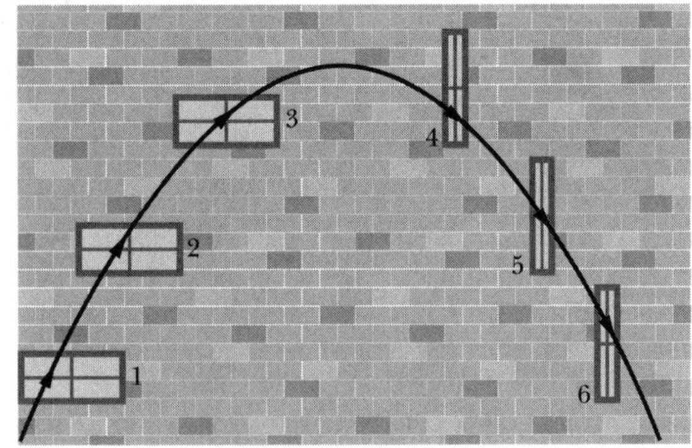

FIGURE 4-47 Questions 22 and 23.

tically. Rank those three windows according to (a) the time the cream tangerine takes to pass them and (b) the average speed of the cream tangerine during the passage, greatest first.

23. Continuation of Question 22: The cream tangerine then moves down past windows 4, 5, and 6, which are identical in size and irregularly spaced horizontally. Rank those three windows according to (a) the time the cream tangerine takes to pass them and (b) the average speed of the cream tangerine during the passage, greatest first.

24. You throw a ball with a launch velocity of $\mathbf{v}_i = (3\text{ m/s})\mathbf{i} + (4\text{ m/s})\mathbf{j}$ toward a wall, where it hits at height h_1 in time t_1 after the launch (Fig. 4-48). Suppose that the launch velocity were, instead, $\mathbf{v}_i = (5\text{ m/s})\mathbf{i} + (4\text{ m/s})\mathbf{j}$. (a) Would the time taken by the ball to reach the wall be greater than, less than, or equal to t_1, or is the question unanswerable without more information? (b) Would the height at which the ball hits be greater than, less than, or equal to h_1, or is the question unanswerable?

Suppose, instead, that the launch velocity were $\mathbf{v}_i = (3\text{ m/s})\mathbf{i} + (5\text{ m/s})\mathbf{j}$. (c) Would the time taken by the ball to reach the wall be greater than, less than, or equal to t_1, or is the question unanswerable? (d) Would the height at which the ball hits be greater than, less than, or equal to h_1, or is the question unanswerable?

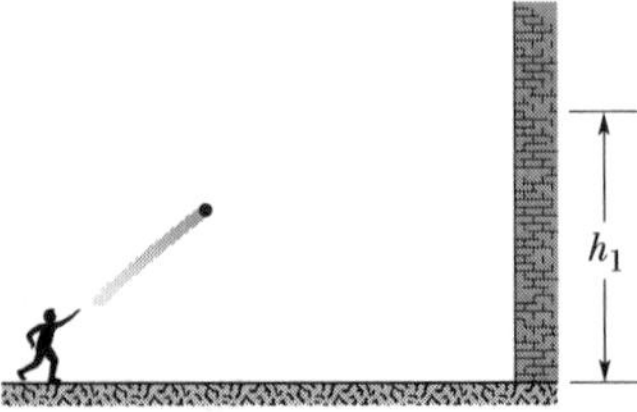

FIGURE 4-48 Question 24.

25. *Organizing question:* A marble is launched from the edge of a high cliff with a velocity of 20 m/s at an angle of 30° above the horizontal. (a) Set up equations, complete with known data, to find the horizontal displacement of the marble from the launch point when the vertical displacement is 2 m below the launch level. (b) Next, set up equations, complete with known data, to find the marble's velocity components v_x and v_y just then.

26. A rock can be launched at ground level with an initial velocity of $\mathbf{v} = (3\text{ m/s})\mathbf{i} + (4\text{ m/s})\mathbf{j}$ toward a wall. You can adjust the horizontal distance d between the launch site and the wall. With that adjustment, what are (a) the least speed and (b) the greatest speed at which the rock can hit the wall?

Suppose that at a certain value of d the rock's velocity is horizontal as it hits the wall. If you move the launch site closer to the wall, do (c) the height at which the rock hits the wall and (d) the speed with which it hits increase, decrease, or remain the same? (e) Does the angle between the velocity vector and the horizontal increase upward or increase downward?

27. Figure 4-49 shows four tracks (either half or quarter circles) that can be taken by a train, which moves at a constant speed. Rank the tracks according to the magnitude of the train's acceleration on the curved portion, greatest first.

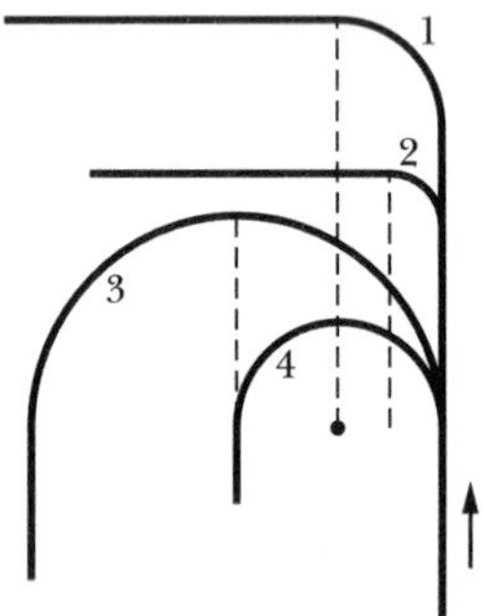

FIGURE 4-49 Question 27.

28. Figure 4-50 shows four paths along which a mechanical rabbit can move from point i to point f. Paths 1 and 3 are different portions of the same circle; path 2 is half of a different circle; and path 4 is straight. (a) If the rabbit must make the move in a certain time, then rank the paths according to the average velocity of the rabbit, greatest first. (b) If, instead, the rabbit must travel at a certain constant speed, then rank the paths according to the magnitude of the acceleration the rabbit will have, greatest first.

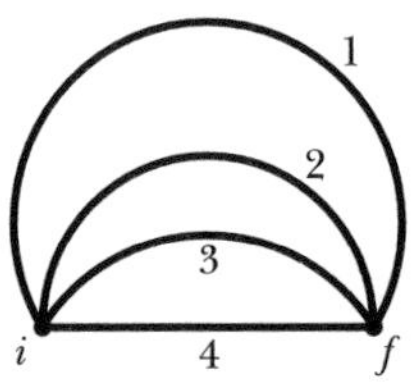

FIGURE 4-50 Question 28.

EXERCISES & PROBLEMS

98. A baseball is hit at ground level. The ball reaches its maximum height above ground level 3.0 s after being hit. Then 2.5 s after reaching its maximum height, the ball barely clears a fence that is 97.5 m from where it was hit. Assume the ground is level. (a) What maximum height above ground level is reached by the ball? (b) How high is the fence? (c) How far beyond the fence does the ball strike the ground?

99. The position $\mathbf{r}$ of a particle moving in the xy plane is given by $\mathbf{r} = 2t\mathbf{i} + [2\sin(\pi/4)(\mathrm{rad/s})t]\mathbf{j}$, where $\mathbf{r}$ is in meters and t is in seconds. (a) Calculate the x and y components of the particle's position at $t = 0$, 1.0, 2.0, 3.0, and 4.0 s and sketch the particle's path in the xy plane for the interval $0 \leq t \leq 4.0$ s. (b) Calculate the components of the particle's velocity at $t = 1.0$, 2.0, and 3.0 s. Show that the velocity is tangent to the path of the particle and in the direction the particle is moving at each time by drawing the velocity vectors on the plot of the particle's path in part (a). (c) Calculate the components of the particle's acceleration at $t = 1.0$, 2.0, and 3.0 s.

100. A 200 m wide river flows due east at a uniform speed of 2.0 m/s. A boat with a speed of 8.0 m/s relative to the water leaves the south bank pointed in a direction of 30° west of north. (a) What is the velocity of the boat relative to the ground? (b) How long does the boat take to cross the river?

101. Two seconds after being projected from ground level, a projectile is displaced 40 m horizontally and 53 m vertically above its point of projection. (a) What are the horizontal and vertical components of the initial velocity of the projectile? (b) At the instant the projectile achieves its maximum height above ground level, how far is it displaced horizontally from its point of projection?

102. A car travels around a flat circle on the ground, at a constant speed of 12 m/s. At a certain instant the car has an acceleration of 3 $\mathrm{m/s^2}$ toward the east. What are its distance and direction from the center of the circle at that instant if it is traveling (a) clockwise around the circle and (b) counterclockwise around the circle?

103. After flying for 15 min in a wind blowing 42 km/h at an angle of 20° south of east, an airplane pilot is over a town that is 55 km due north of the starting point. What is the speed of the airplane relative to the air?

104. Oasis A is 90 km west of oasis B. A camel leaves oasis A and during a 50 h period walks 75 km in a direction 37° north of east. The camel then walks toward the south a distance of 65 km in a 35 h period, after which it rests for 5.0 h. (a) What is the camel's displacement with respect to oasis A after resting? (b) What is the camel's average velocity from the time it leaves oasis A until it finishes resting? (c) What is the camel's average speed from the time it leaves oasis A until it finishes resting? (d) If the camel is able to go without water for five days (120 h), what must its average velocity be after resting if it is to reach oasis B just in time?

105. A hang glider is 7.5 m above ground level with a velocity of 8.0 m/s at an angle of 30° below the horizontal and a constant acceleration of 1.0 $\mathrm{m/s^2}$, up. (a) Assume $t = 0$ at the instant just described and write an equation for the elevation y of the hang glider as a function of t, with $y = 0$ at ground level. (b) Use the equation to determine the value of t when $y = 0$. (c) Explain why there are two solutions to part (b). Which one represents the time it takes the hang glider to reach ground level? (d) How far does the hang glider travel horizontally during the interval between $t = 0$ and the time it reaches the ground? (e) For the same initial position and velocity, what constant acceleration will cause the hang glider to reach ground level with zero velocity? Express your answer in terms of the unit vectors $\mathbf{i}$ (horizontal, in the direction of travel) and $\mathbf{j}$ (upward).

106. A frightened rabbit runs onto a large area of level, frictionless ice, with an initial velocity of 6.0 m/s toward the east. As the rabbit slides across the ice, the force of the wind causes it to have a constant acceleration of 1.4 $\mathrm{m/s^2}$, directed due north. Choose a coordinate system with the origin at the rabbit's initial position on the ice and the positive x axis directed toward the east. In unit-vector notation, what are the rabbit's (a) velocity and (b) position when it has slid for 3.0 s?

107. The magnitude of the velocity of a projectile when it is at its maximum height above ground level is 10 m/s. (a) What is the magnitude of the velocity of the projectile 1.0 s before it achieves its maximum height? (b) What is the magnitude of the velocity of the projectile 1.0 s after it achieves its maximum height? If we take $x = 0$ and $y = 0$ to be at the point of maximum height and positive x to be in the direction of the velocity there, where is the projectile (c) 1.0 s before and (d) 1.0 s after it achieves its maximum height?

108. A high diver pushes off horizontally with a speed of 2.00 m/s from the edge of a platform that is 10.0 m above the surface of the water. (a) What horizontal distance from the edge of the platform is the diver 0.800 s after pushing off? (b) What vertical distance above the surface of the water is the diver 0.800 s after pushing off? (c) How long after pushing off does the diver strike the water? (d) What horizontal distance from the edge of the platform does the diver strike the water?

109. The pitcher in a women's slow-pitch softball game releases the ball at a point 3.0 ft above ground level. A strobe plot of the position of the ball is shown in Fig. 4-51, where the flashes are 0.25 s apart and the ball is released at $t = 0$. (a) What is the initial speed of the ball? (b) What is the speed of the ball at the instant it reaches its maximum height above ground level? (c) What is that maximum height?

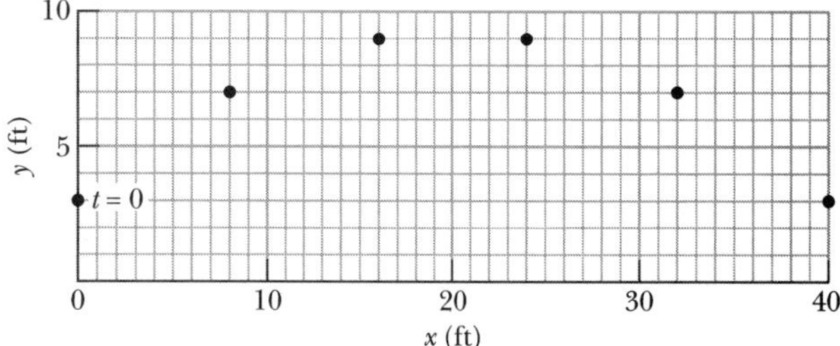

FIGURE 4-51 Problem 109.

110. A baseball is hit at Fenway Park in Boston at a point 0.762 m above home plate with an initial velocity of 33.53 m/s directed 55.0° above the horizontal. The ball is observed to clear the 11.28 m high wall in left field (known as the "green monster") 5.0° s after it is hit at a point just inside the left-field foul-line pole. Find (a) the horizontal distance down the left-field foul line from home plate to the wall; (b) the vertical distance by which the ball clears the wall; (c) the horizontal and vertical displacements of the ball with respect to home plate 0.50° s before it clears the wall.

111. For women's volleyball the top of the net is 2.24 m above the floor and the court measures 9.0 m by 9.0 m on each side of the net. Using a jump serve, a player strikes the ball at a point that is 3.0 m above the floor and a horizontal distance of 8.0 m from the net. If the initial velocity of the ball is horizontal, (a) what minimum magnitude must it have if the ball is to clear the net and (b) what maximum magnitude can it have if the ball is to strike the floor inside the back line on the other side of the net?

112. A track meet is held on a planet in a distant solar system. A shot-putter releases a shot at a point 2.0 m above ground level. A strobe plot of the position of the shot is shown in Fig. 4-52, where the flashes are 0.50 s apart and the shot is released at $t = 0$. (a) What is the initial velocity of the shot in terms of **i** (in the positive direction of x) and **j** (in the positive direction of y)? (b) What is the magnitude of the acceleration due to gravity on the planet? (c) How long after it is released does the shot reach the ground? (d) If an identical throw of the shot is made on the surface of Earth, how long after it is released will it reach the ground?

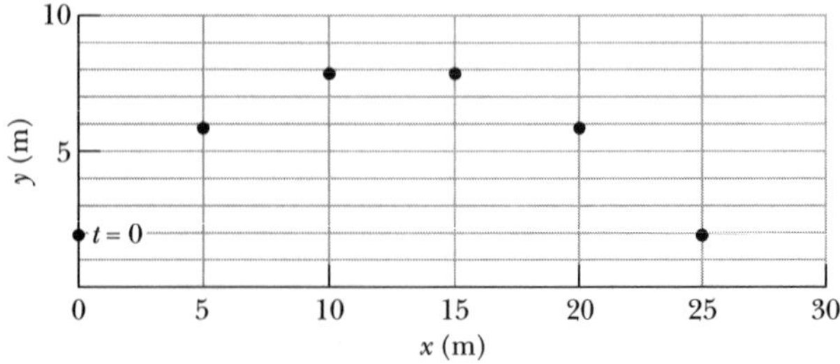

FIGURE 4-52 Problem 112.

113. A golf ball is struck at ground level. The speed of the golf ball as a function of the time is shown in Fig. 4-53, where $t = 0$ at the instant the ball is struck. (a) How far does the golf ball travel horizontally before returning to ground level? (b) What is the maximum height above ground level attained by the ball?

114. A carnival merry-go-round rotates about a vertical axis at a constant rate. A passenger standing on the edge of the merry-go-round has a constant speed of 3.66 m/s. For each of the following situations state where the passenger is with respect to the center of the merry-go-round. (a) The passenger has an acceleration of 1.83 m/s², east. (b) The passenger has an acceleration of 1.83 m/s², south.

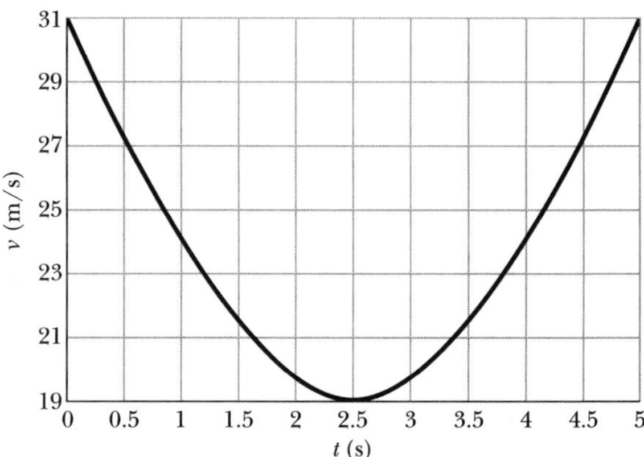

FIGURE 4-53 Problem 113.

115. A pilot maintains a constant speed while flying in a circular path (radius = 500 m) that lies in a horizontal plane. For each of the following situations, give the acceleration of the pilot. Unless there is sufficient information to determine both the magnitude and direction of the acceleration, you should answer "not possible to determine." (a) The airplane is directly north of the circle's center and its speed is 40 m/s. (b) The velocity of the airplane is 40 m/s, east. (c) The airplane is traveling clockwise around the circle as viewed from the ground and its velocity is 40 m/s, north. (d) The airplane is traveling counterclockwise around the circle as viewed from the ground and its speed is 40 m/s.

116. The pilot of an aircraft flies due east relative to the ground in a wind blowing 20 km/h toward the south. If the speed of the aircraft in the absence of wind is 70 km/h, what is the speed of the aircraft relative to the ground?

117. A 200 m wide river has a uniform flow speed through the jungle of 1.1 m/s toward the east. An explorer wishes to leave a small clearing on the south bank and cross the river in a powerboat that moves at a constant speed of 4.0 m/s with respect to the water. There is a clearing on the north bank 82 m upstream from a point directly opposite the clearing on the south bank. (a) In what direction must the boat be pointed in order to travel in a straight line and land in the clearing on the north bank? (b) How long will the boat take to cross the river and land in the clearing?

118. Ship A is located 4.0 km north and 2.5 km east of ship B. Ship A has a velocity of 22 km/h toward the south and ship B has a velocity of 40 km/h in a direction 37° north of east. (a) What is the velocity of A relative to B? (Express your answer in terms of the unit vectors **i** and **j**, where **i** is toward the east.) (b) Write an expression (in terms of **i** and **j**) for the position of A relative to B as a function of t, where $t = 0$ for the positions just described. (c) At what time is the separation between the ships least? (d) What is that least separation?

Clustered Problems

Cluster 1

119. In Fig. 4-54*a*, a ball is fired horizontally through a horizontal distance of 150 m, hitting the ground in 3.00 s. (a) At what height h above the ground is the ball fired? (b) What is the ball's speed just as it hits the ground?

120. In Fig. 4-54*b*, a ball is fired at an angle of 30.0° from the horizontal, landing 3.00 s later after traveling a horizontal distance of 100 m. (a) At what height h above the firing level does the ball land? (b) At what speed is it fired? (c) At what speed does it land?

121. In Fig. 4-54*c*, a ball is thrown up onto a building's roof, landing 4.00 s later, 20.0 m above where it was released. Its path just before landing is angled at 60.0° with the roof. (a) What horizontal distance d does it travel? (*Hint:* One way to the answer is to reverse the motion, as if a videotape of the throw is reversed. Then the landing, for which the angle is given, becomes the launch.) (b) What are the magnitude and direction of the velocity at which the ball is thrown?

122. In Fig. 4-54*d*, a ball that is thrown from the edge of a building hits the ground at an angle of 60.0° with the horizontal, 25.0 m from the building and 1.50 s after it is thrown. (a) From what height h is the ball thrown? (*Hint:* One way to the answer is to reverse the motion, as if a videotape of the throw is reversed. Then the landing, for which the angle is given, becomes the launch.) (b) What are the magnitude and direction of the velocity at which the ball is thrown?

123. In Fig. 4-54*e*, a baseball is hit at a height of 1.0 m and then caught at the same height. It travels alongside a wall, moving up past the top level 1.00 s after it is hit and then down past the top level 4.00 s later, 50.0 m farther along the wall. (a) What horizontal distance is traveled by the ball from the hit to the catch? (b) What are the magnitude and direction of the ball's velocity just after being hit? (c) How high is the wall?

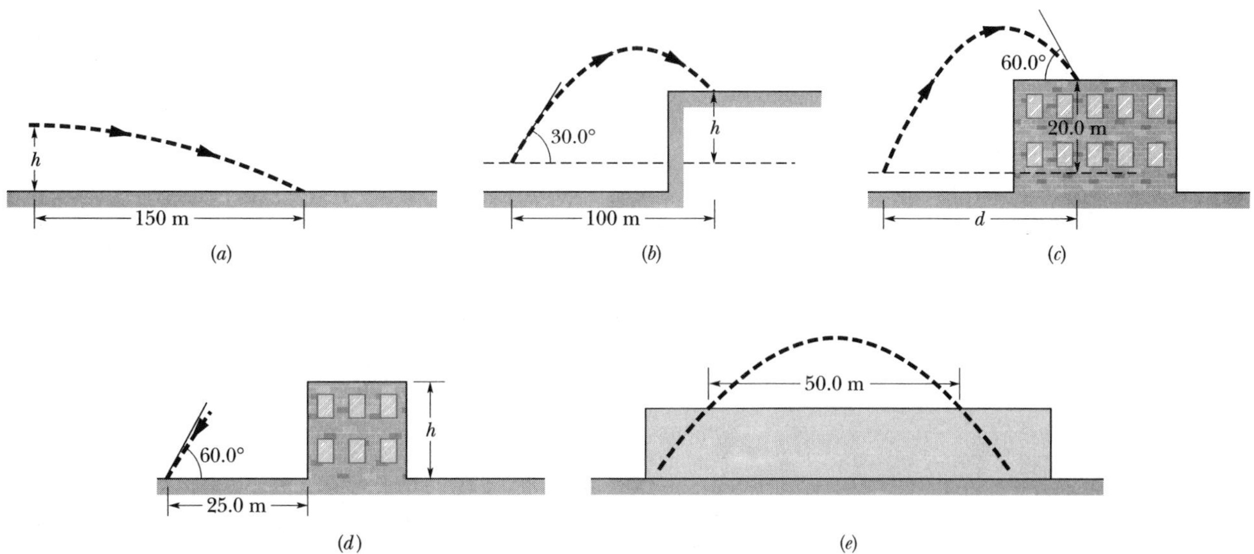

FIGURE 4-54 Problems 119 through 123.

Cluster 2

124. In Fig. 4-55*a*, a ball is thrown horizontally from a height of 1.50 m at a fence that is 10.0 m away, barely passing through a hole that is centered 0.500 m below the throwing level. (a) What is the initial velocity of the ball? (b) How far horizontally has the ball traveled when it hits the ground on the far side of the fence? (c) What is the speed of the ball when it hits the ground?

125. In Fig. 4-55*b*, a ball is thrown at an angle of 30.0° above the horizontal and toward a window of a building that is 20.0 m away. The ball barely clears the bottom of the window, which is 5.00 m above the level from which the ball was thrown. (a) What is the initial velocity of the ball? (b) The floor on the other side of the window is 1.00 below the window. What is the horizontal distance between the window and the point at which the ball hits the floor?

126. In Fig. 4-55*c*, a golf ball is hit toward a building such that just as it reaches the edge of the roof, its velocity is horizontal and it goes rolling or sliding along the roof (rather than bouncing). The ball is launched 30.0 m from the building and 20.0 m below the roof's surface. What are the speed and angle of the launch?

127. In a golfing challenge, you must hit a golf ball through a pair of small, identical holes in vertical walls that are 6.00 m apart (Fig. 4-55*d*). Both holes are 10.0 m from the ground; the nearest wall is 20 m from the ball's launch point. What are the required

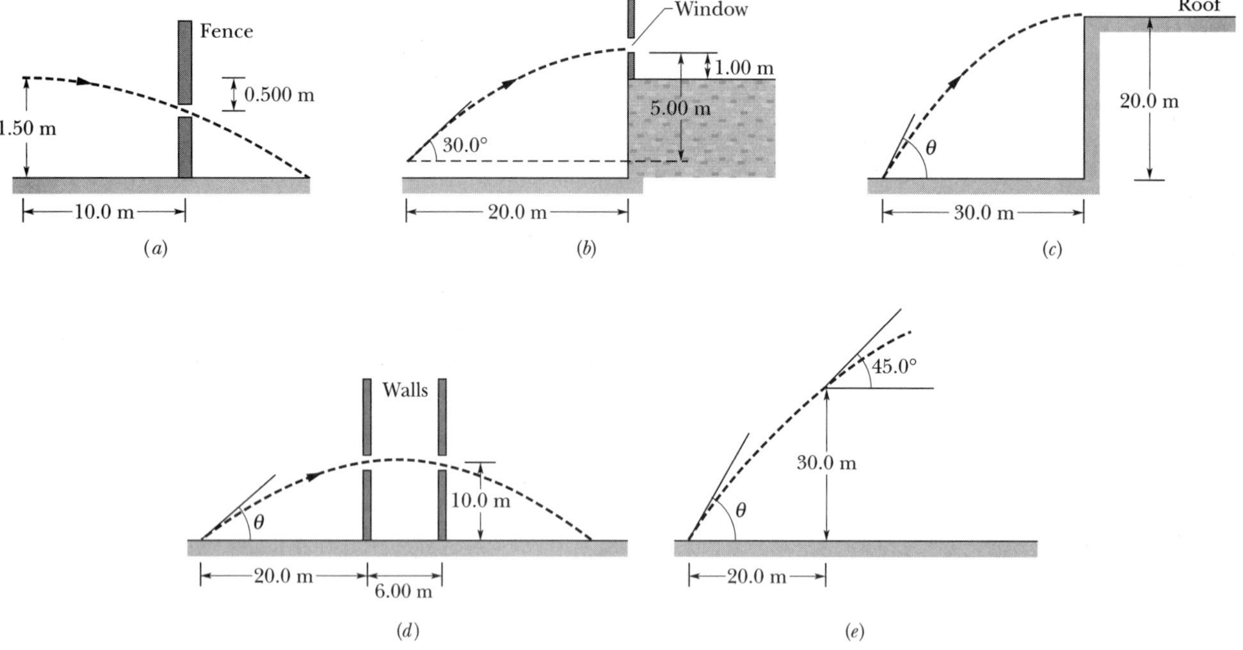

FIGURE 4-55 Problems 124 through 128.

speed and angle of launch? (*Hint:* How far horizontally does the ball travel to reach its highest point? How long does it take to reach there, in terms of the vertical component of the launch velocity and then in terms of the horizontal component of the launch velocity? In terms of that horizontal component, how long does the ball take to reach the first hole?)

128. During the flight of a golf ball, the ball passes a point at a vertical distance of 30.0 m and a horizontal distance of 20.0 m from where it was hit (Fig. 4-55*e*). At that point, the ball's trajectory is angled upward by 45.0° from the horizontal. What are (a) the magnitude and direction of the ball's initial velocity, (b) the maximum height reached by the ball, and (c) the range of the ball? (*Hint:* What does the angle of 45.0° imply about the horizontal and vertical components of the ball's velocity at that point?)

Cluster 3

129. You are to throw a ball with a speed of 12.0 m/s at a target that is 5.00 m above the level at which you release the ball (Fig. 4-56*a*). You want the ball's velocity to be horizontal at the instant it reaches the target. (a) At what angle θ must you throw the ball? (b) What is the horizontal distance from the release point to the target? (c) What is the speed of the ball just as it reaches the target?

130. You are to throw a ball with a speed of 12.0 m/s. When the ball passes through point *A* in Fig. 4-56*b*, 3.00 m above the point at which you release it, its trajectory is to be angled downward by 45.0° from the horizontal. (a) What is the horizontal distance between point *A* and the release point? (*Hint:* What does the angle of 45.0° imply about the horizontal and vertical components of the ball's velocity at point *A*?) (b) What is the speed of the ball at point *A*?

131. In a game of American football, a kicker must kick a field goal by sending the football over a crossbar that is horizontally 30.0 m and vertically 5.00 m from the point of the kick (Fig. 4-56*c*). Let v_0 represent the speed at which the football is launched. (a) Set up an equation, complete with v_0 and known data, to find the minimum angle $\theta_{0,\min}$ and the maximum angle $\theta_{0,\max}$ at which the football must be launched if it is to just barely clear the crossbar. Use Eq. 4-19 and the trigonometric identity

$$\frac{1}{\cos^2\theta} = \sec^2\theta = 1 + \tan^2\theta.$$

(b) Plot $\theta_{0,\min}$ and $\theta_{0,\max}$ as a function of v_0 for the values of v_0 from 18 m/s to 25 m/s.

132. In the situation of Problem 131, what is the minimum initial velocity (magnitude and direction) at which the football can be kicked and still pass over the crossbar?

(*Hint:* The minimum initial velocity occurs when $\theta_{0,\min} = \theta_{0,\max}$; that is, only one launch angle then allows the football to barely pass over the crossbar. To find the answer graphically, examine the curves in part (b) of Problem 131. To find the answer analytically, use the equation requested in part (a) for the launch angle. First, set the expression inside the square root equal to zero and solve that expression for the launch speed v_0 [a polynomial

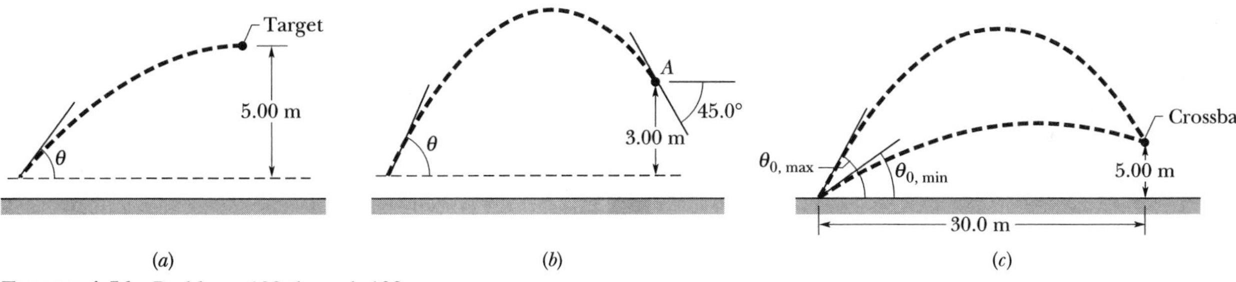

FIGURE 4-56 Problems 129 through 133.

solving program on a calculator helps]. Then substitute that value for v_0 into the equation for the launch angle and solve for the launch angle.)

133. (a) If the kicker of Problem 131 can kick a football with an initial speed of only 15.0 m/s, what is the maximum horizontal distance x_{max} at which the football can be kicked over the crossbar? (b) At what angle must the ball be kicked? (*Hint:* Use the hint of Problem 132.)

Graphing Calculators

SAMPLE PROBLEM 4-17

Parametric Plotting. Graphically determine what speed is required by the stuntman in Sample Problem 4-7 to reach the second building.

SOLUTION: We can plot the flight of the stuntman and then adjust his initial horizontal velocity until we find that he reaches the second building. We mentally superimpose an *xy* coordinate system on Fig. 4-15, placing the origin at the right side of the first building, level with the roof of the second building, with the *x* axis extending rightward and the *y* axis extending upward. From Eqs. 4-15 and 4-16, with $\theta_0 = 0$ and $v_0 = 4.5$ m/s, the stuntman's flight between buildings is then given by

$$x = 4.5t \quad \text{and} \quad y = 4.8 - 4.9t^2,$$

with *x* and *y* in meters and *t* in seconds.

We want to make a parametric plot of the flight; that is, we want to plot *y* versus *x*, where both coordinates depend on *t*. To do so on a TI-85/86, we first check the mode menu by pressing in 2nd MODE. If Param is not highlighted, move the cursor down to it and then press ENTER. Then exit the menu.

Press GRAPH and then choose E(t)= from the menu. On the screen we enter the motion along the *x* axis by completing the first line as

xt1 = 4.5t

with the F1 key used to enter the t symbol. Press ENTER and then enter the motion along the *y* axis by completing the next line as

yt1 = −4.9t² + 4.8

where the negation sign is required for the first term and the square can be obtained with the x² key or by pressing the ^ key and then the 2 key.

We then set the range or window of the graph: Press 2nd M2 to choose RANGE from the higher menu line. Set tMin = 0, tMax = 2 (to graph from time $t = 0$ to $t = 2$ s), and tStep = .01 (in steps of 0.01 s). Next set xMin = 0, xMax = 6.5 (slightly farther than the 6.2 m to the second building), xScl = 1 (tick-marks every 1 m). Then set yMin = −1 (lower than the second roof), yMax = 5 (slightly higher than the initial height of 4.8 m), and yScl = 1. Then choose the GRAPH option from the screen menu. Note that the plot passes through $y = 0$ (the level of the second roof) well before the second building is reached, which would alarm the stuntman.

Let us insert other values for v_0 to see if the stuntman can reach the second roof if he runs somewhat faster. Return to the xt1 and yt1 statements by choosing the E(t)= option. (If the menu is not present, press GRAPH to see it.) Place the cursor on the 4 in the first line and then press DEL three times to erase the 4.5 there. Press 2nd INS to be able to insert into the xt1 statement.

To insert several values for v_0 to be plotted on the same graph, we use a list. Press 2nd LIST, choose the symbol { to start a list, and then press in the v_0 values

4.8,5.2,5.8,6.3

where the comma key is used to separate the values. Next, close the list with the symbol } and exit the list menu to return to the graph menu. Choosing the GRAPH option, we soon see a graph of four curves that correspond to the four v_0 values in the list. We find that the last speed (6.3 m/s) barely allows the stuntman to reach the second roof. Thus

$$v_0 = 6.3 \text{ m/s} \qquad \text{(Answer)}$$

gives a successful stunt.

To change the format of the graph, press MORE and then choose FORMT. The format menu is identical to that described in Sample Problem 2-15 for functional plotting. We can also store a list under a name, such as V, and then insert that name into the xt1 statement (see Sample Problem 2-13). If we want to return to the calculations later, we can store the graph database by choosing STGDB in the menu (press MORE to find that option) and then later recall the database by finding and choosing RCGDB. (See Sample Problem 2-15(d).)

134. When balls are juggled through a high arc from one hand to the other, launching each ball with the same speed and angle is crucial, otherwise catching them becomes improbable. Assume that a ball is launched with a speed of 6.3 m/s and then caught at the launch height. (a) At what optimum angle θ_{opt} relative to horizontal (in degrees) should it be launched if the hands are to have a comfortable horizontal separation of 40 cm? (Is the answer displayed by the calculator the angle? See Sample Problem 4-8(a).) (b) Using the parametric graphing capability on a calculator, plot the ball's flight.

Using the list capability of the calculator, set up list L_1 of the values of five angles: θ_{opt}, $\theta_{opt} \pm 1°$, and $\theta_{opt} \pm 2°$. Then substitute this list for the symbol of the launch angle that you used in step (b) and replot the ball's flight. (c) For each of the angles in the list, estimate from the plots how far horizontally the ball travels to return to the launch height. (d) Also from the plots, estimate the error in catching the ball, that is, the horizontal distance between the returning ball and the catching hand. (e) Using list L_1 in a calculation line, generate a list of the errors for the five launch angles. You will find that even a small error in the launch angle can throw the ball off the mark.

135. A shaving razor with twin blades is designed to remove hair below the skin surface to provide a longer-lasting clean-shaven look. As with a single-blade razor, the first blade catches, drags, and then cuts off the exposed portion of a hair. As the hair is then retracted by the skin, a short section briefly emerges. If the second blade reaches the hair before that section disappears into the skin, it can cut off all or part of the section.

Figure 4-57*a* shows the initial arrangement as the first blade reaches a hair with length *a* below the skin surface. Figure 4-57*b* shows the arrangement just as the first blade makes its cut: θ_0 is the angle between the remaining shaft and a perpendicular to the skin and y_0 is the distance through which the hair's root has been raised. As the remaining shaft is retracted, the angle θ with the perpendicular and the height y of the root both decrease. The decrease in θ causes a section of height h to briefly protrude from the skin (Fig. 4-57*c*). (a) Find h in terms of a, y, and θ.

Assume that $a = 2.1$ mm, $\theta_0 = 1.0$ rad, and $y_0 = 1.0$ mm and that the decreases in θ and y are given by

$$\theta = \theta_0 e^{-30t} \quad \text{and} \quad y = y_0 e^{-30t},$$

where t (in seconds) is the time after the first blade's cut. From a graph of $h(t)$, determine (b) the maximum exposed height h_{max} and (c) the time t_{max} at which it occurs. (d) Assuming the blade separation is the standard 1.5 mm, at what speed V should the razor be drawn across the skin for the second blade to cut off the height h_{max}? (Adapted from "On the Optimum Hand Speed for Two-Blade Razor Shaving," by A. D. Fitt, A. A. Lacey, and P. Wilmott, *Teaching Mathematics and Its Applications,* 1991, Vol. 10, No. 3, pp. 122–126.)

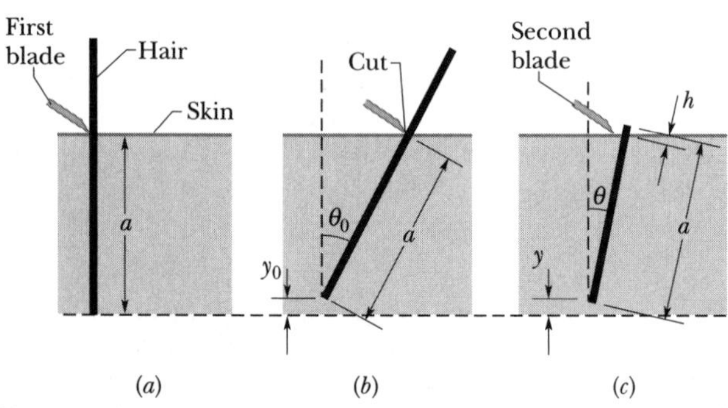

FIGURE 4-57 Problem 135.

SAMPLE PROBLEM 4-18

Using the Equation Solver. A projectile is to be launched at ground level with an initial speed $v_0 = 25.0$ m/s. It must pass through a point that is 12.0 m above the ground when it has traveled 5.00 m horizontally. At what initial angle must it be launched?

SOLUTION: We can solve the problem by using Eq. 4-19 with the Equation Solver on a TI-85/86. To store the equation so that we can solve other problems with it without having to rekey the equation, press in

YP = XP*tan θ0 − 9.8*XP²/(2(V0*cos θ0)²)

while in the home screen. Here we let YP represent the vertical position y, XP the horizontal position x, and θ0 the launch angle θ_0. Pressing ENTER stores the expression at the right of the equals symbol under the variable name YP.

The variables XP, V0, and θ0 might have old assigned values that are left over from previous work on the calculator. We could wait until we turn on the Equation Solver to change the values. However if those variables have been used to store expressions, we shall have a mess in the Equation Solver. To avoid that possibility, we assign zeros to the variables by pressing in the following sequence: 0 STO→ XP and then ENTER; 0 STO→ V0 and then ENTER; and finally 0 STO→ θ0 and then ENTER.

Next press 2nd SOLVER. If anything appears to the right of the colon, press CLEAR. Find and then choose the menu option YP and then press ENTER. The top line then shows the name YP of the stored expression. The second line gives the generic name exp for the expression. The next three lines show the three variables in the expression. (If we had neglected to assign zero to any of these variables, their old values would be shown. We could then move to those variables, in turn, and change their values as explained below.) The last line gives the range over which an answer will be sought. To speed up a solution or to find a particular answer when there are several, we can narrow that range. We shall not do so here.

Press in the given height of 12 for exp, the given horizontal distance of 5 for XP, and the given speed of 25 for V0. Then move to the line $\theta 0$ (which shows a value of zero) and choose SOLVE from the menu. When the busy signal at the upper right of the screen disappears, the calculator tells us that $\theta 0 = 69.89$, which means that the launch angle should be

$$\theta_0 = 69.9^\circ. \qquad \text{(Answer)}$$

(To see the power of the Equation Solver, try solving this problem with the usual methods.) We could have speeded up the solution if we had inserted a guess of, say, 50 for $\theta 0$. To do this, press in the guess on the $\theta 0$ line and then, with the cursor still there, choose the SOLVE option. A guess also helps when there is more than one answer.

Instead of storing an expression before going to the Equation Solver, you can compose one after the colon in the first screen of the solver. (However, first be sure that none of the variables have old expressions stored in them.) Press ENTER to then go to the second screen. If the calculator signals an error, it might mean that there is no answer.

Chapter Five
Force and Motion—I

QUESTIONS

16. Two horizontal forces,

$$\mathbf{F}_1 = (3\text{ N})\mathbf{i} - (4\text{ N})\mathbf{j} \quad \text{and} \quad \mathbf{F}_2 = -(1\text{ N})\mathbf{i} - (2\text{ N})\mathbf{j},$$

pull a banana split across a frictionless luncheon counter. Without using a calculator, tell which of the vectors in the free-body diagram of Fig. 5-63 best represent (a) $\mathbf{F}_1$ and (b) $\mathbf{F}_2$. What is the net-force component along (c) the x axis and (d) the y axis? Into which quadrants do (e) the net-force vector point and (f) the acceleration vector point?

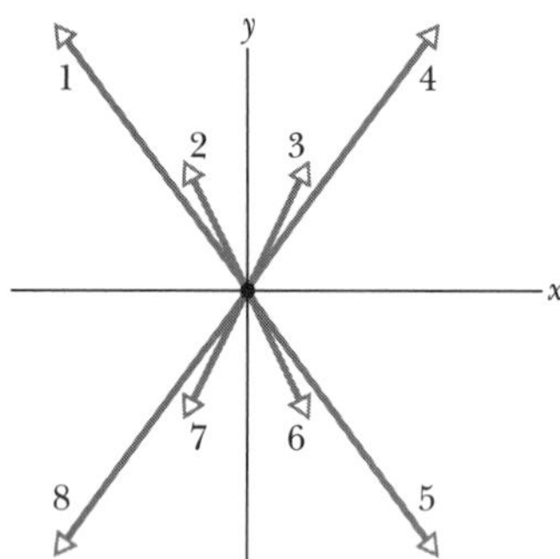

FIGURE 5-63 Question 16.

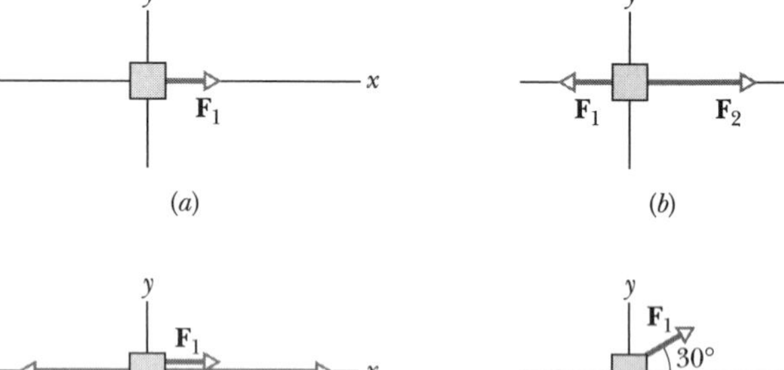

(c)

(d)

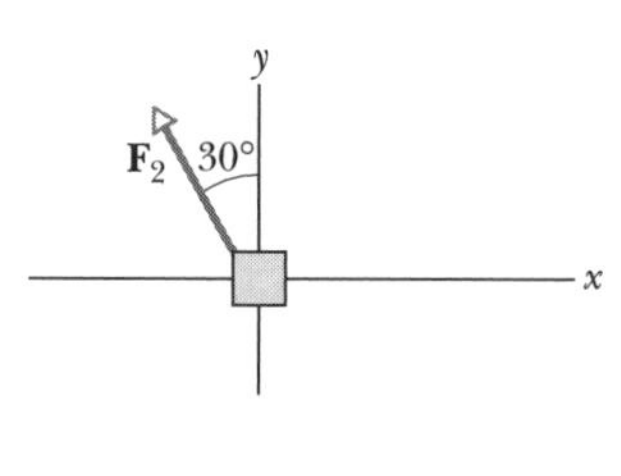

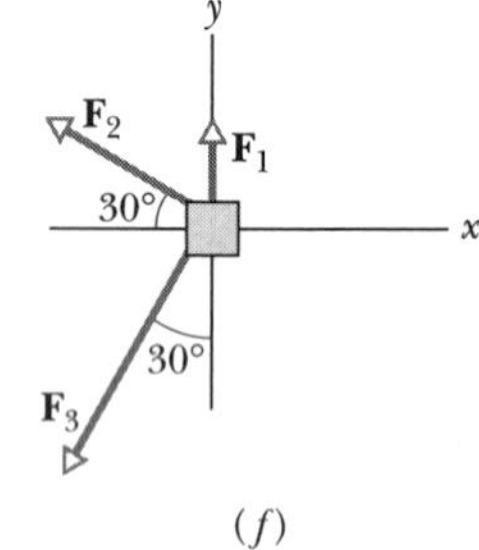

FIGURE 5-64 Question 17.

17. *Organizing question:* Figure 5-64 shows overhead views of six situations in which a 2 kg box is pulled over a frictionless floor by one or more forces. The magnitudes of the forces are $F_1 = 1$ N, $F_2 = 2$ N, $F_3 = 3$ N, and $F_4 = 4$ N. For each situation, set up equations, complete with known data, to find the acceleration components a_x and a_y; use the first two scalar equations of Eq. 5-2 ($\Sigma F_x = ma_x$ and $\Sigma F_y = ma_y$).

18. *Organizing question:* Figure 5-65 gives overhead views of six situations in which a 2 kg box is accelerated at a magnitude of 1 m/s^2 along a frictionless floor by one or more forces. The magnitude of force $\mathbf{F}_1$ differs for each situation, but the magnitudes of the other forces are always $F_2 = 2$ N, $F_3 = 3$ N, and $F_4 = 1$ N. For each situation, set up the first two scalar equations of Eq. 5-2 ($\Sigma F_x = ma_x$ and $\Sigma F_y = ma_y$), complete with known data.

19. Figure 5-66 shows overhead views of six situations in which a pirate's chest is pulled over a frictionless surface by pirates who apply horizontal forces $\mathbf{F}_1$ and $\mathbf{F}_2$ in the directions indicated. No other horizontal forces act on the chest in these situations. An acceleration vector is also indicated. In which of the situations, if any, is the direction of that vector physically impossible to achieve for any possible values of the magnitudes F_1 and F_2 (and, in part (f), the angle θ)?

20. Figure 5-67 gives the free-body diagram for four situations in which an object is pulled by several forces across a frictionless floor, seen from an overhead perspective. In each situation, determine if the object is in equilibrium or is accelerating along (a) the x axis and (b) the y axis. In each, and without calculation, determine the direction of the acceleration, giving either a quadrant or a direction along an axis.

21. Figure 5-68 gives three graphs of velocity component $v_x(t)$ and three graphs of velocity component $v_y(t)$. The graphs are not

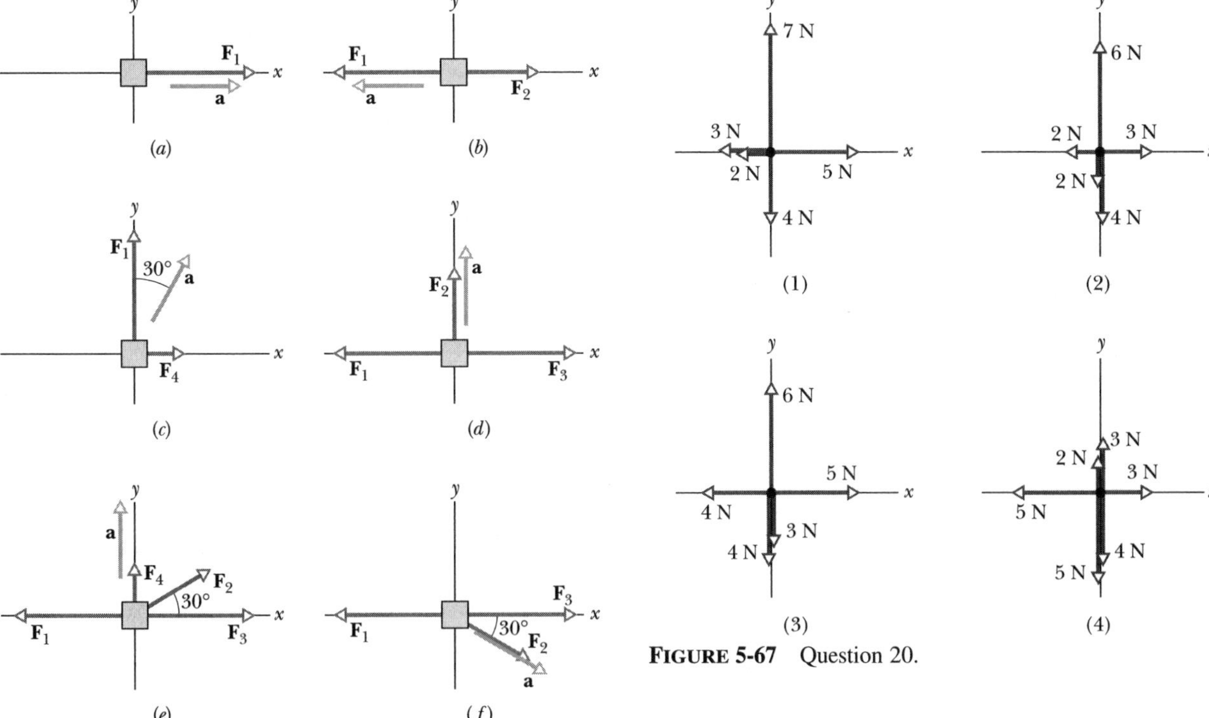

FIGURE 5-65 Question 18.

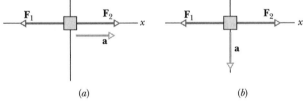

FIGURE 5-67 Question 20.

to scale. Which $v_x(t)$ graph and which $v_y(t)$ graph best correspond to which of the four situations in Question 20 and Fig. 5-67?

22. In Fig. 5-69, two forces $\mathbf{F}_1$ and $\mathbf{F}_2$ act on a Rocky-and-Bullwinkle lunch box as the lunch box slides at constant velocity over a frictionless lunchroom floor. We are to decrease the angle θ of $\mathbf{F}_1$ without changing the magnitude of $\mathbf{F}_1$. To keep the lunch box sliding at constant velocity, should we increase, decrease, or maintain the magnitude of $\mathbf{F}_2$?

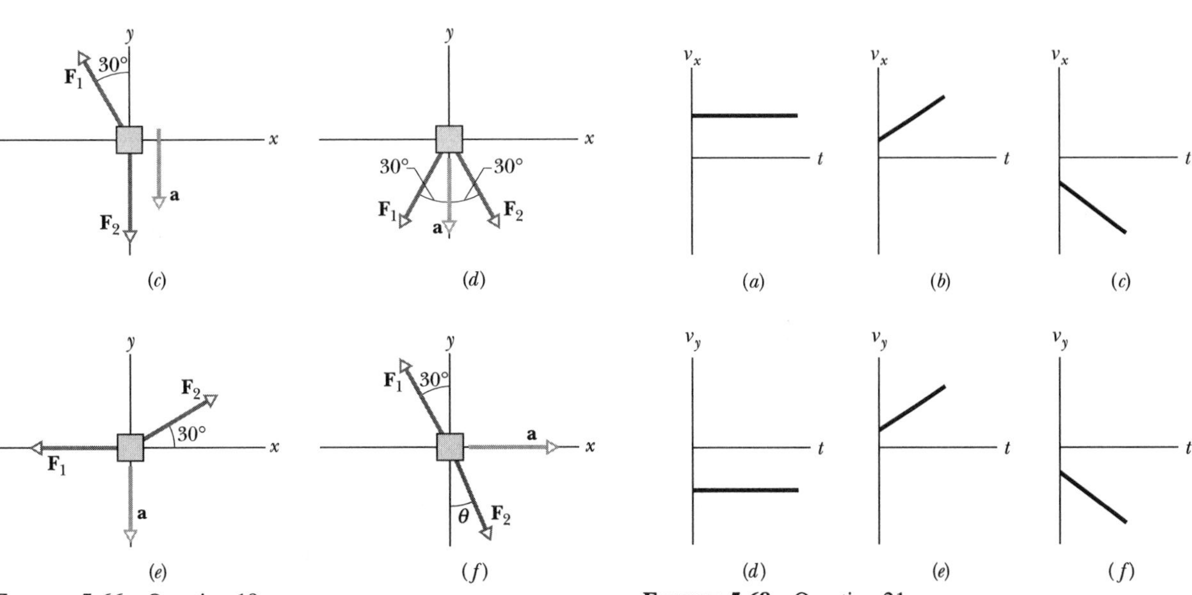

FIGURE 5-66 Question 19.

FIGURE 5-68 Question 21.

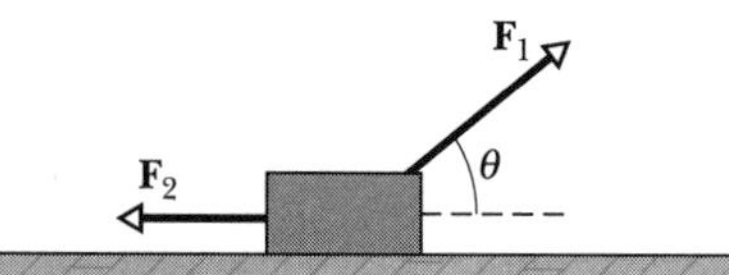

FIGURE 5-69 Question 22.

23. Figure 5-70 shows an applied force **F** pushing a parade of objects: a clown's suitcase *s*, a costume box *b*, and a railroad trunk *t*. Assume that the floor is slippery enough for the friction between the floor and these objects to be negligible. (a) Name the action–reaction pairs involved in the suitcase–box–trunk interactions. (b) Which forces in those pairs determine the rate at which the speed of the costume box changes?

FIGURE 5-70 Question 23.

24. Figure 5-71 shows three groups of four identical videocassettes that are pushed, as a group, across a frictionless video-store counter by a horizontal 10 N force. Rank the groups according to (a) the magnitude of their acceleration and (b) the magnitude of the force between their two stacks, greatest first.

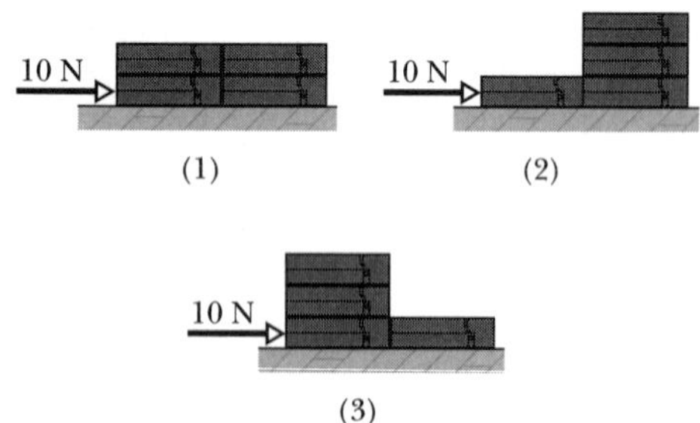

FIGURE 5-71 Question 24.

25. Seven identical dominoes are to be stacked in three columns (Fig. 5-72 gives an example) and pushed across a frictionless ice rink by a horizontal 10 N force. How many dominoes should be in each column, with a minimum of one, (a) to maximize the acceleration of the dominoes, (b) to maximize the force on column *C* due to column *B*, (c) to maximize the *net* force on column *B* due to columns *A* and *C*, and (d) to maximize the force on column *B* due to column *A*?

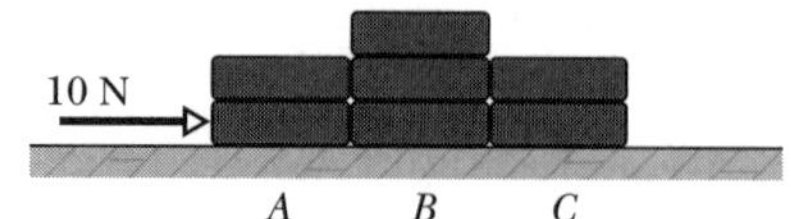

FIGURE 5-72 Question 25.

26. At time $t = 0$, a single force **F** begins to act on a particle-like cookie that is moving along an x axis through deep space. The cookie continues to move along that axis. (a) For time $t > 0$, which of the following is a possible function $x(t)$ for the cookie's position: (1) $x = 4t - 3$, (2) $x = -4t^2 + 6t - 3$, (3) $x = 4t^2 + 6t - 3$? (b) For which function is **F** directed opposite the cookie's initial direction of motion?

27. During a space walk, three astronauts race one another by using their jet backpacks. The race begins at $t = 0$ and with them at rest relative to the racecourse alongside their ship. During the race, their position functions $x(t)$ along the racecourse are the following, with x in meters and t in seconds: (a) $x = 2t^2$, (b) $x = 8t - 3$, (c) $x = 4t^2 + 2$. Assuming that the astronauts have the same mass, rank them according to the magnitudes of the forces on them from the jet backpacks, greatest first.

28. *Organizing question:* In each situation of Fig. 5-73, a force **F** of magnitude 5 N is applied to a 0.2 kg tin of sardines that can slide along a frictionless plane tilted at 30°. For each situation, set up equations, complete with known data, to find the acceleration a of the tin and the magnitude N of the normal force on the tin; use the tilted xy coordinate system shown in Fig. 5-22 and the first two scalar equations of Eq. 5-2 ($\Sigma F_x = ma_x$ and $\Sigma F_y = ma_y$).

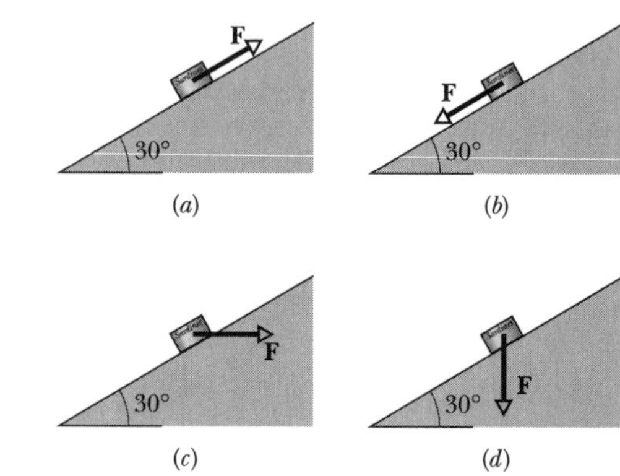

FIGURE 5-73 Question 28.

EXERCISES & PROBLEMS

77. A motorcycle and 60.0 kg rider accelerate at 3.0 m/s² up a ramp inclined 10° above the horizontal. (a) What is the magnitude of the net force acting on the rider? (b) What is the magnitude of the force exerted by the motorcycle on the rider?

78. A block weighing 3.0 N is at rest on a horizontal surface. A 1.0 N upward force is applied to the block by means of an attached vertical string. What are the magnitude and the direction of the force of the block on the horizontal surface?

79. A 1.0 kg mass on a frictionless inclined surface is connected to a 2.0 kg mass as shown in Fig. 5-74. The pulley is massless and frictionless. The 2.0 kg mass is acted on by an upward force $\mathbf{F} = 6.0$ N and has a downward acceleration of 5.5 m/s^2. (a) What is the tension in the connecting cord? (b) What is the angle β?

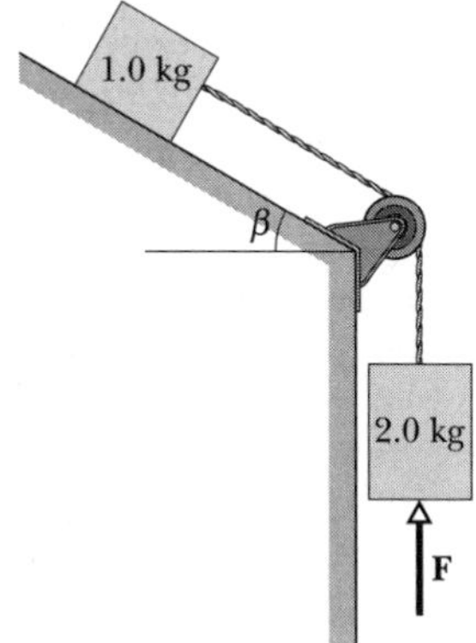

FIGURE 5-74 Problem 79.

80. Only two forces act on a 3.0 kg object that moves with an acceleration of 3.0 m/s^2 in the positive y direction. If one of the forces acts in the positive x direction and has a magnitude of 8.0 N, what is the magnitude of the other force?

81. A 0.20 kg hockey puck has a velocity of 2.0 m/s toward the east as it slides over the frictionless surface of a frozen lake. What are the magnitude and direction of the average force that must act on the puck during a 0.50 s interval to change its velocity to (a) 5.0 m/s, due west, and (b) 5.0 m/s, due south?

82. Only two forces act on a 3.0 kg mass. One force is 9.0 N, acting due east, and the other is 8.0 N, acting in a direction 62° north of west. What is the magnitude of the acceleration of the mass?

83. When the system in Fig. 5-75 is released from rest, the 3.0 kg mass has an acceleration of 1.0 m/s^2 to the right. The surfaces and the pulley are frictionless. (a) What is the tension in the connecting cord? (b) What is the value of M?

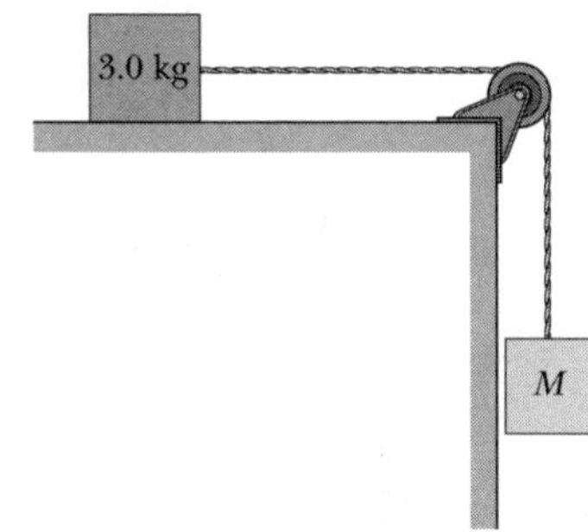

FIGURE 5-75 Problem 83.

84. A 0.20 kg hockey puck has a velocity of 2.0 m/s toward the east as it slides over the frictionless surface of an ice hockey rink. What constant net force (magnitude and direction) must act on the puck during an 0.40 s interval to change its velocity to (a) 5.0 m/s toward the west and (b) 5.0 m/s toward the south?

85. The only two forces acting on a body have magnitudes of 20 N and 35 N and directions that differ by 80°. The resulting acceleration has a magnitude of 20 m/s^2. What is the mass of the body?

86. A 12 kg penguin runs onto a large area of level, frictionless ice with an initial velocity of 6.0 m/s toward the east. As the penguin slides across the ice, it is pushed by the wind with a force that is constant in magnitude and direction. Fig. 5-76 shows an overhead strobe photograph of the penguin as it slides on the frictionless ice surface; the positive direction of the x axis is toward the east. The penguin first makes contact with the ice at $t = 0$, and the flashes on the strobe camera record the position of the penguin at 1.0 s intervals. What are the magnitude and direction of the force of the wind on the penguin as it slides across the ice?

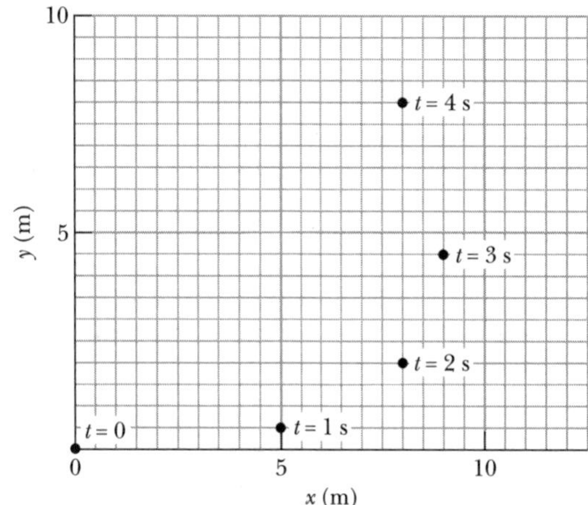

FIGURE 5-76 Problem 86.

87. A 45 kg woman is ice-skating toward the east on a frictionless frozen lake when she collides with a 90 kg man who is ice-skating toward the west. The maximum force exerted on the woman by the man during the collision is 180 N, west. (a) What is the maximum force exerted on the man by the woman during the collision? (b) What is the maximum acceleration experienced by the woman during the collision? (c) What is the maximum acceleration experienced by the man during the collision?

88. A 29.0 kg child, who has a 4.50 kg backpack strapped to him, is waiting patiently for the school bus when he becomes restless and jumps in the air. Give the magnitude and direction of the following forces: (a) the force of the child–backpack on the sidewalk while the child waits, (b) the force of the child–backpack on the sidewalk while in the air, (c) the force of the child alone on Earth while waiting patiently, and (d) the force of the child alone on Earth while in the air.

89. A 50 kg passenger rides in an elevator that starts from rest on the ground floor of a building at $t = 0$ and rises to the top floor during a 10 s interval. The acceleration of the elevator as a function of the time is shown in Fig. 5-77, where positive values of the acceleration mean that it is directed upward. Give the magnitude and direction of the following forces: (a) the maximum force exerted by the floor on the passenger, (b) the minimum force exerted by the floor on the passenger, and (c) the maximum force exerted by the passenger on the floor.

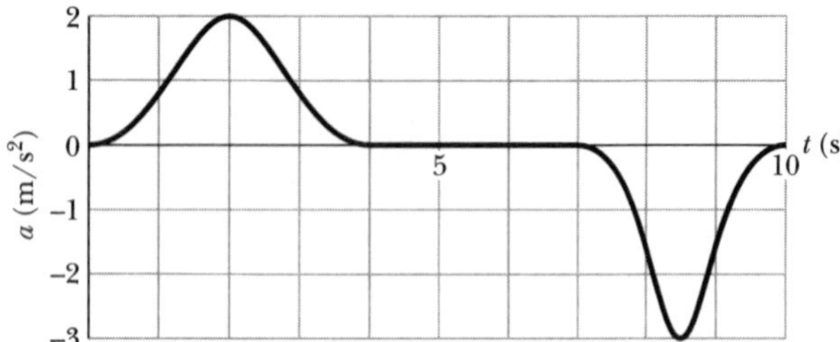

FIGURE 5-77 Problem 89.

90. A 100 kg crate sits on the floor of a freight elevator that starts from rest on the ground floor of a building at $t = 0$ and rises to the top floor during an 8.0 s interval. The velocity of the elevator as a function of time is shown in Fig. 5-78. For the following questions give both the magnitude and direction of the force. (a) What is the force of the floor of the elevator on the crate at $t = 1.8$ s? (b) What is the force of the floor of the elevator on the crate at $t = 4.4$ s? (c) What is the force of the floor of the elevator on the crate at $t = 6.8$ s?

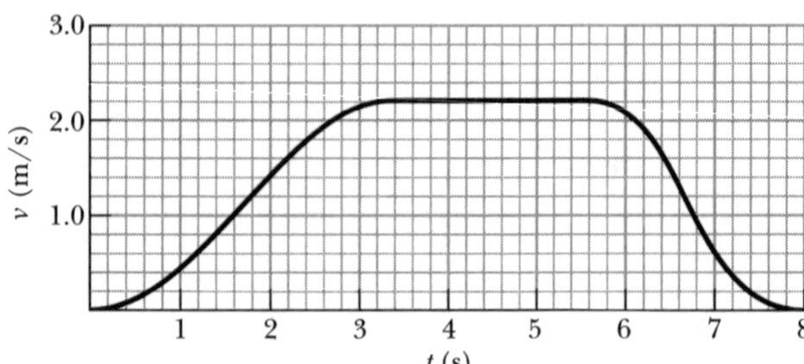

FIGURE 5-78 Problem 90.

91. A 40 kg skier comes directly down a frictionless ski slope that is inclined at an angle of 10° with the horizontal on a day when a strong wind is blowing parallel to the slope. Determine the magnitude and direction of the force of the wind on the skier if (a) the magnitude of the velocity of the skier is constant, (b) the magnitude of the velocity of the skier is increasing at a rate of 1.0 m/s², and (c) the magnitude of the velocity of the skier is increasing at a rate of 2.0 m/s².

92. Just before she releases a 5.44 kg shot, a shot-putter exerts a force on the shot of 280 N at an angle of 45° above the horizontal. What is the acceleration of the shot at this instant? Give the direction in terms of the angle it makes with the horizontal.

93. Tarzan, who weighs 820 N, swings from a cliff at the end of a 20 m vine that hangs from a high tree limb and initially makes an angle of 22° with the vertical. Immediately after Tarzan steps off the cliff, the tension in the vine is 760 N. Choose a coordinate system for which the x axis points horizontally away from the edge of the cliff and the y axis points upward. (a) What is the force of the vine on Tarzan in unit-vector notation? (b) What is the net force acting on Tarzan in unit-vector notation? (c) What are the magnitude and direction of the net force acting on Tarzan? (d) What are the magnitude and direction of Tarzan's acceleration?

94. A 50 kg skier is pulled up a frictionless ski slope that makes an angle of 8.0° with the horizontal by holding onto a tow rope that moves parallel to the slope. Determine the magnitude of the force of the rope on the skier at an instant when (a) the rope is moving with a constant speed of 2.0 m/s and (b) the rope is moving with a speed of 2.0 m/s that is increasing at a rate of 0.10 m/s².

95. In Fig. 5-79, a 3.0 kg mass on a 30° frictionless incline is connected to a 2.0 kg mass on a 60° frictionless incline. The pulley is frictionless and massless. What is the tension in the connecting cord?

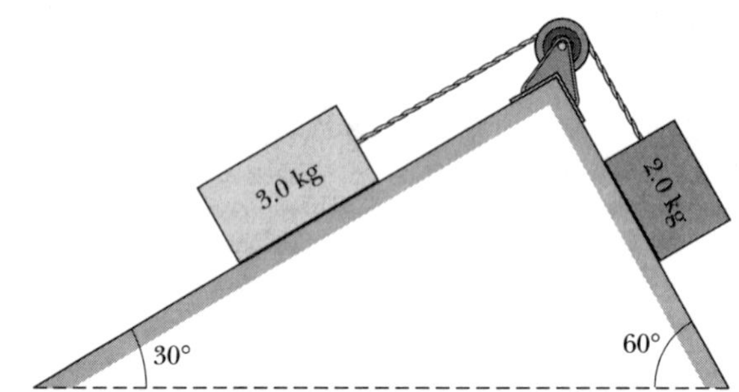

FIGURE 5-79 Problem 95.

96. In Fig. 5-80, a 1.0 kg mass on a 30° frictionless incline is connected to a 3.0 kg mass on a horizontal frictionless surface. The pulley is frictionless and massless. (a) If the magnitude of **F** is 2.3 N, what is the tension in the connecting cord? (b) What is the largest value that the magnitude of **F** may have without the connecting cord becoming slack?

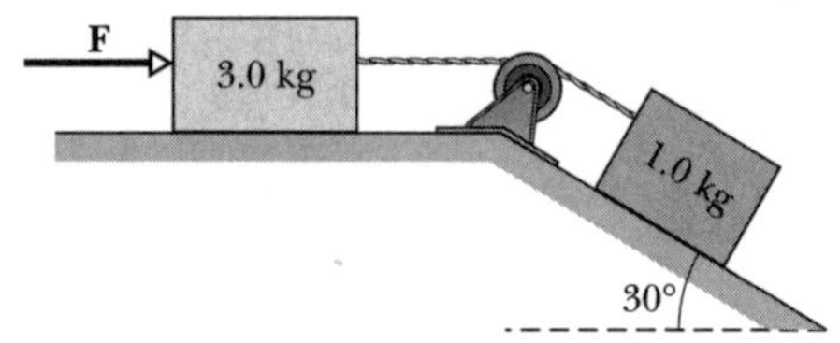

FIGURE 5-80 Problem 96.

Clustered Problems

Cluster 1

In each problem of this cluster, a 10.0 kg block is on a frictionless surface inclined by 30.0° to the horizontal.

97. In Fig. 5-81*a*, the block slides down the plane. What are (a) the magnitude of the block's acceleration and (b) the magnitude of the normal force on the block from the inclined surface?

98. In Fig. 5-81*b*, force **F** is applied to the block, directed upward along the inclined surface. What are the magnitude and direction of the block's acceleration if the magnitude of **F** is (a) 40.0 N and (b) 60.0 N?

99. In Fig. 5-81*c*, a 60.0 N force **F** is applied directly upward to the block. (a) What are the magnitude and direction of the block's acceleration? (b) What is the magnitude of the normal force on the block from the inclined surface? (c) What must the magnitude of the applied force be if the block is to be stationary? (d) Prove your answer to (c) using only a free-body diagram of the block.

100. In Fig. 5-81*d*, a 40.0 N force is applied horizontally to the block. (a) What are the magnitude and direction of the block's acceleration? (b) What is the magnitude of the normal force on the block from the inclined surface? (c) What must the magnitude of the applied force be if the block is to be stationary?

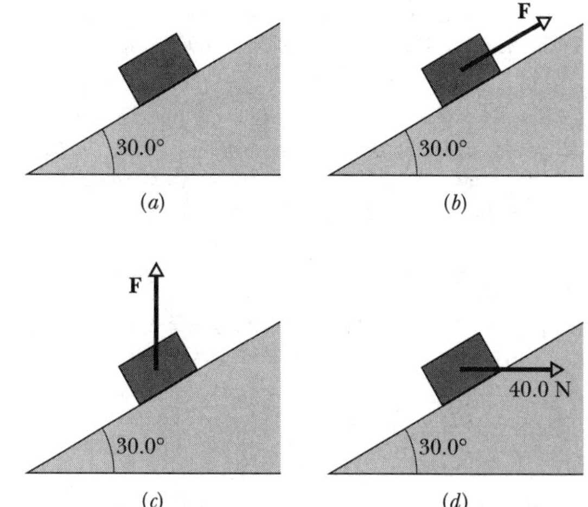

FIGURE 5-81 Problems 97 through 100.

Cluster 2

In each problem of this cluster, block 1 has mass $m_1 = 1.00$ kg and block 2 has mass $m_2 = 2.00$ kg. The blocks are connected by a massless string that wraps over a massless, frictionless pulley, and the surfaces are frictionless.

101. In Fig. 5-82*a* (a) what is the magnitude of the acceleration of the blocks? (b) In what direction does block 1 accelerate? (c) What is the tension in the string?

102. In Fig. 5-82*b* (a) what is the magnitude of the acceleration of the blocks? (b) What is the tension in the string?

103. In Fig. 5-82*c* (a) what is the magnitude of the acceleration of the blocks? (b) In what direction does block 1 accelerate? (c) What is the tension in the string?

104. In Fig. 5-82*d* (a) what is the magnitude of the acceleration of the blocks? (b) What is the tension in the string?

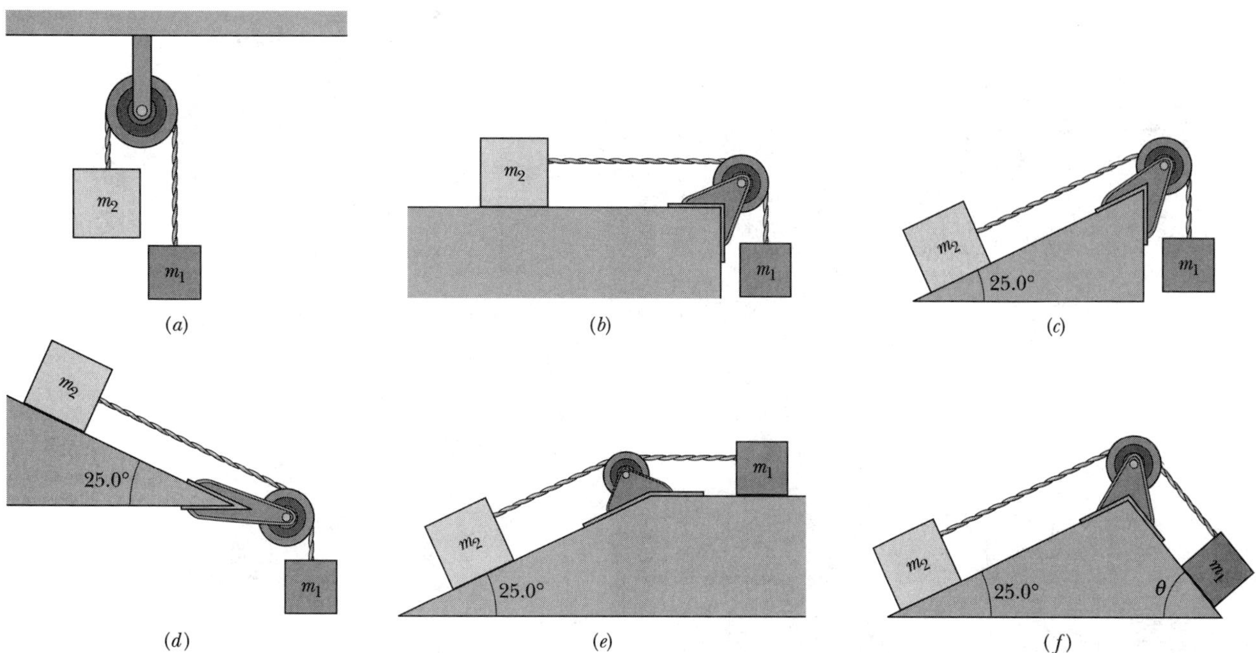

FIGURE 5-82 Problems 101 through 106.

105. In Fig. 5-82*e* (a) what is the magnitude of the acceleration of the blocks? (b) What is the tension in the string?

106. In Fig. 5-82*f*, with $\theta = 50.0°$, (a) what is the magnitude of the acceleration of the blocks? (b) In what direction does block 1 accelerate? (c) What is the tension in the string? What value must angle θ have if the blocks are to be (d) stationary and (e) moving at a constant speed?

Cluster 3

In each problem of this cluster, block 1 has mass $m_1 = 1.00$ kg, block 2 has mass $m_2 = 2.00$ kg, and block 3 has mass $m_3 = 3.00$ kg. The blocks are connected by massless strings that wrap over massless, frictionless pulleys. The surfaces are frictionless.

107. In Fig. 5-83*a* (a) what is the magnitude of the acceleration of the blocks? (b) In what direction does block 1 accelerate? What are the tensions in (c) the left-hand string and (d) the right-hand string?

108. In Fig. 5-83*b* (a) what is the magnitude of the acceleration of the blocks? (b) In what direction does block 1 accelerate? What are the tensions in (c) the left-hand string and (d) the right-hand string? (e) What value of m_1 would allow the blocks to be stationary?

109. In Fig. 5-83*c* (a) what is the magnitude of the acceleration of the blocks? (b) In what direction does block 1 accelerate? What are the tensions in (c) the left-hand string and (d) the right-hand string?

110. In Fig. 5-83*d*, with $\theta = 30°$, (a) what is the magnitude of the acceleration of the blocks? (b) In what direction does block 1 accelerate? What are the tensions in (c) the left-hand string and (d) the right-hand string? What value must angle θ have if the blocks are to be (e) stationary and (f) moving at a constant speed?

111. In Fig. 5-83*e*, with $\theta = 30.0°$, (a) what is the magnitude of the acceleration of the blocks? What are the tensions in (b) the left-hand string and (c) the right-hand string? (d) What is the minimum value of θ such that block 1 moves independently of blocks 2 and 3 (that is, the tension in the left-hand string is zero)?

Cluster 4

In each problem of this cluster, block 1 of mass m_1 and block 2 of mass m_2 are connected by massless strings that are wrapped around or attached to massless, frictionless pulleys. Assume that a y axis is directed upward in these figures; take upward force vectors as positive and downward force vectors as negative. When a string loops halfway around a pulley as in some of the situations given, the string pulls on the pulley with a net force that is twice the tension in the string. (This feature is actually true only when a pulley is stationary, but we shall assume it to be also true here when a massless pulley accelerates.)

112. In Fig. 5-84*a*, the blocks are released from rest. If block 2 moves downward by distance d_2 in a certain time, how far and in which direction do (a) pulley *A* and (b) block 1 move? (c) What, then, is the relationship between acceleration a_1 of block 1 and acceleration a_2 of block 2? (d) What is the relationship between tensions T_1 and T_2? What is Newton's second law written for (e)

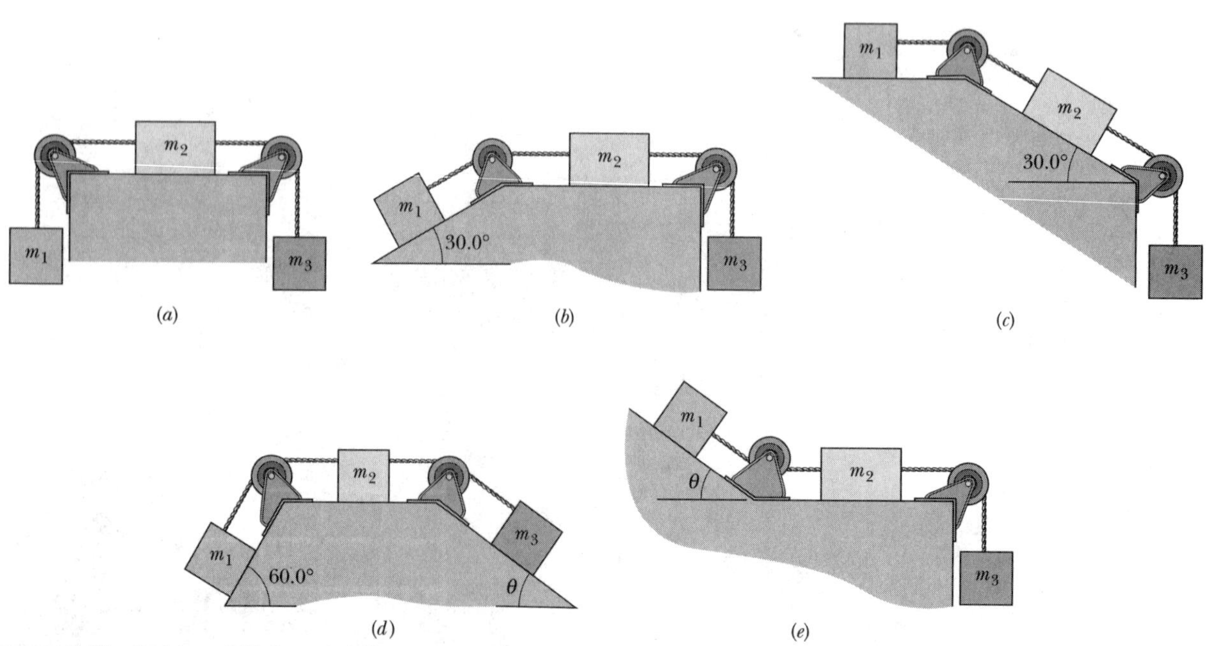

FIGURE 5-83 Problems 107 through 111.

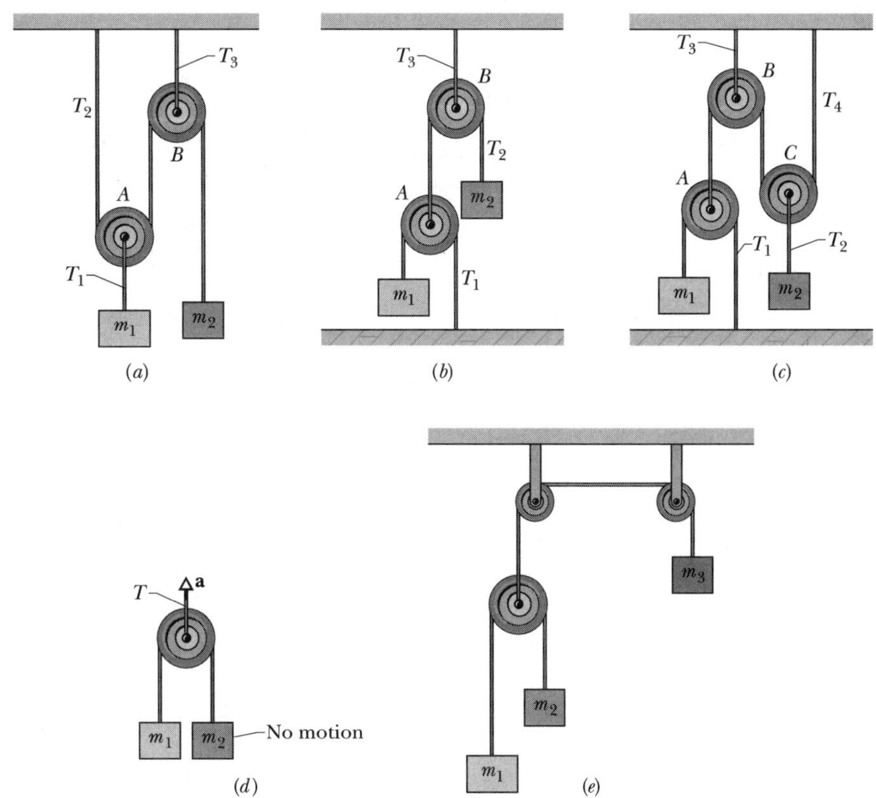

FIGURE 5-84 Problems 112 through 116.

block 1 and (f) block 2? In terms of the masses and the free-fall acceleration g, find expressions for (g) a_1, (h) a_2, and (*i*) T_2.

If $m_1 = 2.00$ kg and $m_2 = 3.00$ kg, what are the magnitudes and directions of (j) a_1 and (k) a_2, and (1) what is the value of T_2? (m) For what value of the ratio m_1/m_2 will the boxes be stationary? (n) Verify that in such a case $T_2 + T_3 = (m_1 + m_2)g$.

113. Repeat parts (a) through (m) of Problem 112 for the situation of Fig. 5-84*b*. For part (n) verify that $T_3 - T_1 = (m_1 + m_2)g$.

114. Repeat parts (a) through (m) of Problem 112 for the situation of Fig. 5-84*c*. For part (n) verify that $T_3 + T_4 - T_1 = (m_1 + m_2)g$.

115. In Fig. 5-84*d*, with $m_1 < m_2$, the pulley is accelerated upward at the rate a via a string fixed at its center. However, that rate has been chosen so that block 2 can be stationary. (a) If the pulley moves upward by distance d, how far and in what direction does block 1 move? (b) What is the acceleration a_1 of block 1 in terms of a? In terms of the masses and the free-fall acceleration g, what are (c) the tension T_{12} in the string connecting the two blocks, (d) the acceleration a_1 of block 1, (e) the acceleration a, and (f) the tension T in the string fixed to the pulley?

116. In Fig. 5-84*e*, the blocks are released from rest and tend to accelerate. (a) In terms of the masses m_1 and m_2, find the value of the mass m_3 of block 3 for which block 2 can remain stationary. (b) Verify your answer for the case $m_1 = m_2$. In that case what are the magnitudes of the accelerations of (c) block 1 and (d) block 3?

Graphing Calculators

SAMPLE PROBLEM 5-12

Solving Multiple-Force Problems Quickly. Repeat Sample Problem 5-3 using the vector capabilities of a graphing calculator. (Doing so greatly speeds up the solution and avoids common errors.)

SOLUTION: From the problem we know that

$$\mathbf{F}_1 + \mathbf{F}_2 + \mathbf{F}_3 = m\mathbf{a},$$

from which we write

$$\mathbf{F}_3 = m\mathbf{a} - \mathbf{F}_1 - \mathbf{F}_2.$$

On a TI-85/86, set the mode menu for degrees. Then, exiting that menu, press in the information for m, **a**, $\mathbf{F}_1$, and $\mathbf{F}_2$ in the manner explained in Sample Problem 3-9 in these pages. The screen should then show

$$2[8\angle 270+30]-[10\angle 180+60]-[12\angle 90]$$

where we have not even bothered to calculate two of the angles mentally. For an answer in magnitude-angle notation, choose SphereV in the mode menu and then exit that menu. Pressing ENTER then gives us, depending on the style of display and number of decimal places set in the menu,

$$[22\angle -53]$$

which means that

$$\text{magnitude of } \mathbf{F}_3 = 22 \text{ N}$$

and
$$\text{angle of } \mathbf{F}_3 = 53°$$

clockwise from positive x direction. (Answer)

For an answer in unit-vector notation, choose RectV in the mode menu. Exit the menu and press ENTER. We find

$$[13 \ \ -17]$$

which means

$$\mathbf{F}_3 = (13 \text{ N})\mathbf{i} - (17 \text{ N})\mathbf{j},$$

exactly as we found with considerable more effort in Sample Problem 5-3.

Chapter Six
Force and Motion—II

QUESTIONS

15. *Organizing question:* Figure 6-49 shows six situations in which one or two forces are applied to a 2 kg box on a floor. Assume that the box remains stationary due to the friction that also acts on it. The magnitudes of the applied forces are $F_1 =$ 10 N and $F_2 = 2$ N, and the coefficient of static friction between the box and the floor is 0.6. For each situation, set up equations, complete with known data, to find the magnitude f_s of the static frictional force and the magnitude N of the normal force on the box; use the first two scalar equations of Eq. 5-2 ($\Sigma F_x = ma_x$ and $\Sigma F_y = ma_y$).

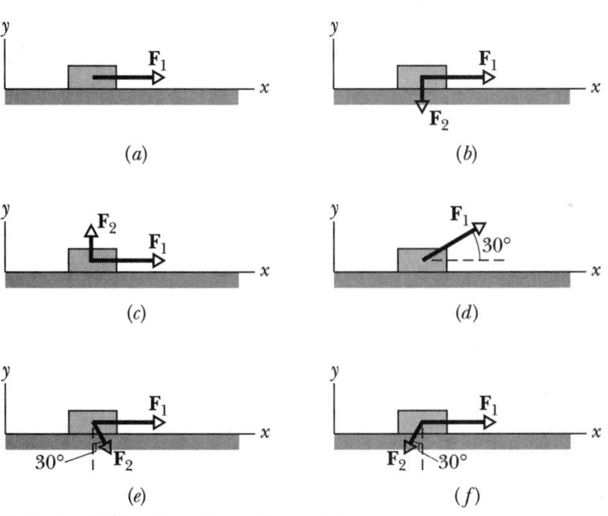

FIGURE 6-49 Questions 15 and 16.

16. *Organizing question:* For the situations of Question 15 and Fig. 6-49, assume that the box is sliding to the right and the coefficient of kinetic friction between the box and the floor is 0.3. For each situation, set up equations, complete with known data, to find the magnitude a of the box's acceleration and the magnitude N of the normal force on the box; use the first two scalar equations of Eq. 5-2 ($\Sigma F_x = ma_x$ and $\Sigma F_y = ma_y$).

17. In Fig. 6-19 of Question 7, if the angle θ of force **F** on the sliding box is increased, does the magnitude of the frictional force on the box increase, decrease, or remain the same?

18. Figure 6-50 shows five drag-racing boxes (a new sport) in which an applied force is directed rightward and a kinetic frictional force is directed leftward. The boxes, which have identical masses, cross the starting line with the same speed. Rank the boxes according to their arrival at the finish line, first box first.

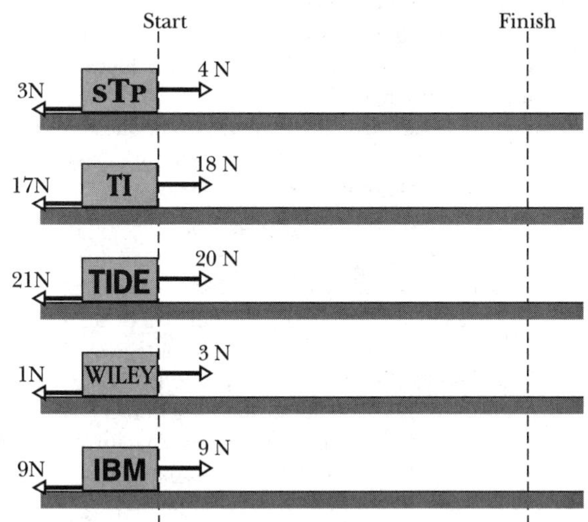

FIGURE 6-50 Question 18.

19. In Fig. 6-51, a horizontal force of 100 N is to be applied to a 10 kg slab that is initially stationary on a frictionless floor, to accelerate the slab. A 10 kg block lies on top of the slab; the coefficient of friction μ between the block and slab is not known, and the block might slip. (a) Considering that possibility, what is the possible range of values for the slab's acceleration a_{slab} (*Hint:* You don't need written calculations; just consider extreme values for μ). (b) What is the possible range for the block's acceleration a_{block}?

20. Figure 6-52 shows a block of mass m on a slab of mass M

FIGURE 6-51 Question 19.

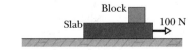

and a horizontal force **F** applied to the block, causing it to slide over the slab. There is friction between the block and slab (but not between the slab and the floor). (a) What mass determines the magnitude of the frictional force between the block and the slab? (b) At the block–slab interface, is the magnitude of the frictional force acting on the block greater than, less than, or equal to that of the frictional force acting on the slab? (c) What are the directions of those two frictional forces? (d) If we write Newton's law for the slab, what mass should be multiplied by the acceleration of the slab? (Warm-up for Problem 39.)

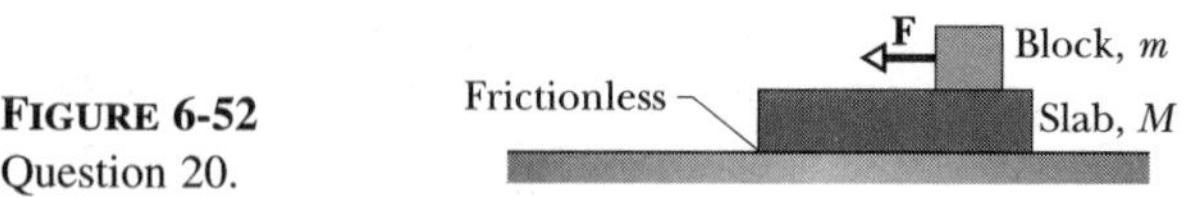

FIGURE 6-52 Question 20.

21. Follow-up to Problem 38: Suppose the larger block in Fig. 6-42 is, instead, fixed to the surface below it. Again, the smaller block is not to slip down the face of the larger block. (a) Is the magnitude of the frictional force between the blocks then greater than, less than, or the same as in the original problem? (b) Is the required minimum magnitude of the horizontal force **F** then greater than, less than, or the same as in the original problem?

22. *Organizing question:* In the two situations of Fig. 6-53, forces are applied to a stationary 2 kg tin of tuna that can slide along a plane tilted at 30°. The applied forces have magnitudes of $F_1 = 20$ N and $F_2 = 2$ N, and the coefficient of static friction is 0.8. In Fig. 6-53*a*, the tin is close to sliding up the plane; in Fig. 6-53*b*, it is close to sliding down the plane. For each situation, set up equations, complete with known data, to find the magnitude f_s of the frictional force on the tin and the magnitude N of the normal force on the tin; use the tilted *xy* coordinate system shown in Fig. 5-22 and the first two scalar equations of Eq. 5-2 ($\Sigma F_x = ma_x$ and $\Sigma F_y = ma_y$).

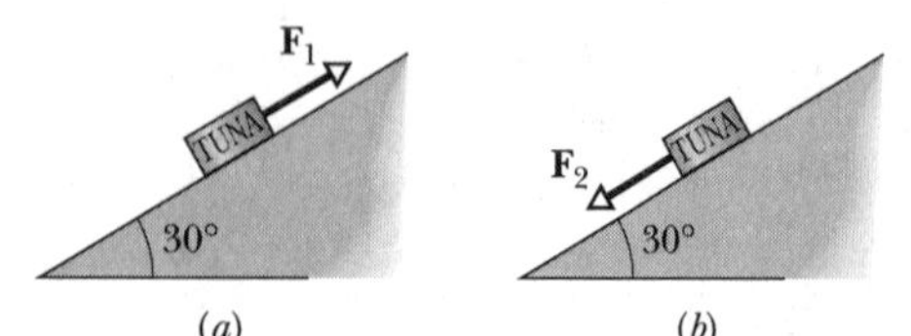

FIGURE 6-53 Question 22.

23. *Organizing question:* In the two situations of Fig. 6-54, forces are applied to a 2 kg tin of tuna that slides along a plane tilted at 30°. The applied forces have a magnitude of $F_1 = 20$ N, the coefficient of kinetic friction is 0.6, and the directions of sliding are indicated by the motion arrows. For each situation, set up equations, complete with known data, to find the acceleration a of the tin and the magnitude N of the normal force on the tin; use the tilted *xy* coordinate system shown in Fig. 5-22 and the first two scalar equations of Eq. 5-2 ($\Sigma F_x = ma_x$ and $\Sigma F_y = ma_y$).

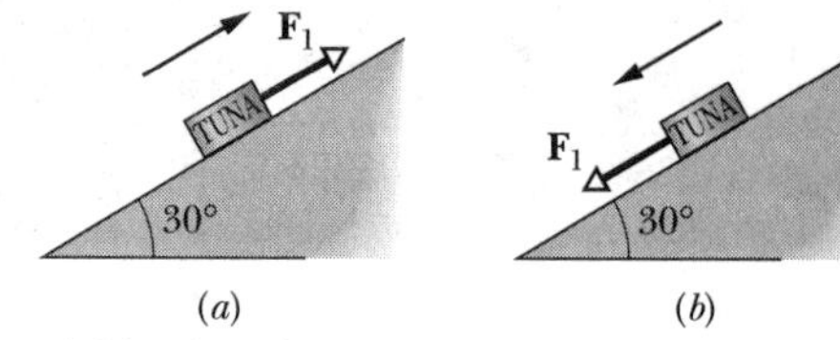

FIGURE 6-54 Question 23.

24. During a routine flight on September 21, 1956, test pilot Tom Attridge put his Grumman F11F-1 jet fighter into a 20° dive for a test of the aircraft's 20-mm machine cannons. While traveling faster than sound at 4000 m, he shot a burst of rounds. Then, after allowing the cannons to cool, he shot another burst at 2000 m; his speed was then 344 m/s and the speed of the rounds relative to him was 730 m/s.

Almost immediately the canopy around Attridge was shredded and his right air intake was damaged. With little power capability left, the jet crashed into a wooden area, but Attridge managed to escape the resulting explosion by crawling from the fuselage (in spite of four fractured vertebrae). Explain what apparently happened just after the second burst of cannon rounds.

25. In 1987, as a Halloween stunt, two sky divers passed a pumpkin back and forth between them while they were in free fall just west of Chicago. The stunt was great fun until the last sky diver with the pumpkin opened his parachute. The pumpkin broke free from his grip, plummeted about 0.5 km, ripped through the roof of a house, slammed into the kitchen floor, and splattered all over the newly remodeled kitchen. From the sky diver's viewpoint and from the pumpkin's viewpoint, why did the sky diver lose control of the pumpkin?

26. A person riding a Ferris wheel moves through positions at (1) the top, (2) the bottom, and (3) midheight. If the wheel rotates at a constant rate, rank these three positions according to (a) the magnitude of the person's centripetal acceleration, (b) the magnitude of the net centripetal force on the person, (c) the apparent weight of the person, and (d) the magnitude of the normal force on the person, greatest first.

27. A ball is hung by a cord from a porch ceiling. Initially, the cord is vertical, as shown in Fig. 6-55*a*, but then a steady horizontal wind moves the ball so that the cord makes an angle with the vertical. Two such positions of the ball are shown in Figs. 6-55*b* and *c*. Rank those two positions and the initial one according to the tension in the cord, greatest first.

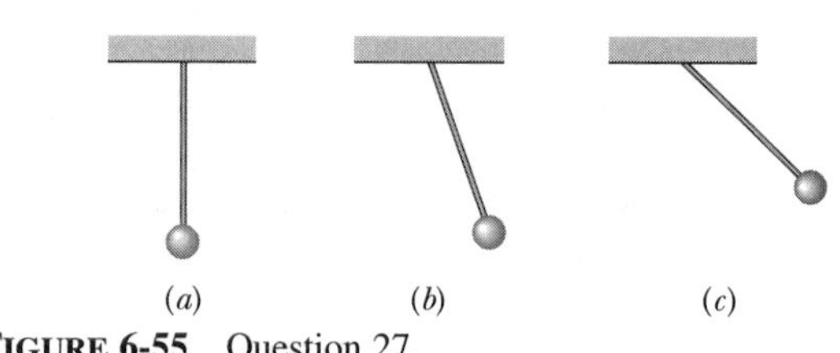

FIGURE 6-55 Question 27.

EXERCISES & PROBLEMS

73. A force **P**, parallel to a surface inclined 15° above the horizontal, acts on a 45 N block, as shown in Fig. 6-56. The coefficients of friction for the block and surface are $\mu_s = 0.50$ and $\mu_k = 0.34$. If the block is initially at rest, determine the magnitude and direction of the frictional force acting on the block for magnitudes of **P** of (a) 5.0 N, (b) 8.0 N, and (c) 15 N.

FIGURE 6-56 Problem 73.

74. An amusement park ride consists of a car moving in a vertical circle on the end of a rigid boom of negligible mass. The combined weight of the car and riders is 5.0 kN, and the radius of the circle is 10 m. What are the magnitude and direction of the force of the boom on the car at the top of the circle if the car's speed there is (a) 5.0 m/s and (b) 12 m/s?

75. Two blocks are accelerated across a horizontal surface by a horizontal force applied to one of the blocks as shown in Fig. 6-57. The magnitude of the frictional force on the smaller block is 2.0 N, and the magnitude of the frictional force on the larger block is 4.0 N. If the magnitude of **F** is 12 N, what is the magnitude of the force exerted on the larger block by the smaller block?

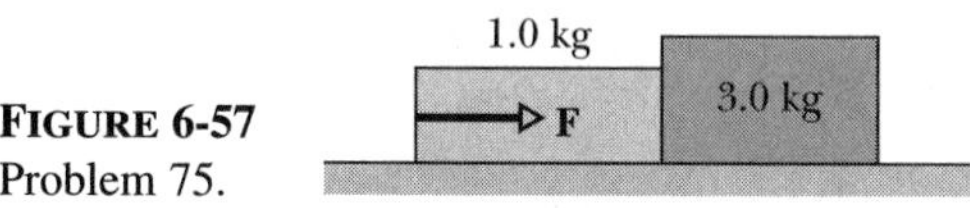

FIGURE 6-57 Problem 75.

76. A 2.5 kg block is initially at rest on a horizontal surface. A 6.0 N horizontal force and a vertical force **P** are applied to the block as shown in Fig. 6-58. The coefficients of friction for the block and surface are $\mu_s = 0.40$ and $\mu_k = 0.25$. Determine the magnitude and direction of the frictional force acting on the block if the magnitude of **P** is (a) 8.0 N, (b) 10 N, and (c) 12 N.

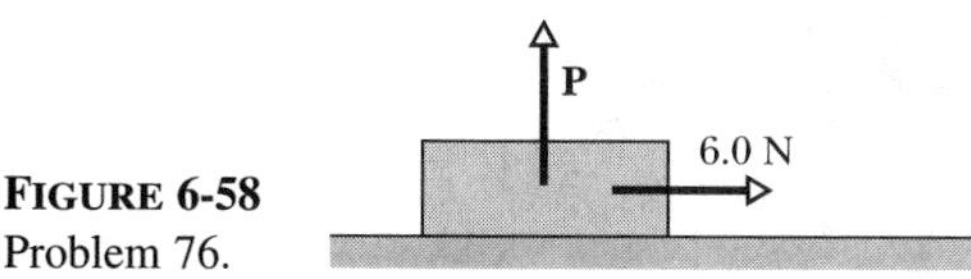

FIGURE 6-58 Problem 76.

77. As a 40 N block slides down a plane that is inclined at 25° to the horizontal, its acceleration is 0.80 m/s^2, directed up the plane. What is the coefficient of kinetic friction between the block and the plane?

78. A 45 kg skier skis over a frictionless circular hill of radius 15 m and then down the hill to a frictionless circular dip of 25 m, as shown in Fig. 6-59. At the top of the hill the ground exerts an upward force of 320 N on the skier, and at the bottom of the dip the ground exerts an upward force of 1.1 kN on the skier. What is the speed of the skier (a) at the top of the hill and (b) at the bottom of the dip?

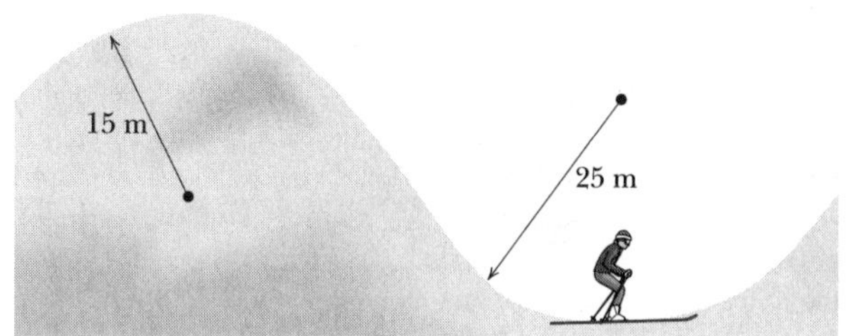

FIGURE 6-59 Problem 78.

79. The three blocks in Fig. 6-60 are released from rest and accelerate at the rate of 1.5 m/s^2. If $M = 2.0$ kg, what is the magnitude of the frictional force on the block that slides horizontally?

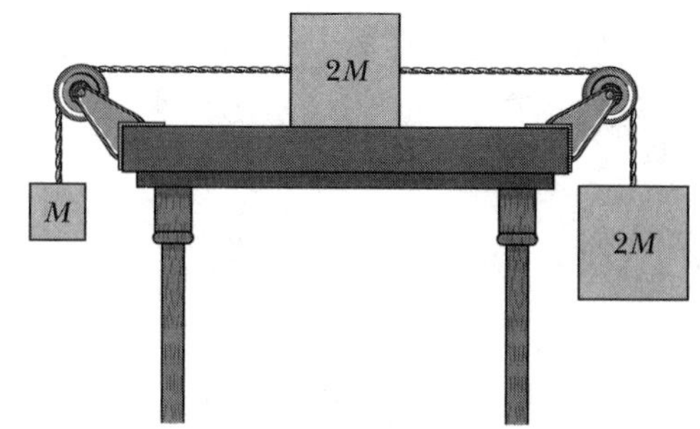

FIGURE 6-60 Problem 79.

80. A warehouse worker exerts a constant horizontal force of magnitude 85 N on a 40 kg box that is initially at rest on the horizontal floor of the warehouse. After moving a distance of 1.4 m, the speed of the box is 1.0 m/s. What is the coefficient of kinetic friction between the box and the warehouse floor?

81. Luggage is transported from one location to another in an airport by a conveyor belt. At a certain location, the belt moves down an incline that makes an angle of 2.5° with the horizontal. Assume that there is no slipping of the luggage. Determine the magnitude and direction of the frictional force exerted by the belt on a box weighing 69 N when the box is on the inclined portion of the belt for the following cases: (a) The belt is temporarily at rest. (b) The belt has a speed of 0.65 m/s that is constant. (c) The belt has a speed of 0.65 m/s that is increasing at a rate of

0.20 m/s ^{2}. (d) The belt has a speed of 0.65 m/s that is decreasing at a rate of 0.20 m/s^2. (e) The belt has a speed of 0.65 m/s that is increasing at a rate of 0.57 m/s^2.

82. A child weighing 140 N sits at rest at the top of a playground slide that makes an angle of 25° with the horizontal. The child keeps from sliding by holding onto the sides of the slide. After letting go of the sides, the child has a constant acceleration of 0.86 m/s^2 (down the slide, of course). (a) What is the coefficient of kinetic friction between the child and the slide? (b) What maximum and minimum values for the coefficient of static friction between the child and the slide are consistent with the information given here?

83. A 1.5 kg box is initially at rest on a horizontal surface when at $t = 0$ a horizontal force, $\mathbf{F} = (1.8)t\mathbf{i}$ N (where t is in seconds), is applied to the box. The acceleration of the box as a function of t is given by $\mathbf{a} = 0$ for $0 \leq t \leq 2.8$ s and $\mathbf{a} = (1.2t - 2.4)\mathbf{i}$ m/s^2 for $t > 2.8$ s. (a) What is the coefficient of static friction between the box and the surface? (b) What is the coefficient of kinetic friction between the box and the surface?

84. A block weighing 22 N is held at rest against a vertical wall by a horizontal force **F** of magnitude 60 N as shown in Fig. 6-61. The coefficient of static friction between the wall and the block is 0.55 and the coefficient of kinetic friction is 0.38. A second force **P** acting parallel to the wall is applied to the block. For the following values of **P**, determine whether the block moves, the direction of motion, and the magnitude and direction of the frictional force acting on the block: (a) 34 N, up, (b) 12 N, up, (c) 48 N, up, (d) 62 N, up, (e) 10 N, down, and (f) 18 N, down.

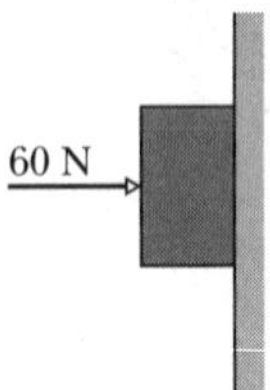

FIGURE 6-61 Problem 84.

85. Some years ago, the bookstore at Georgia Tech was in the basement of a building. As boxes were unloaded from delivery trucks at ground level, they were allowed to slide down a wooden ramp to the basement area. A strobe photograph of a box sliding down the 2.5 m high ramp is represented in Fig. 6-62. Assume that the box is released from rest at the top of the ramp at $t = 0$ and slides with constant acceleration. The flashes on the strobe camera record the position of the box at 0.5 s intervals. (a) What is the magnitude of the acceleration of the box? (b) If the box weighs 240 N, what is the magnitude of the force of kinetic friction exerted by the ramp on the box?

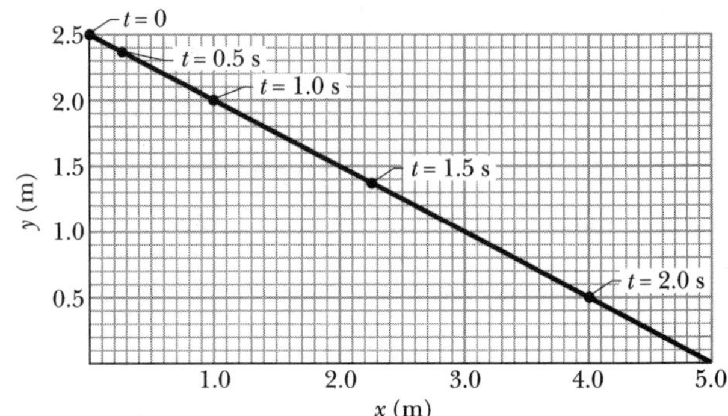

FIGURE 6-62 Problem 85.

86. A four-person bobsled (total mass = 630 kg) comes down a straightaway at the start of a bobsled run. The straightaway is 80.0 m long and is inclined at a constant angle of 10.2° with the horizontal. Assume that the combined effects of friction and air resistance produce a constant force on the bobsled of 62.0 N that acts parallel to the incline and up the incline. For the following questions, express your answers to three significant digits. (a) If the magnitude of the velocity of the bobsled at the start of the run is 6.20 m/s, how long does the bobsled take to come down the straightaway? (b) Suppose through practice the crew is able to reduce the effects of friction and air resistance to a constant force of 42.0 N on the bobsled. For the same initial velocity, how long will the bobsled now take to come down the straightaway?

87. A toy chest and its contents weigh 180 N. The coefficient of static friction between the toy chest and the floor is 0.42. A child attempts to move the chest across the floor by pulling on an attached rope, as shown in Fig. 6-63. (a) If the angle θ that the rope makes with the horizontal is 42°, what is the magnitude of the force **F** that the child must exert on the rope in order to put the chest on the verge of moving? (b) Write an expression for the magnitude of the force **F** required to put the chest on the verge of moving as a function of angle θ. Determine (c) the value of θ for which the magnitude of **F** is a minimum and (d) that minimum magnitude.

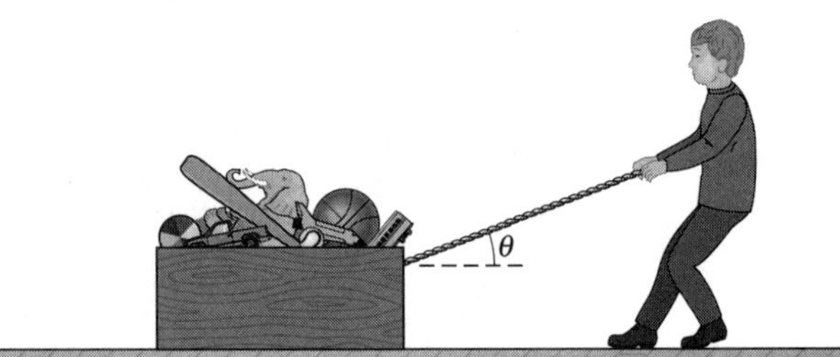

FIGURE 6-63 Problem 87.

88. A 3.0 kg block on a 30° incline is connected by a string to a 2.0 kg block on a horizontal surface, as shown in Fig. 6-64. The 30° incline is frictionless and the coefficient of kinetic friction between the 2.0 kg block and the horizontal surface is 0.25. The

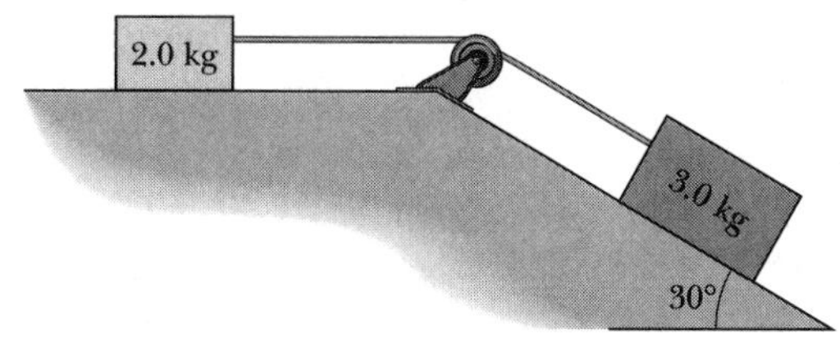

FIGURE 6-64 Problem 88.

pulley is massless and frictionless. What is the tension in the connecting string after the blocks are released from rest?

89. A 2.0 kg block and a 1.0 kg block are connected by a string and are pushed across a horizontal surface by a force applied to the 1.0 kg block as shown in Fig. 6-65. The coefficient of kinetic friction between the blocks and the surface is 0.20. If the magnitude of **F** is 20 N, what is the tension in the connecting string?

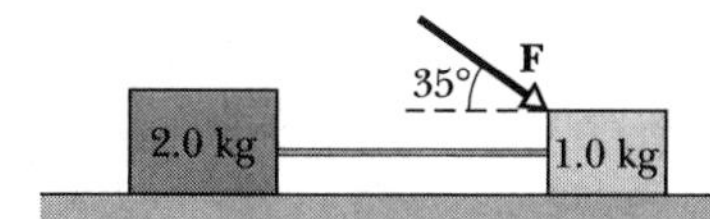

FIGURE 6-65 Problem 89.

90. A 26 kg child on a steadily rotating Ferris wheel moves at a constant speed of 5.5 m/s around a vertical circular path of radius 12 m. What are the magnitude and direction of the force of the seat on the child (a) at the highest point of the path and (b) at the lowest point of the path?

91. A roller-coaster car has a mass of 1200 kg when fully loaded with passengers. As the car passes over the top of a circular hill of radius 18 m, its speed is not changing. What are the magnitude and direction of the force of the track on the car at the top of the hill if the car's speed is (a) 11 m/s and (b) 14 m/s?

92. A high-speed railway car goes around a flat, horizontal circle of radius 470 m at a constant speed. The magnitudes of the horizontal and vertical components of the force of the car on a 51.0 kg passenger are 210 N and 500 N, respectively. (a) What is the magnitude of the net force (of *all* the forces) on the passenger? (b) What is the speed of the car?

93. An airplane flies in a horizontal circle of radius 4000 m at a constant speed of 100 m/s. The mass of the pilot is 52 kg. Answer the following in terms of the horizontal and vertical components of the quantity in question. (a) What is the net force on the pilot when the airplane is directly east of the circle's center? (b) What is the force of the airplane on the pilot when the airplane is directly east of the circle's center?

Clustered Problems

Cluster 1

94. In Fig. 6-66*a*, a 10 kg block remains stationary on a horizontal surface when a horizontal force of magnitude 25 N is applied to it. The coefficient of static friction between the block and the surface is 0.4. What are the magnitude and direction of the frictional force on the block due to the surface?

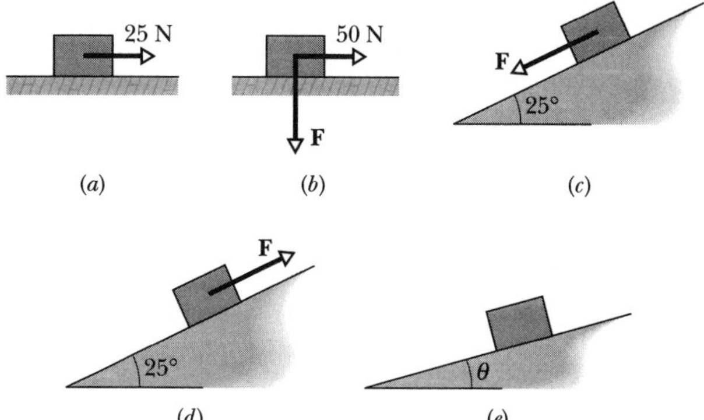

FIGURE 6-66 Problems 94 through 98.

95. In Fig. 6-66*b*, a 50 N force is to be applied horizontally to a 10 kg block on a horizontal surface while a second force, of magnitude F, is to be applied directly downward on the block. If the coefficient of static friction between the block and the surface is 0.4, what is the least value of F that will keep the block stationary?

96. In Fig. 6-66*c*, a 10 kg block is stationary on a plane inclined at 25° to the horizontal. The coefficients of static friction and kinetic friction between the block and the plane are 0.60 and 0.20, respectively. A force of magnitude F is to be applied to the block downward along the plane just long enough to start the block moving in that direction. What are (a) the maximum magnitude of the static frictional force on the block, (b) the magnitude of the block's weight component along the plane, and (c) the least value required of F to start the motion? (d) What are the magnitude and direction of the block's acceleration after the applied force is removed?

97. In Fig. 6-66*d*, a 10 kg block is stationary on a plane inclined at 25° to the horizontal. The coefficients of static friction and kinetic friction between the block and the plane are 0.60 and 0.20, respectively. A force of magnitude F is to be applied to the block upward along the plane. (a) What is the least value of F required to make the block move up along the plane? What are the magnitude and direction of the block's acceleration if (b) the applied force is maintained on the block and (c) it is removed?

98. In Fig. 6-66*e*, a 10 kg block is stationary on a plane inclined at angle θ to the horizontal. The coefficients of static friction and kinetic friction between the block and the plane are 0.60 and 0.20, respectively. (a) With $\theta = 15°$ and the block stationary, what are the magnitude and direction of the frictional force acting on the block? (b) If θ is then increased, at what angle does the block begin to slide? (c) Should θ then be increased or decreased if the sliding is to be at constant speed? (d) What θ is required for that constant-speed sliding?

Cluster 2

In each problem of this cluster, blocks are connected by massless srings that are wrapped around massless, frictionless pulleys.

When a string loops halfway around a pulley as in Fig. 6-67a, the string pulls on the pulley with a net force that is twice the tension in the string. (This feature is actually true only when a pulley is stationary, but we shall assume it to be also true here when a massless pulley accelerates.)

99. In Fig. 6-67*a*, block 2 (mass 2 kg) on a horizontal surface is connected to block 1 (mass 1 kg) via a string and two pulleys. The coefficients of static friction and kinetic friction between block 2 and the surface are 0.60 and 0.40, respectively. A force of magnitude F is applied directly downward on block 1. (a) What is the least value of F that will cause the blocks to move?

A greater value of F is actually used to get the blocks moving, and then the applied force is removed. (b) If block 2 moves a distance d_2 in a certain time t, how far does block 1 move? (c) What, then, is the relationship between the acceleration a_1 of block 1 and the acceleration a_2 of block 2? What are the values of (d) a_1 and (e) a_2?

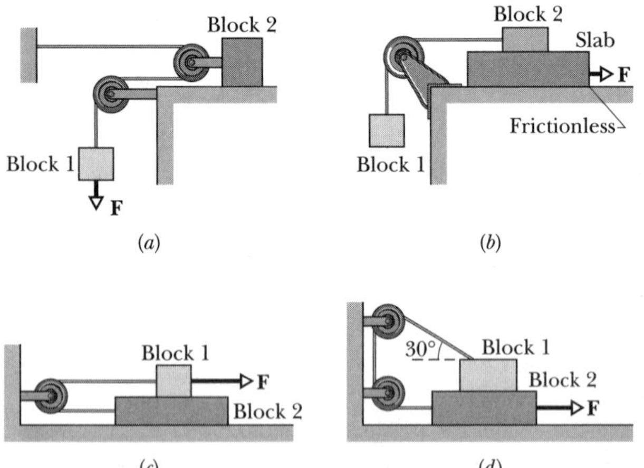

FIGURE 6-67 Problems 99 through 102.

100. In Fig. 6-67*b*, block 2 (mass 2.0 kg) lies on a slab (mass 3.0 kg) that lies on a frictionless horizontal surface. Also, block 2 is connected by a string and pulley to block 1 (mass 1.0 kg). A force of magnitude F is applied horizontally to the slab. The coefficients of static friction and kinetic friction between block 2 and the slab are 0.60 and 0.45, respectively.

(a) What is the least value of F that keeps the slab–blocks system stationary? (b) What is the greatest value of F that allows block 2 and the slab to move rightward without them slipping past each other? With that value of F, what are the magnitudes of the accelerations of (c) block 2 and (d) the slab? If the value of F is then increased by 10%, what are the magnitudes and directions of the accelerations of (e) block 2 and (f) the slab as they slide past each other? (g) Repeat the question using a coefficient of kinetic friction of 0.55.

101. In Fig. 6-67*c*, a force of magnitude F is applied horizontally to block 1 (mass 1 kg), which not only lies on block 2 (mass 2 kg) but is also connected to it by a string. The coefficients of kinetic friction are 0.30 for the contact between the blocks and 0.20 for the contact between block 2 and the floor across which it moves. (a) What value of F results in the two blocks traveling past each other at a constant speed? (b) For that value of F, what is the tension in the connecting string?

102. In Fig. 6-67*d*, the same blocks and floor of Problem 101 and Fig. 6-67*c* are used except that, at block 1, the string now makes a 30° angle with the horizontal and the applied force now acts on block 2. What value of F now results in the two blocks traveling past each other at a constant speed?

Graphing Calculators

PROBLEM SOLVING TACTICS

TACTIC 2: *Storing Equations or Notes about Friction*

In Problem Solving Tactics 19 and 21 of Chapter 2 we discussed how to store equations and notes as a program that can function as a notebook. If you now wish to store the equations involving friction or notes about friction, you need to put the Greek letter mu on the screen. You can find it by pressing 2nd CHAR, choosing GREEK, and then pressing MORE. Choose the mu by pressing F4.

Chapter Seven
Kinetic Energy and Work

QUESTIONS

18. Figure 7-42 gives four plots of position versus time of a cat carrier set in motion along an x axis on a frictionless floor by an applied force. Plots 1 and 2 are symmetric about the time axis with plots 4 and 3, respectively. Rank the plots according to the kinetic energy of the carrier, greatest first.

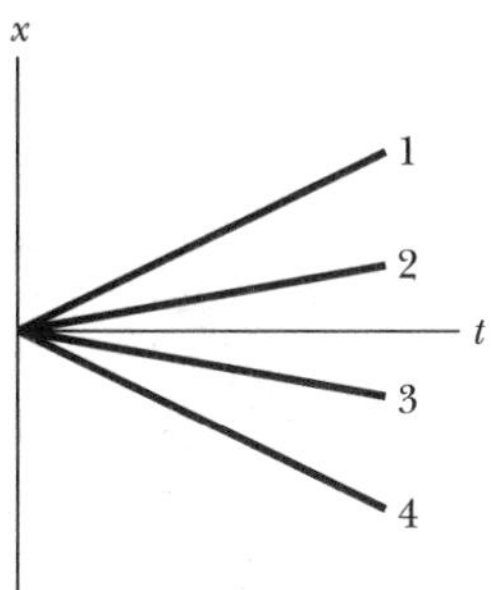

FIGURE 7-42 Question 18.

19. *Organizing question:* Figure 7-43 shows four situations in which a 2 kg box slides over a frictionless floor while the forces $\mathbf{F}_1$ and $\mathbf{F}_2$ are applied to it. The magnitudes of those forces are $F_1 = 10$ N and $F_2 = 2$ N. In each situation, the box begins with speed $v_i = 3$ m/s and slides through a distance of 4 m in the positive direction along an x axis. For each situation, set up an equation, complete with known data, for the kinetic energy K_f of the box at the end of the 4 m.

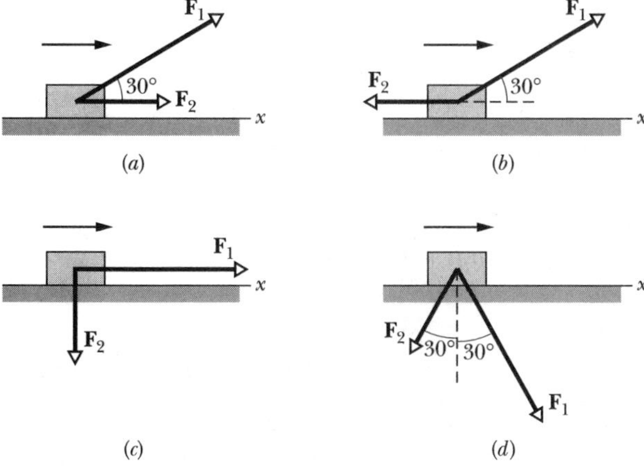

FIGURE 7-43 Question 19.

20. Figure 7-44 shows four situations in which the same can of corned beef hash is pulled by an applied force up frictionless ramps through (and then past) the same vertical distance h. In each situation that force has a magnitude of 10 N. In situations 2 and 4, the force is directed along the plane; in situations 1 and 3, it is directed at an angle $\theta = 20°$ to the plane, as drawn. Rank the situations according to the work done on the can in the vertical distance h by (a) the applied force and (b) the can's weight, greatest first.

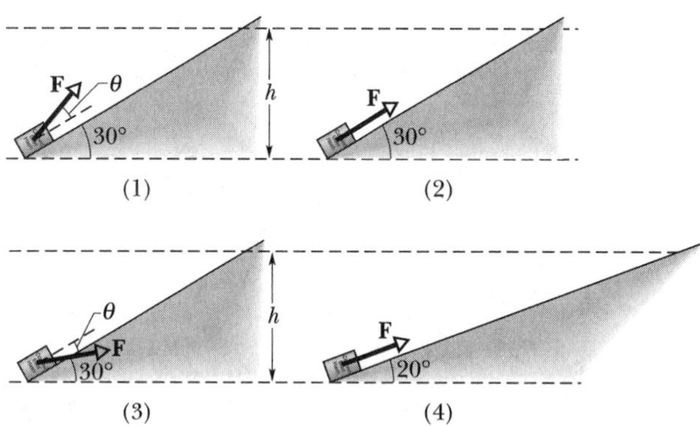

FIGURE 7-44 Question 20.

21. *Organizing question:* Figure 7-45 shows three situations in which a 2 kg box moves downward through a distance of 4 m while forces $\mathbf{F}_1$ and $\mathbf{F}_2$ are applied to it. The magnitudes of those forces are $F_1 = 10$ N and $F_2 = 2$ N. In each situation, the box begins with a speed $v_i = 3$ m/s. For each situation, set up an

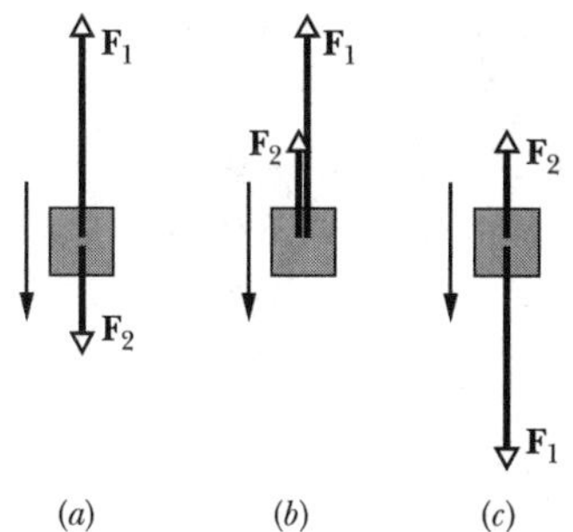

FIGURE 7-45 Question 21.

equation, complete with known data, for the kinetic energy K_f of the box at the end of the 4 m.

22. In four situations, a briefly applied horizontal force changes the velocity of a hockey puck that slides over frictionless ice. The overhead views of Fig. 7-46 indicate, for each situation, the puck's initial speed v_i, the final speed v_f, and the directions of the corresponding velocity vectors. Rank the situations according to the work done on the puck by the applied force, most positive first and most negative last.

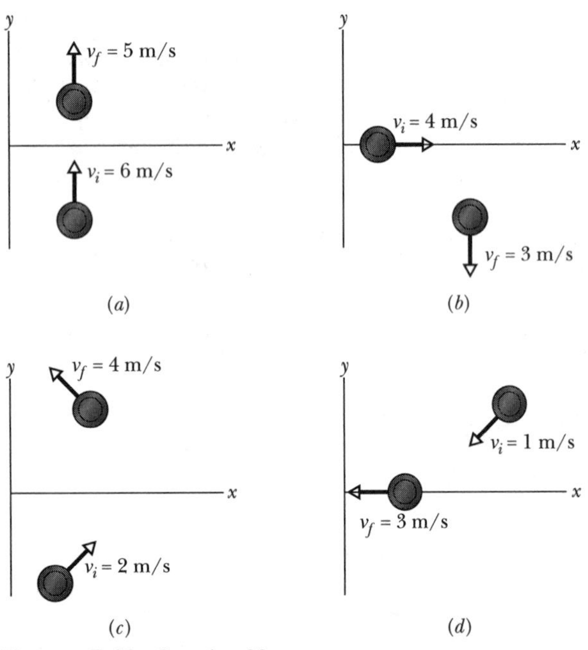

FIGURE 7-46 Question 22.

23. Figure 7-47 shows five situations in which this book is pulled by a force along a frictionless surface through a distance of 4 m. In Fig. 7-47*a* the force is directed along the surface and its magnitude is a constant 3 N. In Fig. 7-47*b* the force is directed along the surface and its magnitude varies with time t as (3 N/s)t. In Fig. 7-47*c* the force is directed along the surface and its magnitude varies with the distance x along the surface as (3 N/m)x. In Fig. 7-47*d* the force is directed at a constant angle $\phi = 0.3$ rad relative to the surface and its magnitude is a constant 3 N. In Fig. 7-47*e* the force has a constant magnitude of 3 N but its angle ϕ relative to the surface varies with time as (0.3 rad/s)t. In which of the situations can we find the work done on the book by the force with (a) Eq. 7-11 and (b) Eq. 7-31?

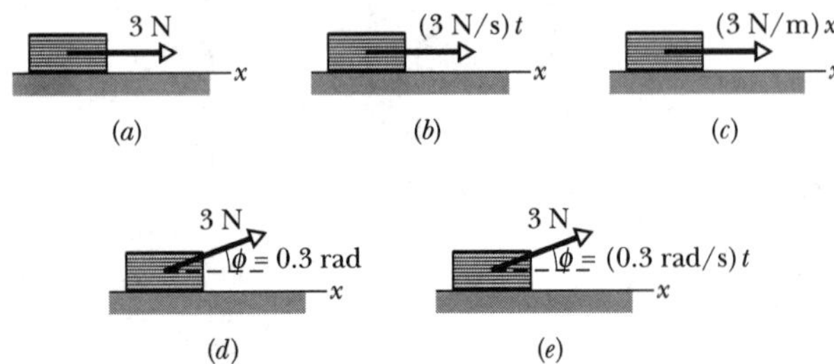

FIGURE 7-47 Question 23.

24. Figure 7-48 gives, for three situations, the position versus time of a box of contraband pulled by applied forces directed along an x axis on a frictionless surface. In each situation, the box starts from rest. Line B is straight; the others are curved. Rank the situations according to the kinetic energy of the box at (a) time t_1 and (b) time t_2, greatest first. (c) Now rank them according to the net work done on the box by the applied forces during the time period 0 to t_2, greatest first.

(d) For each situation, which of the following best describes the net work done by the applied forces during the period 0 to t_2:

(1) transfers energy to the box at a constant rate
(2) transfers energy to the box at an increasing rate
(3) transfers energy to the box at a decreasing rate
(4) transfers energy from the box and then later to the box
(5) transfers energy to the box and then later from the box
(6) transfers energy to the box initially but not later

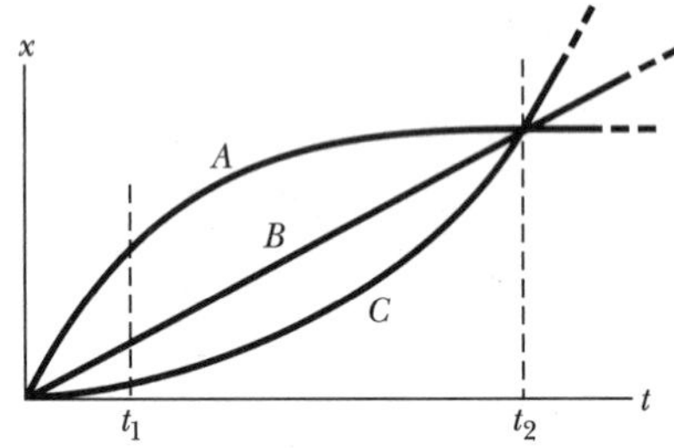

FIGURE 7-48 Question 24.

25. In Fig. 7-49, an initially stationary load of *Rolling Stone* magazines is lifted through (and then past) a height h via a cable in three ways. In Fig. 7-49*a*, the load is accelerated to a speed of 3 m/s through the first $h/3$ and then moved at that speed through the rest of the distance. In Fig. 7-49*b* it is accelerated to 3 m/s through the first $h/2$ and then moved at 3 m/s through the rest of the distance. In Fig. 7-49*c* the acceleration to 3 m/s takes the full height h. Rank the three ways of lifting according to (a) the work done on the load by the cable during the lift of height h and (b) the rate at which that work is done, greatest first.

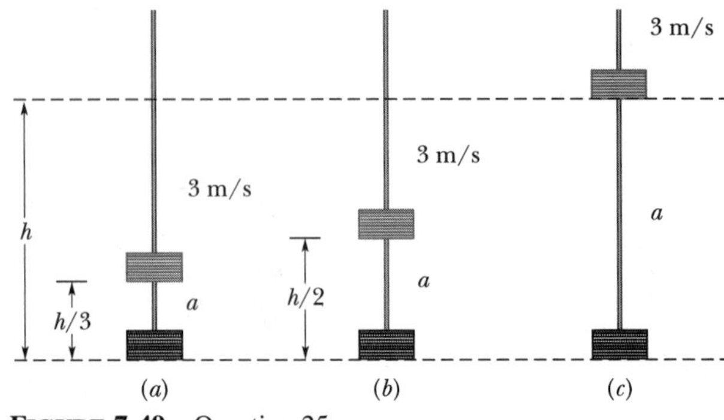

FIGURE 7-49 Question 25.

26. A glob of Slime is launched or dropped from the edge of a cliff. Which of the graphs in Fig. 7-50 could possibly show how the kinetic energy of the blob changes during the blob's flight?

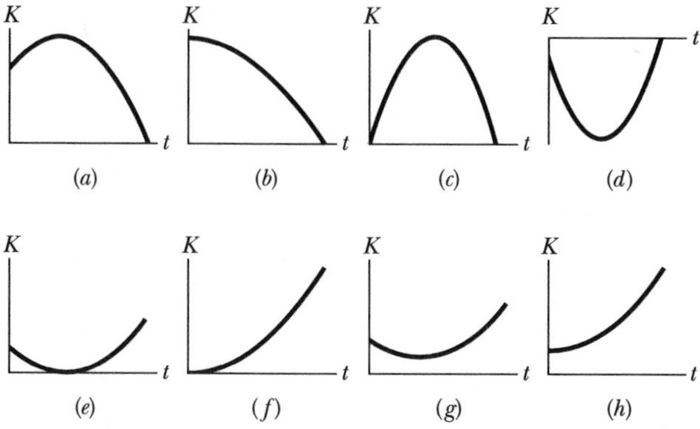

FIGURE 7-50 Question 26.

27. Figure 7-51 gives the velocity versus time of a scooter car being moved along an axis by a varying applied force. (a) During which of the numbered time periods is energy transferred *from* the scooter car by the applied force? (b) Rank the time periods according to the work done on the scooter car by the applied force during the period, most positive work first, most negative work last. (c) Rank the periods according to the rate at which the applied force transfers energy, greatest rate of transfer *to* the scooter car first, greatest rate of transfer *from* it last.

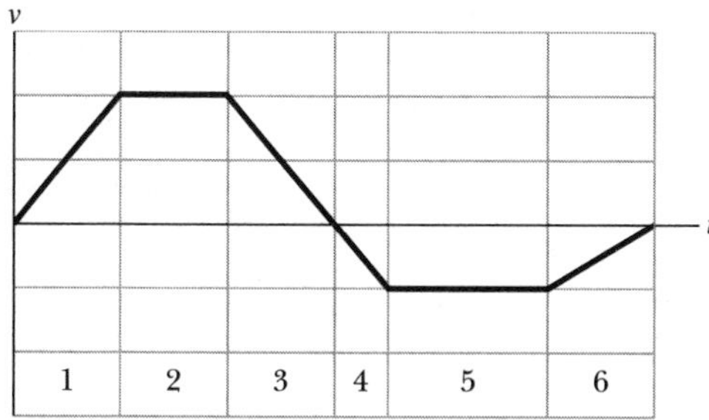

FIGURE 7-51 Question 27.

28. Figure 7-52 gives the velocity versus time of a carriage being moved along an axis by an applied force. Rank the numbered time periods according to (a) the work done by that applied force during that time period and (b) the rate at which the work is done, greatest first.

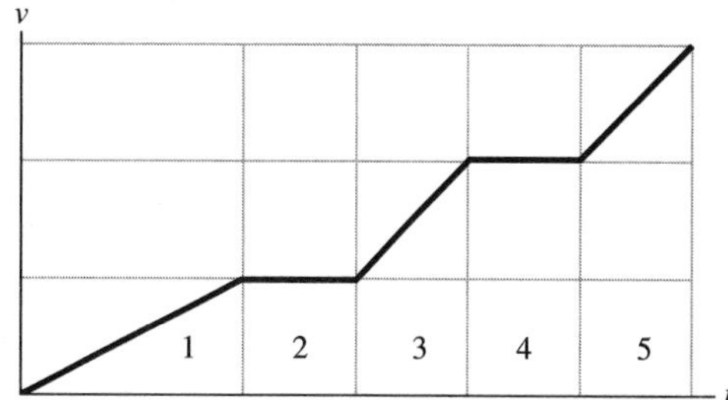

FIGURE 7-52 Question 28.

29. In three situations, an initially stationary can of crayons is sent sliding over a frictionless floor by a different applied force. Plots of the resulting acceleration versus time for the three situations are given in Fig. 7-53. Rank the plots according to the work done by the applied force during the acceleration period, greatest first.

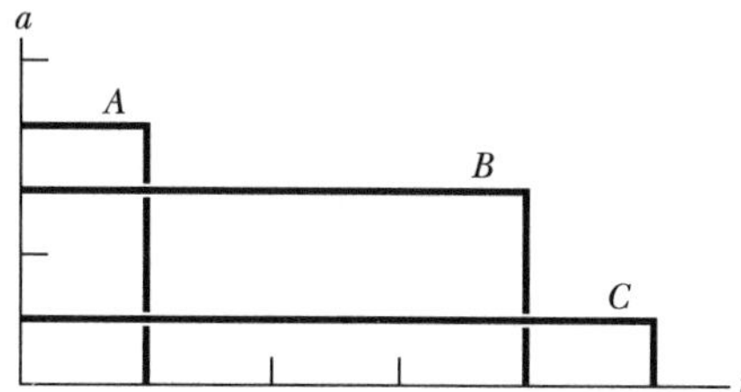

FIGURE 7-53 Question 29.

30. Here, for three situations, is the force acting on a particle and the (instantaneous) velocity of the particle: (a) $\mathbf{F} = -2\mathbf{i}$, $\mathbf{v} = -3\mathbf{i}$; (b) $\mathbf{F} = 3\mathbf{i}$, $\mathbf{v} = 2\mathbf{i}$; (c) $\mathbf{F} = 4\mathbf{i}$, $\mathbf{v} = -2\mathbf{i}$; (d) $\mathbf{F} = 3\mathbf{i}$, $\mathbf{v} = 4\mathbf{j}$. Rank the situations according to the rate at which the force is transferring energy to or from the particle at that instant, with the greatest rate *to* the particle first and the greatest rate *from* the particle last.

EXERCISES & PROBLEMS

56. A constant force of magnitude 10 N makes an angle of 150° (measured counterclockwise) with the direction of increasing x as it acts on a 2.0 kg object. How much work is done on the object by the force as the object moves from the origin to the point with position vector $(2.0 \text{ m})\mathbf{i} - (4.0 \text{ m})\mathbf{j}$?

57. An iceboat is at rest on a frictionless frozen lake when a sudden gust of wind exerts a constant force of 200 N, toward the east, on the boat. Due to the angle of the sail, the gust of wind causes the boat to slide in a straight line for a distance of 8.0 m in a direction 20° north of east. What is the kinetic energy of the iceboat after it slides that 8.0 m?

58. The only force acting on a 2.0 kg canister that is moving in an xy plane has a magnitude of 5.0 N. The canister initially has a velocity of 4.0 m/s in the positive x direction and sometime later has a velocity of 6.0 m/s in the positive y direction. How much work is done on the canister by the 5.0 N force during this interval?

59. A 50 kg ice-skater is moving toward the east across a frictionless frozen lake with a constant velocity when a sudden gust of wind exerts a constant force toward the west on the skater. A plot of position x versus time t for the skater's motion is shown in Fig. 7-54, where $t = 0$ is taken to be the instant the wind starts to blow and the positive direction of the x axis is toward the east. (a) What is the kinetic energy of the ice-skater at $t = 0$? (b) How much work does the force of the wind do on the ice-skater between $t = 0$ and $t = 3.0$ s?

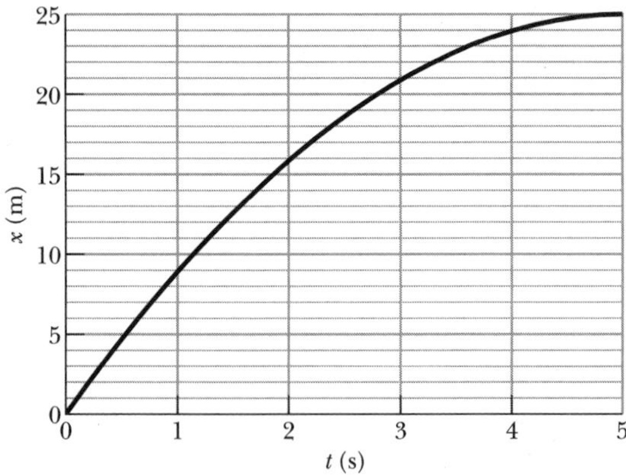

FIGURE 7-54 Problem 59.

60. A 1.0 kg standard body is at rest on a frictionless horizontal air track when a constant horizontal force **F** acting in the positive direction of an x axis along the track is applied to the body. A strobe photograph of the body as it slides to the right is shown in Fig. 7-55. The force **F** is applied to the body at $t = 0$, and the flashes on the strobe camera record the position of the body at 0.50 s intervals. How much work is done on the body by the applied force **F** between $t = 0$ and $t = 2.0$ s?

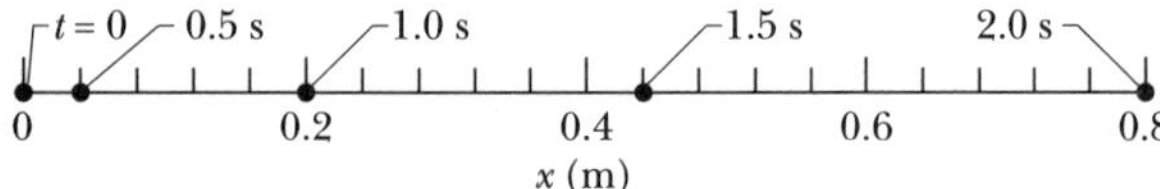

FIGURE 7-55 Problem 60.

61. A frightened child is restrained by her mother as the child comes down a frictionless playground slide. If the mother exerts a force of 100 N up the slide on the child, the child's kinetic energy increases by 30 J as she comes down along the slide by a distance of 1.8 m. (a) How much work is done on the child by the child's weight during the descent? (b) If the child is not restrained by her mother, how much will the child's kinetic energy increase as she comes down the slide by that distance of 1.8 m?

62. A 1.5 kg block is initially at rest on a horizontal frictionless surface when a horizontal force in the positive direction of an x axis is applied to the block. The force is given by $\mathbf{F}(x) = (2.5 - x^2)\mathbf{i}$ N, where x is in meters and the initial position of the block is $x = 0$. (a) What is the kinetic energy of the block as it passes through $x = 2.0$ m? (b) What is the maximum kinetic energy of the block between $x = 0$ and $x = 2.0$ m?

63. In Fig. 7-11a, a box of mass m lies on a horizontal frictionless surface and is attached to one end of a horizontal spring (with spring constant k) whose other end is fixed. The box is initially at rest at the position where the spring is unstretched ($x = 0$) when a constant horizontal force **P** in the direction of increasing x is applied to it. A plot of the resulting kinetic energy of the box versus its position x is shown in Fig. 7-56. (a) What is the magnitude of **P**? (b) What is the value of the spring constant k?

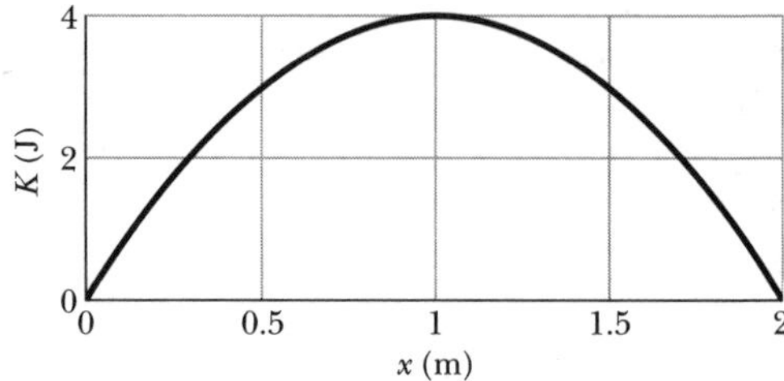

FIGURE 7-56 Problem 63.

64. Boxes are transported from one location to another in a warehouse by means of a conveyor belt that moves with a constant speed of 0.50 m/s. At a certain location the conveyor belt moves for 2.0 m up an incline that makes an angle of 10° with the horizontal and then for 2.0 m horizontally and finally for 2.0 m down an incline that makes an angle of 10° with the horizontal. Assume that a 2.0 kg box rides on the belt without slipping. At what rate is the force of the conveyor belt doing work on the box (a) as the box moves up the 10° incline, (b) as the box moves horizontally, and (c) as the box moves down the 10° incline?

65. A skier is pulled by a tow rope up a frictionless ski slope that makes an angle of 12° with the horizontal. The rope moves parallel to the slope with a constant speed of 1.0 m/s. The force of the rope does 900 J of work on the skier as the skier moves a distance of 8.0 m up the incline. (a) If the rope moves with a constant speed of 2.0 m/s, how much work will the force of the rope do on the skier as the skier moves a distance of 8.0 m up the incline? At what rate is the force of the rope doing work on the skier when the rope moves with a speed of (b) 1.0 m/s and (c) 2.0 m/s?

Graphing Calculators

SAMPLE PROBLEM 7-11

Functional and Graphical Integration. A force acting on a particle is directed along an x axis and is given by $F = 10.0xe^{-x/10.0}$, with F in newtons and x in meters. How much work is done on the particle by the force as the particle moves from $x = 5.00$ m to $x = 15.0$ m?

SOLUTION: Equation 7–27 tells us that the work done by a one-dimensional force on a particle is given by

$$W = \int_{x_i}^{x_f} F(x)\, dx.$$

Here the function $F(x)$ is $10.0xe^{-x/10.0}$, the variable of integration is x, the lower limit x_i is 5.00, and the upper limit $x_f = 15.0$. To enter this information on a TI-85/86, press 2nd CALC and then choose the fnInt option. Next complete the line that the calculator begins on the screen by pressing in

10xe^(−x/10),x,5,15)

using the x-VAR key to put in x and the e^x key to put in the exponential function e^. (Be sure to use the negation key.)

Pressing ENTER we find that the integration equals 352. Thus the work done on the particle by the force is

$$W = 352 \text{ J.} \qquad \text{(Answer)}$$

We can also find the answer by graphing the function $10.0xe^{-x/10.0}$ and integrating on the graph. To do so, first choose Func in the mode menu to ready the calculator for functional graphing. Next press GRAPH, choose the y(x) = option, and then press in the expression for $F(x)$ as y1 (and deactivate any other functions). Next choose the RANGE (TI-85) or WIND (TI-86) option and set the graphing parameters as xMin = 0, xMax = 15, xScl = 5, yMin = 0, yMax = 50, and yScl = 10, and then graph the function.

To integrate on the graph, first press MORE, then choose the FRMT option. If CoordOn is not highlighted, move to it and then press ENTER. Next, return to the first graphing menu by pressing GRAPH. To go to the math menu, press MORE and choose the MATH option. Next choose the ∫ f(x) option for integration, which causes a cursor to appear on the plotted curve.

Move the cursor leftward until it reaches x = 5, as indicated by the location coordinates printed on the screen. (The coordinates appear because we chose CoordOn in the format menu.) Press ENTER to mark this location as the lower limit of the integration. Then move the cursor rightward to x = 15 and press ENTER to mark this location as the upper limit of integration. The calculator soon displays the result of the integration. Again we find 352. In addition, a TI-86 shades in the region of integration on the graph.

66. A can of sardines is made to move along an x axis from $x = 0.25$ m to $x = 1.25$ m by a force with a magnitude given by $F = \exp(-4x^2)$, with x in meters and F in newtons. How much work is done on the can by the force?

Chapter Eight
Potential Energy and Conservation of Energy

QUESTIONS

14. In Problem 10 and Fig. 8-33, the block is released from rest at height h above the bottom of the circular loop; it experiences a certain normal force of magnitude N_1 when it reaches the bottom of the loop and a certain centripetal acceleration of magnitude a_2 when it reaches point Q. (a) If we increase the release height, will the magnitude of the normal force on the block at the bottom of the loop then be greater than, less than, or equal to N_1? (b) Will the magnitude of the block's centripetal acceleration at point Q then be greater than, less than, or equal to a_2?

15. In Fig. 8-65, a block slides along a track that descends by distance h. The track is frictionless except for the lower section. There the block slides to a stop in a certain distance D because of friction. (a) If we decrease h, will the block now slide to a stop in a distance that is greater than, less than, or equal to D? (b) If, instead, we increase the mass of the block, will the stopping distance now be greater than, less than, or equal to D?

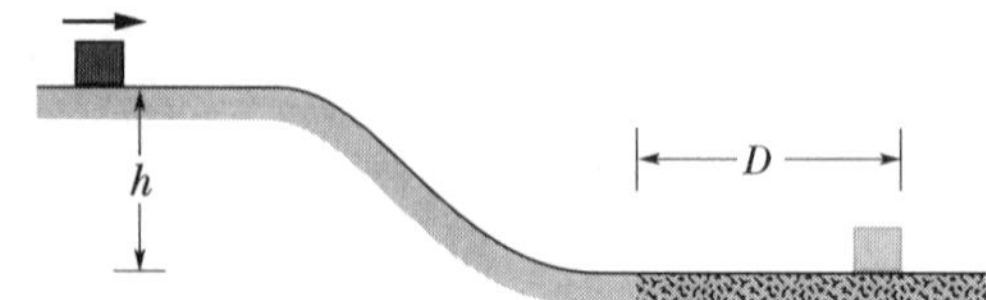

FIGURE 8-65 Question 15.

16. *Organizing question:* Figure 8-66 shows 10 situations in which a block of mass m with an initial speed of v_i is brought to a stop. For each situation, what are the initial total mechanical energy E_i of the block, the change ΔE_d in the block's mechanical energy due to dissipation by a frictional force, and the final total mechanical energy E_f of the block or, in some situations, of the block–spring system? For the 10 situations of Fig. 8-66, the block slides

(*a*) into a region of frictional force f, where it stops in distance d;
(*b*) up a frictionless ramp at angle θ until it (momentarily) stops at height h;
(*c*) up a ramp at angle θ against a frictional force f until it stops at height h;

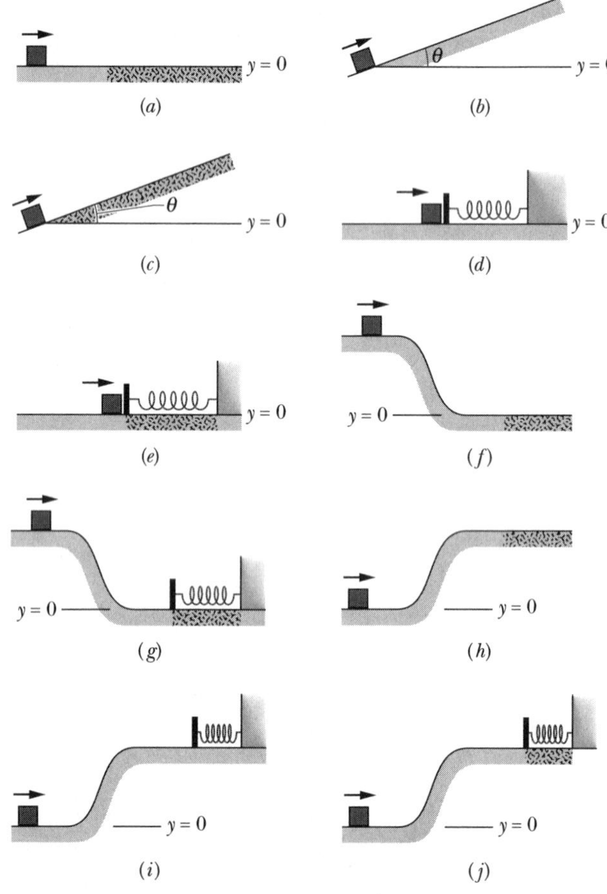

FIGURE 8-66 Question 16.

(*d*) across a frictionless floor and onto a spring of spring constant k, stopping (momentarily) when the spring is compressed by distance d;
(*e*) onto a spring of spring constant k and simultaneously into a region of frictional force f, stopping when the spring is compressed by distance d;
(*f*) downward by height h onto a lower floor and into a region of frictional force f, stopping in that region in distance d;

(*g*) downward by height h onto a lower floor and then simultaneously onto a spring of spring constant k and into a region of frictional force f, stopping when the spring is compressed by distance d;
(*h*) upward by height h onto a higher floor and then into a region of frictional force f, stopping in distance d in that region;
(*i*) upward by height h onto a higher floor and then onto a spring of spring constant k, stopping when the spring is compressed by distance d; and
(*j*) upward by height h onto a higher floor and then simultaneously onto a spring of spring constant k and into a region of frictional force f, stopping when the spring is compressed by distance d.

17. Repeat Checkpoint 3 but now assume that the three ramps have the same coefficient of friction (the block still reaches points B).

18. A spring that lies along an x axis is attached to a wall at one end and a block at the other end, as in Fig. 7-11*a*. The block rests on a frictionless surface at $x = 0$, as shown. Then an applied force of constant magnitude begins to compress the spring, displacing the block by a distance x, until the block comes to a maximum displacement x_{max}. During this displacement, which of the curves in Fig. 8-67 best represents (a) the elastic potential energy of the spring, (b) the kinetic energy of the block, and (c) the work done on the spring–block system by the applied force? (d) What is the sum of curves 2 and 3? In what range of displacement does the applied force (e) increase the block's kinetic energy and (f) decrease it? (g) At what displacement is the block's kinetic energy maximum? (h) What is the kinetic energy when the block reaches x_{max}? Does the applied force transfer more energy to the block's kinetic energy or the spring's potential energy during (i) the first half of the compression and (j) the second half?

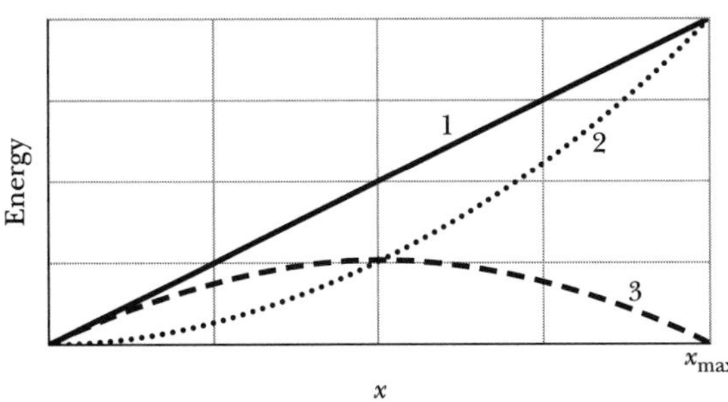

FIGURE 8-67 Question 18.

19. In Question 12 and Fig. 8-26, what is the least mechanical energy the particle can have in regions (a) BC, (b) DE, and (c) FG?

20. In Fig. 8-68, a block is released from rest on a track with an initial gravitational potential energy E_i. The curved portions on the track are frictionless, but the horizontal portion, of length L, produces a frictional force f on the block. (a) How much mechanical energy is dissipated by the friction if the block passes once through length L? How many times does the block pass through that length if the initial potential energy E_i is equal to (b) $0.50fL$, (c) $1.25fL$, and (d) $2.25fL$? (e) For those three situations, does the block come to a stop at the center of the horizontal portion, to the left of center, or to the right of center. (Warm-up for Problem 85.)

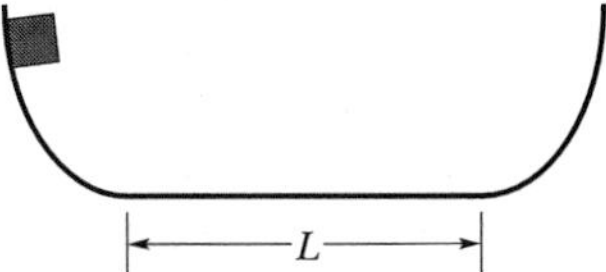

FIGURE 8-68 Question 20.

21. Figure 8-69 is a sketch (not to scale) of a sky diver's kinetic energy K versus time t during a fall. Which of the following descriptions of the fall best corresponds to the sketch: the sky diver accelerates throughout the fall; the sky diver first reaches terminal speed in a spread-eagle orientation and then switches to a head-down orientation; the sky diver first reaches terminal speed in a head-down orientation and then switches to a spread-eagle orientation?

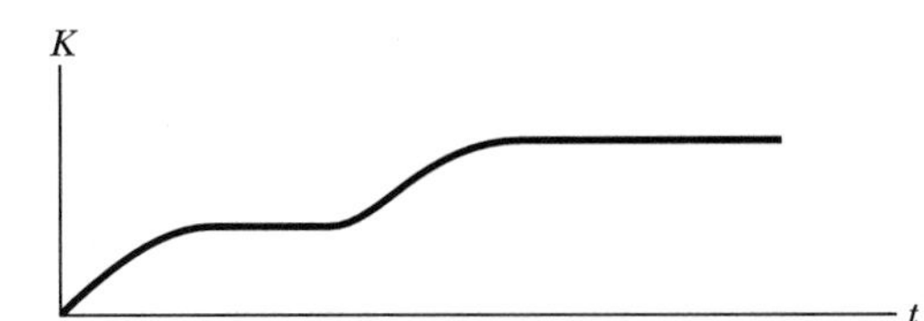

FIGURE 8-69 Question 21.

22. A stationary elementary particle K-zero (K^0) can spontaneously decay to (suddenly transform into) neutral pions π^0 via the reaction $K^0 \rightarrow \pi^0 + \pi^0 + \pi^0$. The pions produced by the decay fly away from the decay site. Is the mass of the K^0 greater than, less than, or equal to the sum of the masses of the pions?

EXERCISES & PROBLEMS

101. A spring with spring constant $k = 200$ N/m is suspended vertically with its upper end fixed to the ceiling and its lower end at position $y = 0$ (Fig. 8-70). A block of weight 20 N is attached to the lower end, held still for a moment, and then released. What are the kinetic energy and the changes in the gravitational potential energy and the elastic potential energy of the spring–block system when the block is at y values of (a) -5.0 cm, (b) -10 cm, (c) -15 cm, and (d) -20 cm?

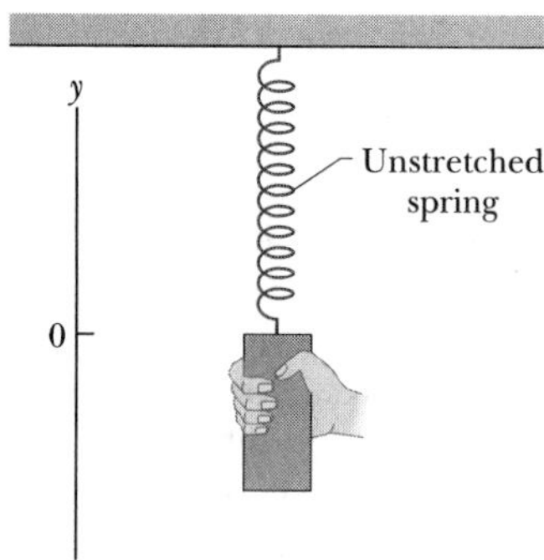

FIGURE 8-70 Problem 101.

102. A 60 kg skier leaves the end of a ski-jump ramp with a velocity of 24 m/s directed 25° above the horizontal. Suppose that as a result of air resistance the skier returns to the ground with a speed of 22 m/s and lands at a point that is 14 m vertically below the end of the ramp. How much energy is dissipated because of air resistance from the time the skier leaves the ramp until she returns to the ground?

103. A 0.40 kg particle moves under the influence of a single conservative force. At point A, where the particle has a speed of 10 m/s, the associated potential energy is 40 J. As the particle moves from A to B, the force does 25 J of work on the particle. What is the value of the potential energy at point B?

104. A spring (k = 200 N/m) is fixed at the top of a frictionless plane that makes an angle of 40° with the horizontal, as shown in Fig. 8-71. A 1.0 kg mass is projected up the plane, from an initial position that is 0.60 m from the end of the uncompressed spring, with an initial kinetic energy of 16 J. (a) What is the kinetic energy of the mass at the instant it has compressed the spring 0.20 m? (b) With what kinetic energy must the mass be projected up the plane if it is to stop momentarily when it has compressed the spring by 0.40 m?

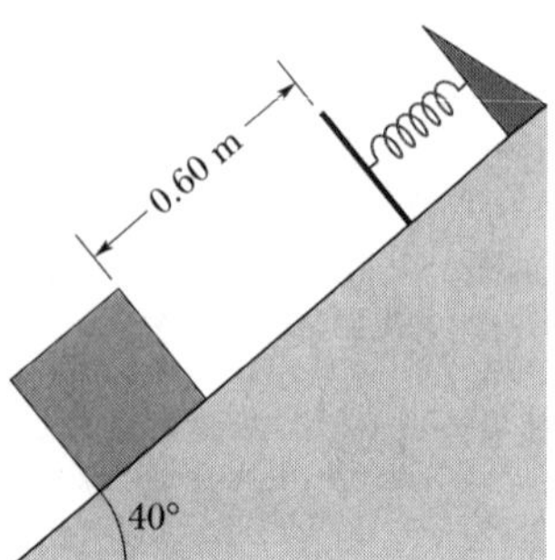

FIGURE 8-71 Problem 104.

105. At t = 0 a 1.0 kg ball is thrown from the top of a tall tower with velocity $\mathbf{v}$ = (18 m/s)$\mathbf{i}$ + (24 m/s)$\mathbf{j}$. What is the change in the potential energy of the ball between t = 0 and t = 6.0 s?

106. Two blocks, of masses M and $2M$ where M = 2.0 kg, are connected to a spring of spring constant k = 200 N/m that has one end fixed, as shown in Fig. 8-72. The horizontal surface and the pulley's axle are frictionless, and the pulley is massless. The system is released from rest with the spring unstretched. (a) What is the combined kinetic energy of the two masses after the hanging mass has fallen 0.090 m? (b) What is the kinetic energy of the hanging mass after it has fallen 0.090 m? (c) What maximum distance does the hanging mass fall before momentarily stopping?

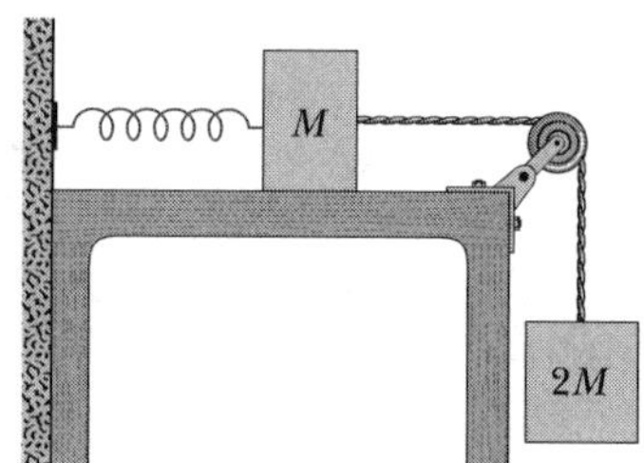

FIGURE 8-72 Problem 106.

107. A 20 kg block on a horizontal surface is attached to a horizontal spring of spring constant k = 4.0 kN/m. The block is pulled to the right so that the spring is extended 10 cm beyond its unstretched length, and the block is then released from rest. The frictional force between the sliding block and the surface has a magnitude of 80 N. (a) What is the kinetic energy of the block when it has moved 2.0 cm from its point of release? (b) What is the kinetic energy of the block when it first slides back through the point where the spring is unstretched? (c) What is the maximum kinetic energy attained by the block as it slides from its point of release to the point where the spring is unstretched?

108. In a circus act, a 60 kg clown is shot from a cannon with an initial velocity of 16 m/s at some unknown angle above the horizontal. A short time later the clown lands in a net that is 3.9 m vertically above the clown's initial position. Disregard air resistance. What is the kinetic energy of the clown as he lands in the net?

109. A skier weighing 600 N goes over a frictionless circular hill of radius 20 m (Fig. 8-73). Assume that the effects of air resistance on the skier are negligible. As she comes up the hill, her speed is 8.0 m/s at point B. (a) What is the skier's speed at the top of the hill (point A) if she coasts over the hill without using her poles? (b) What minimum speed can the skier have at point B and still coast to the top of the hill? (c) Do the answers to these two questions increase, decrease, or remain the same if the skier weighs, instead, 700 N?

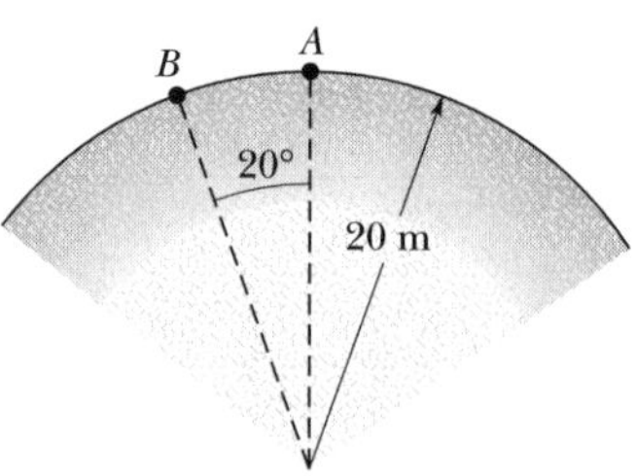

FIGURE 8-73 Problem 109.

110. A 60 kg skier starts from rest at a height of 20 m above the end of a ski-jump ramp as shown in Fig. 8-74. As the skier leaves the ramp, his velocity makes an angle of 28° with the horizontal. Neglect the effects of air resistance and assume the ramp is frictionless. What is the maximum height h of his jump above the end of the ramp?

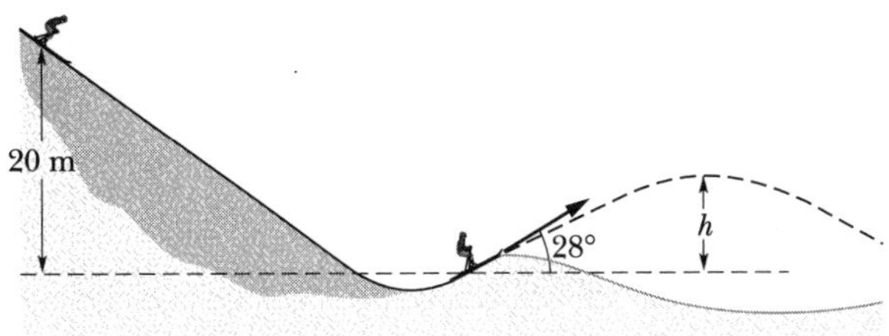

FIGURE 8-74 Problem 110.

111. A single conservative force $\mathbf{F} = (6.0x - 12)\mathbf{i}$ N, where x is in meters, acts on a particle moving along an x axis. The potential energy associated with this force is assigned a value of 27 J at $x = 0$. (a) Write an expression for the potential energy U as a function of x. (b) What is the maximum positive potential energy? (c) At what values of x will the potential energy be equal to zero?

112. A spring with spring constant $k = 170$ N/m is at the top of a 37.0° frictionless incline. The end of the incline is 1.00 m from the end of the uncompressed spring as shown in Fig. 8-75. A 2.00 kg canister is pushed against the spring until the spring is compressed 0.200 m and released from rest. (a) What is the speed of the canister at the instant the spring returns to its uncompressed length (which is when the canister loses contact with the spring)? (b) What is the speed of the canister at the instant it reaches the bottom of the incline?

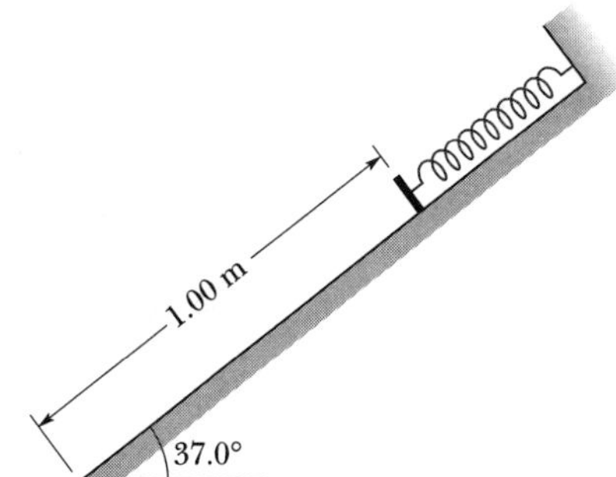

FIGURE 8-75 Problem 112.

113. A spring with spring constant $k = 400$ N/m is placed in a vertical orientation with its lower end supported by a horizontal surface. The upper end is depressed 25.0 cm, and a block of weight = 40.0 N is placed (unattached) on the depressed spring. The system is then released from rest. Assume the gravitational potential energy of the block is zero at the release point ($y = 0$) and calculate the gravitational potential energy, the elastic potential energy, and the kinetic energy of the block for y equal to (a) 0, (b) 5.00 cm, (c) 10.0 cm, (d) 15.0 cm, (e) 20.0 cm, (f) 25.0 cm, and (g) 30.0 cm. Also, (h) how far above its point of release does the block rise?

114. A 2.0 kg bread box on a frictionless 40° incline is connected to a light spring of spring constant $k = 120$ N/m, as shown in Fig. 8-76. The box is released from rest when the spring is unstretched. Assume that the pulley is massless and frictionless. (a) What is the magnitude of the velocity of the box when it has moved 10 cm down the incline? (b) How far down the incline from its point of release does the box slide before coming momentarily to rest? (c) What are the magnitude and direction of the acceleration of the box at the instant it is momentarily at rest?

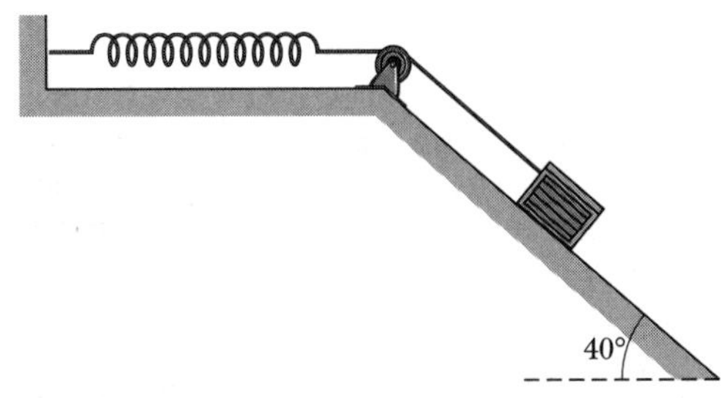

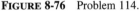

FIGURE 8-76 Problem 114.

115. A single conservative force $F(x)$ acts on a 1.0 kg particle that moves along an x axis. The potential energy $U(x)$ associated with $F(x)$ is given by

$$U(x) = -4x\,e^{-x/4}\text{ J},$$

where x is in meters. At $x = 5.0$ m the particle has a kinetic energy of 2.0 J. (a) What is the mechanical energy of the system? (b) Make a plot of $U(x)$ as a function of x for $0 \leq x \leq 10$ m and on the same graph draw a line that represents the mechanical energy of the system. (c) Use part (b) to determine the values of x between which the particle can move. (d) Use part (b) to determine the maximum kinetic energy of the particle and the value of x at which it occurs. (e) Determine the equation for $F(x)$ as a function of x. (f) For what (finite) value of x is $F(x) = 0$?

116. A 70 kg firefighter slides, from rest, 4.3 m down a vertical pole. (a) If the firefighter holds onto the pole lightly, so that the frictional force of the pole on her is negligible, what is her speed just before reaching the ground? (b) If the firefighter more firmly grasps the pole as she slides, so that the average frictional force of the pole on her is 500 N upward, what is her speed just before reaching the ground floor?

117. A large fake cookie sliding on a horizontal surface is attached to one end of a horizontal spring with spring constant $k = 400$ N/m; the other end of the spring is fixed in place. The cookie has a kinetic energy of 20.0 J as it passes through the position where the spring is unstretched. As the cookie slides, a frictional force of magnitude 10.0 N acts on it. (a) How far will the cookie slide from the position where the spring is unstretched before coming momentarily to rest? (b) What will be the kinetic energy of the cookie as it slides back through the position where the spring is unstretched?

118. An 0.42 kg shuffleboard disk is initially at rest when a player uses a cue to accelerate the disk at a constant value to a speed of 4.2 m/s. The acceleration takes place over a 2.0 m distance, at the end of which the cue loses contact with the disk. Then the disk slides an additional 12 m before stopping. Assume that the shuffleboard court is level and that the force of friction on the disk is constant. How much work is done on the disk by the force of friction (a) for that additional 12 m and (b) for the entire 14 m distance? (c) How much work is done on the disk by the cue?

Clustered Problems

Cluster 1

In problems involving friction, the coefficient of friction is usually given with only two significant figures. So, answers are usually rounded off to two significant figures. However, such rounding off should be done only as a last step in a solution. And if the answer to, say, part (a) of a problem is needed in part (c), you should use the answer before *rounding off.*

119. A 5.0 kg block is projected at 5.0 m/s up a plane that is inclined at 30° with the horizontal. How far up along the plane will the block go (a) if the plane is frictionless and (b) if the coefficient of kinetic friction between the block and the plane is 0.40? (c) In the latter case, how much mechanical energy is dissipated by the frictional force during the block's ascent? (d) If the block then slides back down against the frictional force, what is the block's speed when it reaches the original projection point?

120. A 1500 kg car begins sliding down a 5.0° inclined road with a speed of 30 km/h. The engine is turned off, and the only forces acting on the car are its weight and a net frictional force from the road. After traveling 50 m along the road, the car's speed is 40 km/h. (a) How much mechanical energy has been dissipated by the net frictional force? (b) What is the magnitude of that net force?

121. A 15 kg block is accelerated at 2.0 m/s² along a horizontal frictionless surface, with the speed increasing from 10 m/s to 30 m/s. Assume that energy dissipation by frictional forces can be neglected. What are (a) the change in the block's mechanical energy and (b) the average rate at which energy is transferred to the block? What is the instantaneous rate of that transfer when the block's speed is (c) 10 m/s and (d) 30 m/s?

122. Repeat Problem 121, but now have the block accelerated up along a frictionless plane that is inclined at 5.0° to the horizontal.

Cluster 2

123. A spring with spring constant k is hung with one end attached to a ceiling. A block of mass M is attached at the other end when the spring is at its rest length; then the block is allowed to fall. How far does the block fall to reach (a) the point where it momentarily stops and (b) the point where it has its greatest speed? (c) What is that greatest speed?

124. In Fig. 8-77*a*, a block of mass M falls from rest a distance H to a vertical spring with spring constant k. (a) What is the distance x_c of maximum compression of the spring? Suppose the block sticks to the spring. As the block moves back upward, with the spring eventually being extended, what is the distance x_e of maximum extension of the spring?

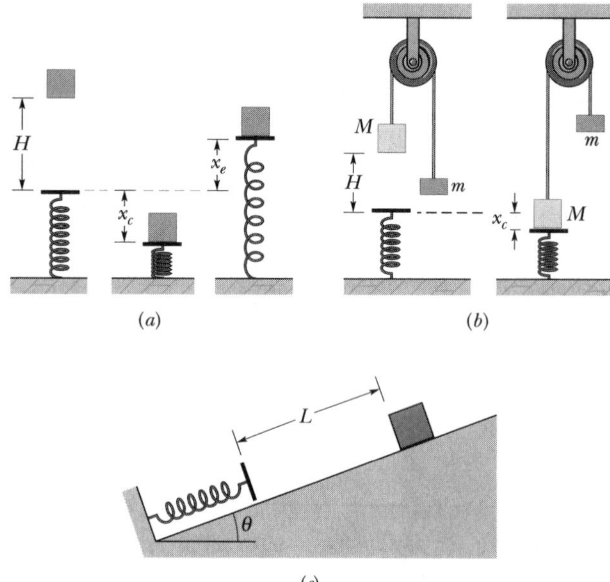

FIGURE 8-77 Problems 124 through 126.

125. In Fig. 8-77*b*, two blocks are connected by a massless string wrapping around a massless, frictionless pulley. The blocks have masses M and m, with $M > m$, and are initially held in place. Then the blocks are released and the block of mass M falls a distance H down onto a vertical spring with spring constant k. What then is the distance x_c of maximum compression of the spring?

126. In Fig. 8-77*c*, a block is released from rest on a plane inclined at angle θ with the horizontal. It slides down the plane by a distance L to a spring of spring constant k. The coefficient of friction between the block and the plane is μ. What is the distance x_c of maximum compression of the spring?

Cluster 3

See the caution at the start of Cluster 1.

127. In Fig. 8-78*a*, which is not drawn to scale, a 2 kg block slides along a frictionless track through point A on a plateau, down into a valley through point B, and then onto a second, lower plateau and through point C. Then it travels through a length $L = 10$ m where the coefficient of kinetic friction between the block and the track is 0.30. Thereafter, the track is again frictionless, but the block runs into a spring of spring constant $k =$

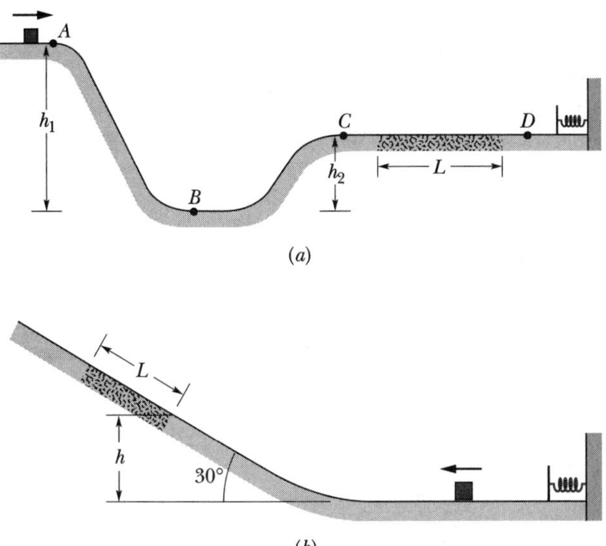

FIGURE 8-78 Problems 127 and 128.

1000 N/m. The plateaus are at heights $h_1 = 5.0$ m and $h_2 = 3.0$ m above the valley floor.

The block has a speed of 10 m/s at point A. What are its speeds at (a) point B, (b) point C, and (c) point D? (d) What is the distance of maximum compression of the spring? (e) Where does the block finally (not just momentarily) stop and what was its final direction of travel?

128. In Fig. 8-78*b*, the 2.0 kg block had been placed in front of a spring (of spring constant 1000 N/m) when the spring was compressed by 0.40 m. Because the spring was released, the block is sliding along a track that is frictionless except for a length L = 2.0 m. That region of friction is on a 30° incline and begins at a height h = 2.0 m from the level track. The coefficients of static and kinetic friction between the block and the track in that region are 0.60 and 0.30, respectively.

Just as the block reaches the region of friction, what are its (a) kinetic energy and (b) speed? (c) How much energy is dissipated by the frictional force acting on the block when the block travels through that region? Just as the block leaves the region, what are its (d) kinetic energy and (e) speed? (f) What maximum height above the level track is reached by the block as it then continues to slide upward?

(g) When the block slides back down, what is its kinetic energy just as it leaves the region of friction? (h) Where does the block finally (not just momentarily) stop, and in what direction is it headed as it stops?

Cluster 4

See the caution at the start of Cluster 1.

129. A 3.0 kg block is pulled along an axis by a conservative force directed along that axis, away from the block's initial point. Figure 8-79*a* gives the magnitude of the force versus the distance the block is moved from its initial point. When the block has been moved by 2.0 m, (a) how much work has been done on it by the force and (b) what is its speed? (c) What is the block's speed when it reaches a distance of 4.0 m from its initial point?

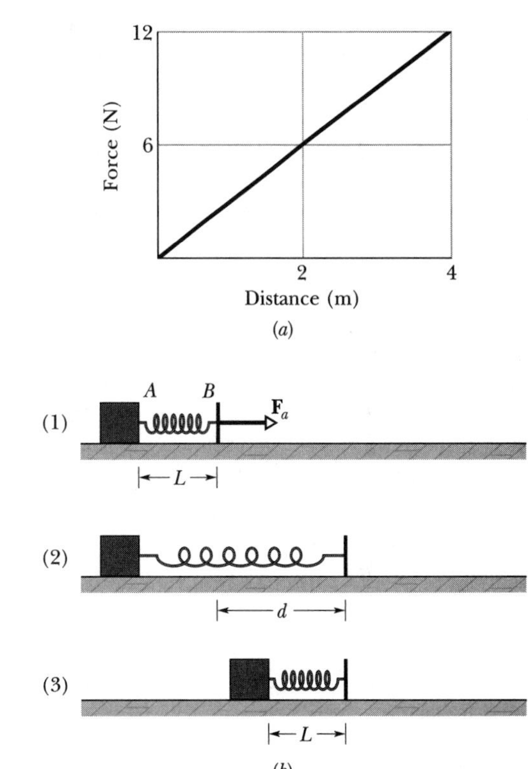

FIGURE 8-79 Problems 129 and 130.

130. End A of a spring (of spring constant 50 N/m and relaxed length L) is attached to a 3.0 kg block that lies on a floor, as shown in Fig. 8-79*b*(1). Then the other end (B) of the spring is pulled parallel to the surface by an applied force $\mathbf{F}_a$ until the block begins to slide over the surface. Just as the sliding begins, end B is held fixed in place, as in Fig. 8-79*b*(2). The coefficients of static and kinetic friction between the block and the floor are 0.40 and 0.12, respectively.

Just as the sliding begins, (a) what is the stretch d of the spring and (b) how much work has been done on the block–spring system by $\mathbf{F}_a$?

The block now slides rightward and through the point at which the spring momentarily has its relaxed length L, shown in Fig. 8-79*b*(3). When it reaches that point, (c) how much work has been done on the block–floor system by the spring force and (d) how much mechanical energy has been dissipated by the frictional force between the block and floor? (e) What is the speed of the block at that point? (f) How much farther toward the right will the block slide before the frictional force and the spring force stop it?

131. A 3 kg block slides along a floor directly toward a spring with spring constant 50 N/m. One end of the spring is fixed to a wall; the block runs into the other end, compressing the spring. When the block is 1.0 m from that end, its speed is 2.0 m/s. The coefficients of static and kinetic friction between the block and the floor are 0.25 and 0.12, respectively. (a) What is the speed of the block just as it reaches the spring? (b) How far is the spring compressed? (c) At the point of maximum compression, which force on the block is greater, the spring force or the frictional force? (d) How far from that point does the block then slide?

132. One end of a spring on a floor is fixed to a wall while a 3.0 kg block is attached to the other end, with the spring at its relaxed length. Then the block is pulled directly away from the wall so that the spring is stretched by 0.20 m. The spring has a spring constant of 50 N/m, and the coefficients of static and kinetic friction between the block and the floor are 0.25 and 0.12, respectively. (a) When we next release the block so that the spring causes it to slide back toward the wall, how far does the block slide before it stops? (*Hint:* It slides past the equilibrium point where it has its relaxed length, but it does not then reverse its motion.) (b) How far does it slide until it reaches its maximum speed, and (c) what is that maximum speed?

Cluster 5

The pulleys shown in Fig. 8-80 are massless and frictionless. When a cord loops halfway around a pulley, the cord pulls on the pulley with a net force that is twice the tension in the cord.

133. In Fig. 8-80*a*, a block of mass M is connected to a spring of spring constant k by means of a cord wrapped over a pulley. (a) When the system is in equilibrium, by how much is the spring stretched? (*Hint:* Imagine doing the following: First place the block on the cord while supporting the block's weight so that the spring is at its relaxed length. Take the gravitational potential energy of the block in this position to be zero. Then gradually allow the block to descend by a distance y until equilibrium is reached. Write an expression for the sum $E(y)$ of the elastic potential energy of the spring and the gravitational potential energy for this state, both written in terms of y. Equilibrium occurs when the sum E is least, that is, when $dE/dy = 0$.)

Next, the block is pulled down from the equilibrium point by a distance H and released. (b) Find the block's speed when it moves through the equilibrium point. (c) How high above that point does the block then travel?

134. In Fig. 8-80*b*, a block of mass M is suspended by a cord that is wrapped over a pulley. The pulley is supported by a spring of spring constant k. (a) When the system is in equilibrium, by how much is the spring stretched? (*Hint:* See the hint of Problem 133.) Next, the block is pulled down from the equilibrium point by a distance H and released. (b) Find the speed of the block when the block moves through the equilibrium point.

135. In Fig. 8-80*c*, a block of mass M is suspended by a cord that is wrapped over a pulley. The pulley is supported by spring 2 of spring constant k_2. The other end of the cord is attached to spring 1 of spring constant k_1. The system is in equilibrium. (a) What is the relation between the tension T in the cord and the stretch y_1 of spring 1? (b) What is the relation between the force on the pulley due to spring 2 and the tension in the cord? (c) What, then, is the relation between the stretch y_2 of spring 2 and that of spring 1? In terms of M, k_1, k_2, and g, what are (d) y_1 and (e) y_2? (*Hint:* To use the hint of Problem 133, first express the total energy of the system in terms of one variable, say, y_1. Then set $dE/dy_1 = 0$.)

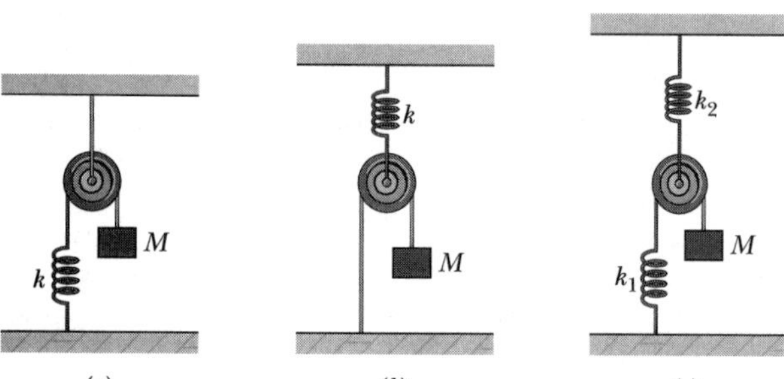

FIGURE 8-80 Problems 133 through 135.

Graphing Calculators

136. *Fly-Fishing and Speed Amplification.* If you throw a loose fishing fly, it will travel horizontally only about 1 m. But if you throw a fly attached to fishing line by casting the line with a rod, the fly will easily travel horizontally to the full length of the line, say, 20 m.

The cast is depicted in Fig. 8-81: Initially (Fig. 8-81*a*) the line of length L is stretched horizontally leftward and moving rightward with a speed v_0. As the fly at the end of the line moves forward, the line doubles over, with the upper section still moving and the lower section stationary (Fig. 8-81*b*). The upper section decreases in length as the lower section increases in length, until the line is horizontally rightward (Fig. 8-81*d*). If air drag is neglected, the initial kinetic energy of the line in Fig. 8-81*a* be-

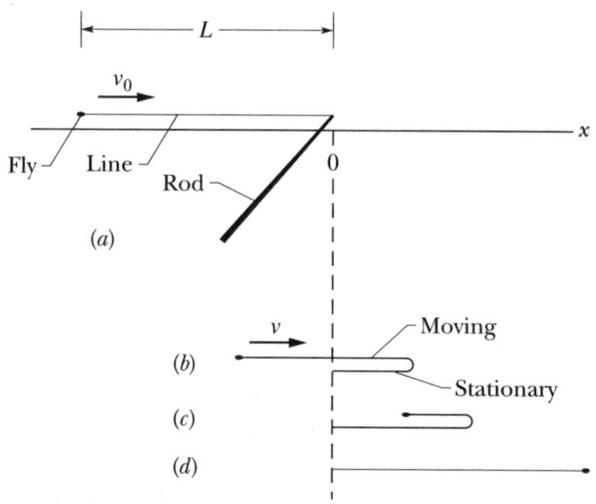

FIGURE 8-81 Problem 136.

comes progressively concentrated in the fly and the decreasing portion of the line that is still moving, resulting in an amplification (increase) in the speed of the fly and that portion.

(a) Using the x axis indicated, show that when the fly position is x, the length of the still moving (upper) section of line is $(L - x)/2$. (b) Assuming that the line is uniform with a linear density ρ (mass per unit length), what is the mass of the still moving section? Let m_f represent the mass of the fly, and assume that the kinetic energy of the moving section is unchanged from its initial value (when the moving section had length L and speed v_0). (c) Find an expression for the speed of the still moving section and the fly.

Assume that initial speed $v_0 = 6.0$ m/s, line length $L = 20$ m, fly mass $m_f = 0.80$ g, and linear density $\rho = 1.3$ g/m. (d) Plot the fly's speed v versus its position x. (e) What is the fly's speed just as the line approaches its final horizontal orientation and the fly is about to flip over and stop? (In more realistic calculations, air drag reduces this final speed.)

Speed amplification can also be produced with a bullwhip and even a curled-up, wet towel that is popped against a victim in a common locker-room prank. (Adapted from "The Mechanics of Flycasting: The Flyline," by Graig A. Spolek, *American Journal of Physics,* Sept. 1986, Vol. 54, No. 9, pp. 832–836.)

Chapter Nine
Systems of Particles

QUESTIONS

13. Some skilled basketball players seem to hang in midair during a jump at the basket, allowing them more time to shift the ball from hand to hand and then into the basket. If a player raises arms or legs during a jump, does the player's time in the air increase, decrease, or stay the same?

14. Figure 9-44 shows, from overhead, the path taken by a toy car moving at a constant speed; the straight sections are either parallel to the x axis, parallel to the y axis, or at 45° to the axes. (a) Rank the curved sections according to the magnitude of the change $\Delta\mathbf{p}$ in linear momentum of the car due to them, greatest first. For which curved sections does $\Delta\mathbf{p}$ have a component in (b) the negative direction of the y axis and (c) the positive direction of the x axis?

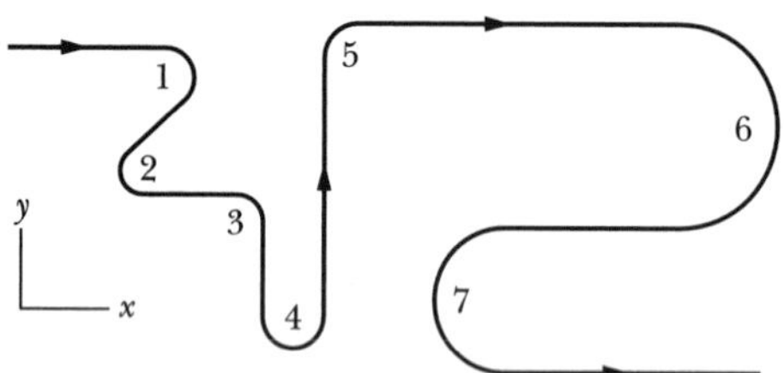

FIGURE 9-44 Question 14.

15. The free-body diagrams in Fig. 9-45 give, from overhead views, the horizontal forces acting on three boxes of chocolates as the boxes move over a frictionless confectioner's counter. For each box, is the linear momentum conserved along the x axis and the y axis?

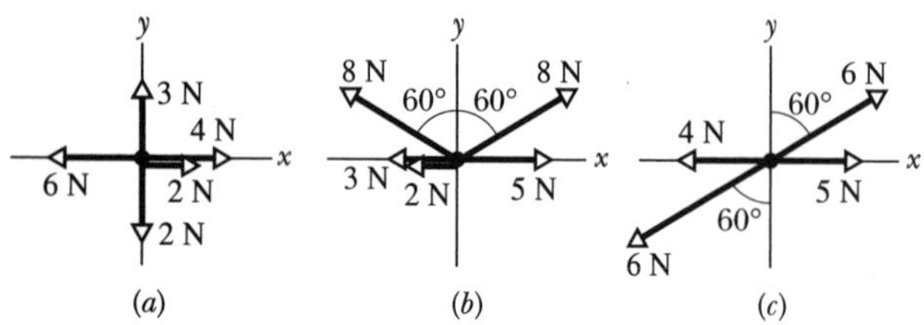

FIGURE 9-45 Question 15.

16. An initially stationary box on a frictionless floor explodes into two pieces: piece A with mass m_A and piece B with mass m_B. These pieces then move across the floor along an x axis. Graphs of the position versus time for the two pieces are given in Fig. 9-46. (a) Which graphs pertain to physically possible explosions? Of those graphs, which best corresponds to the situation where (b) $m_A = m_B$, (c) $m_A > m_B$, and (d) $m_A < m_B$?

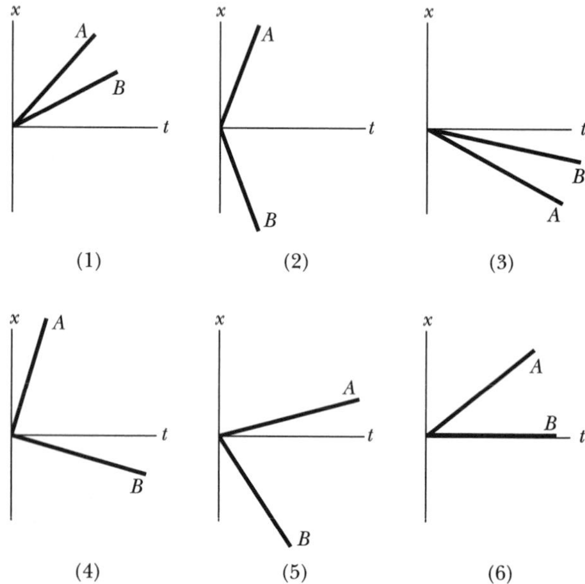

FIGURE 9-46 Question 16.

17. *Organizing question:* A 5 kg block is sliding with a velocity of +4 m/s along an x axis on a frictionless surface when it explodes into two pieces. Piece A, with a mass of 3 kg, and piece B are sent sliding over that surface. Figure 9-47 shows the speed and direction of travel of piece A for three situations. For each situation, set up an equation to find velocity v_{Bf} of piece B, complete with known data.

18. A robot craft R moving through deep space breaks into three pieces A, B, and C. Some of the momentum values (in kilogram-meters per second and expressed in terms of an xyz coordinate

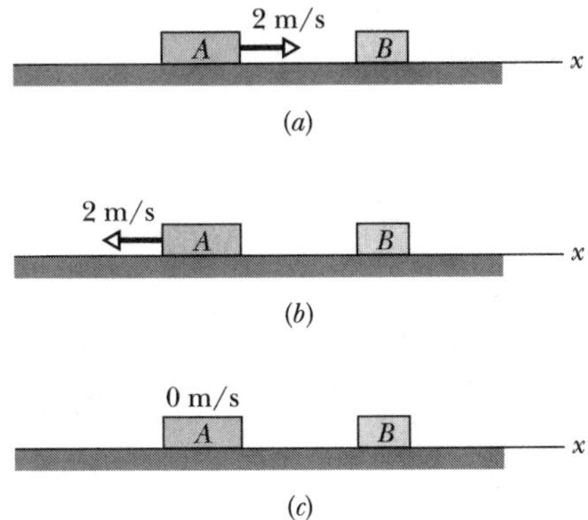

FIGURE 9-47 Question 17.

system) are given in the following table. What are the missing values?

OBJECT	p_x	p_y	p_z
R	600	400	−800
A	200		−1200
B	300	−300	
C		−600	200

19. In the four situations indicated in Fig. 9-48, an object explodes into two equal-mass fragments when the object is at the origin of the coordinate system. The velocity vectors of the fragments are indicated; they are directed either along an axis or at 45° to an axis. For each situation tell the direction of travel of the object before the explosion or indicate that it was stationary.

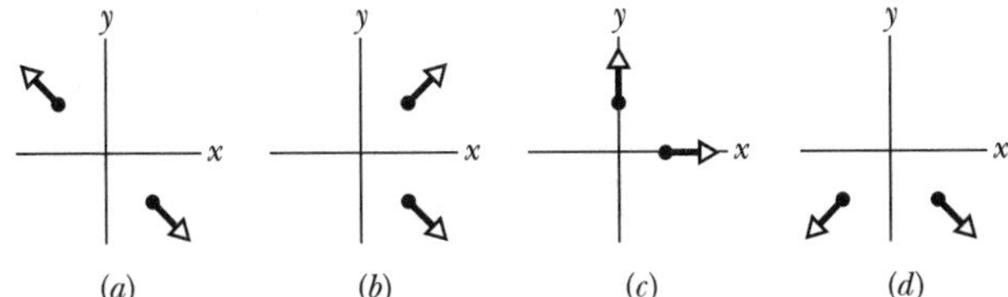

FIGURE 9-48 Question 19.

20. A spaceship that is moving along an x axis separates into two parts, as in Fig. 9-12. (a) Which of the graphs in Fig. 9-49 could possibly give the position versus time for the ship and the two parts? (b) Which of the numbered lines pertains to the trailing part? (c) Rank the possible graphs according to the relative speed between the parts, greatest first.

21. Three rockets, with the same uniformly distributed mass, race along a deep-space raceway. At a certain instant they are even with one another and are traveling at the same speed. But just then they each jettison a rear section, as shown in Fig. 9-50. The resulting relative speeds between the front and rear sections happen to be identical for the rockets. Rank the rockets according to their travel time to the end of the raceway, greatest first.

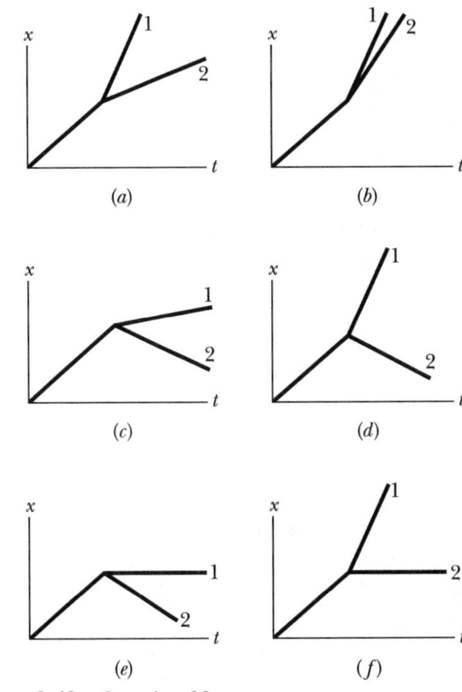

FIGURE 9-49 Question 20.

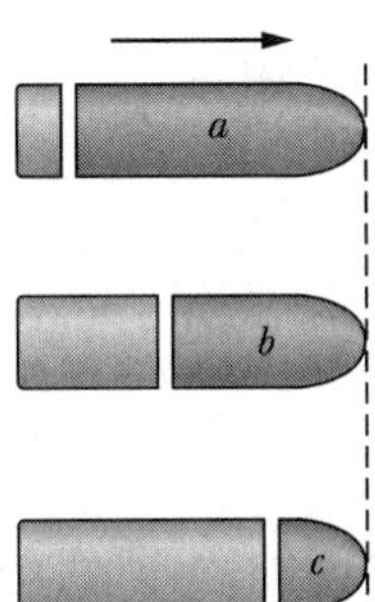

FIGURE 9-50 Question 21.

22. The following table gives data at time T for four rockets racing one another in deep space. Rank the rockets according to (a) their thrusts at time T and (b) the magnitudes of their accelerations just then, greatest first.

ROCKET	RATE OF FUEL CONSUMPTION	SPEED	MASS
1	R	U	M
2	$2R$	U	$2M$
3	R	$2U$	$M/2$
4	$2R$	$2U$	M

EXERCISES & PROBLEMS

78. A 2.0 kg mass sliding on a frictionless surface explodes into two 1.0 kg masses. After the explosion the velocities of the 1.0 kg masses are 3.0 m/s, due north, and 5.0 m/s, 30° north of east. What was the original speed of the 2.0 kg mass?

79. A 4.0 kg particle-like object is located at $x = 0$, $y = 2.0$ m, and a 3.0 kg particle-like object is located at $x = 3.0$ m, $y = 1.0$ m. Where in the xy plane must a 2.0 kg particle-like object be placed in order for the center of mass of the three-particle system to be located at the origin?

80. A 1500 kg car and a 4000 kg truck are moving north and east, respectively, with constant velocities. The center of mass of the car–truck system has a velocity of 11 m/s in a direction 55° north of east. (a) What is the magnitude of the car's velocity? (b) What is the magnitude of the truck's velocity?

81. At $t = 0$, a 1.0 kg jelly jar is projected vertically upward from the base of a 50 m tall building with an initial velocity of 40 m/s. At the same instant and directly overhead, a 2.0 kg peanut butter jar is dropped from rest from the top of the building. (a) How far above ground level is the center of mass of the two-jar system at $t = 3.0$ s? (b) What maximum height above ground level is reached by the center of mass?

82. A big olive ($m = 0.50$ kg) is located at the origin and a big Brazil nut ($M = 1.5$ kg) is located at the point (1.0, 2.0) m in an xy plane. At $t = 0$, a force $\mathbf{F}_o = (2\mathbf{i} + 3\mathbf{j})$ N begins to act on the olive and a force $\mathbf{F}_n = (-3\mathbf{i} - 2\mathbf{j})$ N begins to act on the nut. What is the displacement of the center of mass of the olive–nut system at $t = 4.0$ s with respect to its position at $t = 0$ (answer in terms of **i** and **j**).

83. A 0.20 kg hockey puck is sliding on a frictionless ice surface with a velocity of 10 m/s toward the east just before making contact with a hockey stick. What change in momentum (magnitude and direction) does the puck undergo while in contact with the stick if just afterward the velocity of the puck is (a) 20 m/s toward the east, (b) 5.0 m/s toward the east, and (c) 10 m/s toward the west?

84. A 0.30 kg softball has a velocity of 15 m/s at an angle of 35° below the horizontal just before making contact with the bat. What is the magnitude of the change in momentum of the ball while it is in contact with the bat if the ball leaves the bat with a velocity of (a) 20 m/s, vertically downward and (b) 20 m/s, horizontally away from the batter and back toward the pitcher?

85. A child who is standing in a 95 kg flat-bottom boat is initially 6.0 m from shore. The child sees a frog on the bank and starts to walk along the boat toward the shore. After the child has walked 2.5 m relative to the boat, he is 4.1 m from the shore. Assume there is no friction between the boat and the water. What is the mass of the child?

86. A 40 kg child and her 75 kg father simultaneously dive from a 100 kg boat that is initially motionless. The child dives horizontally toward the east with a velocity of 2.0 m/s, and the father dives toward the south with a velocity of 1.5 m/s at an angle of 37° above the horizontal. Determine the magnitude and direction of the velocity of the boat immediately after their dives.

87. A suspicious package is sliding on a frictionless surface when it explodes into three pieces of equal masses and with the velocities (1) 7.0 m/s, north, (2) 4.0 m/s, 30° south of west, and (3) 4.0 m/s, 30° south of east. (a) Determine the velocity (magnitude and direction) of the package before it explodes. (b) What is the displacement of the center of mass of the three-piece system with respect to the point where the explosion occurred 3.0 s after the explosion?

Clustered Problems

In each problem of this cluster, find the center of mass with the xy coordinate system shown and by using symmetry or Eqs. 9-9.

88. In Fig. 9-51a, a uniform wire forms an isosceles triangle of base B and height H. (a) Find the x and y coordinates of the figure's center of mass by assuming that each side can be replaced with a particle of the same mass as that side and positioned at the center of the side. (*Be careful:* Note that the base and, say, the left-hand side do not have the same mass.) (b) Use Eqs. 9-9 to find the x and y coordinates of the center of mass of the left-hand side.

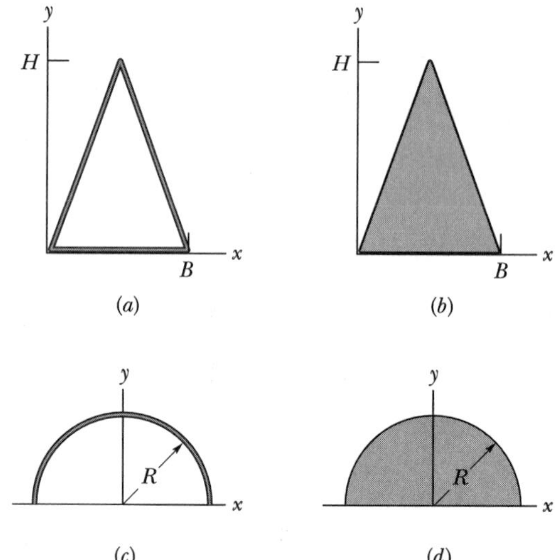

FIGURE 9-51 Problems 88 through 91.

89. Figure 9-51b shows a uniform, solid plate in the shape of an isosceles triangle with base B and height H. What are the x and y coordinates of the plate's center of mass?

90. In Fig. 9-51c, a uniform wire forms a semicircle of radius R. What are the x and y coordinates of the figure's center of mass?

91. Figure 9-51d shows a uniform, solid plate in the shape of a semicircle with radius R. What are the x and y coordinates of the plate's center of mass?

Chapter Ten
Collisions

QUESTIONS

13. *Organizing question:* Figure 10-50 shows four situations in which there is a one-dimensional collision between two blocks of masses $m_1 = 2$ kg and $m_2 = 3$ kg on a frictionless floor. Information about the speeds and directions of travel are given for BEFORE and AFTER the collision except for the final velocity v_{2f} of block 2. For each situation, set up an equation, complete with known data, to find v_{2f}. (*Hint:* Do not assume that the collision is elastic.)

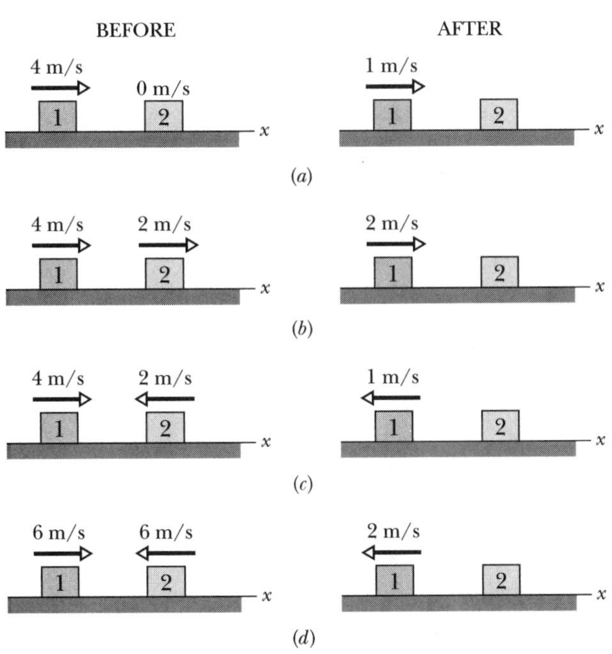

FIGURE 10-50 Questions 13 and 14.

14. *Organizing question:* Set up an equation, complete with known data, to find the change ΔK in the total kinetic energy of the blocks in Question 13.

15. Recall that Eqs. 10-18 and 10-19 correspond to the one-dimensional elastic collision generically depicted in Fig. 10-7. (a) Which of the curves in Fig. 10-51 gives the velocity ratio v_{1f}/v_{1i} versus the mass ratio m_1/m_2? (b) What are the values of the ratios at the intercept of the curve with the vertical axis? (c) What velocity ratio does the curve approach beyond the right side of the graph? (d) What point on the curve corresponds to equal masses of the projectile and the target?

(e) Which of the curves gives the ratio v_{2f}/v_{1i} versus m_1/m_2? (f) What are the values of the ratios at the intercept of the curve with the vertical axis? (g) What velocity ratio does the curve approach beyond the right side of the graph? (h) What point on the curve corresponds to equal masses of the projectile and the target?

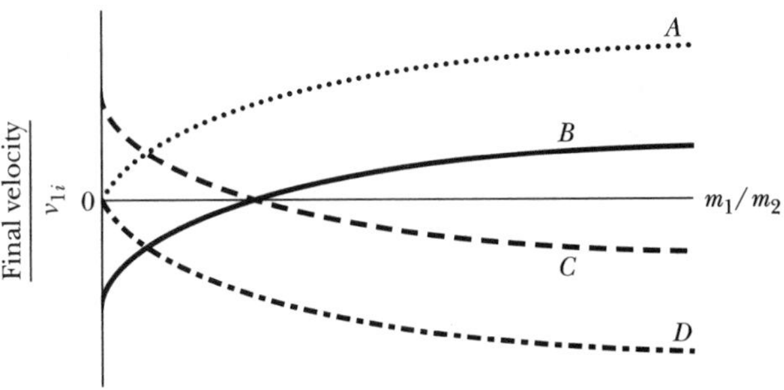

FIGURE 10-51 Question 15.

16. A projectile particle moving along the positive direction of an x axis on a frictionless floor runs into an initially stationary target particle (as in Fig. 10-7) in a one-dimensional collision. Nine choices for a graph of the momenta of the particles versus time (before and after the collision) are given in Fig. 10-52. Determine which choices represent physically impossible situations and explain why.

17. *Organizing question:* Figure 10-53 shows two situations in which a projectile box 1 collides with a stationary target box 2.

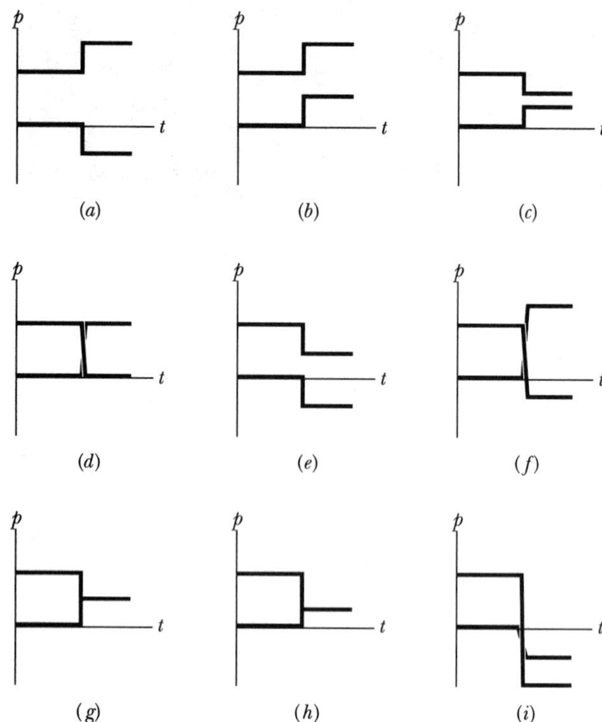

FIGURE 10-52 Question 16.

In both situations the collisions are one-dimensional and elastic; some of the speeds and directions of travel of the boxes are given. For each situation set up equations, complete with known data, to find the velocities whose values are not given.

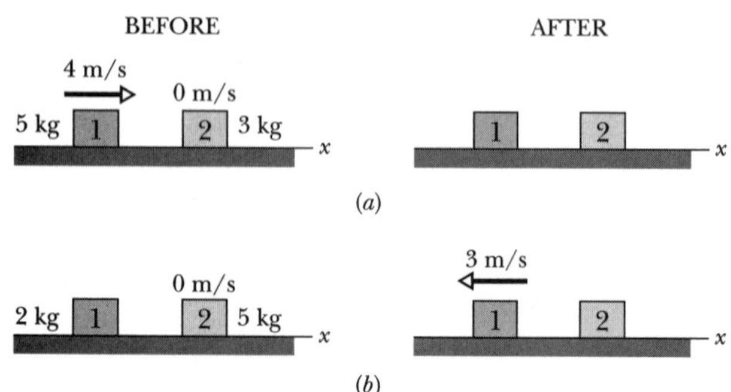

FIGURE 10-53 Question 17.

18. A projectile hockey puck *A*, with initial momentum 5 kg · m/s along an *x* axis, collides with an initially stationary hockey puck *B*. The pucks slide over frictionless ice and are shown in an overhead view in Fig. 10-54. Also shown are three general choices for the path taken by puck *A* after the collision. Which general choice is appropriate if the momentum of puck *B* after the collision has an *x* component of (a) 5 kg · m/s, (b) more than 5 kg · m/s, and (c) less than 5 kg · m/s? Can that *x* component be (d) 1 kg · m/s or (e) −1 kg · m/s?

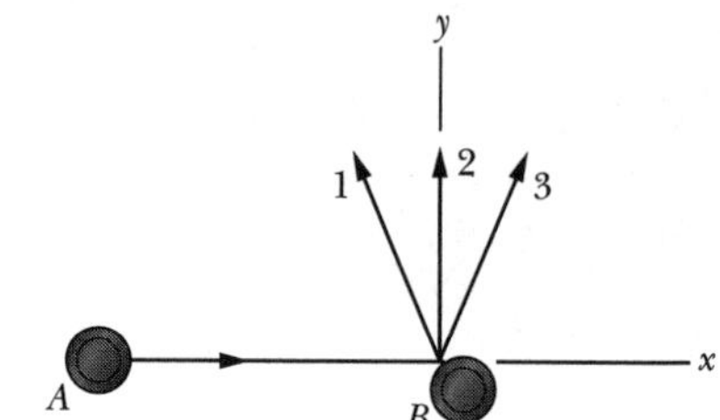

FIGURE 10-54 Question 18.

19. Two bodies undergo a collision along a frictionless floor. Which part of Fig. 10-55 best represents the paths of those bodies and also the path of their center of mass as seen from overhead?

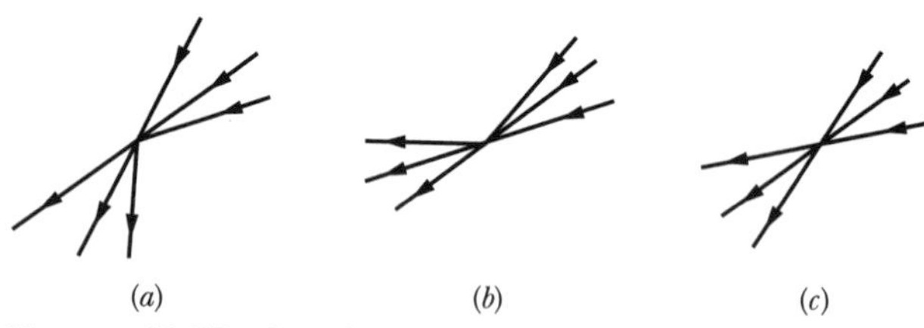

FIGURE 10-55 Question 19.

20. Two objects that are moving along an *xy* plane on a frictionless floor collide. The following table gives some of the momentum components (in kilogram-meters per second) before and after the collision. What are the missing values?

		BEFORE		AFTER	
SITUATION	OBJECT	p_x	p_y	p_x	p_y
1	*A*	3	4	7	2
	B	2	2		
2	*C*	−4	5	3	
	D		−2	4	2
3	*E*	−6			3
	F	6	2	−4	−3

21. A neutron n moving through a room decays (transforms) into a proton p, an electron e, and an antineutrino ($\bar{\nu}$): n → p + e + $\bar{\nu}$. Except for the antineutrino, the following table gives the momentum components of these particles in terms of a basic unit p_0. What are the components for the antineutrino?

PARTICLE	p_x	p_y	p_z
n	$10p_0$	$-5p_0$	$-4p_0$
p	$8p_0$	$-4p_0$	$2p_0$
e	$-6p_0$	$3p_0$	$5p_0$
$\bar{\nu}$			

EXERCISES & PROBLEMS

79. A 6.0 kg mass and a 4.0 kg mass are moving on a frictionless surface, as shown in Fig. 10-56. A spring of spring constant $k = 8000$ N/m is fixed to the 4.0 kg mass. The 6.0 kg mass has an initial velocity of 8.0 m/s toward the right, and the 4.0 kg mass has an initial velocity of 2.0 m/s toward the right. Eventually, the larger mass overtakes the smaller mass. (a) What is the velocity of the 4.0 kg mass at the instant the 6.0 kg mass has a velocity of 6.4 m/s toward the right? (b) What is the elastic potential energy of the system just then?

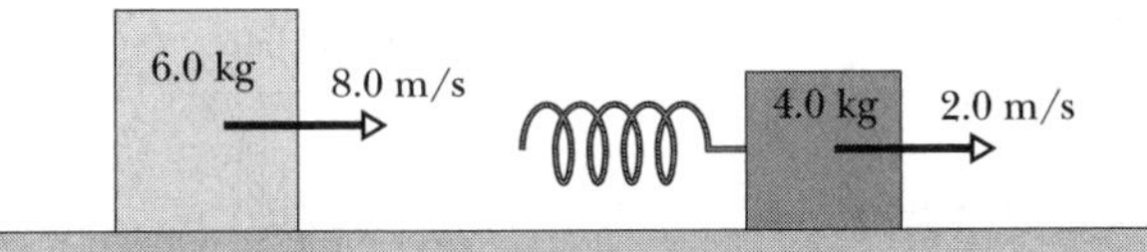

FIGURE 10-56 Problem 79.

80. A completely inelastic collision occurs between a 3.0 kg mass moving upward at 20 m/s and a 2.0 kg mass moving downward at 12 m/s. How high does the combined mass rise above the point of collision?

81. A 5.0 kg mass with an initial velocity of 4.0 m/s, due east, collides with a 4.0 kg mass whose initial velocity is 3.0 m/s, due west. After the collision the 5.0 kg mass has a velocity of 1.2 m/s, due south. (a) What is the magnitude of the velocity of the 4.0 kg mass after the collision? (b) How much energy is dissipated in the collision?

82. A 0.30 kg softball has a velocity of 12 m/s at an angle of 35° below the horizontal just before making contact with a bat. The ball leaves the bat 2.0 ms later with a vertical velocity of 10 m/s as shown in Fig. 10-57. What is the magnitude of the average force of the bat on the ball while the ball is in contact with the bat?

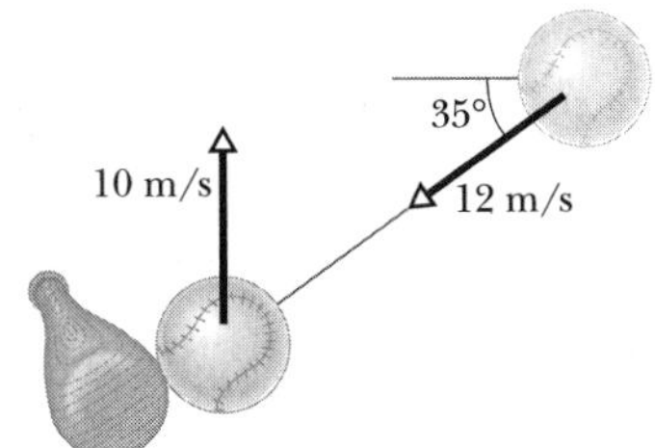

FIGURE 10-57 Problem 82.

83. A 3.0 kg object moving at 8.0 m/s in the direction of increasing x has a one-dimensional, completely elastic collision with an object of mass M, initially at rest. After the collision the object of mass M has a velocity of 6.0 m/s in the direction of increasing x. What is M?

84. A 60 kg man is ice-skating due north with a velocity of 6.0 m/s when he collides with a 38 kg child. The man and child stay together and have a velocity of 3.0 m/s at an angle of 35° north of east immediately after the collision. What were the magnitude and direction of the velocity of the child just before the collision?

85. A 60.0 kg diver has a downward velocity of 3.00 m/s just before making contact with a diving board. The diver leaves the board 1.20 s later with a velocity of 5.00 m/s at an angle of 40.0° with the vertical as shown in Fig. 10-58. (a) What is the magnitude of the average *net* force on the diver while she is in contact with the board? (b) What is the magnitude of the average force of the board on the diver while she is in contact with the board?

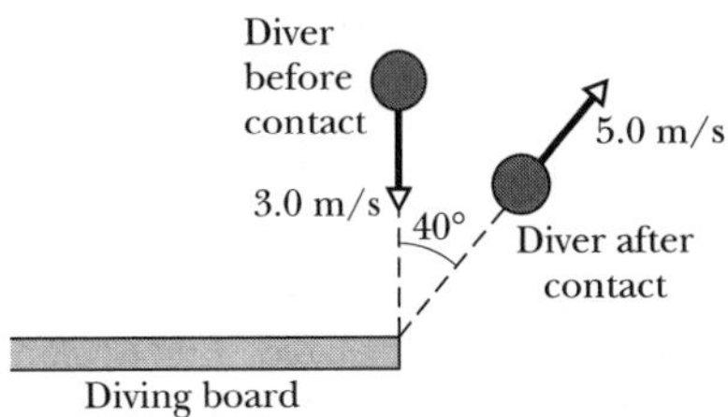

FIGURE 10-58 Problem 85.

86. A remote-controlled toy car of mass 2.0 kg starts from rest at the origin at $t = 0$ and moves in the positive direction of an x axis. The net force on the car as a function of time is shown in Fig. 10-59. (a) What is the time rate of change of the momentum of the car at $t = 3.0$ s? (b) What is the momentum of the car at $t = 3.0$ s?

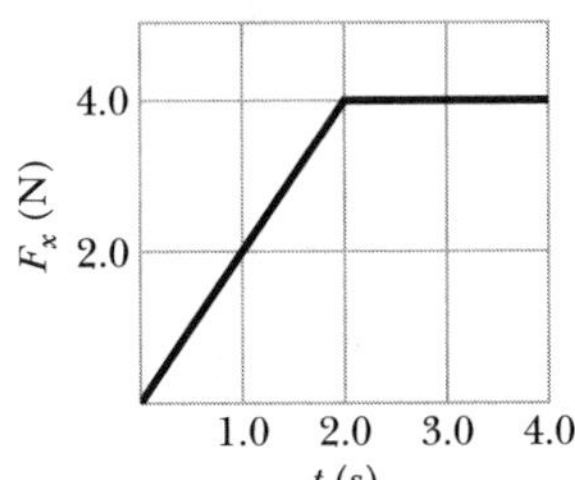

FIGURE 10-59 Problem 86.

87. The only force acting on a 1.6 kg stone moving along an x axis is given by $\mathbf{F} = (16 - t^2)\mathbf{i}$ N for $0 \leq t \leq 8.0$ s. The velocity of the stone at $t = 0$ is zero. (a) What is the velocity of the stone at $t = 3.0$ s? (b) At what value of t will the velocity of the stone again be equal to zero? (c) What maximum velocity in the positive x direction is reached by the stone?

88. A 4.0 kg physics book and a 6.0 kg calculus book, connected by a spring, are stationary on a horizontal frictionless surface. The spring constant is 8000 N/m. The books are pushed together, compressing the spring, and then they are released from rest. When the spring has returned to its unstretched length, the speed of the calculus book is 4.0 m/s. How much energy is stored in the spring at the instant the books are released?

89. A 10 g bullet moving directly upward at 1000 m/s strikes and passes through a 5.0 kg block initially at rest (Fig. 10-60). The bullet emerges from the block moving directly upward at 400 m/s. To what maximum height will the block rise above its initial position?

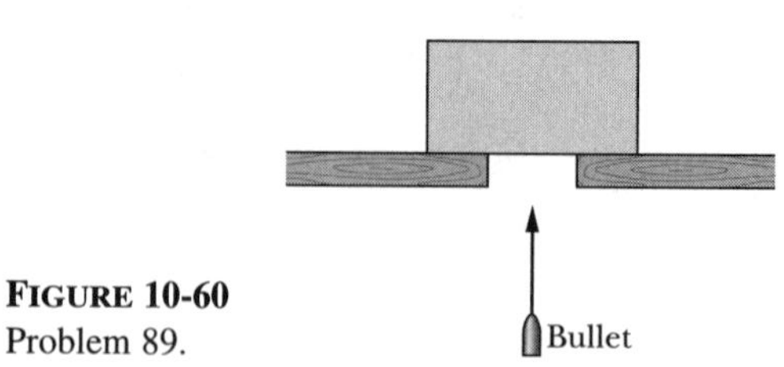

FIGURE 10-60 Problem 89.

90. A 3.2 kg box of running shoes slides on a horizontal frictionless table and collides with a 2.0 kg box of ballet slippers initially at rest on the edge of the table as shown in Fig. 10-61. The speed of the 3.2 kg box is 3.0 m/s just before the collision. If the two boxes stick together because of packing tape on their sides, what is their kinetic energy just before they strike the floor 0.40 m below the table's surface?

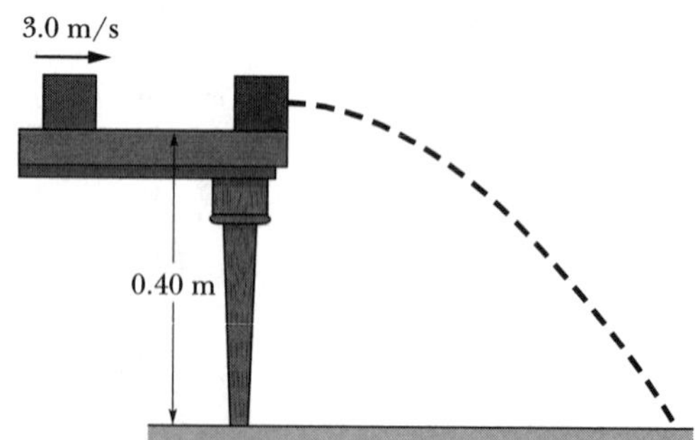

FIGURE 10-61 Problem 90.

91. A 2.0 kg tin cookie with an initial velocity of 8.0 m/s, east, collides with a stationary 4.0 kg cookie tin. Just after the collision, the cookie has a velocity of 4.0 m/s at an angle of 37° north of east. Just then, what are the magnitude and direction of the velocity of the cookie tin?

92. Two 30 kg children, each with a speed of 4.0 m/s, are sliding on a frictionless frozen pond when they collide and stick together due to Velcro straps on their jackets. The two children then collide and stick to a 75 kg man who was sliding at 2.0 m/s. After this collision, the three-person composite is stationary. What was the angle between the initial velocity vectors of the two children?

Clustered Problems

Cluster 1

93. A projectile body of mass m_1 and initial velocity v_{1i} collides with a stationary target body of mass m_2 in a one-dimensional collision. What are the velocities of the bodies after the collision if (a) the bodies stick together (as in Fig. 10-11) and (b) the collision is elastic (as in Fig. 10-7)?

94. A projectile body of mass m_1 and initial velocity $v_{1i} = 10.0$ m/s collides with a stationary target body of mass $m_2 = 2.00m_1$ in a one-dimensional collision. We do not know whether the collision is elastic, but we do know that there is no energy production during the collision (such as with an explosion between the bodies). (a) What is v_{2f} as a function of v_{1f}? (b) Plot this function. What are (c) the greatest possible value of v_{1f}, (d) the corresponding value of v_{2f}, and (e) the circumstance that produces this limiting value of v_{1f}? (f) What is the physical reason that v_{1f} cannot be greater than this limiting value? (g) On your plot of v_{2f} versus v_{1f}, mark the point corresponding to this limiting value.

What are (h) the least (most negative) possible value of v_{1f}, (i) the corresponding value of v_{2f}, and (j) the circumstance that produces this limiting value of v_{1f}? (k) What is the physical reason that v_{1f} cannot be smaller (even more negative) than this limiting value? (l) On your plot of v_{2f} versus v_{1f}, mark the point corresponding to this limiting value, then label the physically possible portion of the plot and indicate the physical reasons for the two impossible portions.

95. Repeat Problem 94 but with $m_2 = 0.500m_1$. Here, however, the least possible value of v_{1f} is positive.

Cluster 2

For the problems in this cluster, consider the two-dimensional collision of Fig. 10-16 with the following data: the projectile body has a mass m_1, an initial velocity of magnitude $v_{1i} = 10.0$ m/s, and a final angle of $\theta_1 = 30.0°$; the initially stationary target body has a mass $m_2 = 2.00m_1$. In solving Problems 97 and 98, you might make use of the trigonometric identity

$$\sin^2 \theta_2 + \cos^2 \theta_2 = 1.$$

96. See the setup for this cluster. We also know that the projectile body has a final velocity $\mathbf{v}_{1f}$ of magnitude 5.00 m/s, but we do not know whether the collision is elastic. (a) After the collision, what are the speed v_{2f} of the target body and the angle θ_2 of its travel? (b) Is the collision elastic? If not, what fraction of the initial kinetic energy of the projectile body is transformed into other types of energy?

97. See the setup for this cluster. We also know that the collision is elastic. After the collision, what are (a) the speed v_{1f} of the projectile body, (b) the speed v_{2f} of the target body, and (c) the angle θ_2 of the target body's travel?

98. See the setup for this cluster. We do not know whether the collision is elastic. (a) What is the final speed v_{2f} of the target body as a function of the final speed v_{1f} of the projectile body? (b) Plot this function. What are (c) the greatest possible value of

v_{1f}, (d) the corresponding value of v_{2f}, and (e) the circumstance that produces this limiting value of v_{1f}? (f) What is the physical reason that v_{1f} cannot be greater than this limiting value? (g) On your plot of v_{2f} versus v_{1f}, mark the point corresponding to this limiting value.

(h) What is the angle θ_2 of the target body's motion as a function of the final speed v_{1f} of the projectile body? (i) Plot this function. (j) What is the value of θ_2 corresponding to the greatest value of v_{1f}? (k) On your plot of θ_2 versus v_{1f}, mark the point corresponding to this limiting value.

What are (l) the least possible value of v_{1f}, (m) the corresponding value of v_{2f}, (n) the corresponding value of θ_2, and (o) the circumstance that produces this limiting value of v_{1f}?

Cluster 3

99. A 50.0 g lump of hash browns slides at 20.0 m/s along a frictionless countertop and into a 20.0 g coffee cup that is poised at the edge of the countertop, 1.00 m above the floor. If the collision is elastic, how far horizontally from the counter do (a) the cup and (b) the hash browns land on the floor? (c) Repeat the question for a completely inelastic collision.

100. In the arrangement of the two metal spheres in Sample Problem 10-3 and Fig. 10-10, sphere 1 of mass m_1 is again raised through a vertical height h_1 and released. It then collides elastically with sphere 2 of mass m_2, which is initially stationary. In terms of these symbols, set up (a) an expression for the maximum height h_1' reached by sphere 1 as it rebounds to the left and (b) an expression for the maximum height h_2' reached by sphere 2 as it moves off to the right.

Assume that the spheres then return to their lowest positions simultaneously so that a second elastic collision occurs there. During the rebound from that second collision, (c) what maximum height h_1'' is reached by sphere 1 and (d) what maximum height h_2'' is reached by sphere 2? (*Hint:* You should calculate the answers, but you might also be able to get them if you imagine videotaping the first collision and then watching the videotape played in reverse.)

101. (a) A 1.00 kg shell is fired from a 25.0 kg cannon that is fixed to the ground. The speed of the shell is 500 m/s relative to the cannon (and thus also the ground). What is the kinetic energy K_1 of the shell?

The shell is again fired from the cannon, but now the cannon can freely recoil over a frictionless surface. If the total kinetic energy of the shell and the recoiling cannon is equal to K_1 of part (a), what are the speeds of (b) the shell and (c) the cannon?

The shell is again fired from the cannon, which can again recoil, but now that recoil is limited to a distance of 10.0 cm by a spring mounted between the cannon and a wall. (d) If again the total kinetic energy of the shell and the recoiling cannon is equal to K_1, what spring constant is required of the spring?

Chapter Eleven
Rotation

QUESTIONS

16. Figure 11-50 gives the angular acceleration $\alpha(t)$ of a disk that rotates like a merry-go-round. In which of the time periods indicated, if any, does the disk move at constant angular speed?

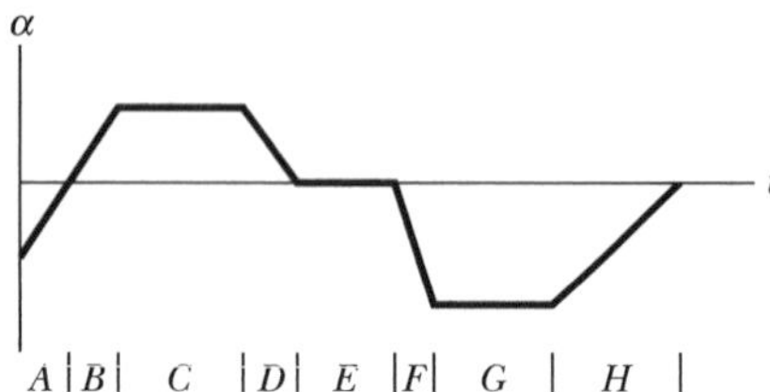

FIGURE 11-50 Question 16.

17. *Organizing question:* A flywheel is spun about its central axis in the four situations given in the table. (a) For situations A and B set up equations, complete with known data, to find the time t the flywheel takes to rotate to the angular position of $+1.0$ rad. (b) Without solving those equations, tell if the time to reach $+1.0$ rad in situation A is greater than, less than, or the same as that in situation B.

(c) Next, for situations C and D set up equations, complete with known data, to find the angular position θ of the flywheel when it has an angular speed of 8 rad/s. (d) Without solving those equations, tell if the magnitude of the angle θ in situation C is greater than, less than, or the same as that in situation D.

Situation	*A*	*B*	*C*	*D*
Initial angular position (rad)	0	0	0	0
Initial angular speed (rad/s)	+5	−5	−5	+5
Constant angular acceleration (rad/s^2)	+2	+2	+2	−2

18. In Fig. 11-51, assume that the vertical axis pertains to the angular velocity ω of a lazy Susan turning like a merry-go-round. Then determine which of the 10 plots of ω versus time t best describes the motion for the following four situations. (Plots 2, 3, 7, and 9 are straight; the others are curved.)

Situation	*a*	*b*	*c*	*d*
Initial θ (rad)	+10	−10	+10	−10
Initial ω (rad/s)	+5	−5	−5	+5
Constant α (rad/s^2)	+2	−2	+2	−2

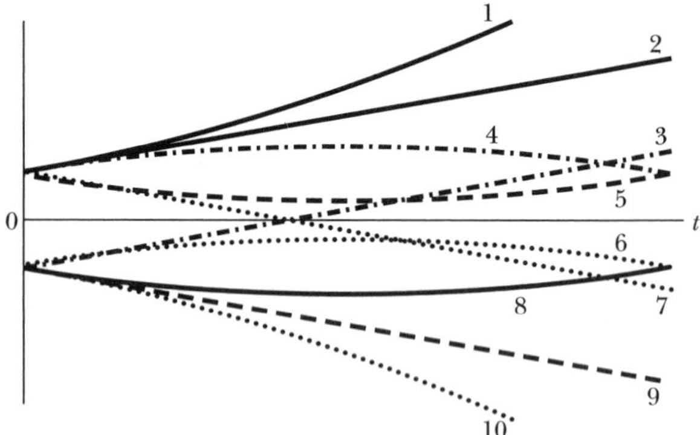

FIGURE 11-51 Questions 18 and 19.

19. Question 19 continued: If, instead, the vertical axis pertains to the angular position θ of the lazy Susan, then which of the 10 plots best describes the motion for the given four sets of values?

20. The overhead view of Fig. 11-52 shows a snapshot of a disk turning counterclockwise like a merry-go-round. The angular speed ω of the disk is decreasing (the disk is turning counterclockwise slower and slower). The figure shows the position of a cockroach that rides the rim of the disk. At the instant of the snapshot, what are the directions of (a) the cockroach's radial acceleration and (b) its tangential acceleration?

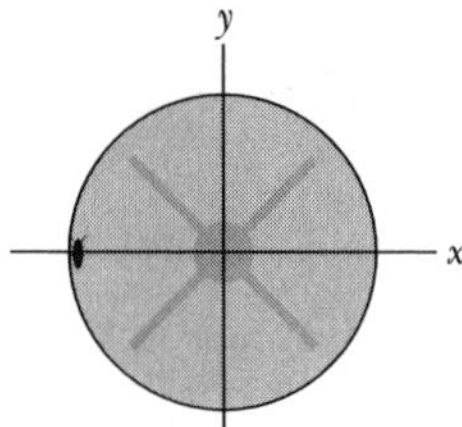

FIGURE 11-52 Question 20.

21. Figure 11-53 shows an assembly of three small spheres of the same mass that are attached to a massless rod with the indicated spacings. Consider the rotational inertia I of the assembly about each sphere, in turn. Then rank the spheres according to the rotational inertia about them, greatest first.

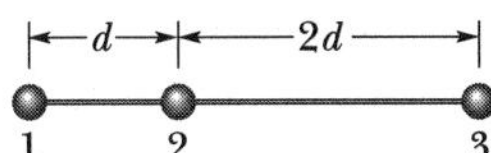

FIGURE 11-53 Question 21.

22. In Fig. 11-54 forces **L** and **R** are applied to an oddly shaped plate of iron that can rotate about a perpendicular axis through point O. The tangential and radial vector components of those applied forces are drawn to the same scale. With respect to O, which applied force has (a) the greater tangential component and (b) the greater radial component? Which of the dashed lines are (c) lines of actions of the applied forces, (d) moment arms to the applied forces, and (e) moment arms to the tangential components? Do forces (f) **L** and (g) **R** tend to rotate the iron plate clockwise or counterclockwise?

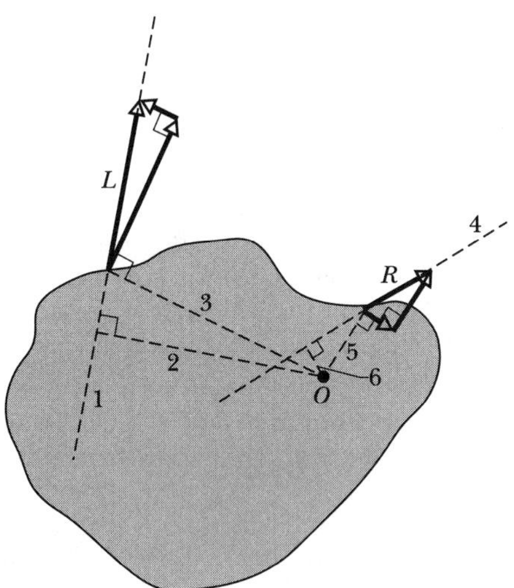

FIGURE 11-54 Question 22.

23. In Fig. 11-55, two forces $\mathbf{F}_1$ and $\mathbf{F}_2$ act on a disk that turns about its center like a merry-go-round. The forces maintain the indicated angles during the rotation, which is counterclockwise and at a constant rate. However, we are to decrease the angle θ of $\mathbf{F}_1$ without changing the magnitude of $\mathbf{F}_1$. (a) To keep the angular speed constant, should we increase, decrease, or maintain the magnitude of $\mathbf{F}_2$? Do forces (b) $\mathbf{F}_1$ and (c) $\mathbf{F}_2$ tend to rotate the iron plate clockwise or counterclockwise?

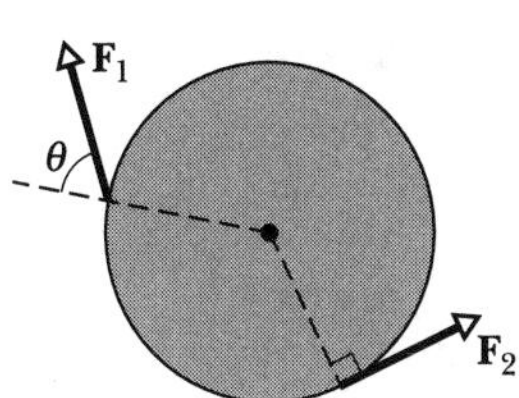

FIGURE 11-55 Question 23.

24. In the overhead view of Fig. 11-56, five forces of the same magnitude act on a merry-go-round for the strange; it is a square that can rotate about point P at midlength along one of the edges. Rank the forces according to the magnitude of the torque they create about point P, greatest first.

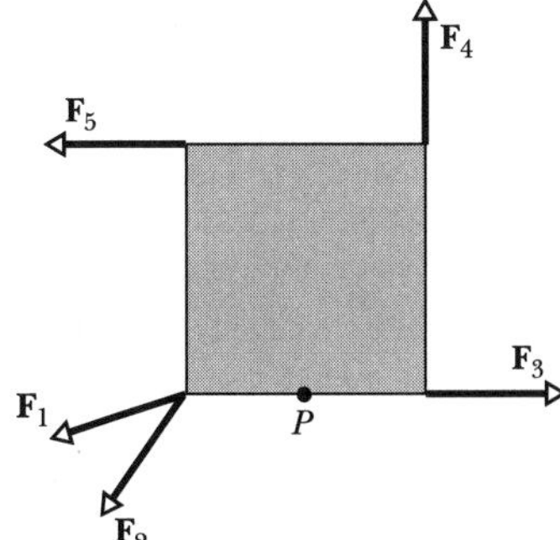

FIGURE 11-56 Question 24.

25. *Organizing question:* Figure 11-57 shows four situations in which a lazy Susan turns like a merry-go-round without friction while one or two forces are applied to its rim. The magnitudes of those forces are $F_1 = 1$ N and $F_2 = 2$ N; during the rotation the forces maintain the angles shown; the radius of the lazy Susan is 0.5 m; the rotational inertia of the lazy Susan is 2 kg · m²; initially the angular velocity is 3 rad/s counterclockwise. (a) For each situation, set up an equation, complete with known data, to find the angular acceleration α of the lazy Susan. (b) Assuming that we then know α, set up an equation, complete with known data, to find the angular displacement θ of the lazy Susan during the first 2 s of rotation.

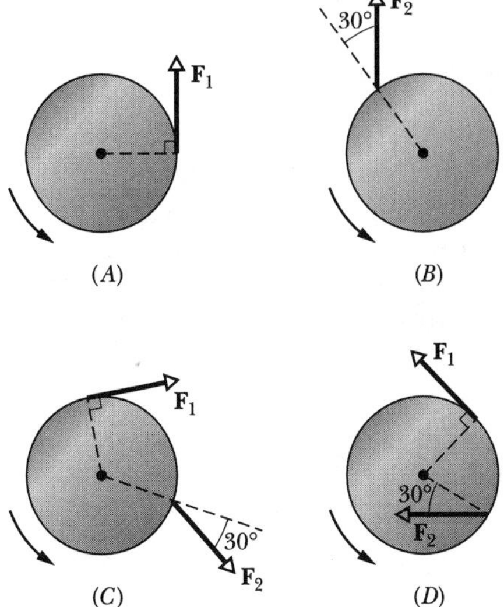

FIGURE 11-57 Questions 25 and 26.

26. *Organizing question:* Figure 11-57 shows four situations in which a lazy Susan turns like a merry-go-round without friction while one or two forces are applied to its rim. The magnitudes of those forces are $F_1 = 1$ N and $F_2 = 2$ N; during the rotation the forces maintain the angles shown; the radius of the lazy Susan is 0.5 m; the rotational inertia of the lazy Susan is 2 kg · m^2. In each situation, the lazy Susan begins with angular speed $\omega_i = 3$ rad/s counterclockwise and rotates through an angular displacement of 1.2 rad counterclockwise.

(a) For each situation, tell if the associated one or two torques tend to transfer energy *to* or *from* the merry-go-round during the 1.2 rad rotation. (b) For each situation, set up an equation, complete with known data, to find the rotational kinetic energy K_f of the lazy Susan at the end of the 1.2 rad rotation.

27. Figure 11-58 gives the angular velocity ω versus time t of a merry-go-round being turned by a varying applied force. (a) During which of the numbered time periods is energy transferred *from* the merry-go-round by the applied force? (b) Rank the time periods according to the work done on the merry-go-round by the applied force during the period, most positive work first, most negative work last. (c) Rank the periods according to the rate at which the applied force transfers energy, greatest rate of transfer *to* the merry-go-round first, greatest rate of transfer *from* it last.

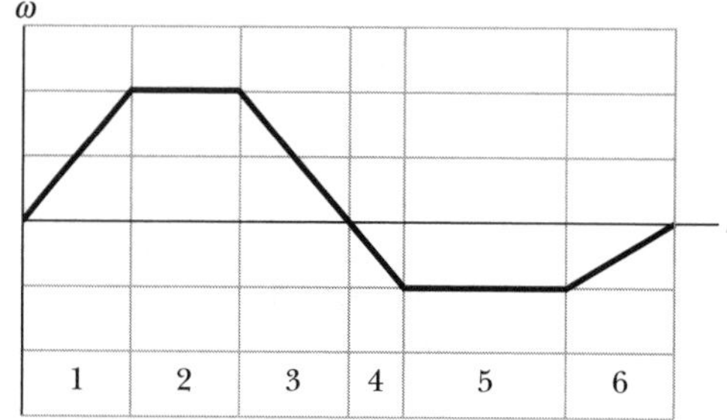

FIGURE 11-58 Question 27.

EXERCISES & PROBLEMS

95. Four identical particles of mass 0.50 kg each are placed at the vertices of a 2.0 m × 2.0 m square and held there by four massless rods, which form the sides of the square. What is the rotational inertia of this rigid body about an axis that (a) passes through the midpoints of opposite sides and lies in the plane of the square, (b) passes through the midpoint of one of the sides and is perpendicular to the plane of the square, and (c) lies in the plane of the square and passes through two diagonally opposite particles?

96. An object rotates about a fixed axis, and a reference line on the object is given by $\theta = 0.40e^{2t}$, where θ is in radians and t is in seconds. Consider a point on the object that is 4.0 cm from the axis of rotation. At $t = 0$, what is the magnitude of (a) the tangential component of the acceleration of the point and (b) the radial component of the acceleration of the point?

97. The turntable of a record player has an angular speed of 8.0 rad/s when it is switched off. Three seconds later, the turntable is observed to have an angular speed of 2.6 rad/s. Through how many radians does the turntable rotate from the time it is turned off until it stops? (Assume constant angular acceleration.)

98. Two thin disks, each of mass 4.0 kg and radius 0.40 m, are attached as shown in Fig. 11-59 to form a rigid body. What is the rotational inertia of this body about an axis A that is perpendicular to the plane of the disks and passes through the center of one of the disks?

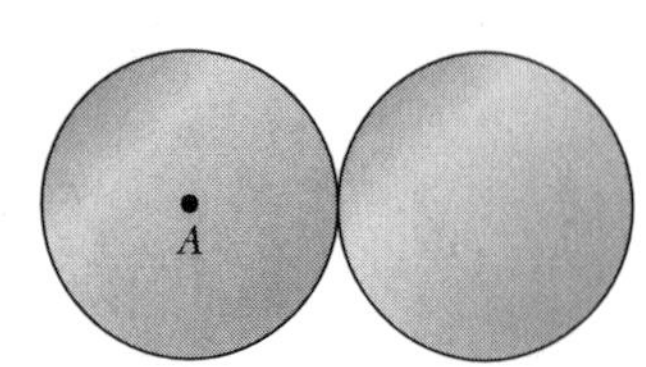

FIGURE 11-59 Problem 98.

99. A wheel rotating about a fixed axis through its center has a constant angular acceleration of 4.0 rad/s^2. In a 4.0 s interval the wheel turns through an angle of 80 rad. (a) What is the angular velocity of the wheel at the start of the 4.0 s interval? (b) Assuming that the wheel started from rest, how long had it been in motion at the start of the 4.0 s interval?

100. The turntable of a record player has an angular velocity of 3.5 rad/s when it is turned off. The turntable comes to rest 1.6 s after being turned off. The radius of the turntable is 15 cm. (a) Assuming constant angular acceleration, through how many radians does the turntable turn after being turned off? (b) What is the magnitude of the linear acceleration of a point on the rim of the turntable 1.0 s after it is turned off? (*Hint:* The linear acceleration has both a radial and a tangential component.)

101. The rigid object shown in Fig. 11-60 consists of three balls and three connecting rods, with $M = 1.6$ kg and $L = 0.60$ m. The balls may be treated as particles, and the connecting rods have negligible mass. Determine the rotational kinetic energy of the object if it has an angular velocity of 1.2 rad/s about (a) an axis that passes through point P and is perpendicular to the plane of the figure and (b) an axis that passes through point P, is perpendicular to the rod of length $2L$, and lies in the plane of the figure.

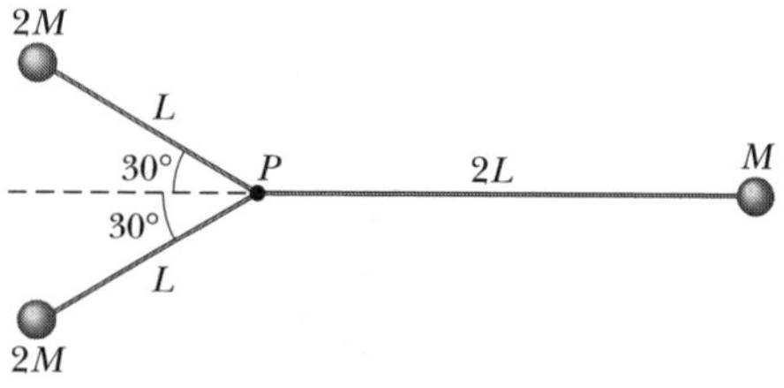

2M
L
30°
P
2L
M
30°
L
2M

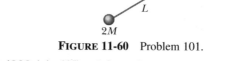

FIGURE 11-60 Problem 101.

102. Three particles, each of mass 0.50 kg, are positioned at the vertices of an equilateral triangle with sides of length 0.60 m. The particles are connected by rods of negligible mass. What is the rotational inertia of this rigid body about (a) an axis that passes through one of the particles and is parallel to the rod connecting the other two, (b) an axis that passes through the midpoint of one of the sides and is perpendicular to the plane of the triangle, and (c) an axis that is parallel to one side of the triangle and passes through the respective midpoints of the other two sides?

103. Two thin rods (each of mass 0.20 kg) are joined together to form a rigid body as shown in Fig. 11-61. One of the rods is 0.40 m long and the other is 0.50 m long. What is the rotational inertia of this rigid body about (a) an axis that is perpendicular to the plane of the paper and passes through the center of the shorter rod and (b) an axis that is perpendicular to the plane of the paper and passes through the center of the longer rod?

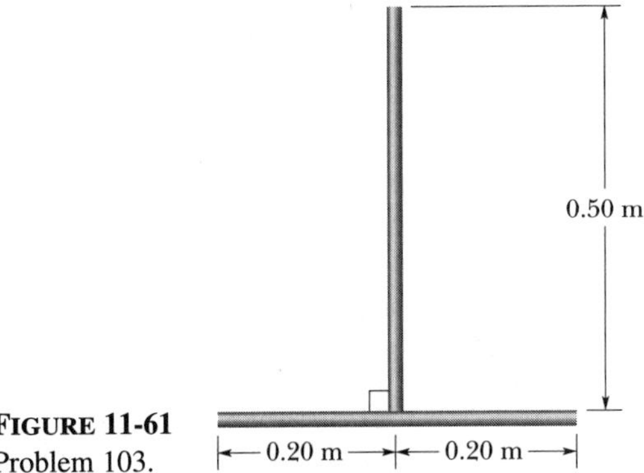

FIGURE 11-61 Problem 103.

104. Figure 11-62 shows a 1.0 kg uniform rod that is 3.0 m long. The rod is mounted to rotate freely about a horizontal axis perpendicular to the rod that passes through a point 1.0 m from one end of the rod. The rotational inertia of the rod about the axis is 1.0 kg · m^2. The rod is released from rest when it is horizontal. What is the angular acceleration of the rod at the instant it is released?

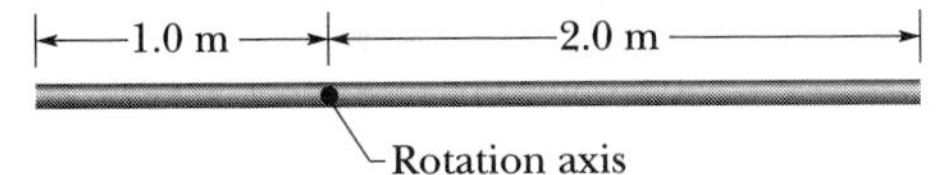

FIGURE 11-62 Problem 104.

105. A wheel of radius 0.20 m is mounted on a frictionless horizontal axis. The rotational inertia of the wheel about the axis is 0.050 kg · m^2. A massless cord wrapped around the wheel is attached to a 2.0 kg block that slides on a horizontal frictionless surface. If a horizontal force of magnitude $P = 3.0$ N is applied to the block as shown in Fig. 11-63, what is the magnitude of the angular acceleration of the wheel?

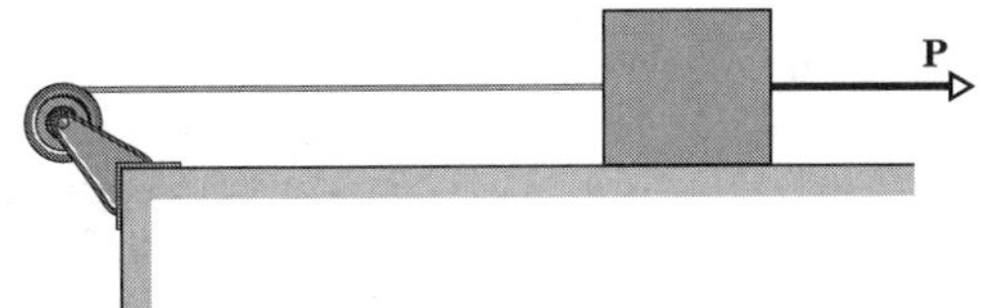

FIGURE 11-63 Problem 105.

106. A yo-yo-shaped device mounted on a horizontal frictionless axis is used to lift a 30 kg box as shown in Fig. 11-64. The outer radius R of the device is 0.50 m and the radius r of the hub is 0.20 m. When a constant horizontal force of 140 N is applied to a rope wrapped around the outside of the device, the box, which is suspended from a rope wrapped around the hub, has an upward acceleration of 0.80 m/s^2. What is the rotational inertia of the yo-yo-shaped device about its axis of rotation?

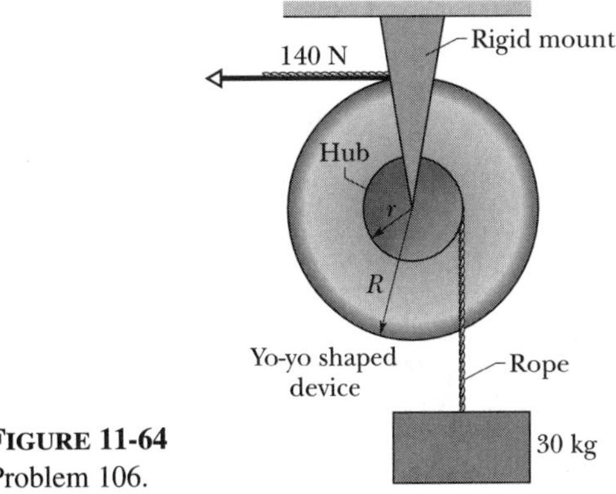

FIGURE 11-64 Problem 106.

107. A wheel of radius 0.20 m is mounted on a frictionless horizontal axis. The rotational inertia of the wheel about the axis is 0.40 kg · m^2. A massless cord wrapped around the wheel's circumference is attached to a 6.0 kg box, as shown in Fig. 11-65. A short time after being released from rest, the box has a kinetic energy of 6.0 J. Just then, what are (a) the rotational kinetic energy of the wheel and (b) the distance the box has fallen from its initial position?

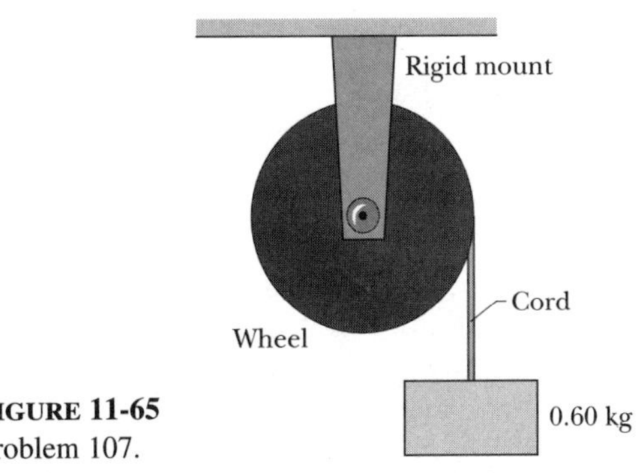

FIGURE 11-65 Problem 107.

108. A thin uniform rod (mass = 3.0 kg, length = 4.0 m) rotates freely about a horizontal axis A that is perpendicular to the rod and passes through a point 1.0 m from the end of the rod, as shown in Fig. 11-66. The kinetic energy of the rod as it passes through the vertical position is 20 J. (a) What is the rotational inertia of the rod about axis A? (b) What is the (linear) speed of the end B of the rod as the rod passes through the vertical position? (c) What angle θ will the rod make with the vertical when it momentarily stops in its upward swing?

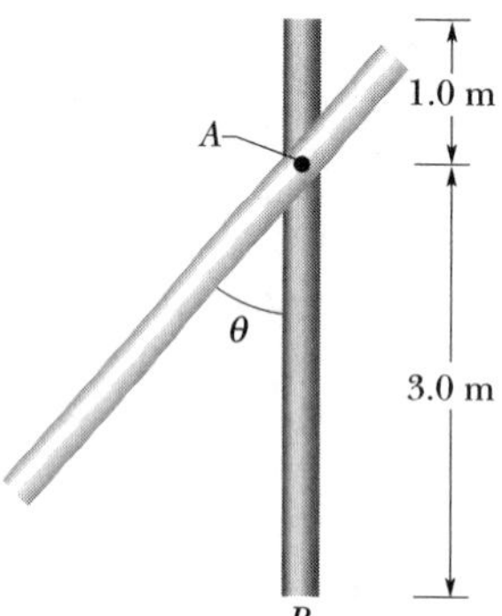

FIGURE 11-66 Problem 108.

109. Four particles, each of mass 0.20 kg, are placed at the vertices of a square with sides of length 0.50 m. The particles are connected by rods of negligible mass. This body can rotate in a vertical plane about a horizontal axis A that passes through one of the particles. The body is released from rest with rod AB horizontal, as shown in Fig. 11-67. (a) What is the rotational inertia of the body about axis A? (b) What is the angular acceleration of the body about axis A immediately after it is released? (c) Use the principle of the conservation of energy to determine the angular speed of the body about axis A at the instant rod AB swings through the vertical position.

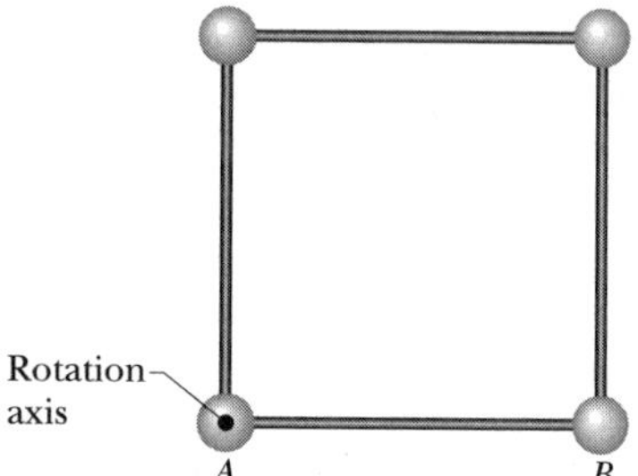

FIGURE 11-67 Problem 109.

Graphing Calculators

PROBLEM SOLVING TACTICS

TACTIC 3: *Storing Equations for Constant Angular Acceleration*

In Problem Solving Tactic 19 of Chapter 2, we discussed how to store equations and notes as a program that can function as a notebook. If you now wish to store the equations of Table 11-1 or notes about angular motion, you need to put Greek letters on the screen. You can find them by pressing 2nd CHAR and then choosing GREEK. Many (but not all) of the Greek letters, some lowercase and others uppercase, can be chosen from the menu that appears. However, omega (ω) is missing; use a lowercase w in its place.

Chapter Twelve
Rolling, Torque, and Angular Momentum

QUESTIONS

16. In Fig. 12-47, a disk rolls smoothly (and without slipping) up an incline at angle θ. (a) What is the direction of the frictional force on the disk at P, the point of contact of the disk with the incline? (b) Which is greater in magnitude, that frictional force or the disk's weight component along the plane?

(c) About the center, which is greater, the torque due to the disk's weight or the torque due to the frictional force on the disk? (d) Repeat the question, but now consider the torques about point P. (e) If we decrease the angle θ, does the torque about P due to the weight increase, decrease, or remain the same?

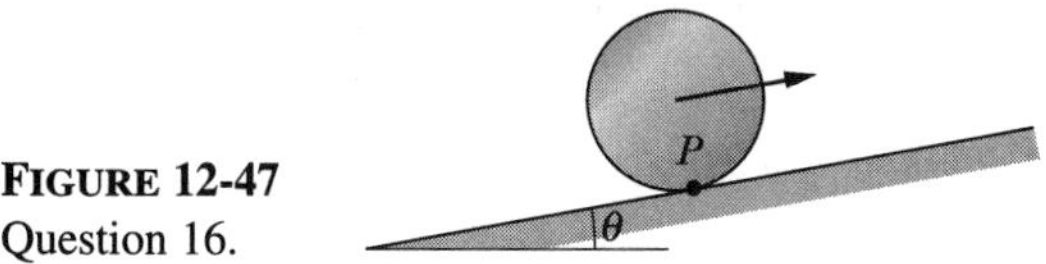

FIGURE 12-47 Question 16.

17. *Organizing question:* Figure 12-48 shows three situations in which an object of mass m, rotational inertia I, and radius r rolls smoothly (and without slipping) up or down a track. For each situation, what are the object's initial total mechanical energy E_i and its final total mechanical energy E_f? In the situations of Fig. 12-48, the object (*a*) has an initial speed of v_i, rolls up a slope and onto a plateau at vertical height h, and there has a final speed of v_f; (*b*) has an initial speed of v_i and rolls up a slope to some maximum vertical height h (take this state as the final state); (*c*) starts from rest at a vertical height h and rolls down into a circular track of radius R and up to the top of that circular track, where it has speed v_f (take this state as the final state).

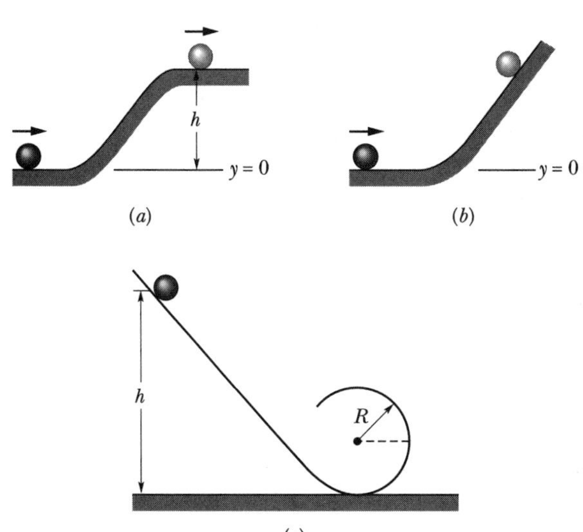

FIGURE 12-48 Question 17.

18. In Fig. 12-49, three forces of the same magnitude are applied to a particle at the origin ($\mathbf{F}_1$ acts directly into the plane of the figure). Rank the forces according to the magnitudes of the torques they create about (a) point P_1, (b) point P_2, and (c) point P_3, greatest first.

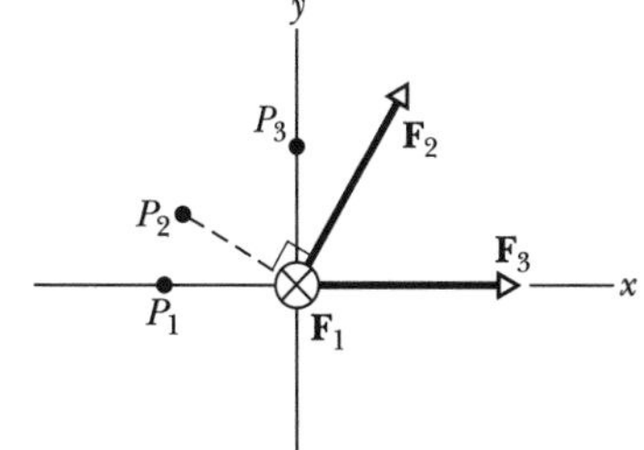

FIGURE 12-49 Question 18.

19. Figure 12-50 shows two particles A and B at xyz coordinates (1 m, 1 m, 0) and (1 m, 0, 1 m). Acting on each particle are three numbered forces, all of the same magnitude and each directed parallel to an axis. (a) Which of the forces produce a torque about the origin that is directed parallel to y? (b) Rank the forces according to the magnitudes of the torques they produce on the particles about the origin, greatest first.

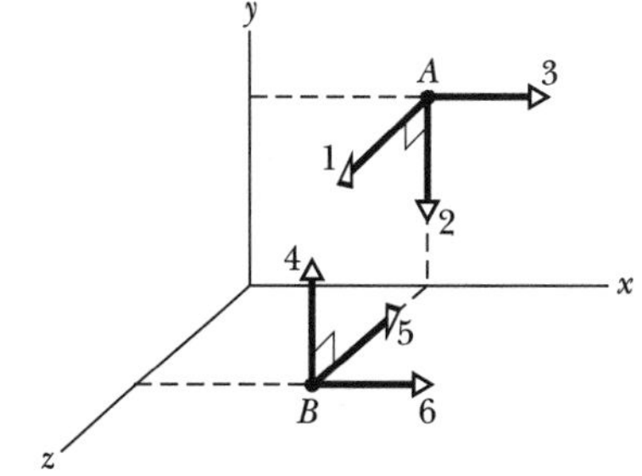

FIGURE 12-50 Question 19.

20. Figure 11-53 shows an assembly of three small spheres of the same mass that are attached to a massless rod with the indicated spacings. The assembly is to be rotated at 3.0 rad/s about an axis through one of the spheres and perpendicular to the plane of the page. There are, of course, three such choices of the axis. Rank those choices according to (a) the angular momentum the assembly will have about the rotation axis and (b) the rotational kinetic energy the assembly will have, greatest first.

21. At a certain instant, the sign of the angular momentum L and the sign of the change dL/dt of a merry-go-round about its center are given for four situations. For each situation, are (a) the angular speed and (b) the rotational kinetic energy of the merry-go-round increasing, decreasing, or staying the same?

	(1)	(2)	(3)	(4)
L	+	+	−	−
dL/dt	+	−	+	−

22. *Organizing question:* In the overhead view of Fig. 11-52, a 0.2 kg Texas cockroach rides the rim of a uniform 3.0 kg disk that is rotating like a merry-go-round. The angular speed is 1.5 rad/s; assume the cockroach is small enough to be a particle. The cockroach is to crawl to either (a) a point halfway to the center of the disk or (b) the center of the disk. For each situation, set up an equation, complete with known data, to find the angular speed ω_f of the disk and cockroach once the move is complete. (c) Reaching which final point, the halfway point or the center, requires more effort by the cockroach? In which situation will (d) ω_f and (e) the final rotational kinetic energy of the cockroach–disk system be greater?

23. If the rotating student in Fig. 12-17 were to drop the dumbbells, would the following, each relative to the rotation axis, increase, decrease, or remain the same: (a) the student's angular momentum, (b) the student's angular speed, (c) the angular momentum of the dumbbells, (d) the speed at which the dumbbells move vertically, (e) the speed at which they move horizontally, (f) the angular speed of the dumbbells, and (g) the angular momentum of the student–dumbbell system?

24. When an astronaut floats in a spacecraft while facing a wall, how can the astronaut turn to face another wall to the left or right without touching anything? Such motion is called *yaw.* How might the astronaut *pitch,* which is to rotate forward or backward around a horizontal axis that runs left and right? Is a *roll,* which is a rotation around a horizontal axis that runs forward and rearward, possible?

25. In ballet's *tour jeté* (or turning jump), the performer leaps from the floor with no apparent spin and then somehow turns on the spin while in midair with a simple rearrangement of arms and legs. Just before the performer lands, the spinning is somehow eliminated. Does this maneuver violate the law about conservation of angular momentum? How is it accomplished?

EXERCISES & PROBLEMS

77. A constant horizontal force of 10 N is applied to a wheel of mass 10 kg and radius 0.30 m as shown in Fig. 12-51. The wheel rolls without slipping on the horizontal surface, and the acceleration of its center of mass is 0.60 m/s^2. (a) What are the magnitude and direction of the frictional force on the wheel? (b) What is the rotational inertia of the wheel about an axis through its center of mass and perpendicular to the plane of the wheel?

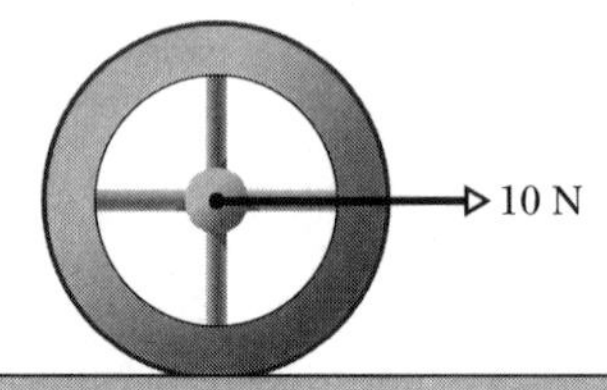

FIGURE 12-51 Problem 77.

78. An 0.80 kg particle is located at the position $\mathbf{r} = (2.0\text{ m})\mathbf{i} + (3.0\text{ m})\mathbf{j}$. The momentum of the particle lies in the xy plane and has a magnitude of 2.4 kg · m/s and a direction of 115° measured counterclockwise from the direction of increasing x. What is the angular momentum of the particle about the origin, in unit-vector form?

79. A phonograph record of mass 0.10 kg and radius 0.10 m rotates about a vertical axis through its center with an angular speed of 4.7 rad/s. The rotational inertia of the record about its axis of rotation is 5.0×10^{-4} kg · m^2. A wad of putty of mass 0.020 kg drops vertically onto the record from above and sticks to the edge of the record. What is the angular speed of the record immediately after the putty sticks to it?

80. A hollow sphere of radius 0.15 m, with rotational inertia $I = 0.040$ kg · m^2 about its center of mass, rolls without slipping up a surface inclined at 30° to the horizontal. At a certain initial position, the sphere's total kinetic energy is 20 J. (a) How much of this initial kinetic energy is rotational? (b) What is the speed of the center of mass of the sphere at the initial position? What are (c) the total kinetic energy of the sphere and (d) the speed of its center of mass after it has moved 1.0 m up along the incline from its initial position?

81. In Fig. 12-52, a constant horizontal force of 12 N is applied to a uniform solid cylinder by fishing line wrapped around the cylinder. The mass of the cylinder is 10 kg, the radius is 0.10 m, and the cylinder rolls without slipping on the horizontal surface. (a) What is the magnitude of the acceleration of the center of mass of the cylinder? (b) What is the magnitude of the angular acceleration of the cylinder about the center of mass?

(c) What are the magnitude and direction of the frictional force acting on the cylinder?

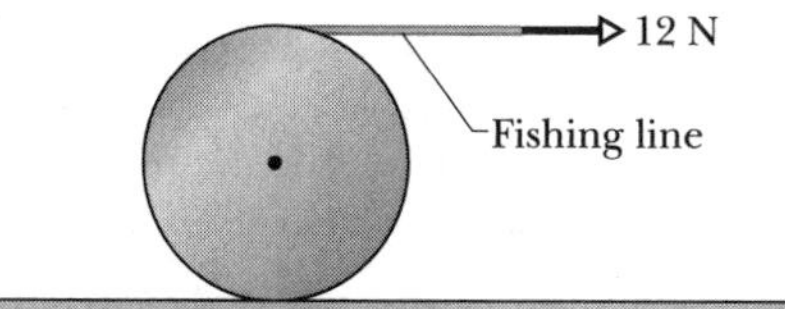

FIGURE 12-52 Problem 81.

82. A uniform wheel of mass 10.0 kg and radius 0.400 m is mounted rigidly on an axle through its center (Fig. 12-53). The radius of the axle is 0.200 m, and the rotational inertia of the wheel–axle combination about its central axis is $0.600 \text{ kg} \cdot \text{m}^2$. The wheel is initially at rest at the top of a surface that is inclined 30.0° with the horizontal; the axle rests on the surface while the wheel extends into a groove in the surface without touching the surface. Once released, the axle rolls down along the surface smoothly and without slipping. When the wheel–axle combination has moved down the surface by 2.00 m, what are (a) its rotational kinetic energy and (b) its translational kinetic energy?

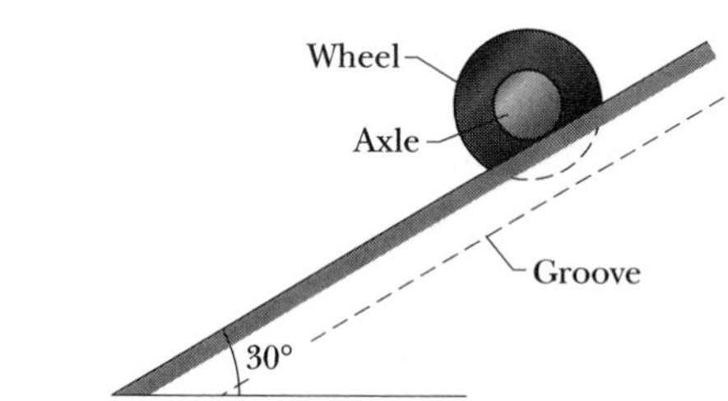

FIGURE 12-53 Problem 82.

83. A 4.0 kg particle moves in an xy plane. At the instant when the particle's position and velocity are $\mathbf{r} = (2.0\mathbf{i} + 4.0\mathbf{j})$ m and $\mathbf{v} = -4.0\mathbf{j}$ m/s, the force on the particle is $\mathbf{F} = -3.0\mathbf{i}$ N. At this instant, determine (a) the particle's angular momentum about the origin, (b) the particle's angular momentum about the point $x = 0$, $y = 4.0$ m, (c) the torque acting on the particle about the origin, and (d) the torque acting on the particle about the point $x = 0$, $y = 4.0$ m.

84. A particle of mass M is dropped from a height h above the ground and a horizontal distance s from an observation point O as shown in Fig. 12-54. What is the magnitude of the angular momentum of the particle with respect to point O when the particle has fallen half the distance to the ground? State your answer in terms of M, h, s, and g.

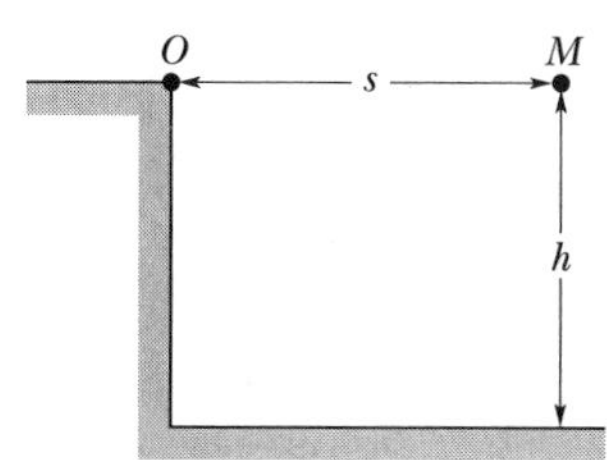

FIGURE 12-54 Problem 84.

85. A small 0.50 kg block has a horizontal velocity of 3.0 m/s when it slides off a 1.2 m high frictionless table as shown in Fig. 12-55. Answer the following in unit vectors for a coordinate system in which the origin is at the edge of the table (point O), the positive x direction is horizontally away from the table, and the positive y direction is up. What are the angular momenta of the block about point A at the foot of the table leg (a) just after the block leaves the table and (b) just before the block strikes the floor? What are the torques on the block about point A (c) just after the block leaves the table and (d) just before the block strikes the floor?

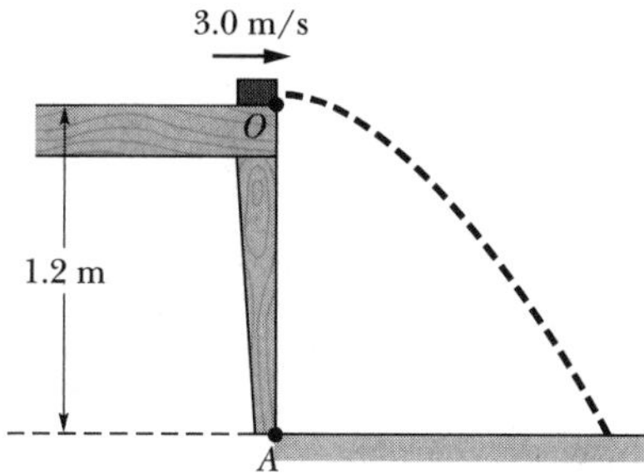

FIGURE 12-55 Problem 85.

86. A 30 kg child stands on the edge of a stationary merry-go-round of mass = 100 kg and radius = 2.0 m. The rotational inertia of the merry-go-round about its axis of rotation is $150 \text{ kg} \cdot \text{m}^2$. The child catches a ball of mass 1.0 kg thrown by a friend. Just before the ball is caught, it has a horizontal velocity of 12 m/s that makes an angle of 37° with a line tangent to the outer edge of the merry-go-round, as shown in the overhead view of Fig. 12-56. What is the angular speed of the merry-go-round just after the ball is caught?

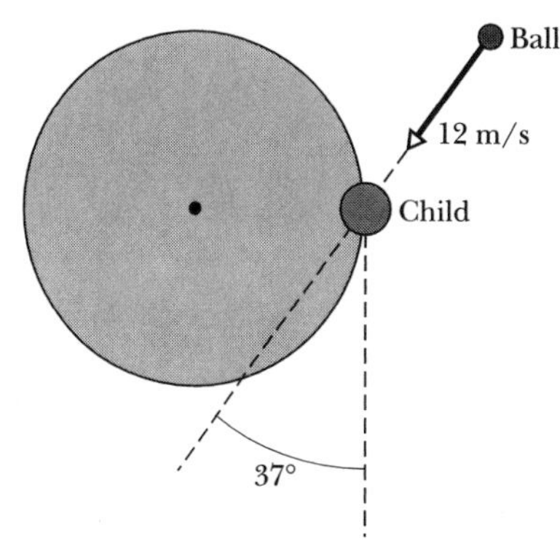

FIGURE 12-56 Problem 86.

87. In Fig. 12-57, a 1.0 g bullet is fired into a 0.50 kg block that is mounted on the end of a 0.60 m nonuniform rod of mass

0.50 kg. The block–rod–bullet system then rotates about a fixed axis at point A. The rotational inertia of the rod alone about A is 0.060 kg · m². Assume the block is small enough to treat as a particle on the end of the rod. (a) What is the rotational inertia of the block–rod–bullet system about point A? (b) If the angular speed of the system about A just after the bullet's impact is 4.5 rad/s, what is the speed of the bullet just before the impact?

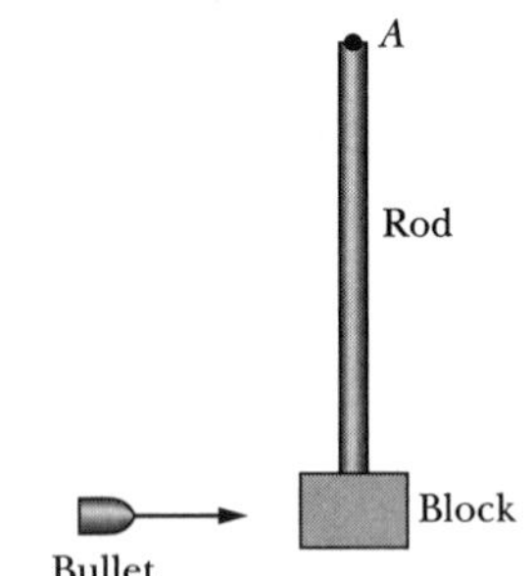

FIGURE 12-57 Problem 87.

88. In Fig. 12-58, a uniform rod (length = 0.60 m, mass = 1.0 kg) rotates about an axis through one end, with a rotational inertia of 0.12 kg · m². As the rod swings through its lowest position, the end of the rod collides with a small 0.20 kg putty wad that sticks to the end of the rod. If the angular speed of the rod just before the collision is 2.4 rad/s, what is the angular speed of the rod–putty system immediately after the collision?

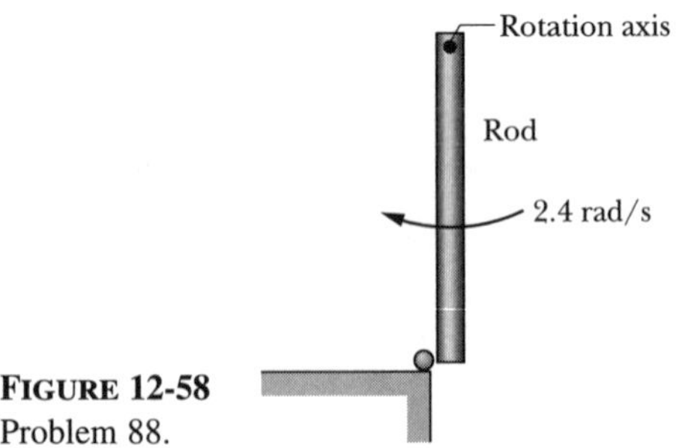

FIGURE 12-58 Problem 88.

Clustered Problems

Cluster 1

The problems in this cluster deal with an object rolling with constant linear acceleration due to an applied force and a frictional force.

89. A horizontal force of magnitude 200 N is applied to the axle of a one-wheel cart with a total mass of 50.0 kg, producing an acceleration of 3.00 m/s² along a level path. The cart's wheel, which does not have uniform distribution of mass, is a long cylinder with a radius of 0.200 m. It rolls smoothly and without sliding, rotating about its axle without friction. What are (a) the magnitude of the frictional force on the wheel from the path, (b) the torque of that force about the rotation axis of the wheel, and (c) the rotational inertia of the wheel about that axis?

90. The cart of Problem 89 is sent rolling down a plane inclined at 30.0° to the horizontal. (a) What is the magnitude of the cart's acceleration? (b) What is the minimum coefficient of static friction needed to keep the cart from sliding? (*Hint:* Use the answer to part (c) of Problem 89; if you cannot get that answer, use the symbol I for the wheel's rotational inertia about the rotation axis.)

91. A disk of rotational inertia I, mass M, and radius R is positioned to roll down a plane inclined at angle θ to the horizontal. The coefficient of static friction between the disk and the inclined plane is μ_s. What is the greatest value of θ at which the disk rolls smoothly down the plane without slipping? (This value is said to be the critical angle for rolling objects.)

92. A uniform disk is upright on its edge on the flat bed of a long truck, ready to roll toward the front or rear. The truck then accelerates at magnitude a in the forward direction, and the disk rolls without slipping until it hits the truck's rear wall. Assume that the bed remains horizontal during the acceleration. During the rolling, what are the magnitudes and directions of the acceleration of the disk relative to (a) the bed and (b) the roadway?

Cluster 2

93. Figure 12-59a shows two pulleys, one of radius R_1 and the other of radius R_2, that are rigidly connected and free to turn about a common, frictionless axle through their centers. The rotational inertia of this two-pulley device about that axle is I. Fishing line that is wrapped around the circumference of the larger pulley extends to block 2 of mass m_2. Similarly, fishing line that is wrapped around the circumference of the smaller pulley extends to block 1 of mass m_1.

Initially we hold the blocks in place. When we release them, what are the magnitudes of the accelerations of (a) block 1 and (b) block 2? Assume that the direction of positive acceleration is upward for block 2 and downward for block 1 and take the direction of positive angular acceleration of the two-pulley device as counterclockwise.

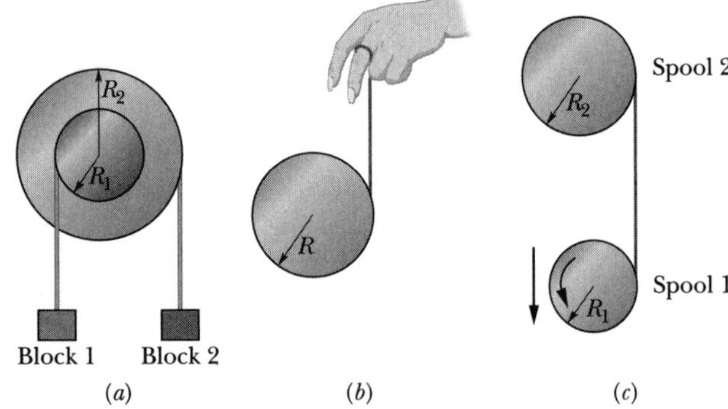

FIGURE 12-59 Problems 93 through 95.

94. In Fig. 12-59*b*, a spool of thread with rotational inertia I, radius R, and mass M falls as thread unwraps from the spool's circumference. A hand holds the upper end of the thread. (a) What is the magnitude of the linear acceleration of the spool's center of mass if the fingers are stationary? (b) What upward acceleration must the hand give the thread if the spool's center of mass is not to fall?

95. In Fig. 12-59*c*, spool 1 falls as thread unwraps from its circumference. At the same time, the upper end of the thread unwraps from the circumference of spool 2, which rotates about a fixed, frictionless axle that coincides with its central axis. Spool 1 has rotational inertia I_1 about its central axis, radius R_1, and mass M. Spool 2 has rotational inertia I_2 about its central axis and radius R_2.

(a) What is the magnitude of the linear acceleration a_1 of the center of mass of spool 1? (*Hint:* Downward acceleration a_1 is equal to the sum of two accelerations: the downward acceleration a_s of the thread that results from its unwrapping from spool 2 and the downward acceleration of spool 1 that results from the thread unwrapping from it. That latter acceleration is like that in Problem 94(a).) (b) Show that if $I_2 \gg I_1$, then the expression for a_1 matches that for the acceleration in Problem 94(a).

Cluster 3

The problems in this cluster deal with torque and the principle of the conservation of energy.

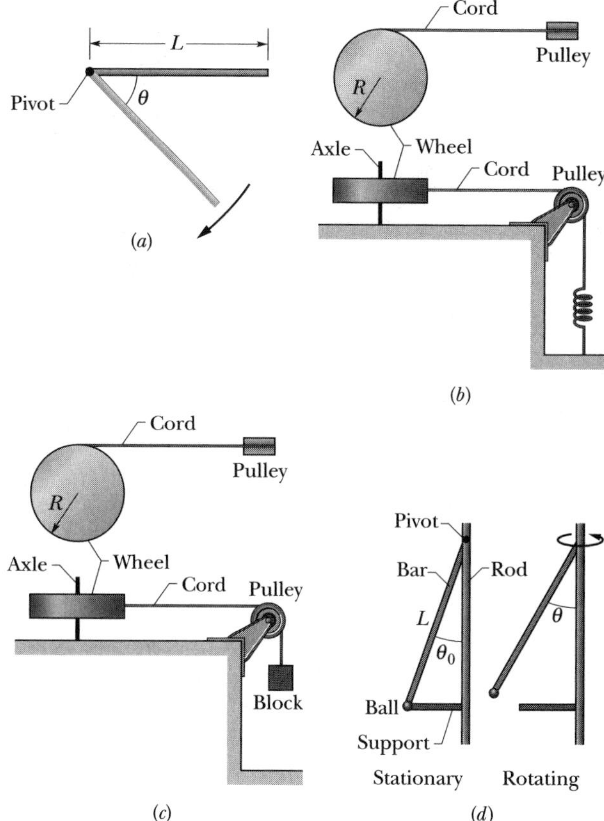

FIGURE 12-60 Problems 96 through 99.

96. In Fig. 12-60*a*, a uniform rod of mass M and length L that can pivot about a rotation axis at one end is held horizontally and then released from rest. What is the linear speed of the end opposite the rotation axis when the rod reaches (a) the vertical orientation and (b) some intermediate orientation with the rod at angle θ with the horizontal?

97. In the overhead and side views of Fig. 12-60*b*, a cord is tied to a spring of spring constant k and is also wrapped around the circumference of a wheel of inertia I and radius R. The other end of the spring is attached to a floor, and the cord passes over a massless, frictionless pulley. We rotate the wheel about a frictionless axle along the wheel's central axis so as to stretch the spring. When the spring is stretched by distance d, we release the wheel from rest. When the spring's stretch is reduced to zero, what are the magnitudes of (a) the angular velocity and (b) the angular acceleration of the wheel?

98. In the overhead and side views of Fig. 12-60*c*, a cord is tied to a block of mass m and is also wrapped around the circumference of a wheel of inertia I and radius R. (The cord also passes over a massless, frictionless pulley. The wheel can rotate about a frictionless axle along the wheel's central axis.) The block is initially held in place. When it is released, what are the magnitudes of (a) its velocity and (b) its acceleration when it has fallen by distance h?

99. In Fig. 12-60*d*, a massless bar of length L can pivot about its upper end where it is connected to the vertical rod. A small ball of mass M is attached to the lower end of the bar and initially rests against a support on the rod such that the bar makes an angle θ_0 with the rod. Initially this assembly is stationary, but then we begin to rotate the vertical rod about its central axis. Thus the ball also rotates about that axis. As we steadily increase the angular speed of the rotation, the bar eventually swings outward from the rod, pivoting about its upper end like some carnival rides with swings hanging from a rotating support. Consequently, the ball, still rotating about the rod, moves outward away from the support.

When the ball is on the verge of moving outward in this way, (a) what is its angular speed about the rod and (b) how much work has been done on it to give it that angular speed? (c) How much more work must be done on it if the rotation increases the angle between the bar and the rod to a value θ?

Assume that the torque causing the ball to rotate about the rod has a magnitude of 1.00 N · m, the ball's mass M is 5.00 kg, the bar's length L is 2.00 m, and the initial angle θ_0 is 20.0°. When the ball is on the verge of moving outward, (d) how many revolutions has it gone through and (e) what is its angular speed? (f) When θ reaches 30.0°, how many more revolutions has the ball gone through and (g) what is its angular speed?

Cluster 4

100. A ball of sticky putty slides over a frictionless air table, colliding with and sticking to the end of a stationary, uniform rod. The rod, which has length L and mass M, can rotate over the table about a rotation axis through its opposite end. The ball has mass m and initial speed v_i, and its initial path is perpendicular to the rod. Are the following properties of the ball–rod system necessarily conserved during the collision: (a) linear momentum, (b) angular momentum, and (c) kinetic energy? (d) What is the angular speed of the rod–ball combination about the rotation axis after the collision?

101. A steel ball slides over a frictionless air table, colliding elastically with the end of a stationary uniform rod. The rod, which has length L and mass M, can rotate over the table about a rotation axis through its opposite end. The ball has mass m and initial speed v_i, and its initial path is perpendicular to the rod. After the collision it is still moving in its original direction. Are the following properties of the ball–rod system necessarily conserved during the collision: (a) linear momentum, (b) angular momentum, and (c) kinetic energy? (d) What is the angular speed of the rod about the rotation axis after the collision? (e) For what mass ratio m/M is all the ball's kinetic energy transferred to the rod during the collision?

102. A ball of sticky putty slides over a frictionless air table, colliding with and sticking to the end of a stationary uniform rod. The rod, which has length L and mass M, has an identical ball of putty already stuck on its other end and is free to move over the table. The sliding ball has mass m and initial speed v_i, and its initial path is perpendicular to the rod.

Are the following properties of the ball–rod system necessarily conserved during the collision: (a) linear momentum, (b) angular momentum, and (c) kinetic energy? After the collision, what are (d) the linear speed and (e) the angular speed of the rod–putty system?

Cluster 5

The problems in this cluster are about a barbell, which, as you probably know, is a weight-lifting apparatus consisting of a bar with disks fastened onto both ends. Here the bar has radius r, the disks have radius R, and the barbell as a whole has a rotational inertia I about the central axis along the bar and a mass m. In each part of Fig. 12-61, where the barbell is seen from one end with the bar's radius exaggerated, a sturdy cord is wrapped around the bar so that a force of magnitude F can be applied tangentially to the bar. The barbell then rolls smoothly and without slipping across a surface (which implies that the surface is not frictionless).

Answer the problems in terms of the given symbols r, R, I, m, g, and, in Problems 106 through 108, θ. In all the Problems, take counterclockwise as the positive direction of rotation. In Problems 103 through 105, determine the positive direction of linear motion using the x axis shown in the figures. In Problems 106 through 108, take up the inclined plane as the positive direction of linear motion.

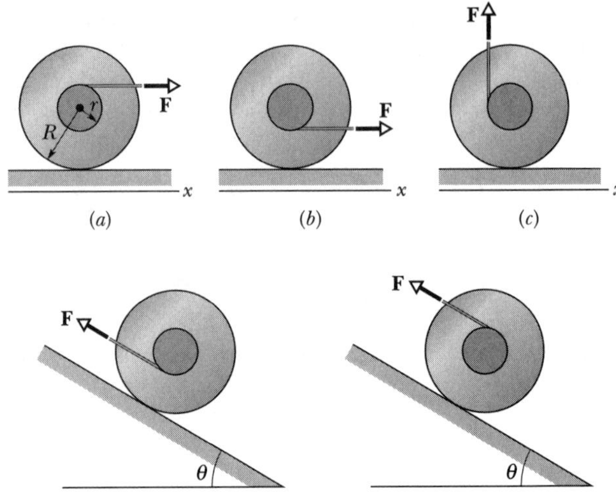

FIGURE 12-61 Problems 103 through 108.

103. See the setup for this cluster. What are (a) the linear acceleration of the barbell and (b) the static frictional force acting on the barbell in Fig. 12-61*a*, where the applied force is horizontal? (c) Making the reasonable assumption that $I > mRr$, determine the direction of the frictional force.

104. See the setup for this cluster. What are (a) the linear acceleration of the barbell and (b) the static frictional force acting on the barbell in Fig. 12-61*b*, where the applied force is horizontal? (c) Determine the direction of the frictional force.

105. See the setup for this cluster. What are the magnitudes of (a) the linear acceleration of the barbell and (b) the static frictional force acting on the barbell in Fig. 12-61*c*, where the applied force is vertical?

106. See the setup for this cluster. In Fig. 12-61*d*, the barbell is on a plane inclined at angle θ with the horizontal and the applied force is directed up along the plane. In what follows, use the given symbols. (a) Find an expression for the magnitude F_{stat} of the applied force that is required to keep the barbell stationary (in place and not rotating). (b) In this situation, is the frictional force acting on the barbell directed up or down the plane? (c) Is F_{stat} greater than, less than, or equal to the component of the barbell's weight along the plane?

Suppose that with a magnitude F, the applied force causes the barbell to accelerate up the plane. In terms of F, find an expression for (d) the barbell's acceleration and (e) the frictional force acting on the barbell. (f) In terms of the given symbols, what magnitude F results in no frictional force? (g) If F is greater than that value, is the frictional force directed up or down the plane?

107. Continuation of Problem 106. Now let μ_s represent the coefficient of static friction between the barbell and the inclined plane. (a) In terms of μ_s, what magnitude of the applied force puts the barbell on the verge of slipping? (b) What is the corresponding acceleration a_{slip}? (c) At what angle θ, said to be the critical angle θ_c, does the barbell slip for any magnitude of the applied force up along the plane? (*Hint:* This condition occurs when $a_{slip} = 0$; that is, the barbell is already on the verge of slipping before we attempt to accelerate it up the plane.)

108. Repeat the parts of Problem 106 for the situation in Fig. 12-61*e*, where the applied force is again directed up along the plane. (In parts (e) and (f), make the reasonable assumption that $I > mRr$.)

Chapter Thirteen
Equilibrium and Elasticity

QUESTIONS

14. In Fig. 13-59, a rigid beam is attached to two posts that are fastened to a floor. A small but heavy safe is placed at the six positions indicated, in turn. Assume that the mass of the beam is negligible compared to that of the safe. (a) Rank the positions according to the force on post A due to the safe, greatest compression first, greatest tension last, and indicate where, if anywhere, the force is zero. (b) Now rank them in the same way according to the force on post B.

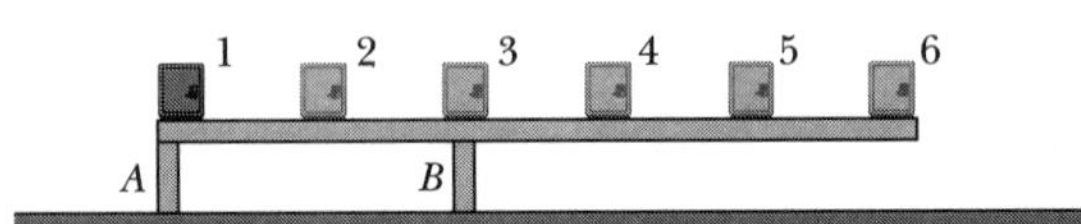

FIGURE 13-59 Question 14.

15. *Organizing question:* Figure 13-60 shows a window washer on his rig, which consists of a uniform beam of mass 200 kg and length 8.0 m, suspended by a cable at each end. The 60 kg washer works 2.0 m from the right end. (a) Set up an equation (one equation), complete with known data, to find the tension T_R in the cable at the right end. (b) Suppose, instead, we want the tension T_L in the cable at the left end. Without relying on the answer to (a), set up an equation, complete with known data, to find T_L.

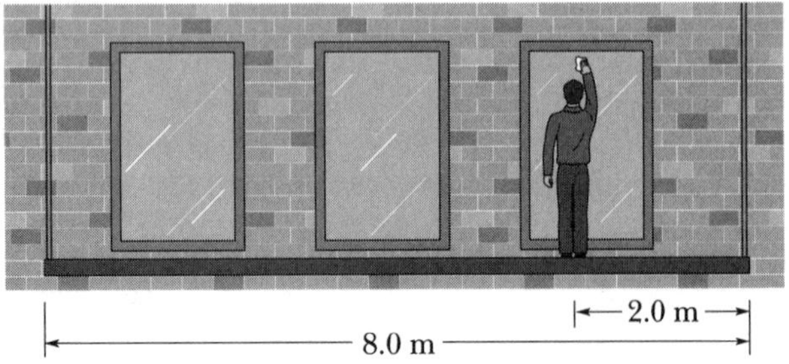

FIGURE 13-60 Question 15.

16. *Organizing question:* Figure 13-61 shows a stationary 50 kg rock climber that is on belay via a climbing harness. The line of action of the force on her from the belay line is at perpendicular distances 2.0 cm from her center of mass, and 80 cm from her feet. The forces on her feet from the rock face are collectively represented by $\mathbf{F}_r$. Set up an equation, complete with known data, to find the tension T in the belay line.

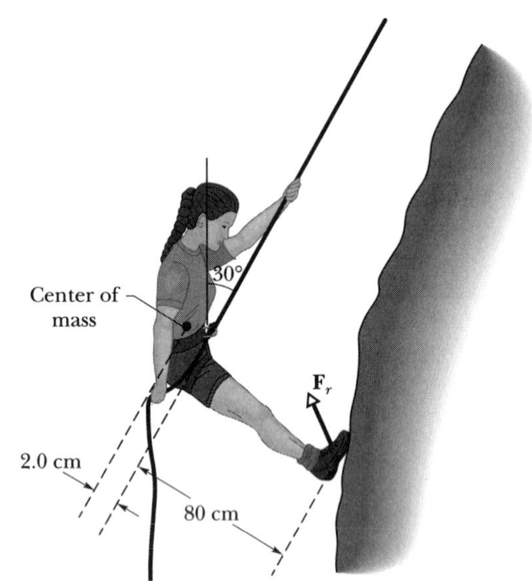

FIGURE 13-61 Question 16.

17. (a) In Sample Problem 13-5 and Fig. 13-9, if the angle θ were made greater (but the cable still kept horizontal), would the tension in the cable be greater than, less than, or the same as in the Sample Problem? (b) Would the magnitude of the net force exerted by the hinge on the beam be greater than, less than, or the same as in the Sample Problem?

18. In Fig. 13-62, a beam of mass M is held horizontal by a massless rod in orientation 1; the beam is attached to a hinge at its other end. Five other orientations of the rod are ghosted in the figure (the tilted orientations have identical tilts). (a) For which orientations is the rod under tension and for which is it under compression? Rank the six orientations according to (b) the magnitude of the tension or compression in the rod, (c) the magnitude of the vertical force on the beam from the hinge, and (d) the magnitude of the horizontal force on the beam from the hinge, greatest first.

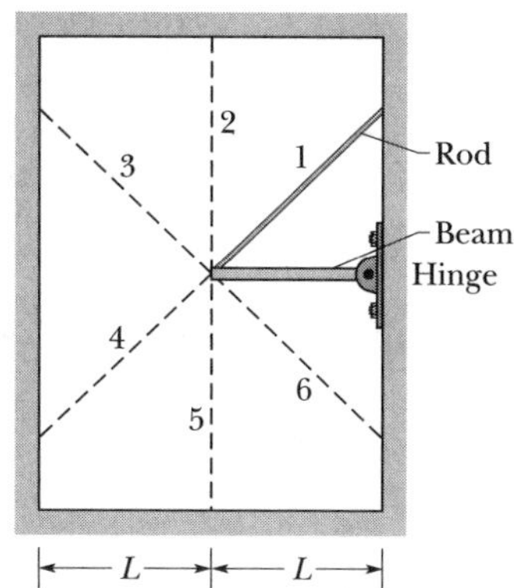

FIGURE 13-62 Question 18.

(b) When an old industrial chimney is taken down, one side of its base is usually knocked out by a bulldozer or blasted out with an explosion. The chimney then rotates around that side toward the ground. During the rotation, the chimney tends to bend backward against the fall until it ruptures at about midheight. Does the rupture begin on the side facing the ground or on the opposite side?

19. (a) As a quick fix to a bridge that has been partially ruined by floodwater, a concrete slab is laid across a gap in the bridge, supported by remaining vertical structures. As a heavy truck is then driven over the slab and the slab sags, is the slab more likely to rupture on its top surface or its bottom surface?

20. Four cylindrical rods are stretched as in Fig. 13-13a. The force magnitudes, the areas of the end faces, the changes in length, and the initial lengths are given here. Rank the rods according to their Young's moduli, greatest first.

Rod	Force	Area	Length Change	Initial Length
1	F	A	ΔL	L
2	$2F$	$2A$	$2\Delta L$	L
3	F	$2A$	$2\Delta L$	$2L$
4	$2F$	A	ΔL	$2L$

EXERCISES & PROBLEMS

57. A uniform beam of length 12 m is supported by a horizontal cable and pin as shown in Fig. 13-63. The tension in the cable is 400 N. What are (a) the weight of the beam and (b) the horizontal and vertical components of the force of the pin on the beam?

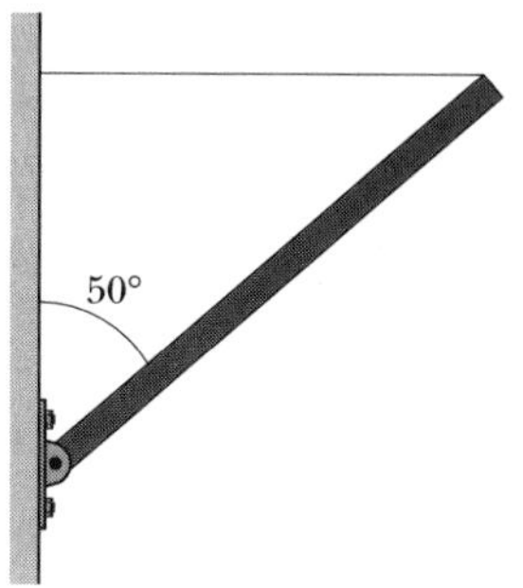

FIGURE 13-63 Problem 57.

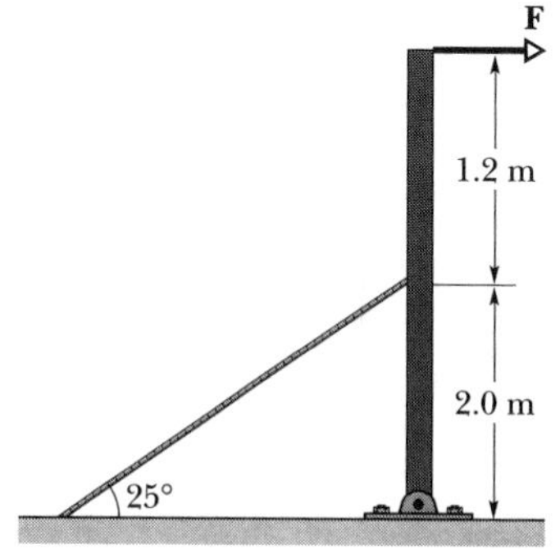

FIGURE 13-64 Problem 58.

58. A uniform beam having a weight of 60 N and a length of 3.2 m is pinned at its lower end and acted on by a horizontal force **F** of magnitude 50 N at its upper end (Fig. 13-64). The beam is held vertical by a cable that makes an angle of 25° with the ground. What are (a) the tension in the cable and (b) the horizontal and vertical components of the force of the pin on the beam?

59. A uniform 10 m ladder weighs 200 N. The ladder leans against a vertical, frictionless wall at a point 8.0 m above the ground, as shown in Fig. 13-65. A horizontal force F is applied to the ladder at a point 2.0 m from the base of the ladder (as measured along the ladder). (a) If $F = 50$ N, what is the force of the ground on the ladder, in terms of the unit vectors shown? (b) If $F = 150$ N, what is the force of the ground on the ladder,

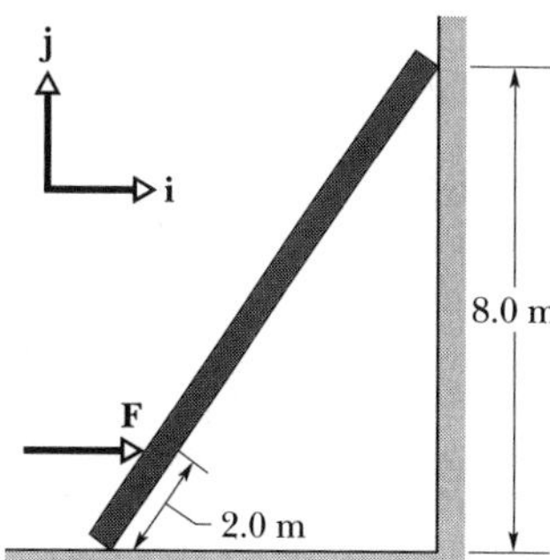

FIGURE 13-65 Problem 59.

in unit-vector form? (c) Suppose the coefficient of static friction between the ladder and ground is 0.38; for what minimum value of F will the base of the ladder just start to move toward the wall?

60. The system shown in Fig. 13-66 is in equilibrium. If $M = 2.0$ kg, what is the tension in (a) string ab and (b) string bc?

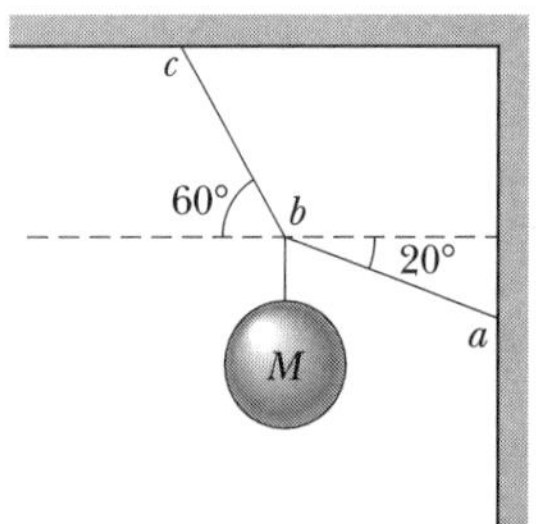

FIGURE 13-66 Problem 60.

61. In an ice plant, 200 kg blocks of ice slide down a frictionless ramp that makes an angle of 10° with the horizontal. In order to keep the blocks of ice from moving too quickly, they are restrained by an attached cable that is parallel to the ramp. If the blocks are temporarily held at rest on the ramp, what is the tension in the cable?

62. A 10 kg sphere is supported on a frictionless plane inclined at 45° from the horizontal as shown in Fig. 13-67. Calculate the tension in the cable.

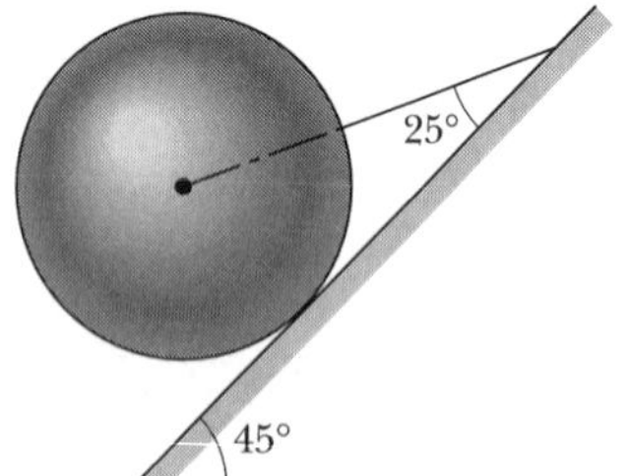

FIGURE 13-67 Problem 62.

63. A makeshift swing is constructed by making a loop in one end of a rope and tying the other end to a tree limb. A child is sitting in the loop with the rope hanging vertically when an adult pulls on the child with a horizontal force and displaces the child to one side. Just before the child is released from rest, the rope makes an angle of 15° with the vertical and the tension in the rope is 280 N. (a) How much does the child weigh? (b) What is the magnitude of the (horizontal) force of the adult on the child just before the child is released? (c) If the maximum horizontal force that the adult can exert on the child is 93 N, what is the maximum angle with the vertical that the rope can make while the adult is pulling horizontally?

64. A uniform beam is 5.0 m long and has a mass of 53 kg. The beam is supported in a horizontal position by a pin and cable as shown in Fig. 13-68. In unit-vector notation, what is the force of the pin on the beam?

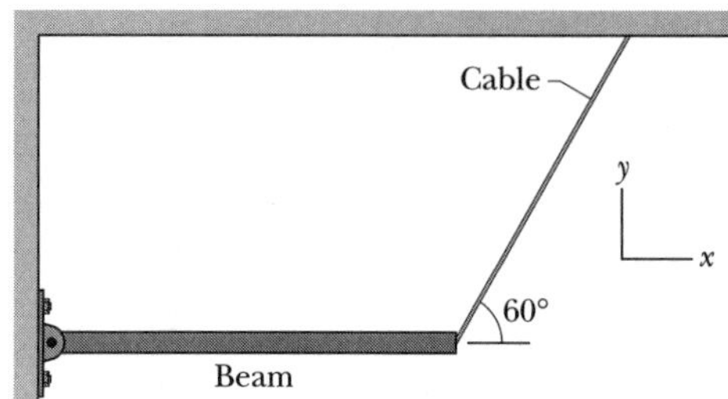

FIGURE 13-68 Problem 64.

65. A gymnast with mass 46.0 kg stands on the end of a uniform balance beam as shown in Fig. 13-69. The beam is 5.00 m long and has a mass of 250 kg (excluding the mass of the two supports). Each of the two supports is 0.540 m from its respective end of the beam. What are the magnitude and direction of (a) the force of support 1 on the beam and (b) the force of support 2 on the beam?

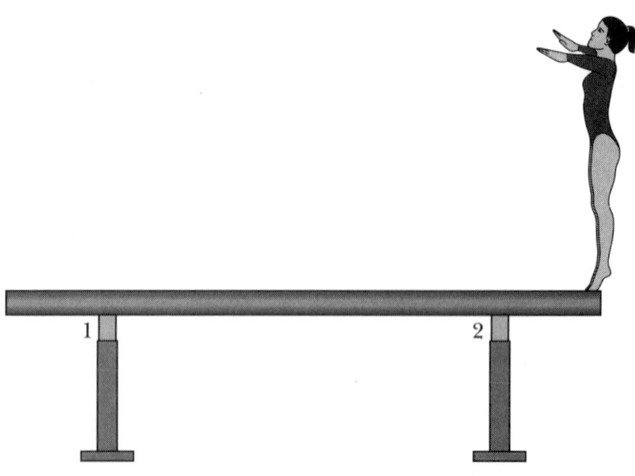

FIGURE 13-69 Problem 65.

66. A uniform ladder whose length is 5.0 m and whose weight is 400 N leans against a frictionless vertical wall. The coefficient of static friction between the level ground and the foot of the ladder is 0.46. What is the greatest distance the foot of the ladder can be placed from the base of the wall without the ladder immediately slipping?

67. A construction worker attempts to lift a uniform beam off the floor and raise it to a vertical position. The beam is 2.5 m long and weighs 500 N. At a certain instant the worker holds the beam momentarily at rest with one end 1.5 m off the floor, as shown in Fig. 13-70, by exerting a force **P** on the beam, perpendicular to the beam. (a) What is the magnitude of the force exerted by the worker? (b) What is the magnitude of the (net) force of the floor on the beam? (c) What is the minimum value that the coefficient of static friction between the beam and the floor can have in order for the beam not to slip at this instant?

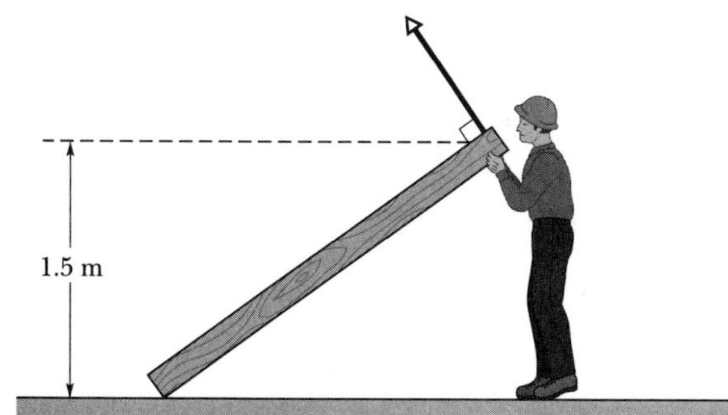

FIGURE 13-70 Problem 67.

68. In Fig. 13-71, a uniform diving board (mass = 40 kg) is 3.5 m long and is attached to two supports. When a diver stands on the end of the board, the support on the other end exerts a downward force of 1200 N on the board. Where on the board should the diver stand in order to reduce that force to zero?

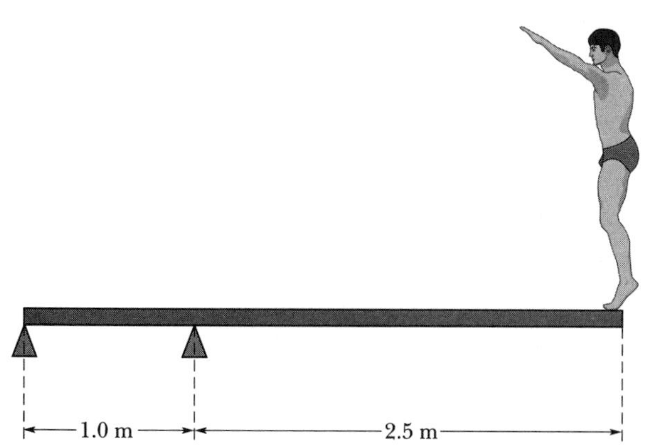

FIGURE 13-71 Problem 68.

Tutorial Problems

69. A uniform wood door has mass m, height h, and width w. It is hanging from two hinges attached to one side; the hinges are located $h/3$ and $2h/3$ from the bottom of the door. (a) Sketch this system, showing the positions of the hinges, the center of mass of the door, h, and w. (b) List the forces acting on the door and draw a free-body force diagram for the door. (*Hint:* Show the true lines of action of the forces instead of making them all act at the center of mass.)

(c) State in complete sentences, without the use of mathematical symbols, the net force and net torque conditions for the static equilibrium of the door. (d) Write all the equations that result from the conditions you stated in part (c). If you had chosen the same notation as one of your classmates, would it be possible for the two of you to write correct but different equations (different in a significant way, not just with a different order of terms)? If you think they could have been different, explain why.

(e) How many equations did you have in part (d) and how many unknowns? Would it be theoretically possible to determine the numerical values of all the forces if you were given the numerical values of m, h, and w? Why or why not?

(f) It is possible to determine an algebraic expression (in terms of m, h, and w) for at least some, if not all, of the forces. Determine the ones you can. If there are any you cannot determine, provide whatever information you can about the unknown forces, then make up a reasonable extra assumption of your own that will enable you to determine all the forces. (g) Suppose that $m = 20.0$ kg, $h = 2.20$ m, and $w = 1.00$ m. What are the numerical values of the forces whose algebraic expressions you determined in part (f)?

Answers

(a) See Fig. 13-72*a*.

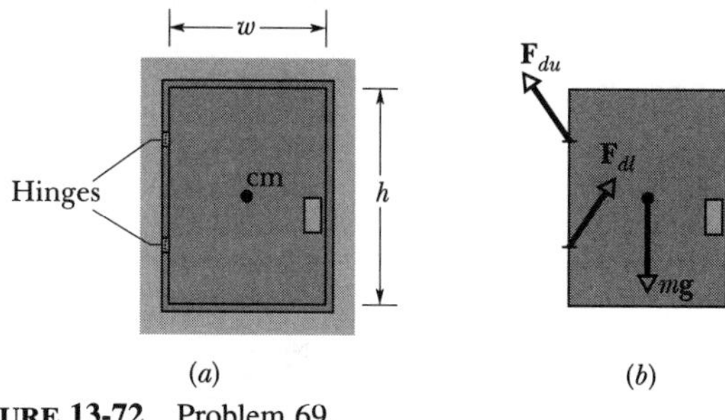

FIGURE 13-72 Problem 69.

(b) The forces acting on the door are: (1) the gravitational force on the door, which acts through the center of mass of the door; (2) the force $\mathbf{F}_{du}$ exerted on the door by the upper hinge; and (3) the force $\mathbf{F}_{dl}$ exerted on the door by the lower hinge (Fig. 13-72*b*).

The forces exerted by the hinges may have both vertical and horizontal components. If we wanted to, we could refer to them as normal and frictional forces, although the term *frictional* may not be appropriate for the vertical component. Let's just denote them as having x and y components, where the x and y axes are horizontal and vertical, respectively.

We can guess that at the hinges the vertical components are probably directed upward, so as to support the weight of the door. The door is expected to pull away from the top hinge and push onto the bottom hinge, so the horizontal components of the reaction forces on the door by the hinges should be to the left for the top hinge and to the right for the bottom hinge. Nevertheless, we shall call the horizontal components F_{dux} and F_{dlx} and expect, in the end, to find $F_{dux} < 0$ and $F_{dlx} > 0$.

(c) For the door to be in static equilibrium, two conditions must be met: (1) the net force acting on the door must be zero, since it is not undergoing linear acceleration and (2) the net torque acting on the door must be zero, since it is not undergoing angular acceleration. This must be true for any one point about which the torques are calculated, and then it will be true for any other point.

(d) For the vertical component of the net force to be zero, we must have

$$F_{duy} + F_{dly} - mg = 0. \qquad (1)$$

For the horizontal component of the net force to be zero, we must have

$$F_{dux} + F_{dlx} = 0 \quad \text{or} \quad F_{dux} = -F_{dlx}. \qquad (2)$$

For the net torque about the center of mass to be zero, we must have

$$-F_{dux}\left(\frac{h}{6}\right) - F_{duy}\left(\frac{w}{2}\right) + F_{dlx}\left(\frac{h}{6}\right) - F_{dly}\left(\frac{w}{2}\right) = 0. \qquad (3)$$

Here we take the counterclockwise torques as positive and the clockwise torques as negative; remember that the force components may be either positive or negative, but only their magnitudes are used to calculate corresponding torques.

Someone else's equations might be different because the net-torque-equals-zero equation might have been determined about a different point. We used the center of mass of the door as our point, so the torque equilibrium condition had no reference to the gravitational force on the door. Had we chosen instead, say, the upper hinge, then the gravitational force on the door, and not the upper hinge forces, would have appeared in the torque equilibrium condition.

(e) There are three equations (two from the net-force-equals-zero equation and one from the net-torque-equals-zero equation), but there are four unknowns (F_{dux}, F_{duy}, F_{dlx}, and F_{dly}). Thus we do not have enough information to determine the numerical values of the forces, even if we are given the values of m, h, and w. We need to know something else.

(f) Equation 2 of part (d) shows that $F_{dux} = -F_{dlx}$. Substituting this into equation 3 for F_{dux} and rearranging, we have

$$-F_{dlx}\left(\frac{h}{6}\right) - F_{dlx}\left(\frac{h}{6}\right) + F_{duy} + F_{dly}\left(\frac{w}{2}\right) = 0.$$

Collecting terms, and using equation 1, which tells us that $F_{duy} + F_{dly} = mg$, we have

$$-2F_{dlx}\frac{h}{6} + mg\frac{w}{2} = 0,$$

so

$$F_{dlx} = \frac{3mgw}{2h}.$$

Also, then,

$$F_{dux} = -F_{dlx} = -\frac{3mgw}{2h}.$$

The sign of F_{dlx} (positive) and the sign of F_{dux} (negative) show that the bottom hinge pushes the door away from the frame, while the upper hinge pulls the door toward the frame.

There is no way to determine the vertical components of the hinge forces, although equation 1 of part (d) shows that they must add up to mg, the weight of the door. A reasonable assumption to make might be that the hinges are identical and identically mounted and that each hinge supports half the weight of the door. In this case we would have

$$F_{duy} = F_{dly} = \frac{mg}{2}.$$

(Another possibility is that the bottom hinge is not screwed into the door frame. That means the door frame could supply a horizontal force component on the door [the same F_{dlx} found above] but not a vertical component, so then the upper hinge would have to support the whole weight of the door, and we would have $F_{duy} = mg$.)

(g) $$F_{dlx} = \frac{3mgw}{2h} = \frac{3(20.0\text{ kg})(9.80\text{ m/s}^2)(1.00\text{ m})}{2(2.20\text{ m})} = 134\text{ N}.$$

$$F_{dux} = -F_{dlx} = -134\text{ N}.$$

$$F_{duy} = F_{dly} = \frac{mg}{2} = \frac{1}{2}(20.0\text{ kg})(9.80\text{ m/s}^2) = 98\text{ N}.$$

In terms of unit-vector notation, then,

$$\mathbf{F}_{du} = (-134\text{ N})\mathbf{i} + (98\text{ N})\mathbf{j}$$

and

$$\mathbf{F}_{dl} = (+134\text{ N})\mathbf{i} + (98\text{ N})\mathbf{j}.$$

Note: In problems of this sort, it is often possible to simplify the equations by choosing the point about which the net-torque-is-zero condition is used. In this problem, if the point about which the torque is measured is the upper hinge, then neither the force at the upper hinge nor the vertical component of the force at the lower hinge will provide a torque. The only torque contributions will be a clockwise torque $mgw/2$ from the weight of the door and a counterclockwise torque $F_{dlx}h/3$ from the horizontal component at the lower hinge. Setting these torques equal gives $F_{dlx} = 3mgw/2h$. Similarly, choosing the torque about the lower hinge immediately leads to $F_{dux} = -3mgw/2h$.

If you find these simpler expressions, great! If you don't, you should still get the right answer if you don't make a mistake. If you wonder why the author of this solution did not do things the easy way, it's to show you that the harder way works and to keep you from asking, "How did he know to take the torque about that point?"

70. A uniform horizontal beam of mass M and length L is supported by two uniform vertical metal rods. Rod 1 has a Young's modulus E_1 and a cross-sectional area A_1 and is located a distance d_1 from the right end of the beam. Rod 2 has a Young's modulus E_2 and a cross-sectional area A_2 and is located a distance d_2 from the right end of the beam. Assume that $0 < d_1 < d_2 < L/2$, so that both rods are on the right half of the beam. Denote the forces on the beam exerted by rods 1 and 2, respectively, by $\mathbf{F}_1$ and $\mathbf{F}_2$.

(a) Make a sketch of this physical situation, labeling the distances involved, and draw a force diagram for the beam. (b) Using the torque condition for static equilibrium, derive algebraic expressions for the forces $\mathbf{F}_1$ and $\mathbf{F}_2$. (c) Check that these two forces you determined by the torque condition for static equilibrium also satisfy the force condition for static equilibrium. (d) Determine for each rod whether the rod is under compression or tension, and derive algebraic expressions for the magnitudes of the stress and strain in the two rods.

(e) Suppose that $M = 5.00$ kg, $L = 80.0$ cm, $A_1 = 4.00$ mm^2, $A_2 = 2.00$ mm^2, $d_1 = 10.0$ cm, $d_2 = 30.0$ cm, and rod 1 is made of steel and rod 2 of aluminum. Determine the numerical values of the forces $\mathbf{F}_1$ and $\mathbf{F}_2$ and of the stress and strain in each rod. If the rods are 50.0 cm long, by how much is each rod elongated or compressed?

Answers

(a) See Figs. 13-73*a* and *b*. The force diagram for the beam shows that there are three forces acting on the beam: (1) the gravitational force on it, of magnitude Mg, directed down; (2) and (3) the forces $\mathbf{F}_1$ and $\mathbf{F}_2$ exerted on the beam by rods 1 and 2, respectively.

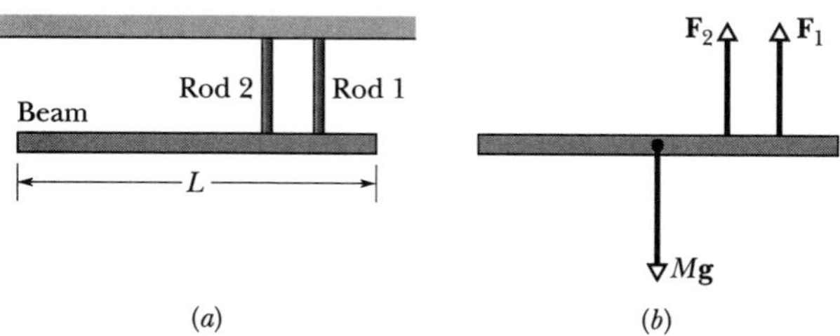

FIGURE 13-73 Problem 70.

(b) We'll represent counterclockwise torques as positive and clockwise torques as negative. First, apply the torque condition about the point where rod 1 supports the beam. Letting F_{2y} represent the upward vertical component of $\mathbf{F}_2$, the torque condition becomes

$$Mg(L/2 - d_1) - F_{2y}(d_2 - d_1) = 0,$$

so

$$F_{2y} = \frac{Mg(L/2 - d_1)}{d_2 - d_1}.$$

Second, apply the torque condition about the point where rod 2 supports the beam. Letting F_{1y} represent the upward vertical component of $\mathbf{F}_1$, the torque condition becomes

$$Mg(L/2 - d_2) + F_{1y}(d_2 - d_1) = 0,$$

so

$$F_{1y} = \frac{Mg(L/2 - d_2)}{d_1 - d_2}.$$

(c) The sum of the upward components of the two forces should equal Mg, and it does:

$$\begin{aligned} F_{1y} + F_{2y} &= \frac{Mg(L/2 - d_1)}{d_2 - d_1} + \frac{Mg(L/2 - d_2)}{d_1 - d_2} \\ &= \frac{Mg(L/2 - d_1)}{d_2 - d_1} - \frac{Mg(L/2 - d_2)}{d_2 - d_1} \\ &= \frac{Mg(L/2 - d_1 - L/2 + d_2)}{d_2 - d_1} = \frac{Mg(d_2 - d_1)}{d_2 - d_1} = Mg. \end{aligned}$$

(d) We see that

$$F_{1y} = \frac{Mg(L/2 - d_2)}{(d_1 - d_2)}$$

must be negative because the numerator is positive while the denominator is negative; so rod 1 is actually pushing down on the beam, and the rod is under compression.

The stress in rod 1 has magnitude

$$\frac{|F_{1y}|}{A_1} = \frac{Mg(L/2 - d_2)}{A_1(d_2 - d_1)},$$

and the strain in it is

$$\frac{\Delta L}{L} = \frac{1}{E_1}\left(\frac{|F_{1y}|}{A_1}\right) = \frac{Mg(L/2 - d_2)}{E_1 A_1(d_2 - d_1)}.$$

Similarly,

$$F_{2y} = \frac{Mg(L/2 - d_1)}{d_2 - d_1}$$

must be positive because both numerator and denominator are positive; so rod 2 is under tension (it is pulling up on the beam).

The stress in rod 2 has magnitude

$$\frac{F_{2y}}{A_2} = \frac{Mg(L/2 - d_1)}{A_2(d_2 - d_1)},$$

and the strain in it is

$$\frac{\Delta L}{L} = \frac{1}{E_2}\left(\frac{F_{2y}}{A_2}\right) = \frac{Mg(L/2 - d_1)}{E_2 A_2(d_2 - d_1)}.$$

(e) The forces are vertical, with y components

$$\begin{aligned} F_{1y} &= \frac{Mg(L/2 - d_2)}{d_1 - d_2} \\ &= \frac{(5.0 \text{ kg})(9.8 \text{ N/kg})(0.40 \text{ m} - 0.30 \text{ m})}{0.10 \text{ m} - 0.30 \text{ m}} \\ &= -24.5 \text{ N} \end{aligned}$$

and

$$\begin{aligned} F_{2y} &= \frac{Mg(L/2 - d_1)}{d_2 - d_1} \\ &= \frac{(5.0 \text{ kg})(9.8 \text{ N/kg})(0.40 \text{ m} - 0.10 \text{ m})}{0.30 \text{ m} - 0.10 \text{ m}} \\ &= +73.5 \text{ N}. \end{aligned}$$

Thus $\mathbf{F}_1 = -(24.5 \text{ N})\mathbf{j}$ and $\mathbf{F}_2 = +(73.5 \text{ N})\mathbf{j}$, using the usual cartesian coordinate system in which the $+y$ direction is vertically upward.

The magnitudes of the stress and strain in the two rods, and their elongations (+ if under tension, − if under compression) are:

Rod 1 (steel): $E_1 = 200 \times 10^9 \text{ N/m}^2$
Stress $= |F_{1y}|/A_1 = (24.5 \text{ N})/(4.00 \times 10^{-6} \text{ m}^2)$
$= 6.13 \times 10^6 \text{ N/m}^2$
Strain $= (\text{stress})/E_1 = (6.12 \times 10^6 \text{ N/m}^2) \div (200 \times 10^9 \text{ N/m}^2) = 3.1 \times 10^{-5}$
Elongation $= -(\text{strain})(\text{length}) = -(3.1 \times 10^5) \times (50.0 \text{ cm}) = -1.6 \times 10^{-3} \text{ cm}$

Rod 2 (aluminum): $E_1 = 70 \times 10^9 \text{ N/m}^2$
Stress $= F_{2y}/A_2 = (73.5 \text{ N})/(2.00 \times 10^{-6} \text{ m}^2)$
$= 36.7 \times 10^6 \text{ N/m}^2$
Strain $= (\text{stress})/E_2 = (36.7 \times 10^6 \text{ N/m}^2) \div (70 \times 10^9 \text{ N/m}^2) = 5.2 \times 10^{-4}$
Elongation $= (\text{strain})(\text{length}) = (5.2 \times 10^4)(50.0 \text{ cm})$
$= 2.6 \times 10^{-2} \text{ cm}$

Chapter Fourteen
Gravitation

QUESTIONS

14. *Organizing question:* Figure 14-46 shows four particles fixed in place in a plane. (a) Set up an equation, complete with known data, to find the x component of the net gravitational force on the 1 kg particle at the origin due to the other three particles. (b) Similarly set up an equation to find the corresponding y component.

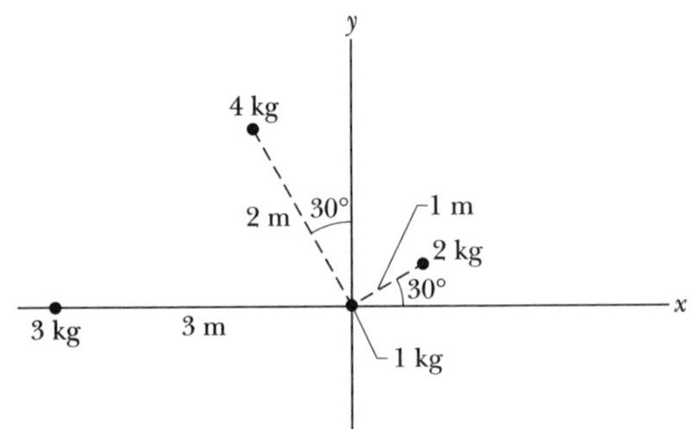

FIGURE 14-46 Question 14.

15. Figure 14-47 shows three uniform spherical planets that are identical in size and mass. The periods of rotation T for the planets are given, and six lettered points are indicated—three points are on the equators of the planets and three points are on the north poles. Rank the points according to the value of the free-fall acceleration g at them, greatest first.

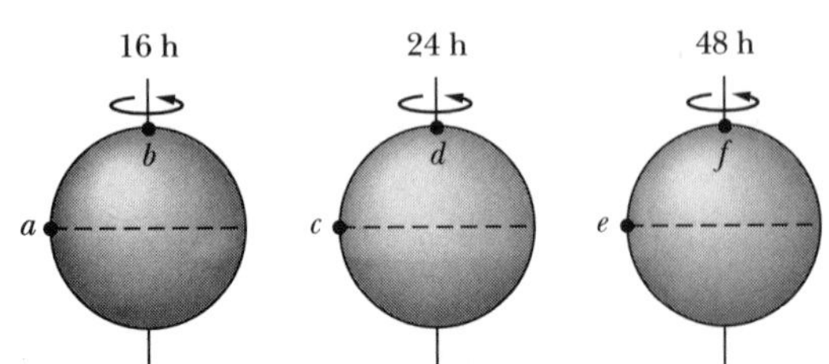

FIGURE 14-47 Question 15.

16. On burning out, a (nonrotating) star collapses onto itself from an initial radius R_i. Which curve in Fig. 14-48 best gives the gravitational acceleration a_g on the surface of the star as a function of the radius of the star during the collapse?

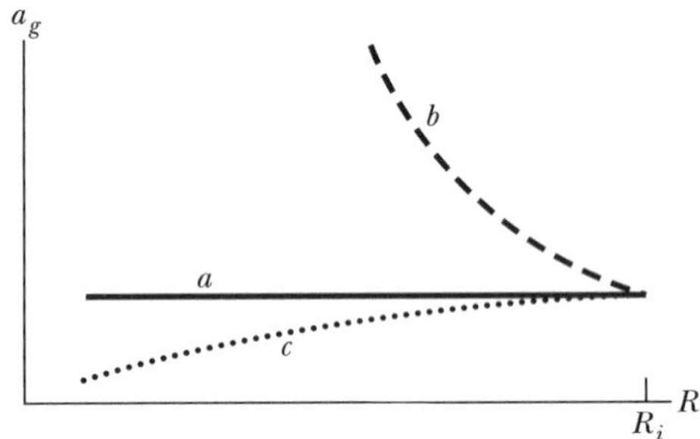

FIGURE 14-48 Question 16.

17. Figure 14-49 gives the gravitational acceleration a_g for four planets as a function of the radial distance r from the center of the planet, starting at the surface of the planet (at radius R_1, R_2, R_3, or R_4). Plots 1 and 2 coincide for $r \geq R_2$; plots 3 and 4 coincide for $r \geq R_4$. Rank the four planets according to (a) their mass and (b) their density, greatest first.

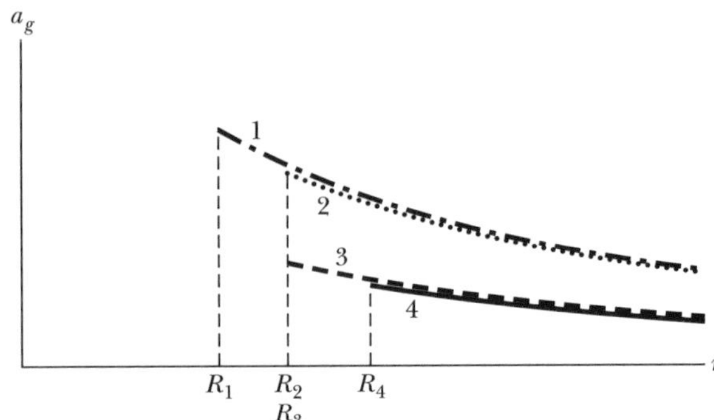

FIGURE 14-49 Question 17.

18. In Fig. 14-50*a*, a stationary spacecraft of mass M is passed first by asteroid A of mass m, then by asteroid B of the same mass

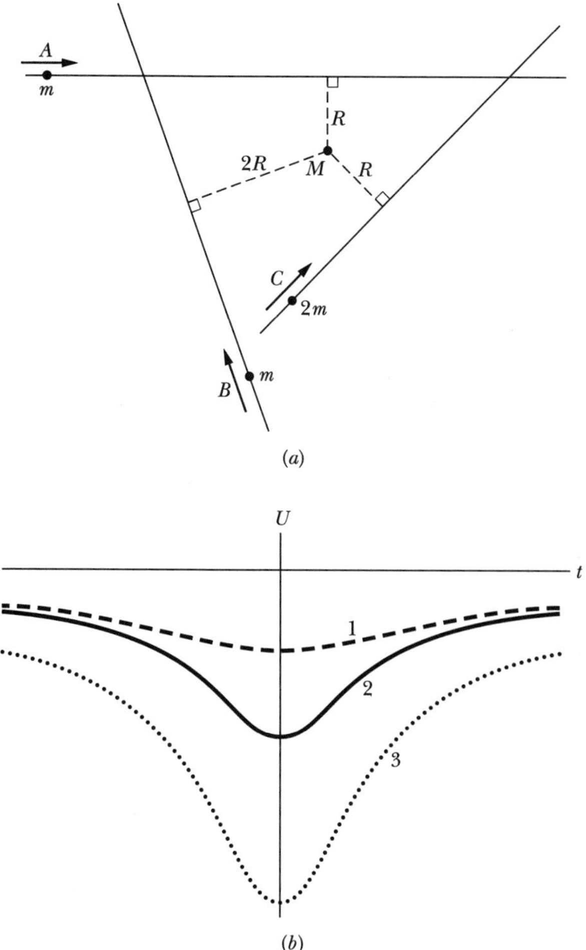

FIGURE 14-50 Question 18.

m, and then by asteroid C of mass $2m$. The asteroids move along the indicated straight paths at the same speed; the perpendicular distances between the spacecraft and the paths are given in terms of R. Figure 14-50b gives the gravitational potential energy $U(t)$ of the spacecraft–asteroid system during the passage of each asteroid. (a) Where is an asteroid along its path at time $t = 0$? (b) Which asteroid corresponds to which plot of $U(t)$?

19. As your spacecraft travels along an x axis through an asteroid belt, the gravitational potential energy $U(x)$ of the spacecraft–asteroid system is given by the curve in Fig. 14-51. From that

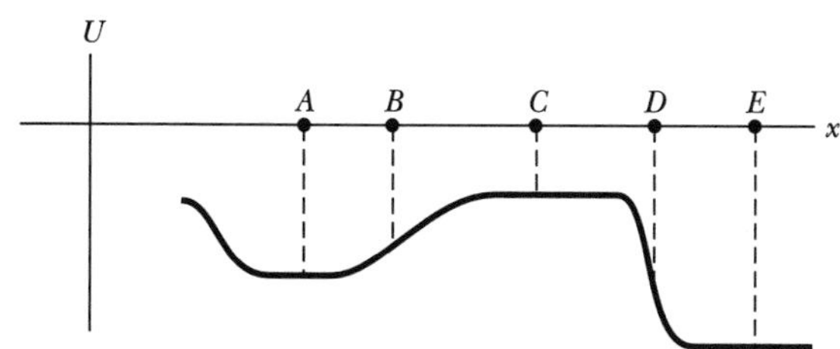

FIGURE 14-51 Question 19.

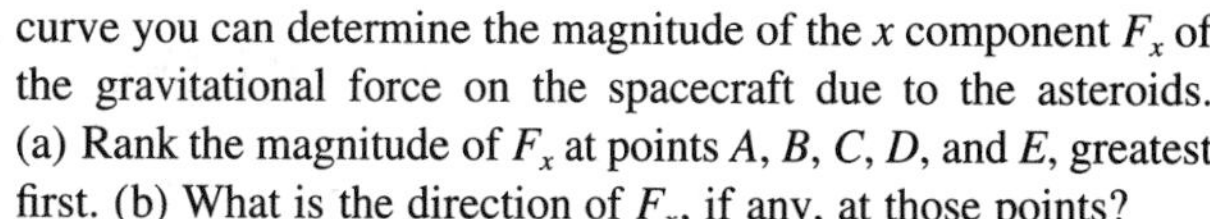
curve you can determine the magnitude of the x component F_x of the gravitational force on the spacecraft due to the asteroids. (a) Rank the magnitude of F_x at points A, B, C, D, and E, greatest first. (b) What is the direction of F_x, if any, at those points?

20. From an inertial frame in space, we watch two identical uniform spheres fall toward one another owing to their mutual gravitational attraction. Approximate their initial speed as zero and take the initial gravitational potential energy of the two-sphere system as U_i. When the separation between the two spheres is half the initial separation, what is the kinetic energy of each sphere?

21. In Fig. 14-14, consider three points along the orbit: point 1 at perihelion, point 2 at aphelion, and point 3 midway between perihelion and aphelion. Rank the three points according to (a) the angular momentum of the planet there and (b) the speed of the planet there, greatest first.

22. Figure 14-52 shows the orbits of three planets about identical stars; the orbits have identical major axes $2a$. Rank the three orbits according to their (a) eccentricities, (b) perihelion distances, (c) aphelion distances, and (d) periods of revolution, greatest first. (*Hint:* Roughly locate the star for each orbit.)

23. Rank the orbits in Question 22 and Fig. 14-52 according to (a) the total energy associated with the orbit of each planet, (b) the orbital speed of the planet at perihelion, and (c) the orbital speed of the planet at aphelion, greatest first.

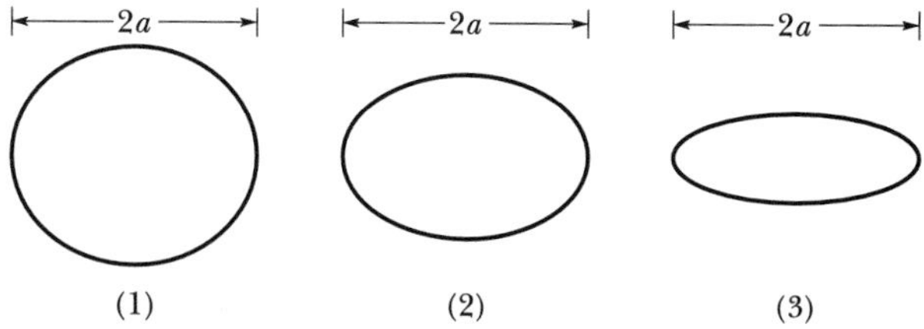

FIGURE 14-52 Questions 22 and 23.

24. Rank the three points in Question 21 according to (a) the total energy of the planet (associated with the planet's orbiting), (b) the kinetic energy of the planet, and (c) the potential energy of the planet (actually, the planet–star system), greatest first.

25. Figure 14-53 gives the masses and separations for three pairs of stars that each form a binary star system. (a) Locate the point about which the stars of each pair orbit. (b) Rank the pairs according to the magnitude of the centripetal acceleration of the stars, greatest first.

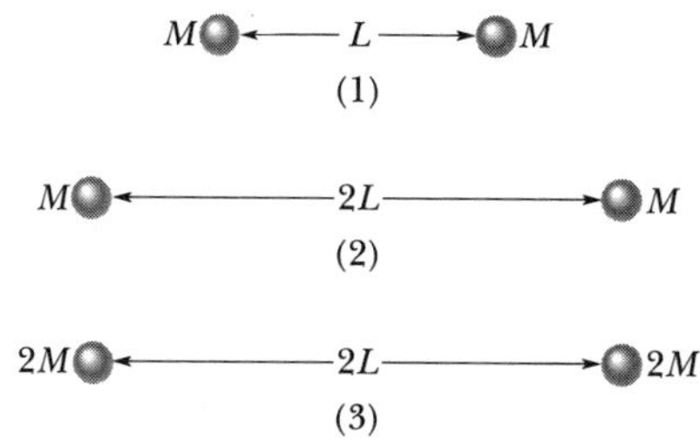

FIGURE 14-53 Question 25.

EXERCISES & PROBLEMS

90. A projectile is launched from the surface of a planet with mass M and radius R; the launch speed is $(GM/R)^{1/2}$. Use the principle of conservation of energy to determine the maximum distance from the center of the planet achieved by the projectile. Express your result in terms of R.

91. A satellite circles a planet of unknown mass in a circular orbit of radius 2.0×10^7 m. The magnitude of the gravitational force exerted on the satellite by the planet is 80 N. (a) What is the kinetic energy of the satellite in this orbit? (b) What would be the magnitude of the gravitational force exerted on the satellite by the planet if the radius of the orbit were increased to 3.0×10^7 m?

92. Three 5.0 kg masses are located in the xy plane as shown in Fig. 14-54. What is the magnitude of the net gravitational force on the mass at the origin due to the other two masses?

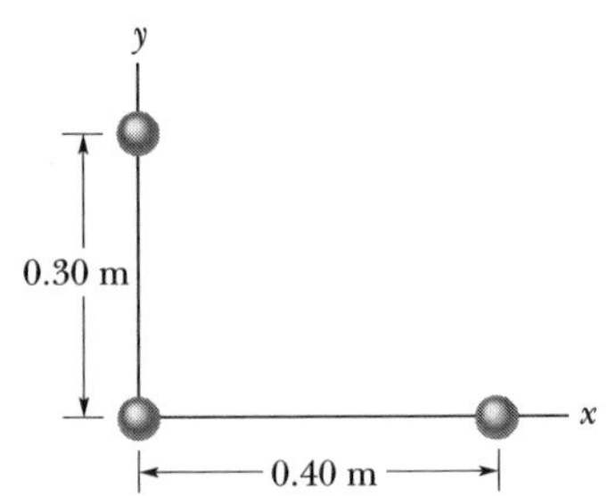

FIGURE 14-54 Question 92.

93. Four identical 1.5 kg particles are placed at the corners of a square with sides equal to 20 cm. What is the magnitude of the net gravitational force exerted on any one of the particles due to the other three?

94. One model for a certain planet has a core of radius R and mass M surrounded by an outer shell of inner radius R, outer radius $2R$, and mass $4M$. If $M = 4.1 \times 10^{24}$ kg and $R = 6.0 \times 10^6$ m, what is the gravitational acceleration of a particle at points (a) R and (b) $3R$ from the center of the planet?

95. Two 20 kg spheres are fixed in place on a y axis, one at $y = 0.40$ m and the other at $y = -0.40$ m. A 10 kg ball is then released from rest at a point on the x axis that is at a great distance (effectively infinite) from the spheres. If the only forces acting on the ball are the gravitational forces due to the spheres, then when the ball reaches the (x, y) point (0.30 m, 0), what are (a) its kinetic energy and (b) the magnitude and direction of the net force on it from the spheres?

96. Planet Roton, with a mass of 7.0×10^{24} kg and a radius of 1600 km, gravitationally attracts a meteorite that is initially at rest relative to the planet, at a great enough distance to take as infinite. The meteorite falls toward the planet. Assuming the planet is airless, find the speed of the meteorite when it reaches the planet's surface.

97. A 150.0 kg rocket moving radially outward from Earth has a speed of 3.70 km/s when its engine shuts off 200 km above Earth's surface. (a) Assuming negligible air drag, find the rocket's kinetic energy when the rocket is 1000 km above the surface. (b) What maximum height above the surface is reached by the rocket?

98. A planet requires 300 (Earth) days to complete its circular orbit about its sun, which has a mass of 6.0×10^{30} kg. What are (a) its orbital radius and (b) its orbital speed?

99. A 20 kg satellite has a circular orbit with a radius of 8.0×10^6 m and a period of 2.4 h around a planet of unknown mass. If the magnitude of the gravitational acceleration on the surface of the planet is 8.0 m/s^2, what is the radius of the planet?

100. A 50 kg satellite circles planet Cruton every 6.0 h. The magnitude of the gravitational force exerted on the satellite by Cruton is 80 N. (a) What is the radius of the orbit? (b) What is the kinetic energy of the satellite? (c) What is the mass of planet Cruton?

Tutorial Problems

101. Let's look at the gravitational forces in a system of four particles arranged in a square with sides of length D. Assume they do not interact with any other objects. Let the masses of the particles be $m_1 = M$, $m_2 = 2.00M$, $m_3 = 3.00M$, and $m_4 = 4.00M$. Use $\mathbf{F}_{41}$ to denote the force on particle 4 due to particle 1; $\mathbf{F}_{42}$ and $\mathbf{F}_{43}$ are similarly defined. Let $\mathbf{F}_4$ denote the total force on particle 4.

(a) Show the forces $\mathbf{F}_{41}$, $\mathbf{F}_{42}$, and $\mathbf{F}_{43}$ on a sketch. (b) Express each of the three gravitational forces on particle 4 in terms of the given quantities (M, D, etc.) using the form $\mathbf{F} = F\hat{F}$, where F is a magnitude and $\hat{F}$ is a unit vector in the direction of $\mathbf{F}$. Each $\hat{F}$ should be expressed in unit vector $(\hat{\imath}, \hat{\jmath})$ notation, using the directions defined by the coordinate axes. (c) Name the principle of physics that can be used to determine the total force $\mathbf{F}_4$ acting on particle 4 in terms of the forces produced by the other particles. (d) Express $\mathbf{F}_4$ in $\hat{\imath}$ and $\hat{\jmath}$ notation. (e) Express $\mathbf{F}_4$ as a magnitude and direction, giving the direction in terms of an angle. (f) Show the vector $\mathbf{F}_4$ on the sketch in part (a).

(g) The gravitational field $\mathbf{g}$ at a point in space is defined as the ratio $\mathbf{F}_{grav}/m$ where $\mathbf{F}_{grav}$ is the total gravitational force on a mass m at that point. Determine the contribution of each particle to the gravitational field $\mathbf{g}$ at the center of the square, expressing each as a magnitude and a unit vector. Add them to determine the total value for $\mathbf{g}$. Show this field on the diagram of part (a).

(h) Consider a point on the positive y axis very far from this system of particles. What is the approximate gravitational field $\mathbf{g}$ as a function of y? Use unit-vector notation. (i) Suppose the four particles in the square are initially at rest and are allowed to move until they have merged into a single object. Assume that no external forces act on the system. Where will that object be located?

Answers

(a) See Fig. 14-55.

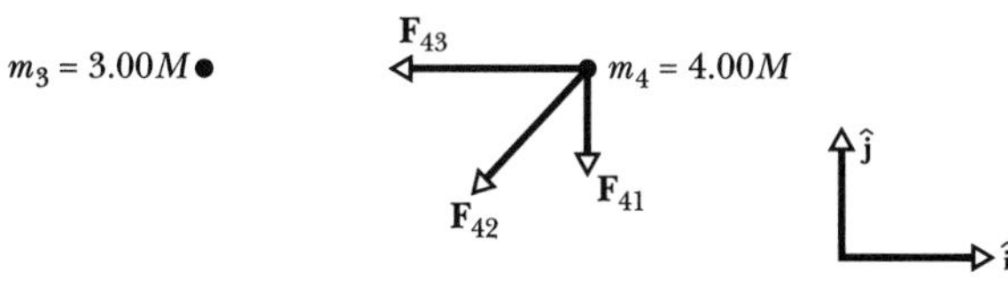

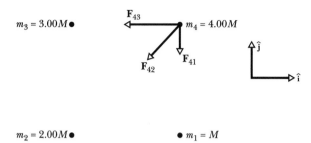

FIGURE 14-55 Question 101.

(b) $\mathbf{F}_{41}$ is directed from m_4 toward m_1, that is, in the direction $-\hat{j}$, so $\hat{F} = -\hat{j}$. The magnitude is

$$F = \frac{Gm_1m_4}{D^2} = \frac{GM(4.00M)}{D^2} = \frac{4.00GM^2}{D^2}.$$

So $$\mathbf{F}_{41} = \frac{4.00GM^2}{D^2}(-\hat{j}).$$

$\mathbf{F}_{42}$ is directed from m_4 toward m_2, that is, at an angle of $-225°$ from the direction of $+x$; so the vector is in the direction

$$\hat{F} = \cos(225°)\,\hat{i} + \sin(225°)\,\hat{j} = -0.707\hat{i} - 0.707\hat{j}.$$

Its magnitude is

$$F = \frac{Gm_2m_4}{(D^2 + D^2)} = \frac{4.00GM^2}{D^2}.$$

Thus $$\mathbf{F}_{42} = F\hat{F} = \frac{4.00GM^2}{D^2}(-0.707\,\hat{i} - 0.707\,\hat{j}).$$

$\mathbf{F}_{43}$ is directed from m_4 toward m_3, that is, in the $-\hat{i}$ direction; so $\hat{F} = -\hat{i}$. The magnitude is

$$F = \frac{Gm_3m_4}{D^2} = \frac{12.0GM^2}{D^2}.$$

So $$\mathbf{F}_{43} = \frac{12.0GM^2}{D^2}(-\hat{i}).$$

(c) To determine the total force on particle 4 we use the principle of superposition: the total force on the particle is the vector sum of all the individual forces acting on it due to the other particles.

(d) In the notation using unit vectors along the axes,

$$\begin{aligned}\mathbf{F}_4 &= \mathbf{F}_{41} + \mathbf{F}_{42} + \mathbf{F}_{43} \\ &= -\frac{4.00GM^2}{D^2}\hat{j} + \frac{2.83GM^2}{D^2}(-\hat{i} - \hat{j}) - \frac{12.0GM^2}{D^2}\hat{i} \\ &= \frac{GM^2}{D^2}(-14.8\hat{i} - 6.83\hat{j}).\end{aligned}$$

(e) The magnitude is

$$F_4 = \frac{GM^2}{D^2}\sqrt{(-14.828)^2 + (-6.828)^2} = \frac{16.3GM^2}{D^2}.$$

The direction has angle (in the third quadrant)

$$\theta = \tan^{-1}[-6.828/(-14.828)] = \tan^{-1}(0.4605) = 24.7°$$

So we can say that the total force on the fourth particle is $16.3(GM^2/D^2)$ in a direction of 24.7° below the $-x$ direction.

(g) The center of the square is a distance $D/\sqrt{2}$ from each of the particles, since that distance is half the length $\sqrt{2}D$ of a diagonal of a square with sides of length D. Thus the factor $1/r^2$ in the gravitational field equals $2/D^2$ in each case. The contribution of m_i to the gravitational field is then $(2Gm_i/D^2)$ times the unit vector that points from the center of the square toward m_i. The contributions are then

$$\begin{aligned} m_1 &= M: && (2.00GM/D^2))(+\hat{i} - \hat{j})/\sqrt{2} \\ m_2 &= 2M: && (4.00GM/D^2)(-\hat{i} - \hat{j})/\sqrt{2} \\ m_3 &= 3M: && (6.00GM/D^2)(-\hat{i} + \hat{j})/\sqrt{2} \\ m_4 &= 4M: && (8.00GM/D^2)(+\hat{i} + \hat{j})/\sqrt{2} \end{aligned}$$

These add up to $\mathbf{g} = (8.00GM/\sqrt{2}\,D^2)\hat{j}$.

(h) Far from the system, the system resembles a single particle of mass

$$m_1 + m_2 + m_3 + m_4 = 10M.$$

So the gravitational field *is*

$$\mathbf{g}(0, y, 0) \approx -\frac{10GM}{y^2}\hat{j}.$$

The negative sign here shows that the field points toward the system of particles.

(i) The object will have to be located at the center of mass of the original particles. That point will be located above the center of the square, about $\frac{2}{5}$ of the way to the top edge of the square.

102. Consider the situation in which Earth, the Moon, and the Sun are positioned in a straight line, as they would be at the time of a solar eclipse. Neglect the effects of all other astronomical objects. (a) Make a rough sketch, not necessarily to scale, of the three astronomical objects. (b) Determine (to two significant figures) the magnitudes and directions of the approximate gravitational forces on Earth due to the other two objects; do the same for the total gravitational force on Earth.

(c) Carry out the same calculation for the Moon. Comment on the direction of the net gravitational force. (d) Carry out the same calculation for the Sun. (e) Suppose, instead, that the Moon was on the opposite side of Earth from the Sun, as it would be during a total lunar eclipse. Which, if any, of the different gravitational forces calculated in parts (b), (c), and (d) would have changed significantly, and how? (f) It has been claimed that the Moon's orbit in space as it revolves around Earth, which is itself revolving around the Sun, is always concave inward with respect to the Sun. How do your numerical calculations support this statement?

Answers

(a) See Fig. 14-56.

FIGURE 14-56 Question 102.

(b) The gravitational force on Earth due to the Moon is directed toward the Moon and has the magnitude

$$\frac{GM_E M_M}{d_{EM}^2} = \frac{(6.67 \times 10^{-11}\ \text{N}\cdot\text{m}^2/\text{kg}^2)(5.98 \times 10^{24}\ \text{kg})(7.36 \times 10^{22}\ \text{kg})}{(3.84 \times 10^{8}\ \text{m})^2} = 2.0 \times 10^{20}\ \text{N}.$$

The gravitational force on Earth due to the Sun is directed toward the Sun and has the magnitude

$$\frac{GM_E M_S}{d_{ES}^2} = \frac{(6.67 \times 10^{-11}\ \text{N}\cdot\text{m}^2/\text{kg}^2)(5.98 \times 10^{24}\ \text{kg})(1.99 \times 10^{30}\ \text{kg})}{(1.496 \times 10^{11}\ \text{m})^2} = 3.5 \times 10^{22}\ \text{N}.$$

The magnitude of the force due to the Sun is more than 100 times that due to the Moon, so it is essentially the magnitude of the net gravitational force as well (to two significant figures).

(c) The gravitational force on the Moon due to Earth is just the negative of the gravitational force on Earth due to the Moon, so it is directed to the left in Fig. 14-56 and has the magnitude 2.0×10^{20} N calculated in part (a).

Since the Sun–Moon distance is very nearly the same as the Sun–Earth distance, the gravitational force on the Moon due to the Sun has the magnitude

$$\frac{GM_S M_M}{d_{EM}^2} = \frac{(6.67 \times 10^{-11}\ \text{N}\cdot\text{m}^2/\text{kg}^2)(1.99 \times 10^{30}\ \text{kg})(7.36 \times 10^{22}\ \text{kg})}{(1.496 \times 10^{11}\ \text{m})^2} = 4.4 \times 10^{20}\ \text{N}.$$

This force is directed to the right in Fig. 14-56.

Thus, the net gravitational force on the Moon is directed toward the Sun and has the magnitude $4.4 \times 10^{20}\ \text{N} - 2.0 \times 10^{20}\ \text{N} = 2.4 \times 10^{20}$ N. Many people expect the gravitational force on the Moon to be due mainly to Earth, since it is commonly understood that the Moon revolves around Earth. From another point of view, however, the Moon revolves around the Sun in an orbit perturbed by the gravitational force of the nearby Earth.

(d) The gravitational forces on the Sun due to Earth and the Moon are both toward the left in Fig. 14-56, and their magnitudes are 3.5×10^{22} N for the force due to Earth and 4.4×10^{20} N for the force due to the Moon. The net force, to two significant figures, is just the gravitational force due to Earth, or magnitude 3.5×10^{22} N.

(e) The magnitudes of the gravitational forces between the objects, taken two at a time, would not change significantly, because the distances have not changed significantly. However, the directions of the gravitational forces on Earth due to the Moon, and on the Moon due to Earth, would have changed. That means that the net force on the Moon, which would still be to the right in Fig. 14-56 (toward the Sun), would have the larger magnitude

$$4.4 \times 10^{20}\ \text{N} + 2.0 \times 10^{20}\ \text{N} = 6.4 \times 10^{20}\ \text{N}.$$

(f) For the Moon's orbit to be always concave inward, the net force acting on the Moon must always have a component pointing toward (rather than away from) the Sun. The case in which the net force is most likely to point away from the Sun is when the Moon is located between Earth and the Sun, so that Earth's gravitational force on the Moon points in that direction. However, the results of part (c) indicate that even in that case, the net force points toward the Sun.

103. Near Earth, the gravitational field (see Problem 101) is mainly due to Earth's mass, with the Sun, Moon, and other astronomical objects making only minor contributions. Earth is approximately spherically symmetric, but its density is not uniform. In this problem we'll use geophysical data to model Earth as a sphere of varying density and determine $g(r)$ both inside Earth ($r < R_E$ = radius of Earth) and outside ($r > R_E$). A key fact to remember is that outside a spherically symmetric object, the gravitational field of the object can be calculated by assuming that the mass is concentrated at the center, while inside the object, the gravitational field is zero. In other words, to determine the gravitational field at any point inside Earth, you need to take into account only that part of Earth's mass that is closer to the center than that point.

From the behavior of seismic waves and other information, geophysicists have shown that Earth does not have a uniform density but can be modeled as consisting of five spherically symmetric regions, each of which is approximately uniform in density. This model is similar to, but a slight simplification of, the density variation shown in Figure 14-6 in the textbook. The five regions of the model are

the inner core ($r < 1221.5$ km), density = 12.9 g/cm^3;

the outer core (1221.5 km $< r <$ 3480 km), density = 10.9 g/cm^3;

the lower mantle (3480 km $< r <$ 5701 km), density = 4.9 g/cm^3;

the upper mantle (5701 km $< r <$ 6346.6 km), density = 3.6 g/cm^3;

the crust/ocean region (6346.6 km $< r <$ 6371 km), density = 2.4 g/cm^3.

(a) First, use Earth's mass (5.98×10^{24} kg) and its mean radius (6371 km) to determine the average density. Which of the five regions are denser than average, and which are less dense than average? (b) Using the information in the preceding list, determine the magnitude $g(r)$ of Earth's gravitational field in the inner core as a function of r. Express your answer as a numerical function of r. What is the functional dependence of $g(r)$ on r (linear, quadratic, inverse square, etc.)?

(c) Determine the value of g at each of the boundaries between the five regions and at Earth's surface. (d) Determine the magnitude $g(r)$ for the region $r > R_E = 6371$ km. (e) Make a plot of $g(r)$ for $0 \le r \le 2R_E = 2 \times 6371$ km, choosing an appropriate scale.

(f) Show that at any value of r inside Earth (or another spherically symmetric object), $g(r)$ will increase with r if and only if the density $\rho(r)$ there exceeds $\frac{2}{3}$ the average density inside the radius r. Check the data in this problem to show that the maximum of $g(r)$ inside Earth occurs at one of the boundaries between the regions, and determine which one.

Answers

(a) Earth's mass is $M = 5.98 \times 10^{24}$ kg and its volume is

$$V = \tfrac{4}{3}\pi R_E^3 = (\tfrac{4}{3})\pi(6.371 \times 10^6 \text{ m})^3 = 1.083 \times 10^{21} \text{ m}^3,$$

so its average density is

$$\rho_{av} = \frac{M}{V} = \frac{5.98 \times 10^{24} \text{ kg}}{1.083 \times 10^{21} \text{ m}^3} = 5500 \text{ kg/m}^3 = 5.5 \text{ g/cm}^3.$$

Only the inner and outer core have a density greater than the average density.

(b) In the inner core, the density has a constant value $\rho = 12.9$ g/cm^3. At a distance r from Earth's center, the mass inside r is

$$\rho V = \rho(\tfrac{4}{3}\pi r^3),$$

so the gravitational field there is

$$g(r) = \frac{GM}{r^2} = \frac{G(4/3\pi\rho r^3)}{r^2} = \tfrac{4}{3}\pi G\rho r$$
$$= (\tfrac{4}{3}\pi)(6.67 \times 10^{-11} \text{ N}\cdot\text{m}^2/\text{kg}^2)(12\,900 \text{ kg/m}^3)r$$
$$= (3.61 \times 10^{-6} \text{ N/kg}\cdot\text{m})r$$

This is a linear function of r; in other words, the gravitational field starts at 0 at $r = 0$ and increases linearly up to the boundary of the inner core with the outer core.

(c) We need to determine the total mass M inside each boundary and then use that mass with the equation $g(r) = GM/r^2$. Apart from the inner core, which is a solid sphere, the regions are spherical shells. The volume of a spherical shell that lies between $r = r_1$ and $r = r_2$ is the difference in volume between spheres of radii r_1 and r_2, namely, $\frac{4}{3}\pi(r_1^3 - r_2^3)$. So the mass of a region is this quantity multiplied by the density of the region. Results are shown in Table 14-4.

(d) For the region $r > R_E$, the region outside Earth, the gravitational field has the form $g(r) = GM/r^2$, where M is the total mass of Earth. At Earth's surface, $r = R_E$ and $g(R_E) = 9.8$ N/kg. We can thus write, for the region $r > R_E$,

$$g(r) = (9.8 \text{ N/kg})\left(\frac{R_E}{r}\right)^2$$

(f) To see this, let $M(r)$ be the mass up to radius r, $\rho(r)$ be the density at radius r, and $\rho_{av}(r)$ be the average density up to radius r. From

$$\rho = \frac{M}{V} = \frac{M}{\frac{4}{3}\pi r^3},$$

we have

$$\frac{dM(r)}{dr} = 4\pi r^2 \rho(r)$$

and

$$M(r) = \tfrac{4}{3}\pi r^3 \rho_{av}(r).$$

The gravitational acceleration at radius r is $g(r) = GM(r)/r^2$; its derivative with respect to r is

$$\frac{dg}{dr} = -\frac{2GM(r)}{r^3} + \frac{G}{r^2}\frac{dM(r)}{dr} = -\frac{2GM(r)}{r^3} + 4\pi G\rho(r)$$
$$= 4\pi G\left(\rho(r) - \frac{2}{3}\rho_{av}(r)\right).$$

This is positive only if $\rho(r) > \frac{2}{3}\rho_{av}(r)$; if $\rho(r) < \frac{2}{3}\rho_{av}(r)$, g will actually decrease with increasing r. The average density of Earth is 5.51 g/cm^3. From the data in this problem we see that the density near the surface is considerably less than $\frac{2}{3}$ of this, which would be 3.67 g/cm^3. The density of the lower mantle,

TABLE 14-4 PROBLEM 103

	INNER CORE	OUTER CORE	LOWER MANTLE	UPPER MANTLE	CRUST/OCEAN
r_1 (km)	0	1221.5	3480	5701	6346.6
r_2 (km)	1221.5	3480	5701	6346.6	6371
Volume (m^3)	7.63×10^{18}	1.69×10^{20}	6.00×10^{20}	2.95×10^{20}	1.24×10^{19}
Density (kg/m^3)	12 900	10 900	4900	3600	2400
Mass (kg)	9.85×10^{22}	1.84×10^{24}	2.94×10^{24}	1.06×10^{24}	2.98×10^{22}
M (kg)	9.85×10^{22}	1.94×10^{24}	4.88×10^{24}	5.94×10^{24}	5.97×10^{24}
$g(r_2)$ (N/kg)	4.40	10.68	10.01	9.83	9.81

4.9 g/cm³, is less than $\frac{2}{3}$ the density of either the inner or the outer cores, so g is still increasing with r even in the lower mantle.

Graphing Calculators

SAMPLE PROBLEM 14-11

Solving Multiple-Force Problems Quickly. In Sample Problem 14-1 we find the net gravitational force $\mathbf{F}_1$ on m_1 due to the other four particles by taking a vector sum of the four gravitational forces they produce. The solution given there involves arguments of symmetry to simplify the calculation.

Instead, we could resolve the four forces into x and y components, sum the x components and then the y components, and then finally find the magnitude of the net force (using the Pythagorean theorem with the net x component and the net y component) and the angle of the net force (using an inverse tangent of the net y component divided by the net x component). This procedure could be called the long way, because it is painfully long, with many chances for error. When symmetry cannot help us, the long way might seem to be our only resort.

However, there is a much easier way to the answer: use the vector capabilities of a graphing calculator. There is even a shortcut in that use.

SOLUTION: Recall that on a TI-85/86, a vector can be entered in the form

[magnitude ∠ angle]

as examined in the Graphing Calculators section of Chapter 3 in this booklet. For example, from Sample Problem 14-1 we know that force $\mathbf{F}_{12}$ has a magnitude of $Gm_1m_2/(2a)^2$ and an angle of 270° − 30°. So, using the given data for the masses and the distance a and putting the calculator into degree mode, we could enter that force as

[6.67E−11*8*2/(2*.02)² ∠ 270 − 30]

on the calculator.

Similarly, we could enter force $\mathbf{F}_{13}$, with magnitude Gm_1m_3/a^2 and angle 90° + 30°, as

[6.67E−11*8*2/.02² ∠ 90 + 30]

and force $\mathbf{F}_{14}$, with magnitude $Gm_1m_4/(2a)^2$ and angle 90° − 30°, as

[6.67E−11*8*2/(2*.02)² ∠ 90 − 30]

and force $\mathbf{F}_{15}$, with magnitude Gm_1m_5/a^2 and angle 90° − 30°, as

[6.67E−11*8*2/.02² ∠ 90 − 30]

Entering all four vectors in this way to sum them involves a lot of keystrokes.

There is a shortcut. All the magnitudes of the vectors include the expression 6.67E−11*8*2/.02². So, we can pull that common expression out front of the sum and enter the sum as

6.67E−11*8*2/.02²([1/2² ∠ 270 − 30]
+ [1 ∠ 90 + 30] + [1/2² ∠ 90 − 30] + [1 ∠ 90 − 30])

in which the calculator will multiply the expression to the left of the parentheses with only the magnitude portion of each vector, not the angle portion. Of course, we could save a few keystrokes by mentally performing some of the easy mathematics (such as 270 − 30).

We can save a few more keystrokes and avoid looking in the book for the value of G by using the value stored in the calculator under the symbol Gc (for gravitational constant). To find it, press 2nd CONS, choose BLTIN (for built-in constants), press MORE, and choose Gc to put the value on the screen. Or, if you happen to remember that Gc is the symbol needed here, press it in with the ALPHA key. Then complete the line so that

Gc*8*2/.02²([1/2² ∠ 270 − 30] + [1 ∠ 90 + 30]
+ [1/2² ∠ 90 − 30] + [1 ∠ 90 − 30])

is our calculation.

If the calculator is in SphereV mode (check the mode menu), pressing ENTER gives us the sum as

[4.62E−6 ∠ 9.00E1]

which tells us that the net force $\mathbf{F}_1$ on particle 1 has

$$\text{magnitude} = 4.6 \times 10^{-6}\ \text{N},$$

and angle = 90° counterclockwise from positive x direction, (Answer)

just as we found in Sample Problem 14-1.

Chapter Fifteen
Fluids

QUESTIONS

12. *The Teapot Effect:* When water is poured slowly from a teapot spout, it can double back under the spout for a considerable distance before detaching and falling. (The water layer is held against the underside of the spout by atmospheric pressure.) In Fig. 15-50, within the water layer in the spout, point a is at the top of the layer and point b is at the bottom of the layer; within the water layer below the spout, point c is at the top of the layer and point d is at the bottom of the layer. Rank those four points according to the gauge pressure in the water there, most positive first, most negative last.

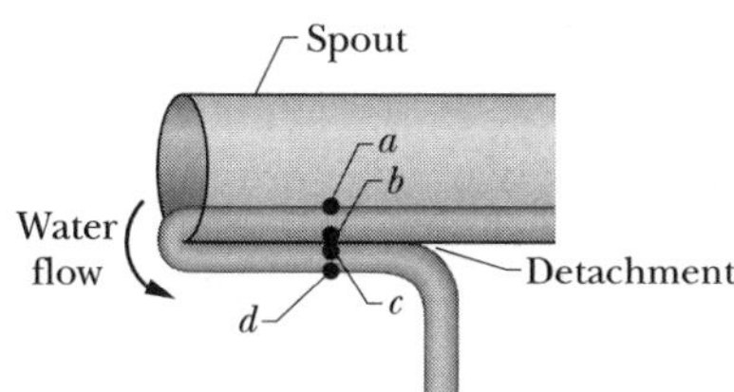

FIGURE 15-50 Question 12.

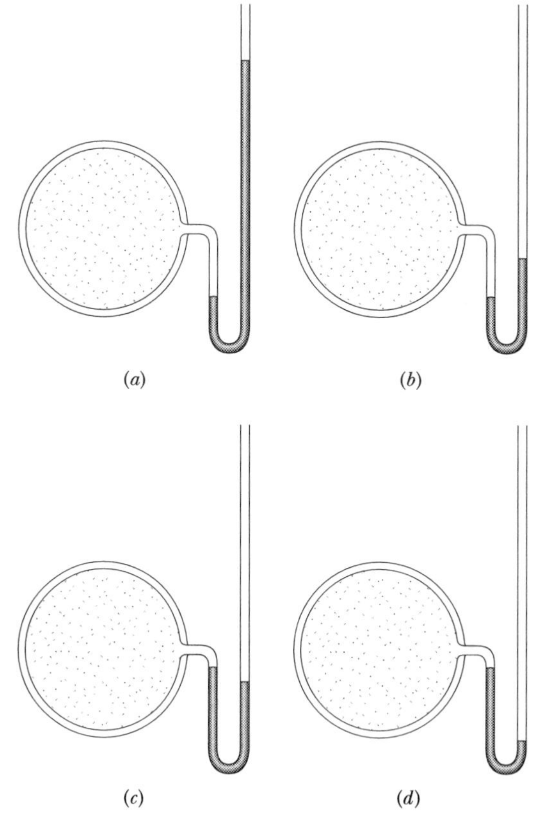

FIGURE 15-51 Question 13.

13. Figure 15-51 shows four situations in which an open-tube manometer, like that in Fig. 15-7, is attached to a tank of gas. Rank the situations according to the gauge pressure of the gas within the tank, most positive first, most negative last.

14. Three hydraulic levers such as that in Fig. 15-9 are used to lift identical loads (at the output side) by identical distances. The levers are identical on the input side but differ in the area of the output piston: lever 1 has a piston of area A, lever 2 has a piston of area $2A$, and lever 3 has a piston of area $3A$. For the lift, rank the levers according to (a) the required work at the input side, (b) the required magnitude of the force (assumed to be constant) at the input side, and (c) the displacement of the piston at the input side, greatest first.

15. Figure 15-52 shows three streamlines in the flow of a fluid. What are the directions of flow of the fluid elements at (a) point B and (b) point C? Rank the fluid elements at points A, B, and C according to (c) the speed of the fluid elements and (d) the pressure on the fluid elements, greatest first.

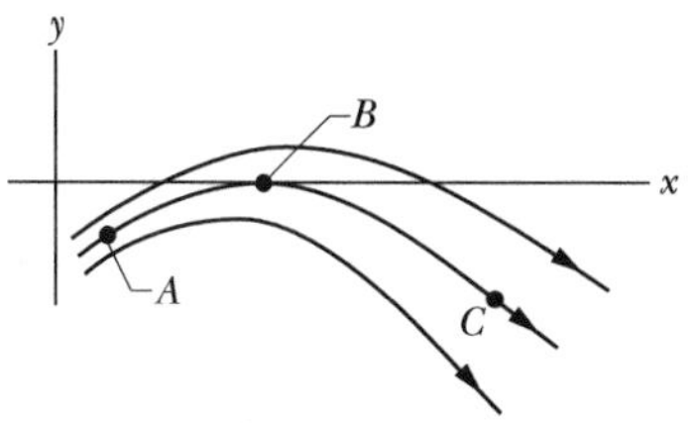

FIGURE 15-52 Question 15.

16. Figure 15-53 shows three straight pipes through which water flows. The figure gives the speed of the water for each pipe and also a cross-sectional area of each pipe. Rank the pipes according to the volume of water that passes through the cross-sectional area per minute, greatest first.

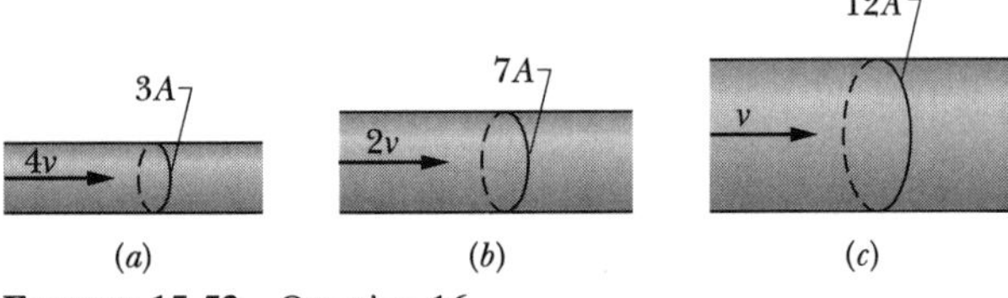

FIGURE 15-53 Question 16.

17. *Organizing question:* Figure 15-54 shows six situations in which milk (density 1030 kg/m^3) flows rightward through a circular pipe that changes radius or elevation. Changes in radius are from 2 cm to either 1 cm or 3 cm, as drawn. Changes in elevation are 0.50 either up or down, as drawn. In all the situations, the initial speed is 4 m/s and the initial pressure is 2×10^5 Pa. For each situation, set up an equation, complete with known data, to find the speed v_2 of the milk after the change in the pipe. Then, assuming we know that speed, set up an equation, complete with known data, to find the pressure p_2 of the milk after the change.

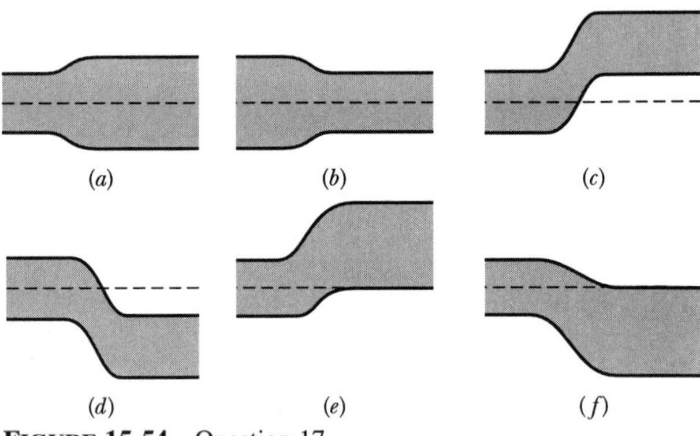

FIGURE 15-54 Question 17.

EXERCISES & PROBLEMS

81. A lead sinker of volume 0.40 cm^3 and density 11.4 g/cm^3 is used in fishing. The sinker is suspended from a vertical string whose other end is attached to the bottom of a spherical cork (of density 0.20 g/cm^3) that is floating on the surface of a lake. Neglecting the effects of the line, hook, and bait, determine what the radius of the cork must be if it is to float with half its volume submerged.

82. What gauge pressure must be produced by a machine for it to suck mud of density 1800 kg/m^3 up a tube by a height of 1.5 m?

83. A 7.00 kg sphere of radius 5.00 cm is at a depth of 1.20 km in seawater that has an average density of 1025 kg/m^3. What are (a) the gauge pressure, (b) the total pressure, and (c) the corresponding total force on the sphere's surface? What are (d) the buoyancy force on the sphere and (e) the magnitude and direction of the sphere's acceleration if it is free to move?

84. In an experiment, a rectangular block with height h is allowed to float in four separate liquids. In the first liquid, which is water, it floats fully submerged. In liquids A, B, and C, it floats with heights $h/2$, $2h/3$, and $h/4$ above the liquid surface, respectively. What are the *relative densities* (the densities relative to that of water) of (a) liquid A, (b) liquid B, and (c) liquid C?

85. A uniform block of length 5.0 cm, width 4.0 cm, and depth 2.0 cm floats in seawater of density 1025 kg/m^3. The block has a broad side facedown, 1.5 cm below the water surface. What is the mass of the block?

86. A cube with a surface area of 24 m^2 floats upright in water. If the density of water is 4.00 times the density of the cube, how far does the cube sink into the water?

87. A laminar stream of water necks down as it falls, as in Sample Problem 15-8 and Fig. 15-18. At one level in the stream, the radius is r_1 and the speed is v_1. At a lower level, the radius is $r_1/2$. (a) What is the speed at that lower level? The *kinetic energy density* of a fluid is measured with the SI unit of joule per cubic meter and is defined as $\frac{1}{2}\rho v^2$, where ρ is the density of the fluid and v is the speed of the fluid. (b) What is the increase in the kinetic energy density of the water as it falls from the higher level to the lower level?

88. Water flows through a horizontal pipe that widens at two points; initial radius = 0.200 m, intermediate radius = 0.400 m, and final radius = 0.600 m. If the initial speed of the water's flow is 9.00×10^{-2} m/s, what is the final speed of flow?

89. A liquid of density 900 kg/m^3 flows through a horizontal pipe that changes in cross-sectional area from 1.90×10^{-2} m^2 to 9.50×10^{-2} m^2. The pressure difference between the liquids on the two sides of the pipe change is 7.20×10^3 Pa. What are (a) the volume flow rate and (b) the mass flow rate through the pipe?

Tutorial Problem

90. A rectangular metal block of height 10.0 cm is suspended by a thin wire of negligible mass. The wire is attached to a spring balance so that the block's weight can be determined; its weight is 353 N in air and its weight is 294 N when it is fully submerged in water. (a) Determine the density, volume, and cross-sectional area of the block.

(b) Suppose the block is suspended in a large container of water, with the top of the block at depth d = 100 cm. Draw a free-body force diagram for the block and determine the magnitude and direction of each of the forces acting on it. (c) Now suppose that the block is raised slowly until its top is level with the water's surface. Determine how much work is done on the block by each of the forces you identified in part (b). Check that the sum of the works is correct. (d) Determine what changes, if any, occur in the gravitational potential energy of the block and the water while the block is raised. How are these changes reflected in the works calculated in part (c)?

Answers

(a) Let ρ be the average density of the block and $\rho' = 1000\ \text{kg/m}^3$ be the density of water. Let the block have height $h = 10.0$ cm, width w, and length ℓ, so its volume is $V = hw\ell$. In air, neglecting the density of air (which is only 1.3 kg/m³), the weight of the block is ρgV. In water its weight, as measured by the spring balance, is $(\rho - \rho')gV$.

The ratio of the weights is

$$\frac{\rho}{\rho - \rho'} = \frac{353}{294} \approx 1.20 = \frac{6}{5},$$

which leads to $\rho/\rho' = 6$. So

$$\rho = 6\rho' = 6(1000\ \text{kg/m}^3) = 6000\ \text{kg/m}^3.$$

The volume of the block must then be

$$V = \frac{\rho gV}{\rho g} = \frac{353\ \text{N}}{(6000\ \text{kg/m}^3)(9.8\ \text{N/kg})} = 6.00 \times 10^{-3}\ \text{m}^3.$$

Since $h = 0.100$ m, the cross-sectional area is

$$w\ell = \frac{V}{h} = \frac{6.00 \times 10^{-3}\ \text{m}^3}{0.10\ \text{m}} = 6.00 \times 10^{-2}\ \text{m}.$$

The exact values of w and ℓ cannot be determined, but they might be, for example, 20 cm and 30 cm.

(b) There are three forces acting on the block: (1) the downward gravitational force of magnitude $\rho gV = 353$ N; (2) the upward buoyant force $\mathbf{F}_b$ of magnitude $\rho' gV = 59$ N; and (3) the upward tension force $\mathbf{T}$ provided by the wire. Since the block is in static equilibrium, the three forces must sum vectorially to zero, so the tension force must have magnitude 353 N − 59 N = 294 N.

(c) The block is raised vertically a distance of 100 cm, either parallel or antiparallel to the directions of the three forces. The work can be computed by multiplying the magnitude of the force by the distance the block is raised, and then including a + or a − sign depending on whether the force and displacement are parallel or antiparallel. The work done by the gravitational force is (353 N)(−1.00 m) = −353 J; that by the buoyancy force is (59 N)(1.00 m) = 59 J; and that by the tension force of the wire is (294 N)(1.00 m) = 294 J. The sum of the works is −353 J + 59 J + 294 J = 0, which is correct, since there is no change in the block's kinetic energy (it is zero initially and finally).

(d) The block is raised 100 cm = 1.00 m; so its gravitational potential energy increases by

$$mg\,\Delta y = \rho gV\,\Delta y = (353\ \text{N})(1.00\ \text{m}) = 353\ \text{J}.$$

This is the negative of the work done by the gravitational force (see part (c)).

The change in the gravitational potential energy of the water can be determined by recognizing that as the block rises 1.00 m, an equal volume of water descends 1.00 m. So its gravitational potential energy decreases. The change in the gravitational potential energy of the water is

$$\rho' gV\,\Delta y = (59\ \text{N})(-1.00\ \text{m}) = -59\ \text{J}.$$

This is the negative of the work done by the buoyancy force.

Chapter Sixteen Oscillations

QUESTIONS

17. *Organizing question:* Figure 16-47 shows the SHM of a block–spring system in four situations. For each, set up an equation, complete with known data, to find the position $x(t)$ of the block.

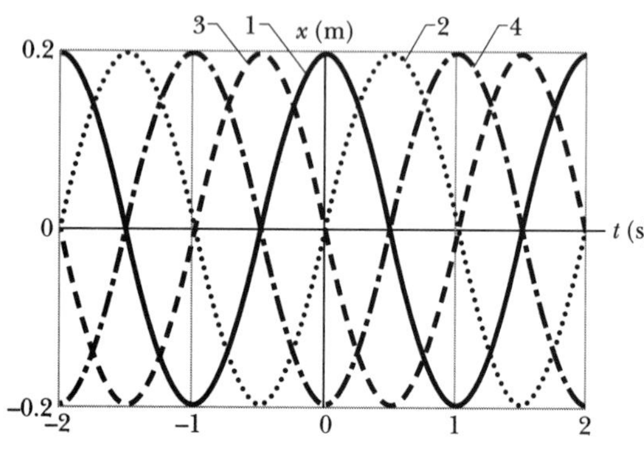

FIGURE 16-47 Question 17.

18. In Fig. 16-48, a spring–block system is put into SHM in two experiments. In the first, the block is pulled from the equilibrium position through a displacement d_1 and then released. In the second, it is pulled from the equilibrium position through a greater displacement d_2 and then released. Are the (a) amplitude, (b) period, (c) frequency, (d) maximum kinetic energy, and (e) maximum potential energy in the second experiment greater than, less than, or the same as those in the first experiment?

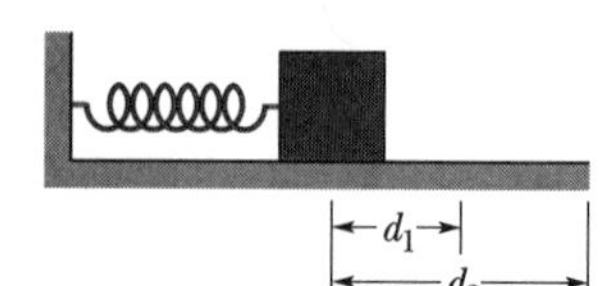

FIGURE 16-48 Question 18.

19. Figure 16-49 shows the $x(t)$ curves for three experiments involving a particular spring–box system oscillating in SHM. Rank the curves according to (a) the system's angular frequency, (b) the spring's potential energy at time $t = 0$, (c) the box's kinetic energy at $t = 0$, (d) the box's speed at $t = 0$, and (e) the box's maximum kinetic energy, greatest first.

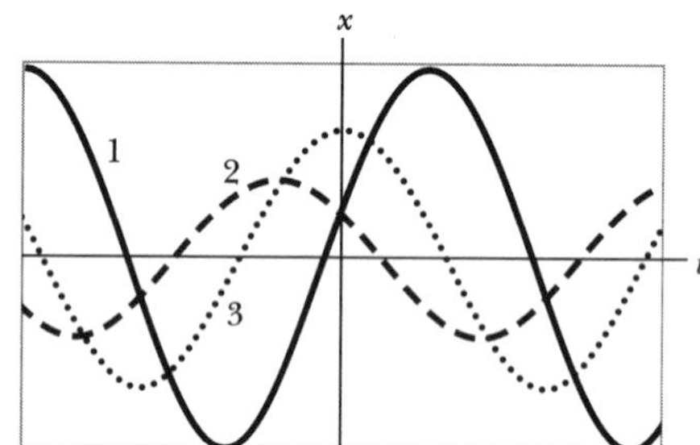

FIGURE 16-49 Question 19.

20. (a) Which curve in Fig. 16-50*a* gives the acceleration $a(t)$ versus displacement $x(t)$ of a simple harmonic oscillator? (b) Which curve in Fig. 16-50*b* gives the velocity $v(t)$ versus $x(t)$?

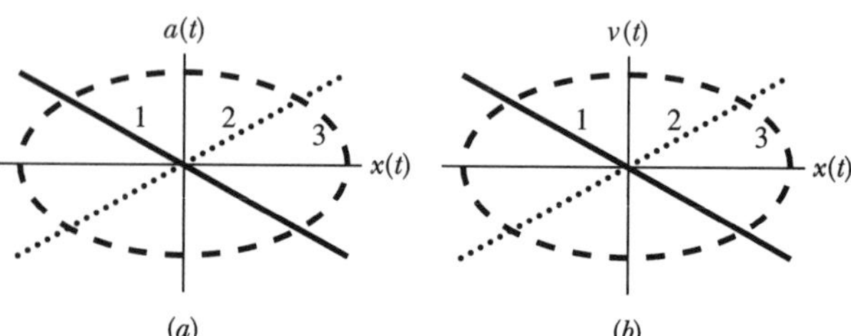

FIGURE 16-50 Question 20.

21. Figure 16-51 gives, for three situations, the displacements $x(t)$ of a pair of simple harmonic oscillators (A and B) that are identical except for phase. For each pair, what phase shift (in radians and in degrees) is needed to shift the curve for A to coincide with the curve for B? Of the many possible answers, choose the shift with the smallest absolute magnitude.

22. Given $x = (2.0\text{ m}) \cos(5t)$ for SHM and needing to find the velocity at $t = 2$ s, should you substitute for t and then differentiate with respect to t or vice versa?

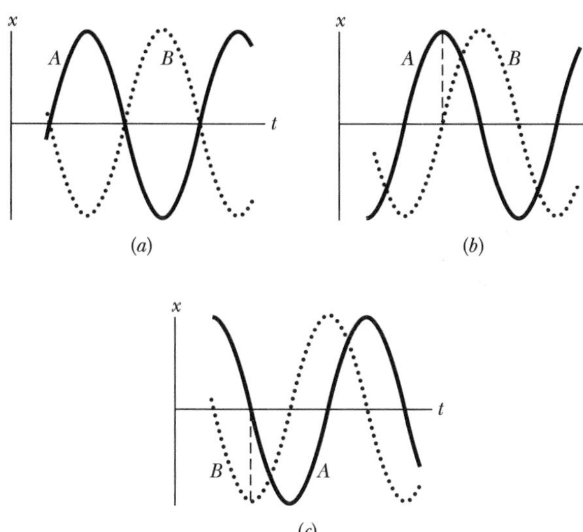

FIGURE 16-51 Question 21.

23. A new type of ride, the SHM Monster, opens up at your local amusement park with three choices of cars (Fig. 16-52). Each car is attached to a large spring that has been pulled from the equilibrium point ($x = 0$) and held fixed at a loading point. Those initial displacements of the identical compartments and the spring constants of the springs are given in the figure in terms of basic units d and k. After you are secured in a car at the loading point, the car is to be released so that its spring can put it and you in SHM along approximately frictionless rails. Rank the three cars according to (a) the time you will take to first reach the equilibrium point, (b) the time you will take to complete 10 cycles of the ride, and (c) the maximum acceleration (and thus fear) that you will experience, greatest first.

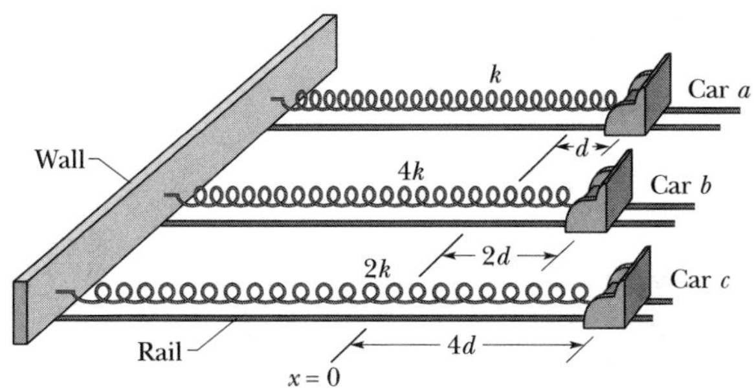

FIGURE 16-52 Question 23.

24. In Fig. 16-53, a small block A sits on a large block B with a certain nonzero coefficient of static friction between the two blocks. Block B, which lies on a frictionless surface, is initially at $x = 0$, with the spring at its relaxed length; then we pull the block a distance d to the right and release it. As the spring–blocks system undergoes SHM, with amplitude x_m, block A is on the verge of slipping over B.

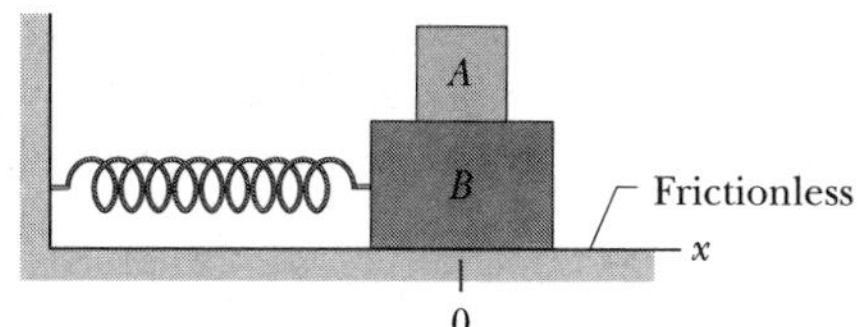

FIGURE 16-53 Question 24.

(a) Is the acceleration of block A constant or does it vary? (b) Is the magnitude of the frictional force accelerating A constant or does it vary? (c) Is A more likely to slip at $x = 0$ or at $x = \pm x_m$? (d) If the SHM had begun with an initial displacement that was greater than d, would slippage then be more likely or less likely? (Warm-up for Problem 25.)

25. Follow-up to Exercise 46: If the speed of the bullet were greater, would the following quantities of the resulting SHM then be greater, less, or the same: (a) amplitude, (b) period, (c) maximum potential energy?

26. Figure 16-54 shows a new variation of a common amusement-park game. A puck is forced to move around a circle of diameter 4 m at a constant angular speed of 2 rad/s. At the right side (at time $t = 0$), it is to bump a pendulum bob from its perch, thus allowing the pendulum to swing to the left. (The pendulum, which is suspended at a high point directly above the center of the circle, swings through a small angle.)

You need to choose a pendulum such that the puck next bumps the bob at the left side of the pendulum's swing. Here, for six pendulums, is the coordinate $x(t)$ of the pendulum's bob during a swing (with x in meters and t in seconds). (a) Which pendulum should you choose? (b) Which should you choose if the next bump is to be at the right side?

(1) $x = 4 \cos (2t)$
(2) $x = 4 \cos (4t)$
(3) $x = 4 \cos (t)$
(4) $x = 2 \cos (2t)$
(5) $x = 2 \cos (4t)$
(6) $x = 2 \cos (t)$

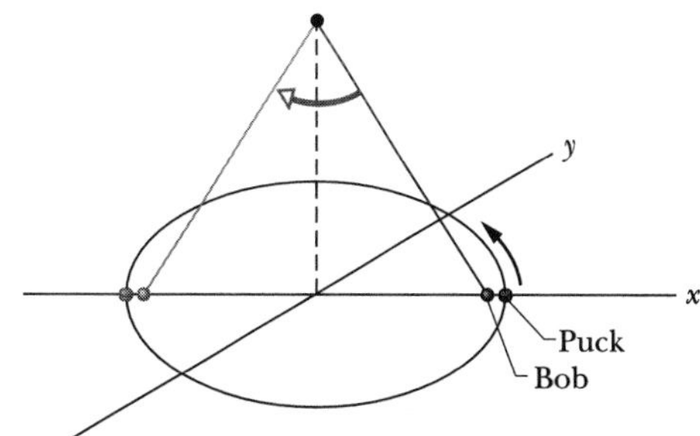

FIGURE 16-54 Question 26.

27. *Math Tool Time.* (a) Which of the following is the derivative of $x = 4.0 \cos(5t)$ with respect to time t: $-4.0 \sin(5t)$, $4.0 \sin t$, $-20 \sin t$, $20 \sin t$, $-20 \sin(5t)$, or $20 \sin(5t)$? (b) Which of the following is the result of $\ln(e^{-at})$: $\ln(-at)$, $\ln(at)$, at, $-at$, $1/at$, or $-1/at$?

EXERCISES & PROBLEMS

94. A simple harmonic oscillator consists of a block attached to a spring of spring constant $k = 200$ N/m. The block slides back and forth along a straight line on a frictionless surface, with equilibrium point $x = 0$ and amplitude 0.20 m. A graph of the velocity v of the block as a function of time t is shown in Fig. 16-55. What are (a) the period of the simple harmonic motion, (b) the mass of the block, (c) the displacement of the block at $t = 0$, (d) the acceleration of the block at $t = 0.10$ s, and (e) the maximum kinetic energy attained by the block?

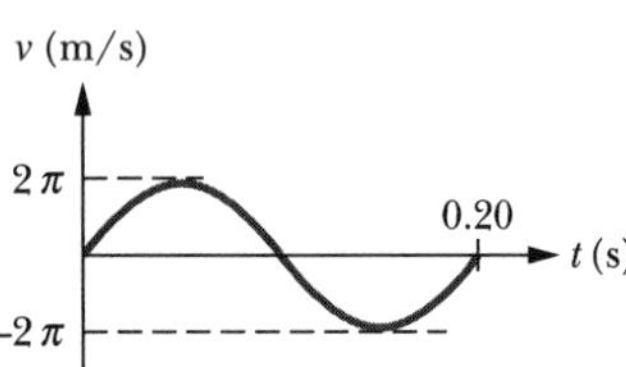

FIGURE 16-55 Problem 94.

95. A simple harmonic oscillator consists of a block of mass 0.50 kg attached to a spring. The block slides back and forth along a straight line on a frictionless surface with equilibrium point $x = 0$. At $t = 0$ the block is at its equilibrium point and is moving in the direction of increasing x. A graph of the magnitude of the net force **F** on the block as a function of its position is shown in Fig. 16-56. What are (a) the amplitude and (b) the period of the simple harmonic motion, (c) the magnitude of the maximum acceleration experienced by the block, and (d) the maximum kinetic energy attained by the block?

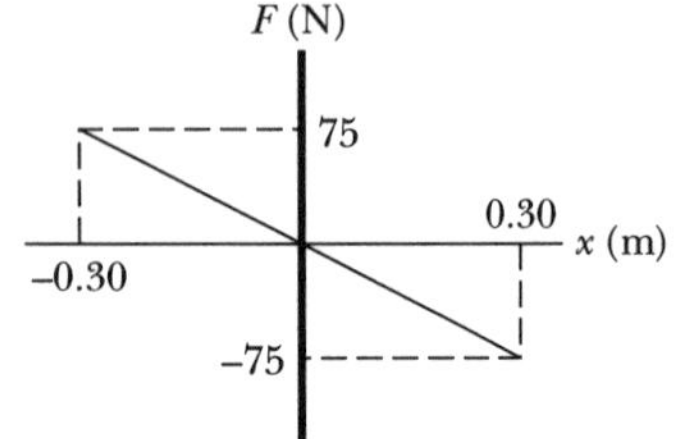

FIGURE 16-56 Problem 95.

96. A block weighing 20 N oscillates at one end of a vertical spring for which $k = 100$ N/m; the other end of the spring is attached to the ceiling. At a certain instant the spring is stretched 0.30 m beyond its unstretched length (the length when no weight is attached) and the block has zero velocity. (a) What is the net force on the block at this instant? What are (b) the amplitude and (c) the period of the resulting simple harmonic motion? (d) What is the maximum kinetic energy of the block as it oscillates?

97. A physical pendulum consists of two meter-long sticks joined together as shown in Fig. 16-57. What is its period of oscillation about a pin inserted through point A?

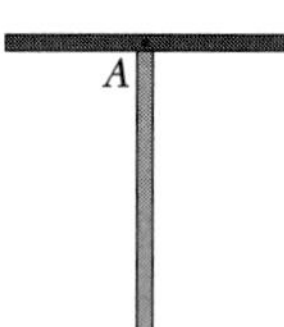

FIGURE 16-57 Problem 97.

98. A particle undergoes simple harmonic motion along an x axis about $x = 0$ with a period of 0.40 s and an amplitude of 0.10 m. At $t = 0$, the particle is at its position of maximum negative displacement, that is, at $x = -0.10$ m. Write expressions, as functions of time, for (a) the particle's position and (b) its velocity.

99. A 1.2 kg block sliding on a horizontal frictionless surface is attached to a horizontal spring with a spring constant of 480 N/m. Let x measure the displacement of the block from the position where the spring is unstretched. At $t = 0$ the block passes through $x = 0$ with a speed of 5.2 m/s in the positive x direction. What are (a) the frequency and (b) the amplitude of the block's motion? (c) Write an expression for the block's displacement x as a function of time.

100. A 2.0 kg tuna can executes SHM while attached to a spring of spring constant 200 N/m. The maximum speed of the can as it slides on a horizontal frictionless surface is 3.0 m/s. What are (a) the amplitude of the block's motion, (b) the magnitude of the block's maximum acceleration, and (c) the magnitude of its minimum acceleration? (d) How long does the can take to complete 7.0 cycles of its motion?

101. A block sliding on a horizontal frictionless surface is attached to a horizontal spring with a spring constant of 600 N/m. The block executes SHM about its equilibrium position with a period of 0.40 s and an amplitude of 0.20 m. As the block slides through its equilibrium position, a 0.50 kg putty wad is dropped vertically onto the block. If the putty wad sticks to the block, determine (a) the new period of the motion and (b) the new amplitude of the motion.

102. A simple harmonic oscillator consists of a block (mass = 0.80 kg) attached to a spring ($k = 200$ N/m). The block slides on a horizontal frictionless surface about the equilibrium point $x = 0$ with a total mechanical energy of 4.0 J. (a) What is the amplitude of the oscillation? (b) How many oscillations does the block complete during a 10 s interval? (c) What is the maximum kinetic energy attained by the block as it oscillates? (d) What is the speed of the block at the instant it is displaced 0.15 m from its equilibrium position?

103. A block weighing 10 N is attached to the lower end of a vertical spring ($k = 200$ N/m), the other end of which is attached to the ceiling. The block oscillates vertically and has a kinetic energy of 2.0 J as it passes through the point where the spring is unstretched. (a) What is the period of the oscillation? (b) Use the law of the conservation of energy to determine the maximum distance the block moves both above and below the point where the spring is unstretched. (These are not necessarily the same.) (c) What is the amplitude of the oscillation? (d) What is the maximum kinetic energy of the block as it oscillates?

104. A grandfather clock has a pendulum that consists of a thin brass disk of radius 15 cm and mass 1.0 kg that is attached to a long thin rod that may be considered massless. The pendulum swings freely about an axis perpendicular to the rod and through the end of the rod opposite the disk as shown in Fig. 16-58. If the pendulum is to have a period of 2.0 s for small oscillations at a point where $g = 9.80$ m/s^2, what must be the length L of the rod? Express your answer to the nearest tenth of a millimeter.

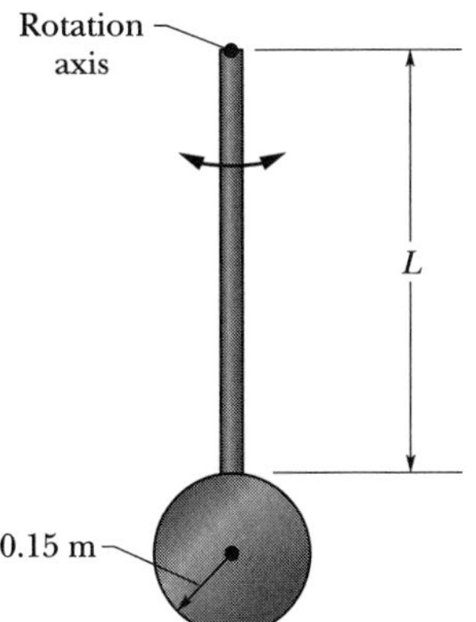

FIGURE 16-58 Problem 104.

105. A thin uniform rod (mass = 0.50 kg) swings about an axis that passes through one end of the rod and is perpendicular to the plane of the swing. The rod swings with a period of 1.5 s and an angular amplitude of 10°. (a) What is the length of the rod? (b) Use the law of the conservation of energy to determine the maximum kinetic energy of the rod as it swings.

Tutorial Problems

106. An object of mass 0.25 kg oscillates in SHM along an x axis with its displacement given by

$$x(t) = 0.25 \text{ m} + (0.50 \text{ m}) \cos [(2.0 \text{ rad/s})t + \pi/6].$$

(a) Make a rough sketch of the displacement of this object as a function of time for several periods. (b) What is the equilibrium position of the object? (c) Determine the amplitude of the motion and the maximum and minimum values of $x(t)$. (d) What is the period of the motion?

(e) Determine numerical expressions for the velocity and acceleration of the object as a function of time. Remember to use notation consistent with the vector nature of these quantities. (f) Determine two numerical expressions for the force acting on the object, one as a function of time and the other as a function of x. (g) What is the total mechanical energy of the object? (h) Determine the position, velocity, and acceleration of the object at time $t = 0.0$ s.

Answers

(a) Sketch a sinusoidal curve between displacements of 0.75 m and -0.25 m, with a displacement of 0.68 at $t = 0$.

(b) The equilibrium position of the object is given by the nonoscillatory term in $x(t)$ and thus $x_{eq} = 0.25$ m.

(c) The amplitude of the motion is the factor in front of the trigonometric function: 0.50 m. The maximum value of $x(t)$ is 0.25 m + 0.50 m = 0.75 m and the minimum value is 0.25 m − 0.50 m = −0.25 m.

(d) The expression for the oscillatory motion should be a trigonometric function of ωt, and from the expression given for $x(t)$ we see that $\omega = 2.0$ rad/s. The period is then $T = 2\pi/\omega = (2\pi \text{ rad})/(2.0 \text{ rad/s}) = 3.1$ s.

(e) The linear velocity of the object has the x component

$$v_x(t) = \frac{dx(t)}{dt} = -(0.50 \text{ m})(2.0 \text{ rad/s}) \sin [(2.0 \text{ rad/s})t + \pi/6]$$
$$= -(1.00 \text{ m/s}) \sin [(2.0 \text{ rad/s})t + \pi/6].$$

The linear acceleration of the object then has the x component

$$a_x(t) = \frac{dv_x(t)}{dt} = -(1.00 \text{ m/s})(2.0 \text{ rad/s}) \cos [(2.0 \text{ rad/s})t + \pi/6]$$
$$= -(2.00 \text{ m/s}^2) \cos [(2.0 \text{ rad/s})t + \pi/6].$$

(f) The force as a function of time is most easily determined by

$$F_x(t) = ma_x(t) = (0.25 \text{ kg})(-(2.00 \text{ m/s}^2) \cos [(2.0 \text{ rad/s})t + \pi/6])$$
$$= -(0.50 \text{ N}) \cos [(2.0 \text{ rad/s})t + \pi/6].$$

To determine the force as a function of x we can use $F_x(x) = -k(x - x_{eq})$, but we would need first to calculate the value of k corresponding to this problem. Another approach is to note that the statement of this problem gave x as a function of t, and it is easy to eliminate the t dependence in $F_x(t)$ by noting that

$$\cos [(2.0 \text{ rad/s})t + \pi/6] = \frac{x - 0.25 \text{ m}}{0.50 \text{ m}}.$$

So

$$F_x(x) = -(0.50 \text{ N}) \left(\frac{x - 0.25 \text{ m}}{0.50 \text{ m}}\right)$$
$$= -(1.00 \text{ N/m})(x - 0.25 \text{ m}).$$

We see that $k = 1.00$ N/m in this situation.

(g) The total mechanical energy of the object can most easily be determined as the value of the maximum kinetic energy of the object:

$$E = K_{max} = \tfrac{1}{2}mv_{max}^2 = (0.5)(0.25 \text{ kg})(1.00 \text{ m/s})^2 = 0.125 \text{ J}.$$

Note that the potential energy of this system will be of the form $U = \frac{1}{2}k(x - x_{eq})^2 = \frac{1}{2}k(x - 0.25 \text{ m})^2$, not $\frac{1}{2}kx^2$.

(h) We can use the general expressions given here for $x(t)$, $v_x(t)$, and $a_x(t)$. At $t = 0.0$ s,

$$\begin{aligned} x &= 0.25 \text{ m} + (0.50 \text{ m}) \cos(\pi/6) \\ &= 0.25 \text{ m} + (0.50 \text{ m})(0.866) = 0.68 \text{ m}. \\ v_x &= -(1.00 \text{ m/s}) \sin(\pi/6) \\ &= -(1.00 \text{ m/s})(0.50) = -0.50 \text{ m/s}. \\ a_x &= -(2.00 \text{ m/s}^2) \cos(\pi/6) \\ &= -(2.00 \text{ m/s}^2)(0.866) = -1.73 \text{ m/s}^2. \end{aligned}$$

107. An object undergoing SHM along an x axis with angular frequency ω has a position function $x(t)$ that satisfies the differential equation

$$\frac{d^2x(t)}{dt^2} + \omega^2 x(t) = 0.$$

This equation is normally obtained by starting from Newton's second law. Let's consider a 2.0 kg object attached to a spring and placed on a horizontal, frictionless track. Suppose that a horizontal force of 20 N is required to hold the object at rest when it is pulled 0.20 m from its equilibrium position (which we can take to be $x = 0.0$ m). The object is released from rest at $x = 0.20$ m and subsequently undergoes SHM.

(a) Starting from the free-body diagram for the object and Newton's second law, determine the differential equation of motion for the object. Show that it has the form given here. Use algebraic, not numerical, expressions in this part. (b) Using the given numerical information, what can you determine about the spring? (c) Determine a complete mathematical expression for the position $x(t)$ of the object and describe in words all the numerical values that appear in that expression.

(d) What is the maximum speed of the object, and where does it occur? (e) What is the maximum magnitude of the acceleration of the object, and where does it occur? (f) What is the total mechanical energy of the system? (g) When the displacement is $\frac{1}{2}$ of its maximum displacement, what are the magnitudes of the velocity and acceleration of the object, and how is the mechanical energy divided between the kinetic energy of the object and the elastic potential energy of the spring?

Answers

(a) The forces acting on the object are the gravitational force $m\mathbf{g} = -mg\hat{\mathrm{j}}$, the normal force $\mathbf{N} = N\hat{\mathrm{j}}$, and the spring force $\mathbf{F} = -kx\hat{\mathrm{i}}$. The vertical forces ($m\mathbf{g}$ and $\mathbf{F}$) must cancel because there is no vertical motion. Along the horizontal x axis, Newton's second law is $F_{net,x} = ma_x$ or

$$-kx = m\left(\frac{d^2x}{dt^2}\right),$$

or

$$m\left(\frac{d^2x}{dt^2}\right) + kx = 0.$$

This gives the differential equation

$$\frac{d^2x}{dt^2} + \left(\frac{k}{m}\right)x = 0,$$

which is the form for SHM, namely,

$$\frac{d^2x}{dt^2} + \omega^2 x = 0$$

if we let $\omega = \sqrt{k/m}$.

(b) From the information that a force of 20 N is necessary to stretch the spring by a distance of 0.20 m, we can determine that the spring constant

$$k = |F/x| = |(20 \text{ N})/(0.20 \text{ m})| = 100 \text{ N/m}.$$

(c) The general form of the position function of the object is $x(t) = x_{eq} + A\cos(\omega t + \phi)$. In this case the equilibrium position of the spring is $x_{eq} = 0.0$ m and the amplitude of the motion is the amount by which the spring is extended at the beginning. So, $A = 0.20$ m. Also, the value of ω is

$$\omega = \sqrt{\frac{k}{m}} = \sqrt{\frac{100 \text{ N/m}}{2.0 \text{ kg}}} = 7.1 \text{ rad/s},$$

and the phase constant ϕ is 0 if we choose $t = 0.0$ s as the time at which the object is released. Then

$$x(t) = (0.20 \text{ m}) \cos[(7.1 \text{ rad/s})\, t].$$

(d) The x component of the velocity of the object is

$$\begin{aligned} v_x(t) &= \frac{dx(t)}{dt} = -(7.1 \text{ rad/s})(0.20 \text{ m}) \sin[(7.1 \text{ rad/s})t] \\ &= (1.4 \text{ m/s}) \sin[(7.1 \text{ rad/s})t]. \end{aligned}$$

So the maximum speed is 1.4 m/s, which occurs at the equilibrium point ($x = 0.0$ m) when the object passes through it in either direction.

(e) The x component of the acceleration of the object is, by Newton's second law,

$$a_x = \frac{F_{net}}{m} = -\frac{kx}{m} = -\omega^2 x.$$

So the maximum value occurs at the same places as the maximum values of $|x|$, namely, at $x = -0.20$ m and $x = 0.20$ m. The maximum magnitude of the acceleration is then

$$a_{max} = \omega^2 |x_{max}| = (7.1 \text{ rad/s})^2(0.20 \text{ m}) = 10 \text{ m/s}^2.$$

(f) The total mechanical energy of the system is

$$E = \tfrac{1}{2}kA^2 = (0.5)(100 \text{ N/m})(0.20 \text{ m})^2 = 2.0 \text{ J}.$$

(g) It is not necessary to determine the quantities in the order requested. Let's try to argue as simply as possible. First, when the displacement is $\frac{1}{2}$ its maximum value, the spring force will be

$\frac{1}{2}$ its maximum value. So the acceleration will be $\frac{1}{2}$ its maximum value. Thus, from part (e), we have $|a_x| = \frac{1}{2}(10\ \text{m/s}^2) = 5.0\ \text{m/s}^2$.

Since the displacement is $\frac{1}{2}$ its maximum value, the elastic potential energy, which is of the form $\frac{1}{2}kx^2$, will be only $\frac{1}{4}$ its maximum value, or $\frac{1}{4}$ of the total mechanical energy E, or 0.5 J. That means the kinetic energy of the object would have to be the other $\frac{3}{4}$ of E, or 1.5 J. So the kinetic energy is three times the elastic potential energy at this position. Since the kinetic energy is $\frac{3}{4}$ of its maximum value, the speed of the object must be $\sqrt{3/4}$ (= 0.866) of its maximum value, or

$$|v_x| = 0.866\ |v_{\text{max}}| = (0.866)(1.4\ \text{m/s}) = 1.2\ \text{m/s}.$$

As a check, this can also be found from $K = \frac{1}{2}mv_x^2$, which gives

$$|v_x| = \sqrt{\frac{2K}{m}} = \sqrt{\frac{2(1.5\ \text{J})}{2.0\ \text{kg}}} = 1.2\ \text{m/s}.$$

108. A simple pendulum with a string of length 66.0 cm has a bob of mass 0.120 kg. At time $t = 0$ s, the string has an angular displacement (from the vertical) of $-9.60°$ (meaning 9.60° to the left) and the bob has a speed of 0.460 m/s and is moving to the right. Neglect frictional effects in this problem. (a) Draw a diagram of this situation, showing the initial conditions. Sketch in the whole path of the bob. Label significant physical quantities. Determine for time $t = 0$ s the height of the bob above the lowest point of its swing.

(b) Determine the initial kinetic energy, potential energy, and mechanical energy of the bob, using the bottom of its swing as the zero of potential energy. (c) Now determine the highest point of the bob's swing and the angular displacement of the string from the vertical at that point. What law or principle of physics are you using to answer this question? (d) Next determine the period and frequency of this pendulum, assuming that the small angle approximation can be used. (e) Write an expression for the angular displacement θ as a function of time. Use the initial conditions to determine the constants in this expression. (f) Sketch the motion of the bob by plotting $\theta(t)$ for several cycles. Indicate the maximum and minimum values of the angle.

Answers

(a) Before starting this problem, we should note that a major part of solving a problem correctly is choosing one's notation carefully. Here, let's use the subscript i for the initial conditions. Looking ahead, we will later use other subscripts for other situations, such as for the physical quantities associated with the highest point of the bob's swing. It is customary to measure angles positive to the right from the vertical (our y axis), and in this case the initial angle has been chosen to be negative (Fig. 16-59).

From the geometry of this physical situation, we see that the initial height y_i of the bob above the bottom of its swing is given by

$$y_i = L - L\cos 9.60° = L(1 - \cos 9.60°)$$
$$= (66.0\ \text{cm})(1 - 0.9860) = 0.924\ \text{cm} = 0.00924\ \text{m}.$$

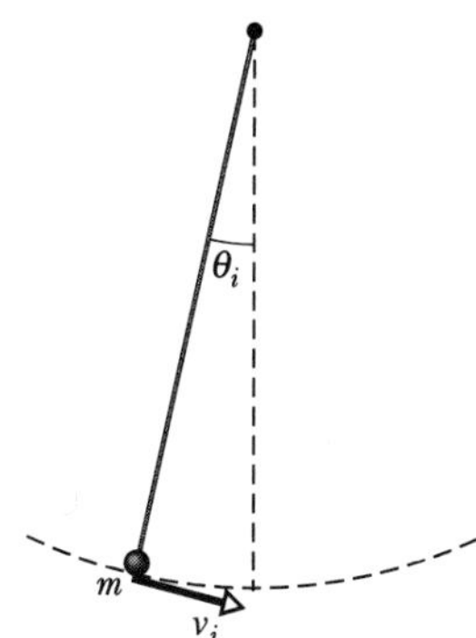

FIGURE 16-59 Problem 108.

Here we have used just the magnitude of θ_i.

(b) The initial kinetic energy of the bob is

$$K_i = \tfrac{1}{2}mv_i^2 = (0.5)(0.120\ \text{kg})(0.460\ \text{m/s})^2 = 0.0127\ \text{J}.$$

Its initial potential energy, with respect to $y = 0$ at the bottom of the swing, is

$$U_i = mgy_i = (0.120\ \text{kg})(9.80\ \text{m/s}^2)(0.00924\ \text{m}) = 0.0109\ \text{J}.$$

Its initial mechanical energy is then

$$E_i = K_i + U_i = 0.0127\ \text{J} + 0.0109\ \text{J} = 0.0236\ \text{J}.$$

(c) We can use the law of the conservation of mechanical energy to determine the highest point of the bob's swing, which will also be the point at which the angular displacement is greatest. At that point, whose height we can call y_{max}, the bob will have zero kinetic energy, so all its mechanical energy is potential energy. Thus $U_{\text{max}} = mgy_{\text{max}} = E$, so

$$y_{\text{max}} = \frac{E}{mg} = \frac{0.0236\ \text{J}}{(0.120\ \text{kg})(9.80\ \text{m/s}^2)} = 0.0201\ \text{m} = 2.01\ \text{cm}.$$

To determine the angular displacement θ_{max} at that point, we make a diagram similar to Fig. 16-59 but for θ_{max}. From it we find that $y_{\text{max}} = L - L\cos\theta_{\text{max}}$.

So, $$\cos\theta_{\text{max}} = \frac{L - y_{\text{max}}}{L} = \frac{66.0\ \text{cm} - 2.01\ \text{cm}}{66.0\ \text{cm}} = 0.96955$$

and $$\theta_{\text{max}} = \cos^{-1}(0.96955) = \pm 14.2°.$$

The $\pm$ sign for θ_{max} means that the maximum angle could be 14.2° to either side.

(d) The period of a simple pendulum, in the small angle approximation, is

$$T = 2\pi\sqrt{\frac{L}{g}} = 2\pi\sqrt{\frac{0.660\ \text{m}}{9.80\ \text{m/s}^2}} = 1.63\ \text{s},$$

so the frequency is

$$f = \frac{1}{T} = \frac{1}{1.63\ \text{s}} = 0.613\ \text{s}^{-1} = 0.613\ \text{Hz}.$$

(e) From the theory of the simple pendulum in the small angle approximation, we can write

$$\theta = \theta_0 \sin(\omega t + \phi) = \theta_0 \sin(\sqrt{g/L}\, t + \phi)$$
$$= \theta_0 \sin[(3.85 \text{ rad/s})t + \phi],$$

where

$$\sqrt{g/L} = \sqrt{(9.8 \text{ m/s}^2)/(0.66 \text{ m})} = 3.85 \text{ rad/s}.$$

From the result of part (c) we know that the amplitude $\theta_0 = |\theta_{max}| = 14.2° = 0.2478$ rad ≈ 0.248 rad. The phase constant ϕ can be found from the fact that the initial value of the angle is $\theta(0) = -9.60° = -0.16755$ rad:

$$-0.16755 \text{ rad} = \theta_0 \sin \phi = (0.2478 \text{ rad}) \sin \phi.$$

So
$$\sin \phi = \frac{-0.16755 \text{ rad}}{0.2478 \text{ rad}} = -0.67616$$

and
$$\phi = \sin^{-1}(-0.67616)$$
$$= -0.7425 \text{ rad} \approx -0.743 \text{ rad}.$$

Finally, we can write

$$\theta(t) = (0.248 \text{ rad}) \sin [(3.85 \text{ rad/s})t - 0.743 \text{ rad}].$$

Graphing Calculators

PROBLEM SOLVING TACTICS

TACTIC 3: *Graphing SHM*

If you have trouble remembering what happens to a graph of simple harmonic motion when you change the phase constant or the angular frequency, why not graph the motion and then make the change on the screen? When you can see the effect of a change, remembering it is much easier.

For example, let's graph the function $x = \cos(5t + \phi)$ for $\phi = 0, 0.5\pi, \pi$, and 1.5π rad to see the effect of increasing ϕ. Go to the primary GRAPH menu, choose y(x) =, and then complete the first line to read

y1 = cos(5*x + {0, .5, 1, 1.5}π)

with x replacing our actual variable t and y1 our actual variable x. Then set the RANGE or WIN values as −2, 2, .5, −2, 2, and 1, respectively. Finally, press GRAPH. As the calculator goes through the values in the list {0, .5, 1, 1.5} one by one, we see the sinusoidal curve march leftward. Thus,

increase ϕ to march the curve leftward,

which we can symbolize as $\overleftarrow{\phi} \uparrow$.

109. In Edgar Allan Poe's masterpiece of terror, "The Pit and the Pendulum," a prisoner who is strapped flat on a floor spies a seemingly motionless pendulum 12 m above him. But then, to his horror, he realizes that the pendulum consists of "a crescent of glittering steel, . . . the under edge as keen as that of a razor. . . ." and that it is gradually descending. As hours go by, the pendulum's motion becomes mesmerizing, with the left–right sweep and the speed at the lowest point of each swing both increasing. The pendulum's intent becomes clear: it is to sweep directly across the prisoner's heart. "Down—steadily down it crept. I took a frenzied pleasure in contrasting its downward with its lateral velocity. To the right to the left—far and wide—with the shriek of a damned spirit! . . . Down—certainly, relentlessly down!"

Assume that the pendulum is ideal, consisting of a particle of mass m on the end of a massless cord of length r. Take the initial cord length r_0 to be 0.80 m and the initial maximum angular speed $\omega_{0,max}$ (when the pendulum passes through $\theta = 0$) to be 1.30 rad/s. Assume also that the pendulum descends only in small steps and only as it passes through $\theta = 0$. (a) Show that this last assumption means that the angular momentum of the pendulum does not change during each step of the descent.

In terms of cord length r during the descent, find (b) the maximum angular speed ω_{max} during a swing and (c) the maximum kinetic energy K_{max} during a swing. (d) As r increases, does K_{max} increase, decrease, or stay the same?

As the pendulum swings upward for any given value of r, assume that its total mechanical energy is conserved. (e) Find the maximum gravitational potential energy U_{max} attained by the pendulum during the upward swing, first in terms of the maximum kinetic energy K_{max} and then in terms of r and the maximum angle θ_{max} reached during a swing.

(f) Using the results of (d) and (e), find θ_{max} as a function of r and given data. (g) Graph θ_{max} versus r; does θ_{max} increase or decrease with the increase in r? (h) For what value of r has θ_{max} changed by a factor of 2 from its initial value (is it twice or half as much)?

(i) Find the horizontal sweep Δx of the pendulum in terms of r and θ_{max}. (j) Graph Δx versus r; does Δx increase or decrease with the increase in r? (k) For what value of r has Δx changed by a factor of 2 from its initial value?

(l) Find the maximum speed v_{max} during a swing in terms of r. (m) Graph v_{max} versus r; does v_{max} increase or decrease with the increase in r? (n) For what value of r has v_{max} changed by a factor of 2 from its initial value? (o) Compare these results with Poe's description.

Chapter Seventeen
Waves—I

QUESTIONS

16. Figure 17-34*a* gives a snapshot of a wave traveling in the direction of increasing x along a string under tension. Four string elements are indicated by the lettered points. For each of those elements, determine whether, at the instant of the snapshot, the element is moving upward or downward or is momentarily at rest. (*Hint:* Imagine the wave as it moves through the four string elements.)

Figure 17-34*b* gives the displacement of a string element at, say, $x = 0$. At the lettered times indicated, is the element moving upward or downward or is it momentarily at rest?

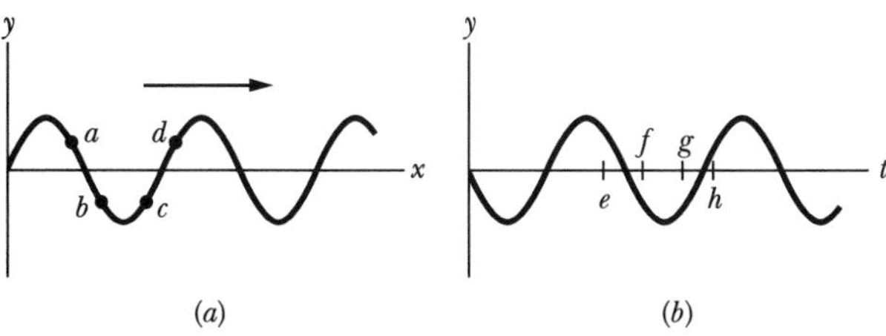

FIGURE 17-34 Question 16.

17. In Fig. 17-35, five points are indicated on the snapshot of a sinusoidal wave. What is the phase difference between point 1 and (a) point 2, (b) point 3, (c) point 4, and (d) point 5? Answer in radians and in terms of the wavelength of the wave. The snapshot shows a point of zero displacement at $x = 0$. In terms of the period T of the wave, when will (e) a crest and (b) the next point of zero displacement reach $x = 0$?

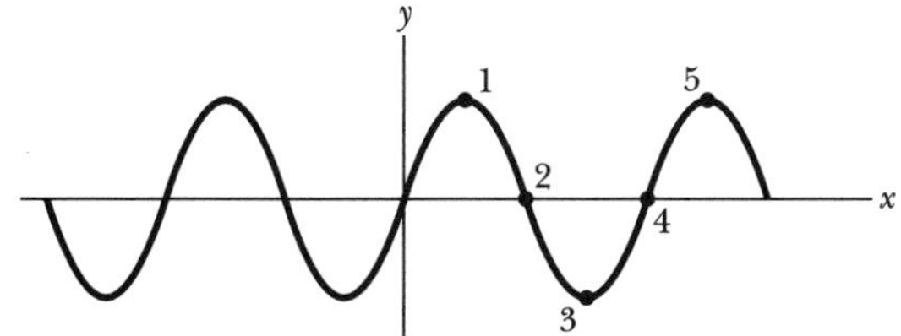

FIGURE 17-35 Question 17.

18. *Organizing question:* Two series of snapshots taken of a wave traveling along a string under tension are shown in Fig. 17-36*a* and *b*. (The snapshots span less than a period of the wave.) For each series, set up an equation, complete with known data, for the traveling wave. (This question continues with Question 25.)

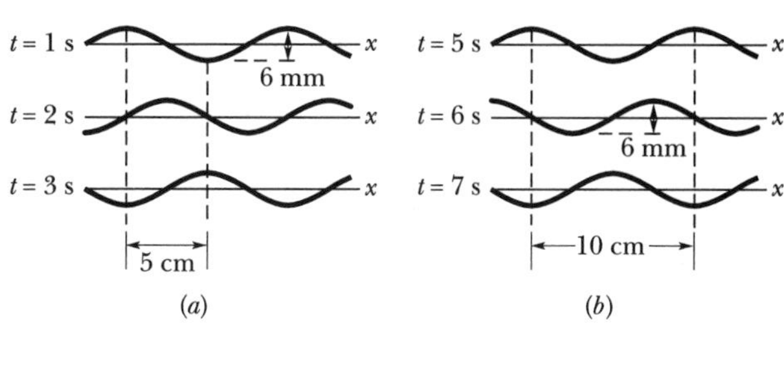

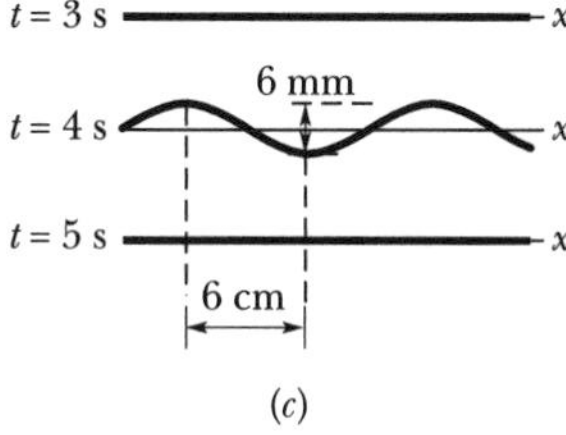

FIGURE 17-36 Questions 18 and 25.

19. The following four waves are sent along strings with the same linear densities (x is in meters and t is in seconds). Rank the waves according to (a) their wave speed and (b) the tension in the strings along which they travel, greatest first:

(1) $y_1 = (3\text{ mm})\sin(x - 3t)$,

(2) $y_2 = (6\text{ mm})\sin(2x - t)$,

(3) $y_3 = (1\text{ mm})\sin(4x - t)$,

(4) $y_4 = (2\text{ mm})\sin(x - 2t)$.

20. Figure 17-37 shows three waves that are *separately* sent along a string that is stretched under a certain tension along an x axis. Rank the waves according to (a) their wavelengths, (b) their speeds, and (c) their angular frequencies, greatest first.

21. Four waves are separately sent along a string that is stretched under a certain tension. The displacement $y(t)$ of a particular

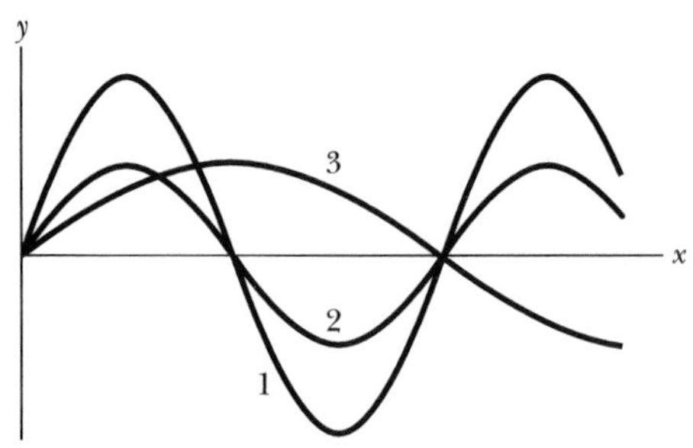

FIGURE 17-37 Question 20.

string element during the passage of each wave is given in Fig. 17-38. Rank the waves according to the average rate at which they transfer energy along the strings, greatest first.

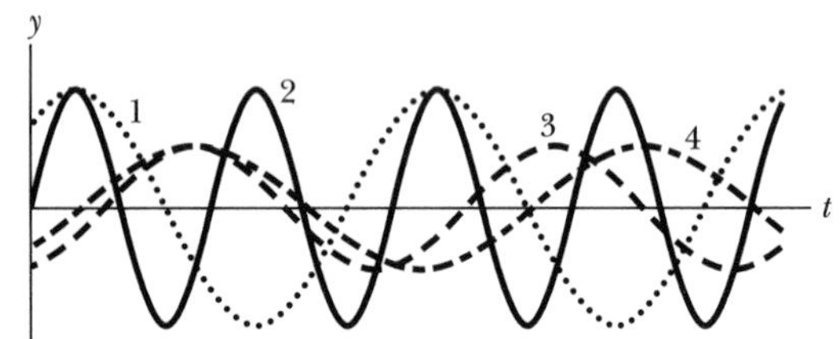

FIGURE 17-38 Question 21.

22. The following table gives the wave speeds (in terms of v_0) and the frequencies (in terms of f_0) for three waves traveling along different strings. The waves have the same amplitudes, and the strings have the same linear densities. Rank the waves according to the average rate at which they transfer energy along the strings, greatest first.

Wave	1	2	3
Wave speed	$2v_0$	$4v_0$	$6v_0$
Frequency	$4f_0$	f_0	$2f_0$

23. Waves of the same amplitude and frequency are sent along three strings. The following table gives the tensions and linear densities of the strings. Rank the strings according to the average rate at which energy is transferred along them, greatest first.

String	1	2	3
Tension	$2\tau_0$	$4\tau_0$	$3\tau_0$
Linear density	$4\mu_0$	$2\mu_0$	$6\mu_0$

24. A wave is sent along a string, which lies along an x axis. Which of the graphs in Fig. 17-39 best represents the kinetic energy K of the wave versus x at a given instant?

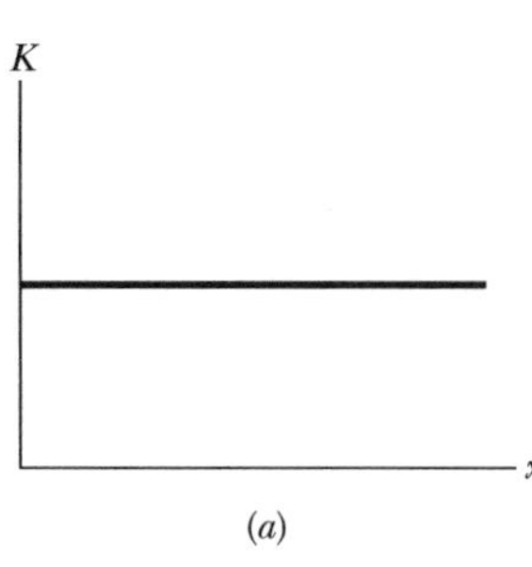

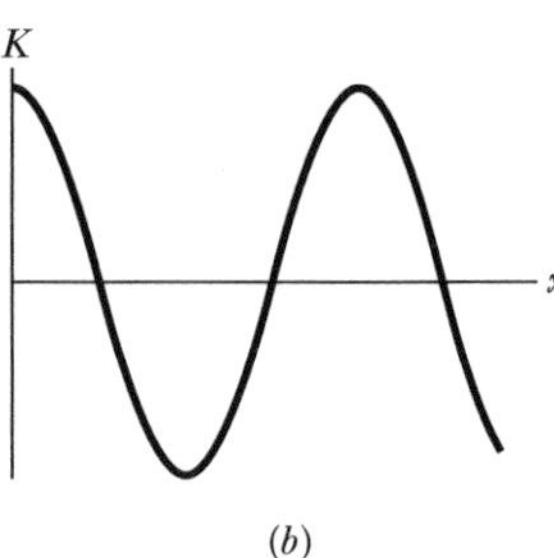

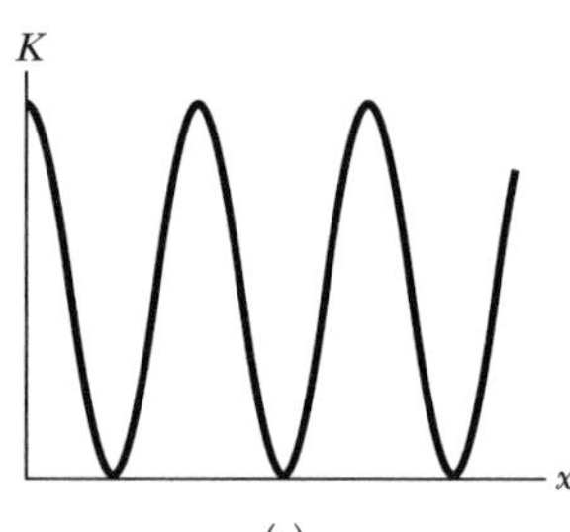

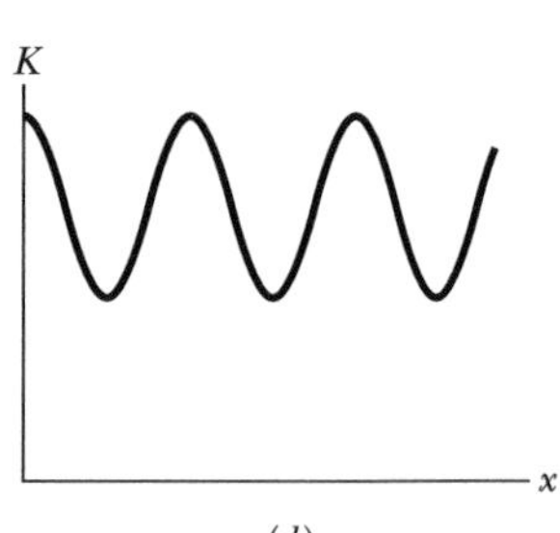

FIGURE 17-39 Question 24.

25. Question 18 continued: Figure 17-36*c* shows three snapshots taken of a string as traveling waves pass along it. Set up equations, complete with known data, for the traveling waves.

26. The first harmonic is set up on the string in Fig. 17-18, and the oscillations of the string have period T. If time $t = 0$ when the string is in its extreme upward configuration (the solid line in Fig. 17-18*a*), at what time, in terms of T, does the string (a) first become horizontal, (b) first reach its extreme downward configuration (the dashed line in Fig. 17-18*a*), and (c) again become horizontal?

27. (a) If a standing wave on a string is given by

$$y'(t) = (3 \text{ mm}) \sin(5x) \cos(4t),$$

is there a node or an antinode of the oscillations of the string at $x = 0$? (b) Repeat the question if the standing wave is given by

$$y'(t) = (3 \text{ mm}) \sin(5x + \pi/2) \cos(4t).$$

28. The nodes of a standing wave on a 10 m string stretched along an x axis happen to be at $x = 0.5, 1.5, 2.5$ m, where $x = 0$ is somewhere near the middle of the string. Is there a node, an antinode, or some intermediate state at (a) $x = -0.5$ m and (b) $x = 3.0$ m?

29. *Math Tool Time.* If $y \sin(3x) = 0$, then for what values of x does $y(x) = 0$?

EXERCISES & PROBLEMS

68. A rope, under a tension of 200 N and fixed at both ends, oscillates in a second-harmonic standing wave pattern. The displacement of the rope is given by

$$y = (0.10 \text{ m})(\sin \pi x/2) \sin 12\pi t,$$

where $x = 0$ at one end of the rope, x is in meters, and t is in seconds. What are (a) the length of the rope, (b) the speed of the waves on the rope, and (c) the mass of the rope? (d) If the rope oscillates in a third-harmonic standing wave pattern, what will be the period of oscillation?

69. A standing wave results from the sum of two transverse traveling waves given by

$$y_1 = 0.050 \cos(\pi x - 4\pi t)$$

and

$$y_2 = 0.050 \cos(\pi x + 4\pi t),$$

where x, y_1, and y_2 are in meters and t is in seconds. (a) What is the smallest positive value of x that corresponds to a node? (b) At what times during the interval $0 \leq t \leq 0.50$ s will the particle at $x = 0$ have zero velocity?

70. A wave on a string is described by

$$y(x, t) = 15 \sin(\pi x/8 - 4\pi t),$$

where x and y are in centimeters and t is in seconds. (a) What is the transverse speed for a point on the string at $x = 6.0$ cm when $t = 0.25$ s? (b) What is the maximum transverse speed of any point on the string? (c) What is the magnitude of the transverse acceleration for a point on the string at $x = 6.0$ cm when $t = 0.25$ s? (d) What is the magnitude of the maximum transverse acceleration for any point on the string?

71. A wave on a string is described by

$$y = 15 \cos(\pi x - 15\pi t),$$

where x and y are in centimeters and t is in seconds. What is the transverse speed for a point on the string at an instant when that point has the displacement $y = 12$ cm?

72. A transverse sinusoidal wave is moving along a string in the positive x direction with a speed of propagation of 80 m/s. At $t = 0$, the string particle at $x = 0$ has a transverse displacement of 4.0 cm from its equilibrium position and is not moving. The maximum transverse speed of the string particle at $x = 0$ is 16 m/s. (a) What is the frequency of the wave? (b) What is the wavelength of the wave? (c) Write an equation describing the wave.

73. A certain transverse sinusoidal wave of wavelength 20 cm is moving to the right. The transverse velocity of the particle at $x = 0$ as a function of the time is shown in Fig. 17-40. (a) What is the speed of propagation of the wave? (b) What is the amplitude of the wave? (c) What is the frequency of the wave? (d) Sketch the wave between $x = 0$ and $x = 20$ cm at $t = 2.0$ s.

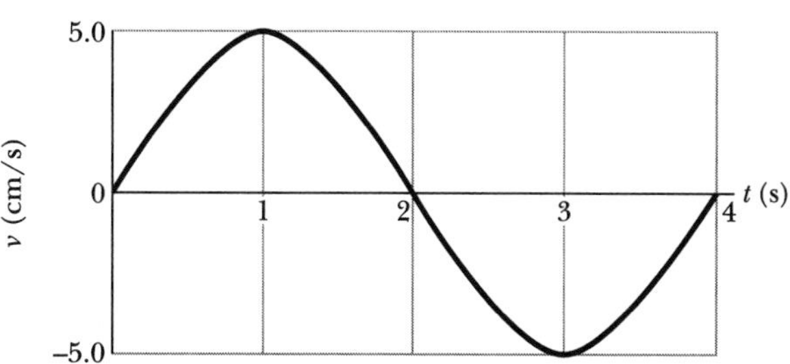

FIGURE 17-40 Problem 73.

74. Two waves are described by

$$y_1 = 0.30 \sin[\pi(5x - 200t)]$$

and

$$y_2 = 0.30 \sin[\pi(5x - 200t) + \pi/3],$$

where y_1, y_2, and x are in meters and t is in seconds. When these two waves are combined, a traveling wave is produced. What are (a) the amplitude, (b) the wave speed, and (c) the wavelength of that traveling wave?

75. A standing wave pattern on a string is described by

$$y(x, t) = 0.040 \sin 5\pi x \cos 40\pi t,$$

where x and y are in meters and t is in seconds. (a) Determine the location of all nodes for $0 \leq x \leq 0.40$ m. (b) What is the period of the oscillatory motion of any (nonnode) point on the string? What are (c) the speed and (d) the amplitude of the two traveling waves that interfere to produce this wave? (e) At what times for $0 \leq t \leq 0.050$ s will all the points on the string have zero transverse velocity?

76. For a certain transverse standing wave on a long string, an antinode is at $x = 0$ and a node is at $x = 0.10$ m. The displacement $y(t)$ of the string particle at $x = 0$ is shown in Fig. 17-41. When $t = 0.50$ s, what are the displacements of the string particles at (a) $x = 0.20$ m and (b) $x = 0.30$ m? At $x = 0.20$ m, what are the transverse velocities of the string particles when (c) $t = 0.50$ s and (d) $t = 1.0$ s? (e) Sketch the standing wave at $t = 0.50$ s for the range $x = 0$ to $x = 0.40$ m.

77. In a demonstration, a 1.2 kg horizontal rope is fixed in place at its two ends ($x = 0$ and $x = 2.0$ m) and made to oscillate up

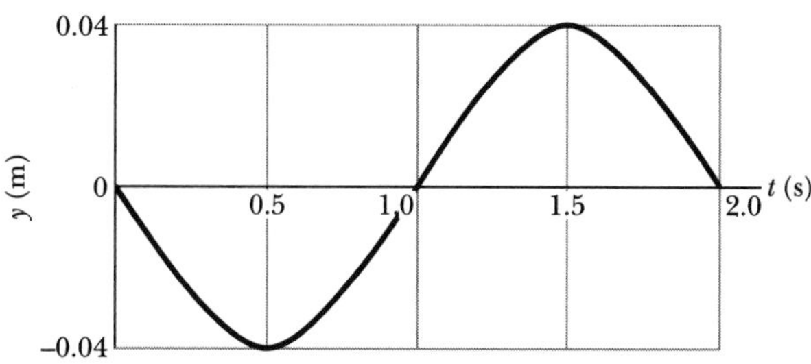

FIGURE 17-41 Problem 76.

and down in a standing wave. The frequency of the fundamental mode is 5.0 Hz. At time $t = 0$, the point at $x = 1.0$ m on the rope has zero displacement and is moving upward (in the positive y direction) with a transverse velocity of 5.0 m/s. What are (a) the amplitude of the motion of that point and (b) the tension in the rope?

The oscillation pattern of the rope might not be due solely to the fundamental mode; other harmonic modes might also be present. (c) With the information given here, determine the lowest frequency of those other harmonic modes that might be present. (d) Write the standing wave equation for the fundamental mode.

Tutorial Problems

78. Figure 17-42 is a graph of a sinusoidal mechanical wave at a particular time, giving displacement y versus position x. (Note the difference in scales on the axes.) (a) The graphed curve gives (at the particular instant) the wave function $y(x)$, that is, y as a function of x. What does this wave function represent physically? (Make no assumption as to whether the wave is a transverse wave or a longitudinal wave.)

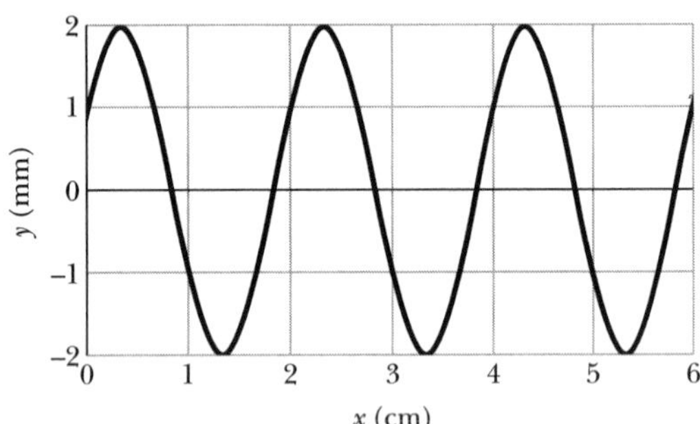

FIGURE 17-42 Problem 78.

(b) Suppose the wave is a transverse wave. Sketch the displacements of the particles (through which the wave moves) at the positions $x = 0$, 1, 2, 3, 4, 5, and 6 cm. (c) Repeat part (b), but now suppose the wave is a longitudinal wave. (d) What are the numerical values of the amplitude and wavelength of the wave? Show these quantities in Fig. 17-42. (e) Graph the wave function for a time that is half a period after the time of Fig. 17-42. Does your graph's appearance depend on whether the wave of Fig. 17-42 is traveling toward increasing x or toward decreasing x?

(f) Now suppose that the wave is traveling toward increasing x with a speed of 4.0 cm/s. On the graph of part (e) include a dashed curve to show the wave function 0.25 s later than the time assumed in drawing that graph. (g) Define in words what is meant by the period of a wave. What is the period of the wave of Fig. 17-42? (h) Define in words what is meant by the frequency of a wave. What is the frequency of the wave of Fig. 17-42? (i) Draw a graph of the wave as a function of time t for a particular position, say, $x = 1.0$ cm. On the graph, indicate the period T.

Answers

(a) The wave function $y(x)$ represents the displacements (from equilibrium) of the particles along which the wave travels. The displacements can be perpendicular to the wave's direction of travel (transverse wave) or parallel to that direction (longitudinal wave).

(b) See Fig. 17-43*a*.

(c) See Fig. 17-43*b*.

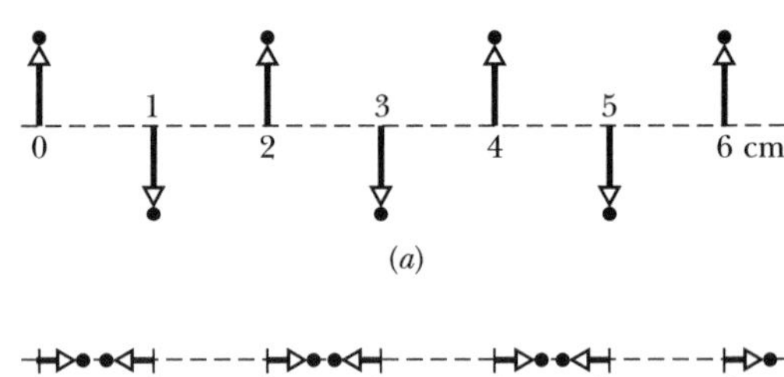

FIGURE 17-43 Problem 78.

(d) The amplitude of the wave is 2.0 mm because the crests and valleys are 2.0 mm from the equilibrium position. In Fig. 17-42, the wavelength is the horizontal distance between successive crests, or twice the horizontal distance between a crest and an adjacent valley. We see that the latter is 1.0 cm, so the wavelength is 2.0 cm.

(e) Crests have become valleys, and vice versa, regardless of whether the wave is traveling toward increasing x or toward decreasing x.

(f) In the 0.25 s the wave moves (4.0)(0.25 s) = 1.0 cm toward increasing x. So, shift the curve of part (e) 1.0 cm to the right.

(g) The period of a wave is the time interval between successive crests at any given position. Here, the wavelength is 2.0 cm while the wave is traveling with a speed of 4.0 cm/s, so this time interval is (2.0 cm)/(4.0 cm/s) = 0.50 s.

(h) The frequency of a wave is the rate at which crests occur at a particular point along the wave. Here, the time between successive crests is 0.50 s, and so the frequency is 1/(0.50 s) = $2.0\ s^{-1}$ = 2.0 Hz.

(i) Sketch an inverted sine function with an amplitude of 2 mm and successive peaks separated by 0.50 s.

79. In this problem we shall consider what happens when two waves are added together. This is partly an exercise in graphing, so make the graphs by hand, instead of with a computer. You may use a calculator or computer, but first figure out what to expect. Let's consider four waves with these wave functions:

$$y_1(x, t) = (1.50 \text{ mm}) \cos[(0.94 \text{ rad/m})x - (4.0 \text{ rad/s})t]$$

$$y_2(x, t) = (1.50 \text{ mm}) \cos[(0.94 \text{ rad/m})x + (4.0 \text{ rad/s})t]$$

$$y_3(x, t) = (1.00 \text{ mm}) \cos[(0.94 \text{ rad/m})x + (4.0 \text{ rad/s})t]$$

$$y_4(x, t) = (1.50 \text{ mm}) \cos[(0.94 \text{ rad/m})x + (4.0 \text{ rad/s})t + \pi]$$

(a) Waves can be characterized by such physical quantities as amplitude, frequency, wavelength, speed, direction of propagation, and phase constant. List three such quantities that are the *same* for all four waves given here. (b) List three physical quantities that are not the same for all four waves. (c) What kind of waves are these? Be as specific as possible. (d) What are the amplitude, frequency, and wavelength of wave 4?

(e) On a graph plot waves 1 and 2 and their superposition at time $t = 0.00$ s, for the range 0 to 10 m. *Suggestion:* If you are using a calculator, begin by calculating the wave every meter and plotting the points, then plot any intermediate points necessary to help to figure out the waves. (f) Repeat (e) for $t = 1.00$ s, plotting waves 1 and 2 and their superposition. (g) On a graph, sketch only the superposition of waves 1 and 2 (not waves 1 and 2 individually), but do so for several different times. You can start with the superpositions plotted on the graphs of parts (e) and (f). What kind of wave is the superposition?

(h) Would the superposition of waves 1 and 3 be the same kind of wave as the superposition of waves 1 and 2? Explain why or why not. If it is the same kind, describe any major differences in the superpositions of the two pairs of waves. (i) Would the superposition of waves 1 and 4 be the same kind of wave as the superposition of waves 1 and 2? Explain why or why not. If it is the same kind, describe any major differences in the superpositions of the two pairs of waves.

Answers

(a) These four waves all have the same wavelength (or wave number), frequency, and speed of propagation.

(b) The four waves don't all have the same amplitude, phase constant, or direction of propagation.

(c) These waves are traveling sinusoidal waves.

(d) Amplitude $A = 1.50$ mm;
frequency $f = \omega/2\pi = (4.0 \text{ rad/s})/2\pi = 0.64$ Hz; and
wavelength $\lambda = 2\pi/k = 2\pi/(0.94 \text{ rad/m}) = 6.68$ m.

(g) The superposition of waves 1 and 2 is a standing wave because those waves have the same amplitude, frequency, and wavelength and are traveling in opposite directions.

(h) Waves 1 and 3 would not give a standing wave because their amplitudes are different. For example, there would be no stationary nodes.

(i) Yes, these would also superpose into a standing wave. Because wave 4 has a different phase from wave 2, the nodes and antinodes would be shifted. In fact, they would be interchanged, since the phase difference is π.

Graphing Calculators

PROBLEM SOLVING TACTICS

TACTIC 3: *Adding Waves; Adding Phasors*

As shown in the next Sample Problem, you can see how two given waves interfere by graphing their sum for several different times within one period of the waves. Although such an exercise is simple, it allows you to see the actual shape of the resultant wave instead of trying to visualize it from the equation for the wave.

To add phasors, keep in mind that they are rotating vectors. So, you can mentally take a snapshot of them, sketch the snapshot, and then add them as vectors on a calculator. Your work is slightly reduced if you take your snapshot with at least one phasor horizontal, as we did in Fig. 17-14.

SAMPLE PROBLEM 17-8

Graphing Waves. Set up a list of times 0, 0.5π, π, and 1.5π under the name TIME. Then make graphs of displacement y versus position x for the wave functions given in (a) through (g). Use the range $x = -8$ m to $x = 8$ m, with tick marks at half-wavelengths, and the range $y = -2$ mm to $y = 2$ mm, with tick mark separations of 0.5 mm. Each of the waves is of the form $y = y_m \sin(kx - \omega t + \phi)$, with y and y_m in millimeters, x in meters, t in seconds, and ϕ in radians. For each graph, describe what occurs as the calculator progresses through the values of the TIME list.

(a) Graph $y1 = \sin(x - \text{TIME})$.

SOLUTION: As time progresses, the sinusoidal wave, with amplitude of 1 mm, travels rightward, in the positive direction of the x axis.

(b) Graph $y1 = \sin(x - \text{TIME})$, $y2 = \sin(x - \text{TIME})$, and $y3 = y1 + y2$. Set the graphing format so that $y1$, $y2$, and $y3$ are graphed simultaneously instead of one after the other. If you have a calculator (such as a TI-86) on which you can draw one curve differently from other curves, do so here and in subsequent parts. For example, you might draw the resultant wave $y3$ with a thicker line. On a TI-86, this change is made with the STYLE option, which is available on the secondary menu under the y(x)= option on the primary menu. Move to the y3= line on the screen and then choose STYLE. The slanted line symbol next to the y3= line thickens. Choose GRAPH by pressing 2nd M5 to see the result.

SOLUTION: Waves $y1$ and $y2$ travel rightward in phase, with their curves on the graph exactly aligned. Resultant wave $y3$ also travels rightward and is in phase with waves $y1$ and $y2$, but has an amplitude of 2 mm.

(c) Graph $y1 = \sin(x - \text{TIME})$, $y2 = \sin(x - \text{TIME} + 0.5\pi)$, and $y3 = y1 + y2$.

SOLUTION: Waves $y1$ and $y2$ are out of phase by 0.5π rad, or 90°. Resultant wave $y3$ peaks between their peaks, with an amplitude of about 1.5 mm. All three waves travel rightward.

(d) Graph $y1 = \sin(x - \text{TIME})$, $y2 = \sin(x - \text{TIME} + \pi)$, and $y3 = y1 + y2$.

SOLUTION: Waves $y1$ and $y2$ are exactly out of phase with each other and travel rightward. Resultant wave $y3$ is flat.

(e) Graph $y1 = \sin(x - \text{TIME})$, $y2 = \sin(x + \text{TIME})$, and $y3 = y1 + y2$.

SOLUTION: Wave $y1$ travels rightward; wave $y2$ travels leftward. They are sometimes exactly aligned and produce a resultant wave $y3$ with an amplitude of 2 mm. At other times they are exactly misaligned and produce a flat resultant wave $y3$. The nodes of the consequent standing wave are separated by 0.5 wavelength; so are the antinodes. One of the nodes is at the origin.

(f) Graph $y1 = \sin(x - \text{TIME})$, $y2 = \sin(x + \text{TIME} + 0.7\pi)$, and $y3 = y1 + y2$.

SOLUTION: The standing wave pattern is shifted from that of part (e).

(g) Graph $y1 = 0.5 \sin(x - \text{TIME})$, $y2 = \sin(x + \text{TIME})$, and $y3 = y1 + y2$. (Note the new amplitude for $y1$.)

SOLUTION: Wave $y1$ travels rightward; wave $y2$ travels leftward. Resultant wave $y3$, however, is not a standing wave because the points of zero displacement oscillate left and right.

Chapter Eighteen
Waves—II

QUESTIONS

15. In Fig. 18-42, three long tubes (A, B, and C) are filled with different gases under different pressures. The ratio of the bulk modulus to the density is indicated for each gas in terms of a basic value B_0/ρ_0. Each tube has a piston at its left end that can send a sound pulse through the tube (as in Fig. 17-2). The pulses are to begin simultaneously in each tube. Rank the tubes according to the arrival of the pulses at the right ends of the tubes, earliest first.

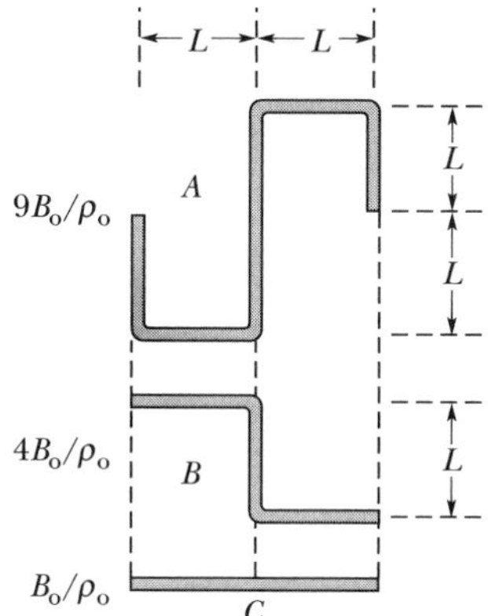

FIGURE 18-42 Question 15.

16. Figure 18-43 is a graph of the displacement functions $s(x)$ of three sound waves that travel along an x axis through air. Rank the waves according to their pressure amplitudes, greatest first.

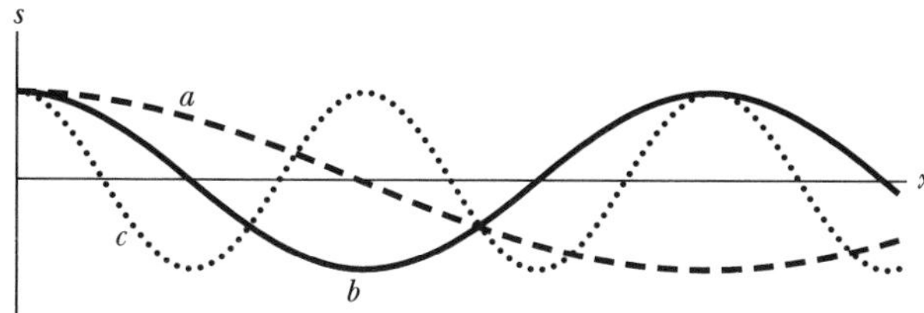

FIGURE 18-43 Questions 16 and 21.

17. The following three pairs of sound waves each produce a resultant wave due to their interference. (Amplitudes are in micrometers, x is in meters, and t is in centiseconds.) Without computation, rank the pairs according to the amplitudes of the resultant waves, greatest first.

(a) $s_1 = 2\sin(3x - 4t)$ and $s_2 = 2\sin(3x - 4t + \pi/2)$
(b) $s_1 = 3\sin(5x - 6t)$ and $s_2 = 3\sin(5x - 6t + \pi/2)$
(c) $s_1 = 7\sin(3x - 5t)$ and $s_2 = 7\sin(3x - 5t + \pi)$

18. *Organizing question:* In Figs. 18-44a through c, two point sources S_1 and S_2 emit sound of wavelength λ uniformly in all directions. Three points P are indicated. For each point, set up an equation, complete with given data, to calculate the phase difference (in terms of wavelength) between the waves arriving at the point from S_1 and S_2.

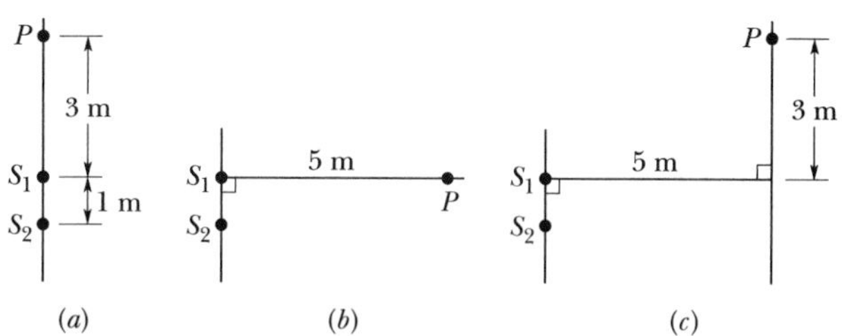

FIGURE 18-44 Question 18.

19. In Fig. 18-45, sound waves A and B, both of wavelength λ, are initially in phase and traveling rightward, as indicated by the two rays. One of the waves reflects from (is redirected by) four surfaces but ends up traveling in its original direction. Expressed in terms of wavelength λ, what is the least value of the distance

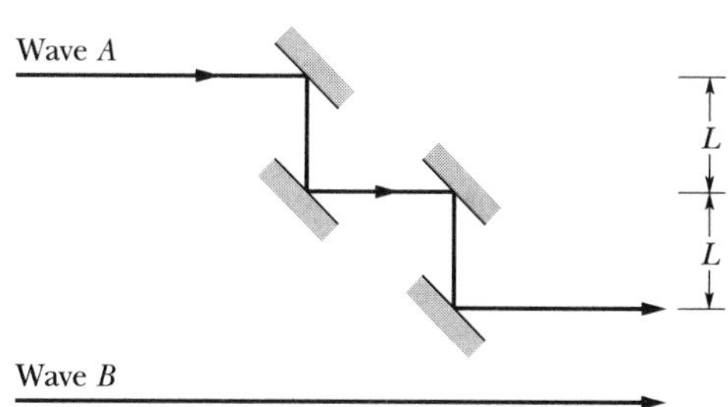

FIGURE 18-45 Question 19.

L in the figure that puts waves A and B exactly out of phase with each other after the reflections? (*Hint:* Consider the path-length difference of the two rays.)

20. (a) In Sample Problem 18-3, suppose that the sources are not in phase, but that instead S_1 emits earlier than S_2 by $0.2T$, where T is the period of the sound waves. Will the points of 0λ in Fig. 18-8 be positioned as they are now, below the horizontal dashed line, or above that dashed line? (b) If S_1 emits earlier than S_2 by $0.5T$, what type of interference then occurs at point P_1 on the perpendicular bisector, and what is the phase difference there? (c) What type occurs at point P_2 on the line extending through the sources, and what is the phase difference there?

21. Figure 18-43 shows graphs of the displacement function $s(x)$ of three sound waves traveling along an x axis through air. Rank the waves according to their intensity on a surface perpendicular to that axis, greatest first.

22. For a particular tube, here are four of the six harmonic frequencies below 1000 Hz: 300, 600, 750, and 900 Hz. What are the two frequencies missing from the list?

23. The third harmonic is set up inside a pipe of length L by a small internal sound source; the pipe has both ends closed and is filled with air (like the external air). Is there a displacement node or antinode (a) across the ends and (b) across the middle? (c) Is the frequency of that third harmonic greater than, less than, or the same as the frequency of the third harmonic of a similarly filled pipe of the same length L with both ends open?

24. If you first set up the third harmonic in a pipe and then switch to the fourth harmonic, does the spacing between adjacent antinodes increase, decrease, or remain the same?

25. In Fig. 18-46, pipe A is made to oscillate in its third harmonic by a small internal sound source. Sound emitted at the right end happens to resonate four nearby pipes, each with only one open end (they are not drawn to scale). Pipe B oscillates in its lowest harmonic, pipe C in its second lowest harmonic, pipe D in its third lowest harmonic, and pipe E in its fourth lowest harmonic. Without computation, rank all five pipes according to their length, greatest first. (*Hint:* Draw the standing waves to scale and then draw the pipes to scale.)

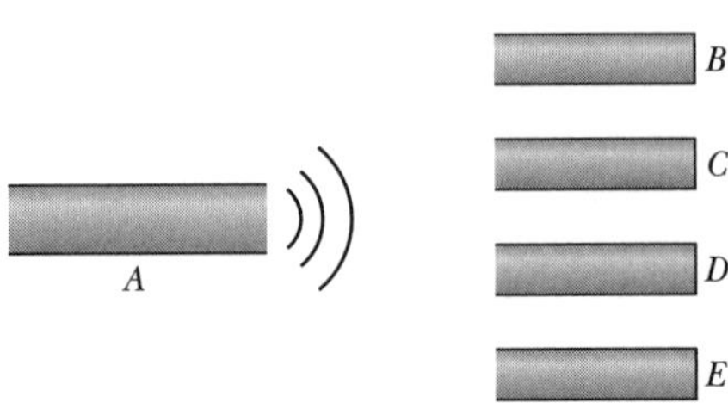

FIGURE 18-46 Question 25.

26. *Organizing question:* A bat flies at speed 9.0 m/s directly toward a moth while emitting sound with a certain frequency f. The moth flies directly away from the bat at speed 7.0 m/s. To determine the frequency f' of the bat's emission heard by the moth with Eq. 18-53, (a) what speed should be substituted for v_D, (b) what sign should be in front of that substitution, (c) what speed should be substituted for v_S, and (d) what sign should be in front of that substitution?

Then, to determine the frequency of the echo heard by the bat from the moth, again with Eq. 18-53, (e) what speed should be substituted for v_D, (f) what sign should be in front of that substitution, (g) what speed should be substituted for v_S, and (h) what sign should be in front of that substitution?

27. A source emitting a sound wave at a certain frequency moves along an x axis (Fig. 18-47a). The source moves directly toward detector A and directly away from detector B. The superimposed three plots of Fig. 18-47b indicate the displacement function $s(x)$ of the sound wave as measured by detector A, by detector B, and by someone (C) in the rest frame of the source. Which plot corresponds to which measurement?

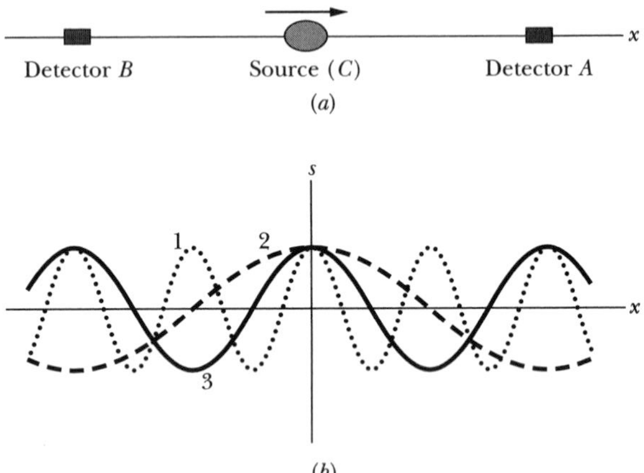

FIGURE 18-47 Questions 27 and 28.

28. A sound wave of a certain frequency is emitted by a source moving along an x axis (Fig. 18-47a). The source moves directly toward detector A and directly away from detector B. Which of the following wave functions best correspond to (a) the wave detected by A, (b) the wave detected by B, and (c) the wave detected by someone (C) in the rest frame of the source and in front of the source? (Amplitudes are in micrometers, x is in meters, and t is in centiseconds.)

$s_1 = 2\sin(2x + 6t)$ $\quad s_2 = 2\sin(x - 3t)$

$s_3 = 2\sin(\frac{2}{3}x - 2t)$ $\quad s_4 = 2\sin(2x - 4t)$

$s_5 = 2\sin(2x - 6t)$ $\quad s_6 = 2\sin(\frac{2}{3}x + 4t)$

$s_7 = 3\sin(\frac{2}{3}x + 2t)$ $\quad s_8 = 2\sin(\frac{2}{3}x + 2t)$

29. Sound waves of frequency f are reflected by a fluid moving through a narrow tube along an x axis (Fig. 18-48a). The tube's diameter varies with x. The change in frequency Δf of the sound, due to the Doppler effect, also varies with x, as shown in Fig.

18-48b. Rank the five indicated regions in terms of the tube's diameter, greatest first. (*Hint:* See Section 15-10.)

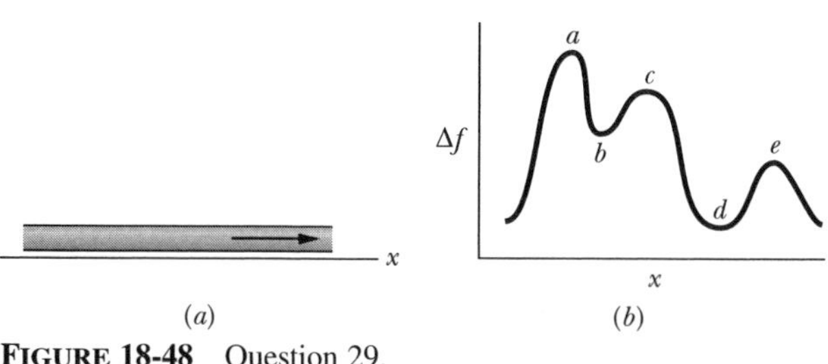

FIGURE 18-48 Question 29.

30. Figure 18-49 indicates the Mach cones that form on an airplane flying at a certain speed at three different altitudes. Rank the altitudes according to (a) the speed of sound there and (b) the Mach number of the airplane there, greatest first. (c) Assuming that the speed of sound in air varies as the square root of the air temperature (in kelvins), rank the altitudes according to the air temperature, greatest first.

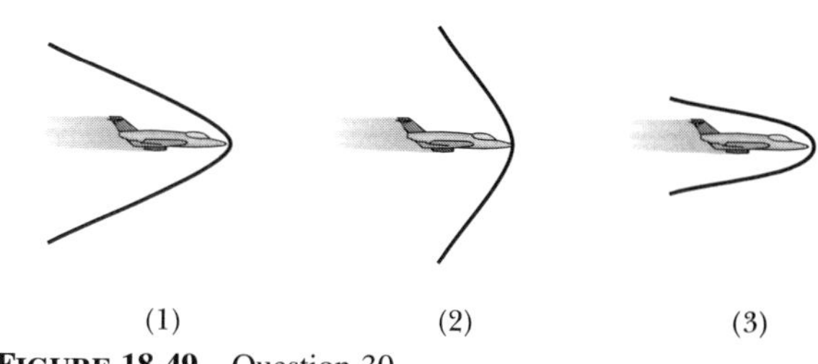

FIGURE 18-49 Question 30.

EXERCISES & PROBLEMS

97. A man strikes one end of a thin rod with a hammer. The speed of sound in the rod is 15 times the speed of sound in air. A woman, at the other end with her ear close to the rod, hears the sound of the blow twice with a 0.12 s interval between; one sound comes through the rod and the other comes through the air alongside the rod. If the speed of sound in air is 343 m/s, what is the length of the rod?

98. The sound intensity is 0.0080 W/m^2 at a distance of 10 m from an isotropic point source of sound. (a) What is the power of the source? (b) What is the sound intensity at a distance of 5.0 m from the source? (c) What is the sound level in decibels at a distance of 10 m from the source?

99. Two pipes, A and B, are of the same length, but A is open at one end only while B is open at both ends. If the ratio of the frequency produced by A to the frequency produced by B is 1/2 and both pipes oscillate in the same harmonic mode, which mode is it? Consider only the first five harmonic modes.

100. A pipe 0.60 m long and closed at one end is filled with an unknown gas. The third lowest harmonic frequency for the pipe is 750 Hz. (a) What is the speed of sound in the unknown gas? (b) What is the fundamental frequency for this pipe when it is filled with the unknown gas?

101. Pipe A, which is 1.2 m long and open at both ends, oscillates at its third lowest harmonic frequency. It is filled with air for which the speed of sound is 343 m/s. Pipe B, which is closed at one end, oscillates at its second lowest harmonic frequency. The frequencies of pipes A and B happen to match. (a) If an x axis extends along the interior of pipe A, with $x = 0$ at one end, where along the axis are the displacement nodes? (b) How long is pipe B? (c) What is the lowest harmonic frequency of pipe A?

102. A guitar player tunes the fundamental frequency of a guitar string to 400 Hz. (a) What will be the next frequency if she then increases the tension in the string by 20%. (b) What is it if, instead, she decreases the length along which the string oscillates by sliding her finger from the tuning key 1/3 of the way down the string toward the bridge at the lower end?

103. A police car is chasing a speeding Porsche 911. The Porsche's maximum speed is 80 m/s (180 mi/h) and the police car's is 54 m/s (120 mi/h). At the moment both cars reach their maximum speed, what frequency will the Porsche driver hear if the frequency of the police car's siren is 440 Hz? The speed of sound in air is 340 m/s.

104. A listener at rest (with respect to the air and ground) hears a signal of frequency f_1 from a source moving toward him with a velocity of 15 m/s, east. If the listener then moves toward the approaching source with a velocity of 25 m/s, west, he hears a frequency f_2 that differs from f_1 by 37 Hz. What is the frequency of the source? The speed of sound in air is 340 m/s.

105. Passengers in an auto traveling 16 m/s toward the east hear a siren frequency of 950 Hz from an emergency vehicle approaching them from behind at a speed (relative to the air and ground) of 40 m/s. The speed of sound in air is 340 m/s. (a) What siren frequency would a passenger riding in the emergency vehicle hear? (b) What frequency would the passengers in the auto hear after the emergency vehicle has passed them?

Tutorial Problem

106. In this problem you will look at the Doppler effect and how motion by the source or detector affects the wavelength of sound in air and the apparent frequency heard at the detector. The Doppler effect is simpler when the relative motion between the source and the detector is along the line between them, and that is the only case that will be discussed quantitatively in this problem.

Consider a police car with a siren emitting sounds at a frequency of 1200 Hz. (a) What does that statement mean? After all, the apparent frequency will depend on the motion of the police car. (b) Draw a qualitative sketch showing the police car moving to the right and how the wavefronts are spaced in front of and behind the car. (c) If the police car were at rest, what would be the wavelength (in air) of the siren sound? Use reasonable values for any physical quantities needed to calculate this wavelength. (d) If the car were moving at a speed of 28 m/s, what would be the wavelength (in air) of the siren sound in front of the car and behind the car? (e) What would be the apparent frequency heard by a person standing behind the police car? Use the results of parts (c) and (d) to make this calculation.

The sound of the siren reflects off a building directly in front of the police car and travels back to the left. The building then acts as a source of the reflected sound. (f) What is the wavelength of the reflected sound and the frequency of this source (the building)? Explain your reasoning thoroughly, using complete sentences. (g) What is the apparent frequency of the reflected sound as heard by the person of part (e)? (h) What is the apparent frequency of the reflected sound as heard by the police in their speeding car? (i) What are (1) the apparent frequency of the direct siren sound and (2) the apparent frequency of the reflected siren sound as heard by someone in a car that is trailing the police car at a speed of 16 m/s?

Answers

(a) The statement means that the siren is oscillating 1200 times a second. The apparent frequency heard by an observer may be equal to, more than, or less than that oscillation frequency. The police in the police car hear a frequency equal to the oscillation frequency.

(b) The wavefronts are closer together in front of the car and farther apart behind the car.

(c) If the police car were at rest, the wavelength of the 1200 Hz sound would be

$$\lambda_0 = \frac{v}{f} = \frac{340 \text{ m/s}}{1200 \text{ Hz}} = 0.283 \text{ m},$$

where 340 m/s is the speed of sound in air.

(d) If the police car were moving, the wavelength in front of it would be reduced and would be

$$\lambda = \lambda_0\left(\frac{v - v_S}{v}\right)$$

$$= (0.283 \text{ m})\left(\frac{340 \text{ m/s} - 28 \text{ m/s}}{340 \text{ m/s}}\right) = 0.260 \text{ m}.$$

The wavelength behind the police car would be increased and would be

$$\lambda = \lambda_0\left(\frac{v + v_S}{v}\right)$$

$$= (0.283 \text{ m})\left(\frac{340 \text{ m/s} + 28 \text{ m/s}}{340 \text{ m/s}}\right) = 0.306 \text{ m}.$$

(e) The apparent frequency f' at the receiver may be found from the wavelength λ by

$$f' = \frac{v}{\lambda} = \frac{340 \text{ m/s}}{0.306 \text{ m}} = 1111 \text{ Hz}.$$

(f) The wavelength in air of the reflected sound is just the wavelength of the sound in front of the police car: 0.260 m, as calculated in part (d). This means that the frequency emitted by the (stationary) building is

$$f = \frac{v}{\lambda} = \frac{340 \text{ m/s}}{0.260 \text{ m}} = 1308 \text{ Hz}.$$

(g) The person hears the sound at its reflected frequency, 1308 Hz, since neither the source (the building) nor the detector (the person) is moving.

(h) The police car acts as a detector that moves toward a stationary source. The apparent frequency heard by the police is then

$$f' = \frac{v + v_{\text{car}}}{\lambda_{\text{reflected}}} = \frac{340 \text{ m/s} + 28 \text{ m/s}}{0.260 \text{ m}} = 1415 \text{ Hz}.$$

(i) (1) The motion of the police car tends to decrease the frequency of the siren's direct sound, but the motion of the trailing car tends to increase it. We can calculate the frequency that is actually heard by using the wavelength we calculated in part (d), that is, the wavelength of the sound behind the police car. The frequency of the direct sound is

$$f_c = \frac{v + v_c}{\lambda} = \frac{340 \text{ m/s} + 16 \text{ m/s}}{0.36 \text{ m}} = 1163 \text{ Hz},$$

where the subscript refers to the trailing car. (2) The sound of the siren as reflected by the building is increased in apparent frequency by the motion of the car; its apparent frequency is then

$$f_{cr} = \frac{v + v_c}{\lambda} = \frac{340 \text{ m/s} + 16 \text{ m/s}}{0.260 \text{ m}} = 1369 \text{ Hz},$$

where we are now using the wavelength (0.260 m) of the reflected sound wave (hence, the extra subscript r).

Graphing Calculators

107. *Coffee-Cup Acoustics.* If you stir water in a ceramic or glass mug with a metal spoon, striking the inner wall with the spoon, you hear a certain resonant frequency f corresponding to a standing sound wave in the water. If you then stir in a powder, such as cocoa or coffee, the resonant frequency quickly and noticeably shifts to a different value f_{shift} before it gradually returns to its former value. The shift in resonant frequency is due to the change in the speed of sound in the water owing to the formation of air bubbles as the powder enters the water. The shift decreases as the air bubbles rise and pop open, until all the bubbles have disappeared and the frequency has returned to f.

(a) Does the frequency increase or decrease? (b) Show that the speed of sound v in the water–bubble mixture is related to the density ρ and volume V of the mixture by

$$\frac{1}{v^2} = \frac{\rho}{V}\frac{dV}{dp},$$

where dV/dp is the change in volume due to the change in pressure of a sound wave.

This volume change consists of a volume change dV_w in the water and a volume change dV_a in the air bubbles: $dV = dV_w + dV_a$. The volume V of the mixture is approximately the volume V_w of the water alone because the volume V_a of the air bubbles is small. That is, the ratio $r = V_a/V_w$ is small. Similarly, the density ρ of the mixture is approximately the density ρ_w of the water because the density ρ_a of air is so small.

(c) Find an expression for v in terms of the variable r and the constants ρ_w, ρ_a, the speed of sound v_a in air, and the speed of sound v_w in water. (d) Substituting appropriate values for the constants in this expression, find the ratio f_{shift}/f in terms of r. In normal circumstances, r can range from 4.0×10^{-3} (many bubbles have just formed) to 0 (the bubbles have dissipated). (e) Graph f_{shift}/f versus r for this range. (f) From the graph, find the value of r that gives $f_{shift}/f = 1/3$. (Adapted from "The Hot Chocolate Effect," by Frank S. Crawford, *American Journal of Physics,* 1982, Vol. 50, pp. 398–404.)

Chapter Nineteen
Temperature, Heat, and the First Law of Thermodynamics

QUESTIONS

16. Three different materials of identical masses are placed, in turn, in a special freezer that can extract energy from a material at a certain constant rate. During the cooling process, each material begins in the liquid state and ends in the solid state; Fig. 19-44 gives the temperature T versus time t for the three materials. (a) For material 1, is the specific heat for the liquid state greater than or less than that for the solid state? Rank the materials according to (b) their freezing-point temperatures, (c) their specific heats in the liquid state, (d) their specific heats in the solid state, and (e) their heats of fusion, all greatest first.

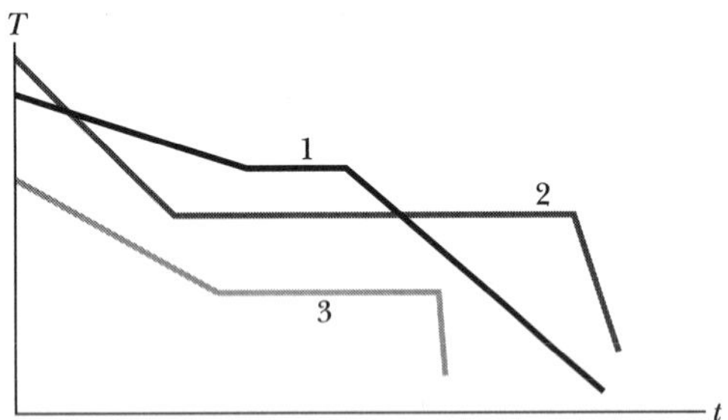

FIGURE 19-44 Question 16.

17. Three different materials of identical masses are placed, in turn, in a special oven where a material absorbs energy at a certain constant rate. During the heating process, each material begins in the liquid state and ends in the gaseous state; Fig. 19-45 gives the temperature T versus time t for the three materials. (a) For material 1, is the specific heat for the liquid state greater than or less than that for the gaseous state? Rank the three materials according to (b) their boiling-point temperatures, (c) their specific heats in the liquid state, and (d) their heats of vaporization, all greatest first.

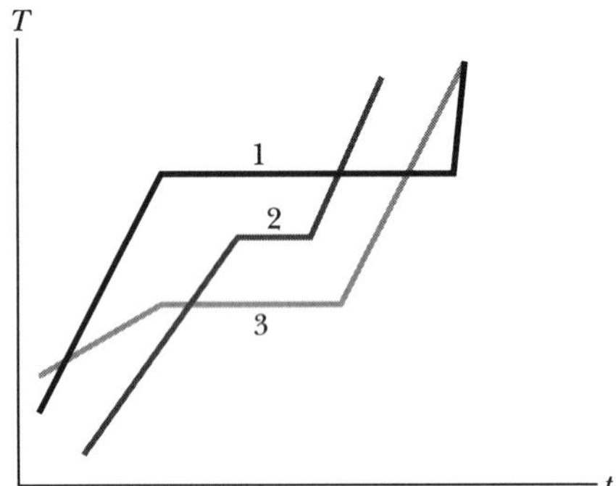

FIGURE 19-45 Question 17.

18. In four experiments, a material A at a particular low temperature T_C and a material B at a particular high temperature T_H are placed in an isolated and insulated container. When they reach thermal equilibrium with each other (no phase change occurs), their final common temperature T is measured. The masses m_A and m_B and specific heats c_A and c_B of the materials are given in the table. Without written calculation, rank the four experiments according to the final temperature T, greatest first.

EXPERIMENT	m_A	c_A	m_B	c_B
1	m	c	m	c
2	m	c	$2m$	c
3	m	$2c$	m	c
4	$2m$	c	m	$2c$

19. In a thermally isolated container, material A of mass m is placed against material B, also of mass m but at a higher temperature. When thermal equilibrium is reached, the temperature changes ΔT_A and ΔT_B of A and B are recorded. Then the experiment is repeated, using A with other materials, all of the same mass m. The results are given in the table. Rank the four materials according to their specific heats, greatest first.

EXPERIMENT	TEMPERATURE CHANGES	
1	$\Delta T_A = +50$ C°	$\Delta T_B = -50$ C°
2	$\Delta T_A = +10$ C°	$\Delta T_C = -20$ C°
3	$\Delta T_A = +\ 2$ C°	$\Delta T_D = -40$ C°

20. Substance A is heated to an initial temperature $T_i = 22$°C and then placed in an equal mass of water at a temperature of 10°C, in a thermally isolated container. When thermal equilibrium is reached, substance A and the water have a final temperature T_f = 12°C. The experiment is repeated with three other substances, each with the same mass as the water, which always has an initial temperature of 10°C. The results are given in the table. Rank the four substances and the water according to their specific heats, greatest first.

SUBSTANCE	T_i	T_f
A	22°C	12°C
B	44°C	14°C
C	33°C	13°C
D	24°C	14°C

21. A sample A of liquid water and a sample B of ice, of identical masses, are placed in a thermally isolated (insulated) container and allowed to come to thermal equilibrium. Figure 19-46a is a sketch of the temperature T of the samples versus time t. (a) Is the equilibrium temperature above, below, or at the freezing point of water? (b) In reaching equilibrium, does the liquid partly freeze or fully freeze, or does it undergo no freezing? (c) Does the ice partly melt or fully melt, or does it undergo no melting?

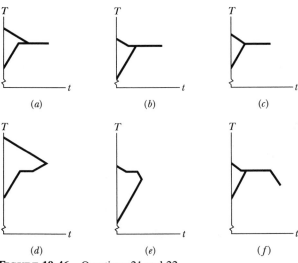

FIGURE 19-46 Questions 21 and 22.

22. Question 21 continued: Figure 19-46 gives additional sketches of T versus t, of which one or more are impossible to produce. (a) What is impossible about any impossible sketch? (b) Of the possible ones, is the equilibrium temperature above, below, or at the freezing point of water? (c) In reaching equilibrium, does the liquid partly freeze or fully freeze, or does it undergo no freezing? And does the ice partly melt or fully melt, or does it undergo no melting?

23. The p-V diagram of Fig. 19-47 indicates four processes that take a gas between state A and state B. Rank the four processes according to (a) the work W done by the gas and (b) the heat Q transferred between the gas and a thermal reservoir, most positive first, most negative last. (c) Rank the four processes according to the magnitude (or absolute value) of $Q - W$.

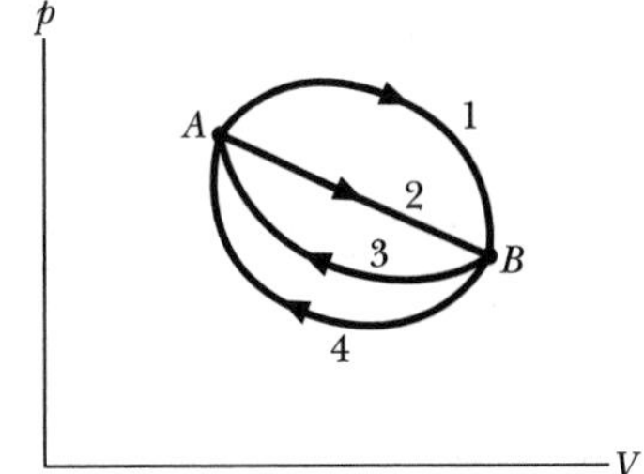

FIGURE 19-47 Question 23.

24. Figure 19-48a gives the temperature along an x axis that extends directly through a wall consisting of three layers. The air temperature on one side of the wall differs from that on the other side; the heat conduction through the wall is constant (steady). (a) Rank the three layers according to their thermal conductivities, greatest first. (b) Figures 19-48b, c, and d are similar in intent

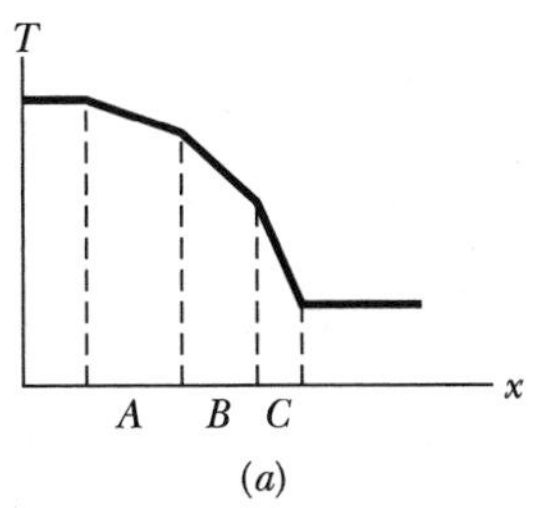

(a)

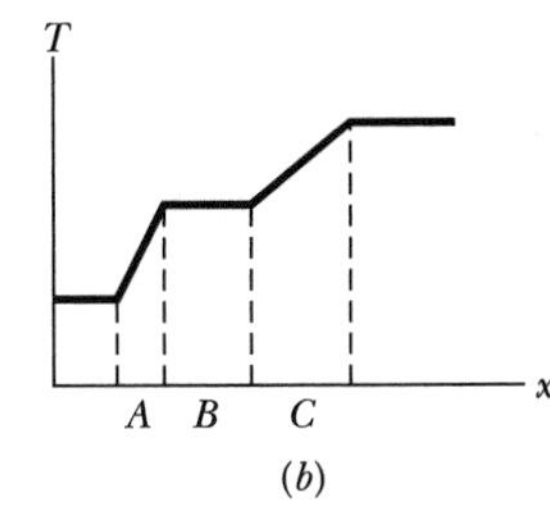

(b)

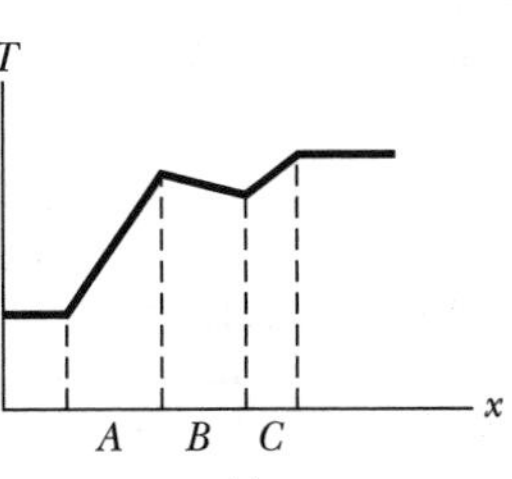

(c)

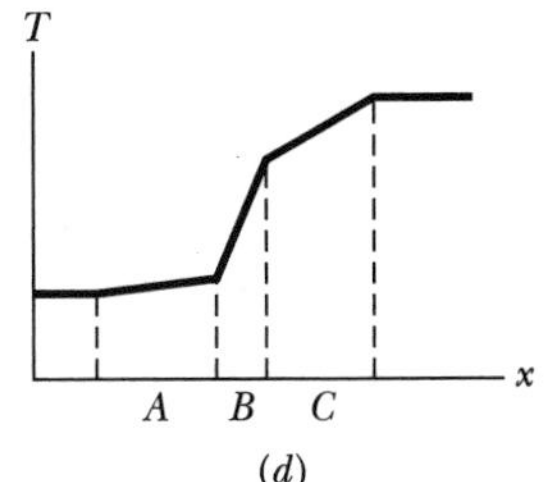

(d)

FIGURE 19-48 Question 24.

but may be impossible. Indicate which are impossible and why. For any that are possible, rank the three layers of the wall according to their thermal conductivities, greatest first.

25. A ball of surface temperature T is in thermal equilibrium with its environment. (a) Which of the curves in Fig. 19-49 gives the energy E radiated by the sphere versus time t? (b) Which curve gives the energy E absorbed by the sphere versus time t?

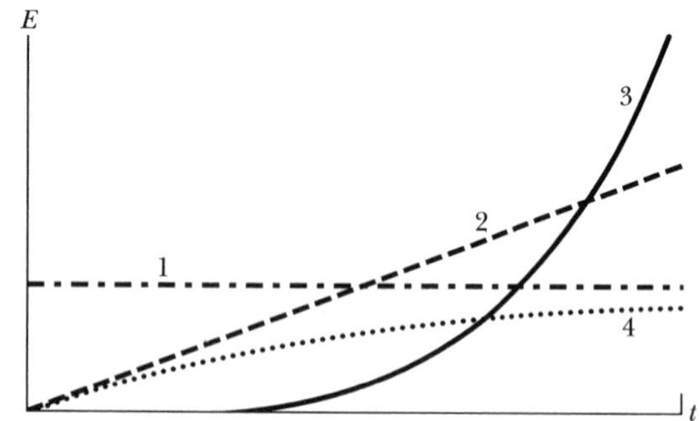

FIGURE 19-49 Question 25.

26. Three spheres of the same radius and emissivity have different (constant) surface temperatures $T_1 > T_2 > T_3$. Three of the curves in Fig. 19-50 pertain to the energy radiated by the surfaces versus time t. Which curve corresponds to which surface temperature?

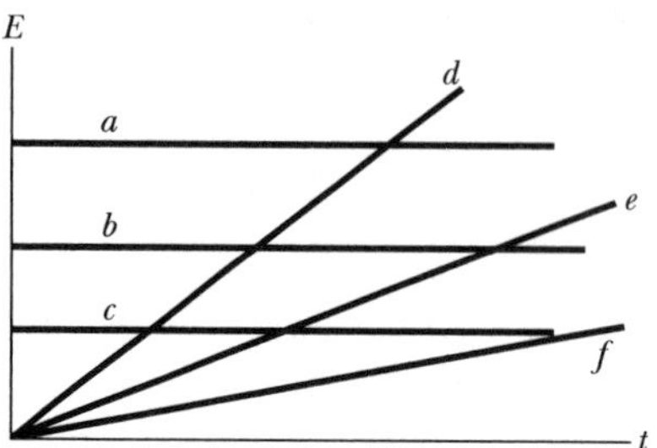

FIGURE 19-50 Question 26.

27. *Organizing question:* A ball at temperature $+20°C$ is in an environment at temperature $-20°C$. The radius of the ball is r and its emissivity is 0.8. Which of the following expressions give the ball's net rate P_n of energy exchange with the environment due to thermal radiation? For each wrong expression, explain why it is wrong.

(a) $\sigma(0.8)\pi r^2[(-20°C)^4 - (20°C)^4]$
(b) $\sigma(0.8)4\pi r^2[(-20°C)^4 - (20°C)^4]$
(c) $\sigma(0.8)\pi r^2(-20°C - 20°C)^4$
(d) $\sigma(0.8)4\pi r^2(-20°C - 20°C)^4$
(e) $\sigma(0.8)\pi r^2[(273°C - 20°C)^4 - (273°C + 20°C)^4]$
(f) $\sigma(0.8)4\pi r^2[(273°C - 20°C)^4 - (273°C + 20°C)^4]$
(g) $\sigma(0.8)\pi r^2[(253\text{ K})^4 - (293\text{ K})^4]$
(h) $\sigma(0.8)4\pi r^2[(253\text{ K})^4 - (293\text{ K})^4]$
(i) $\sigma(0.8)\pi r^2(253\text{ K} - 293\text{ K})^4$
(j) $\sigma(0.8)4\pi r^2(253\text{ K} - 293\text{ K})^4$

EXERCISES & PROBLEMS

98. On an X temperature scale, water freezes at $-125.0°X$ and boils at $375.0°X$. On a Y temperature scale, water freezes at $-70.00°Y$ and boils at $-30.00°Y$. A temperature of $50.00°Y$ corresponds to what temperature on the X scale?

99. A rectangular plate of glass initially has the dimensions of 0.200 m by 0.300 m. The coefficient of linear expansion for the glass is 9.00×10^{-6}/K. What is the change in the plate's area if its temperature is increased by 20 K?

100. A clock pendulum made of Invar (see Table 19-2) has a period of 0.50 s and is accurate at 20°C. If the clock is used in a climate where the temperature averages 30°C, what correction (approximately) is necessary at the end of 30 days to the time given by the clock?

101. The timing of a certain electric watch is governed by a small tuning fork. The frequency of the fork is inversely proportional to the square root of the length of the fork. What is the fractional gain or loss in time for a quartz tuning fork 8.00 mm long at (a) $-40.0°F$ and (b) $+120°F$ if it keeps perfect time at 25.0°F?

102. (a) Show that if the lengths of two rods of different solids are inversely proportional to their respective coefficients of linear expansion at the same initial temperature, the difference in length between them will be constant at all temperatures. (b) What should be the lengths of a steel and a brass rod at 0.00°C so that at all temperatures their difference in length is 0.30 m?

103. In a certain experiment, it was necessary to be able to move a small radioactive source at selected, extremely slow speeds. This was accomplished by fastening the source to one end of an aluminum rod and heating the central section of the rod in a controlled way. If the effective heated section of the rod in Fig. 19-51 is 2.00 cm, at what constant rate must the temperature of the rod be changed if the source is to move at a constant speed of 100 nm/s?

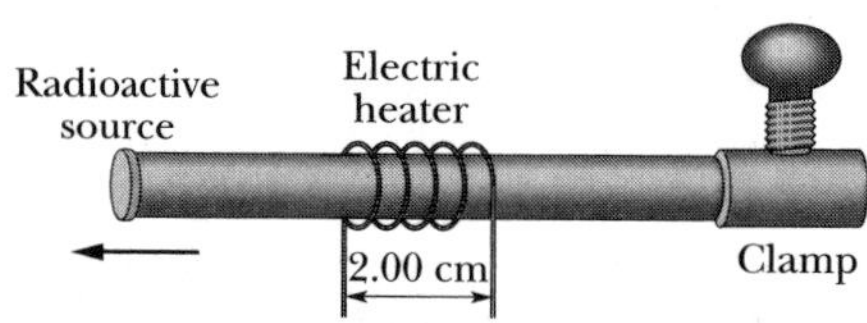

FIGURE 19-51 Problem 103.

104. A 1.28 m long vertical glass tube is half filled with a liquid at 20°C. How much will the height of the liquid column change when the tube is heated to 30°C? Take $\alpha_{\text{glass}} = 1.0 \times 10^{-5}$/K and $\beta_{\text{liquid}} = 4.0 \times 10^{-5}$/K.

105. A thick aluminum rod and a thin steel wire are attached in parallel, as shown in Fig. 19-52. The temperature is 10.0°C. Both the rod and the wire are 85.0 cm long and neither is under stress. The system is heated to 120°C. Calculate the resulting stress in the wire, assuming that the rod expands freely.

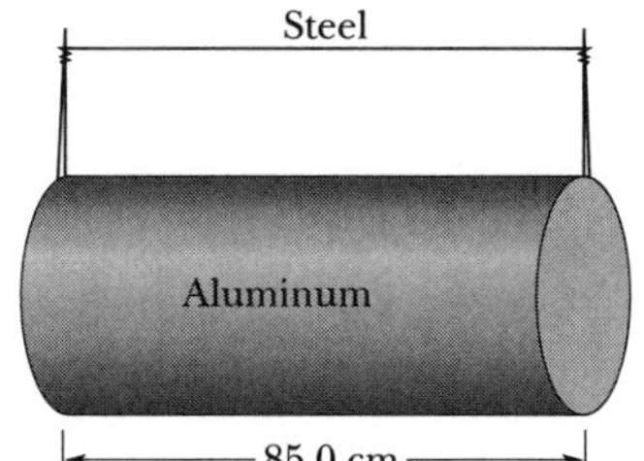

FIGURE 19-52 Problem 105.

106. A thick aluminum cube 20.0 cm on an edge floats on mercury. How much farther will the block sink when the temperature rises from 270 to 320 K? (The coefficient of volume expansion of mercury is 1.80×10^{-4}/K.)

107. Three equal-length straight rods, of aluminum, Invar, and steel, all at 20.0°C, form an equilateral triangle with hinge pins at the vertices. At what temperature will the angle opposite the Invar rod be 59.95°? See Appendix E for needed trigonometric formulas.

108. Two rods of different materials but having the same lengths L and cross-sectional areas A are arranged end to end between fixed, rigid supports, as shown in Fig. 19-53*a*. The temperature is T and there is no initial stress. The rods are heated, so that their temperature increases by ΔT. Show that the rod interface is displaced on heating (Fig. 19-53*b*) by an amount

$$\Delta L = \left(\frac{\alpha_1 E_1 - \alpha_2 E_2}{E_1 + E_2}\right) L\, \Delta T.$$

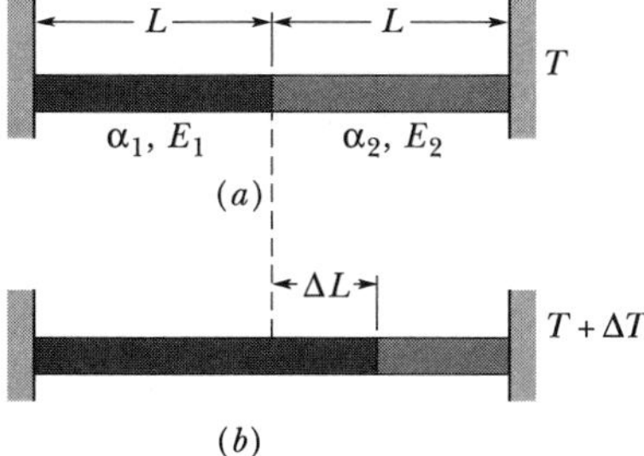

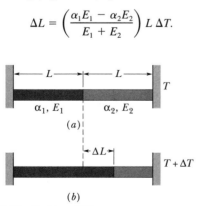

FIGURE 19-53 Problem 108.

109. A 15.0 kg sample of ice is initially at a temperature of −20.0°C. Then 7.0×10^6 J is added as heat to the sample, which is otherwise isolated. What then is the sample's temperature?

110. Ethyl alcohol has a boiling point of 78°C, a freezing point of −114°C, a heat of vaporization of 879 kJ/kg, a heat of fusion of 109 kJ/kg, and a specific heat of 2.43 kJ/kg · K. How much energy must be removed from 0.510 kg of ethyl alcohol that is initially a gas at 78°C so that it is then a solid at −114°C?

111. A 3.00 kg piece of copper at temperature 70.0°C is placed in 4.00 kg of water at temperature 10.0°C. The copper–water system is isolated. What is the final (equilibrium) temperature of the system?

112. Two metal blocks are insulated from their surroundings. The first block, which has mass $m_1 = 3.16$ kg and is at temperature $T_1 = 17.0$°C, has a specific heat four times that of the second block. This second block is at temperature $T_2 = 47.0$°C, and its coefficient of linear expansion is 15.0×10^{-6}/K. When the two blocks are brought together and allowed to come to thermal equilibrium, the area of one face of the second block is found to have decreased by 0.0300%. Find the mass of the second block.

113. Figure 19-54*a* shows a closed cycle for an ideal gas. From *c* to *b*, the heat transfer from the gas is 40 J. From *b* to *a*, the heat transfer from the gas is 130 J and the magnitude of the work done by the gas is 80 J. From *a* to *c*, the heat transfer to the gas is 400 J. What is the work done by the gas from *a* to *c*? (*Hint:* You need to supply the positive and negative signs to the given data.)

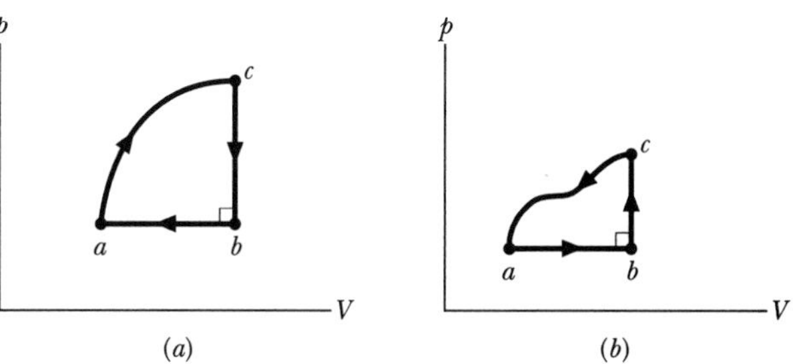

FIGURE 19-54 Problems 113 and 114.

114. Figure 19-54*b* shows a closed cycle for a gas. The change in internal energy along the path from *c* to *a* is −160 J. The heat transferred to the gas is 200 J along the path from *a* to *b* and 40 J along the path from *b* to *c*. How much work is done by the gas along the path (a) from *a* to *c* and (b) from *a* to *b*?

115. A sample of gas undergoes a transition from an initial state *a* to a final state *b* by three different paths (processes), as shown in the *p*-*V* diagram in Fig. 19-55. The heat added to the gas in process 1 is $10p_iV_i$. In terms of p_iV_i, what are (a) the heat added to the gas in process 2 and (b) the change in internal energy that the gas undergoes in process 3?

116. How many 20 g ice cubes, whose initial temperature is −10°C, must be added to 1.0 L of hot tea, whose initial temperature is 90°C, for the final mixture to have a temperature of 10°C? Assume that all the ice is melted in the final mixture and the specific heat of tea is the same as that of water.

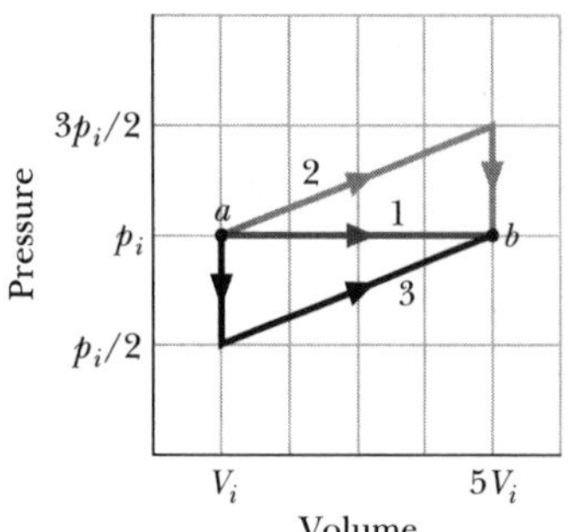

FIGURE 19-55 Problem 115.

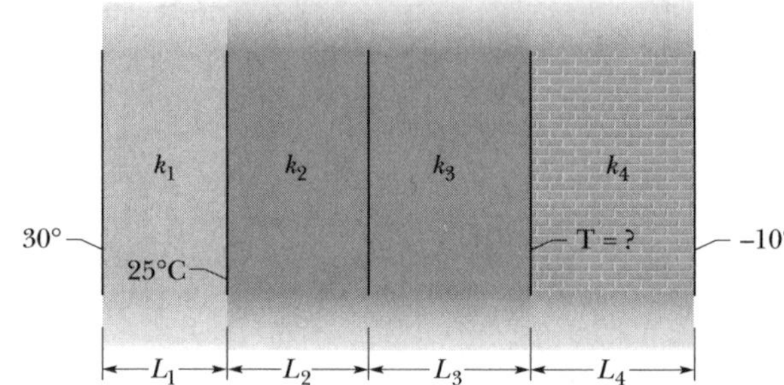

FIGURE 19-57 Problem 119.

117. A sample of gas expands from an initial pressure and volume of 10 Pa and 1.0 m^3 to a final volume of 2.0 m^3. During the expansion, the pressure and volume are related by the equation $p = aV^2$, where $a = 10$ N/m^8. Determine the work done by the gas during this expansion.

118. A cylinder has a well-fitted 2.0 kg metal piston with a cross-sectional area of 2.0 cm^2 (Fig. 19-56). The cylinder contains water and steam at constant temperature. The piston is observed to fall slowly at a rate of 0.30 cm/s because the system loses energy through the cylinder walls as heat. As this happens, some steam condenses in the chamber, where the density of the steam is 6.0×10^{-4} g/cm^3 and the atmospheric pressure is 1.0 atm. (a) Calculate the rate of condensation of steam. (b) At what rate is energy being lost as heat by the chamber? (c) What is the rate of change of internal energy of the system of steam and water inside the chamber?

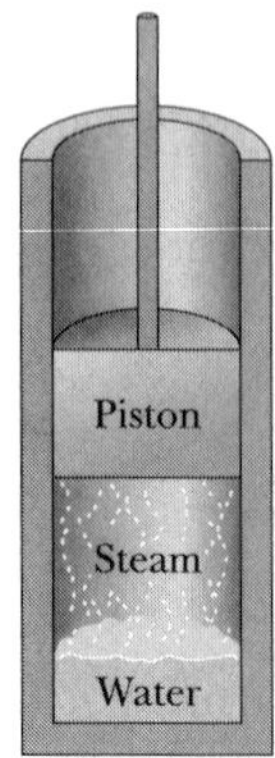

FIGURE 19-56 Problem 118.

119. Figure 19-57 shows (in cross section) a wall that consists of four layers. The thermal conductivities are $k_1 = 0.060$ W/m · K, $k_3 = 0.040$ W/m · K, and $k_4 = 0.12$ W/m · K (k_2 is not known). The layer thicknesses are $L_1 = 1.5$ cm, $L_3 = 2.8$ cm, and $L_4 = 3.5$ cm (L_2 is not known). Energy transfer through the wall is steady. What is the temperature of the interface indicated?

120. A cube with edge length 2.0×10^{-5} m is at a temperature of 50°C and has an emissivity of 0.80. It is suspended in an environment that has a temperature of 20°C. What is the net power at which it loses energy?

121. Suppose that you intercept 5.0×10^{-3} of the energy radiated by a hot sphere that has a radius of 0.020 m, an emissivity of 0.80, and a surface temperature of 500 K. How much energy do you intercept in 2.0 min?

122. An idealized representation of the air temperature as a function of distance from a single-pane window on a calm winter day is shown in Fig. 19-58. The window dimensions are 60 cm $\times$ 60 cm $\times$ 0.50 cm. Assume that heat is conducted along a path that is perpendicular to the window, from points 8.0 cm from the window on one side to points 8.0 cm from it on the other side. (a) At what rate is heat conducted through the window area? (*Hint:* The temperature drop across the window glass is very small.) (b) Estimate the difference in temperature between the inner and outer glass surfaces.

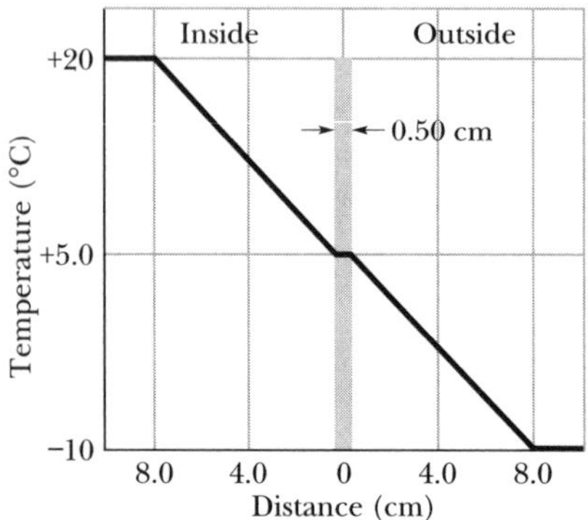

FIGURE 19-58 Problem 122.

Tutorial Problem

123. A system consists of 200 g of water that is initially ice at −20°C. Energy is added to the system until the water is converted to steam at 100°C. (The system is otherwise isolated from its environment.) (a) List the four processes in which energy must be added to the system during this heating. For each, how can the process be described in words, what equation gives the required energy, and how much is that energy? (b) What is the total energy required for the heating, and what fraction of the total is added in each process? (c) Graph the heat versus temperature. Identify which portions of the graph correspond to which process. (d) Suppose that the heat is provided by an electric heater at the rate of 240 W. Graph the temperature of the system versus time.

Answers

(a) ***Process 1:*** warming the ice to the melting temperature (0°C) by a heat in the amount

$$mc_{\text{ice}}\,\Delta T = (0.200\text{ kg})(2220\text{ J/kg}\cdot\text{C°})[0°\text{C} - (-20°\text{C})]$$
$$= 8.88\text{ kJ}.$$

Process 2: melting the ice into liquid water at the melting temperature by a heat in the amount

$$mL_F = (0.200\text{ kg})(333\text{ kJ/kg}) = 66.6\text{ kJ}.$$

Process 3: warming the liquid water from its freezing temperature to the vaporization temperature (100°C) by a heat in the amount of

$$mc_{\text{liq}}\,\Delta T = (0.200\text{ kg})(4190\text{ J/kg}\cdot\text{C°})(100°\text{C} - 0°\text{C})$$
$$= 83.8\text{ kJ}.$$

Process 4: vaporizing the water to form steam at the vaporization temperature by a heat in the amount of

$$mL_V = (0.200\text{ kg})(2256\text{ kJ/kg}) = 451\text{ kJ}.$$

(b) The total heat is (8.88 + 66.6 + 83.8 + 451) kJ = 610 kJ.

Process 1: (8.88 kJ)/(610 kJ) = 0.015 or 1.5%;

Process 2: (66.6 kJ)/(610 kJ) = 0.11 or 11%;

Process 3: (83.8 kJ)/(610 kJ) = 0.014 or 1.4%;

Process 4: (451 kJ)/(610 kJ) = 0.74 or 74%.

(c) From −20°C to 0°C, plot the function given in (a) for process 1. Then plot a vertical line to account for process 2. Then, from 0°C to 100°C, plot the function given in (a) for process 3. Then plot a vertical line to account for process 4.

(d) The time taken by each process is as follows:

Process 1: (8.88 kJ)/(0.240 kW) = 37 s;

Process 2: (66.6 kJ)/(0.240 kW) = 278 s;

Process 3: (83.8 kJ)/(0.240 kW) = 349 s;

Process 4: (451 kJ)/(0.240 kW) = 1880 s.

(The total time is 2540 s, which equals the sum of the individual times and is also found by dividing the total heat 610 kJ by the power 0.240 kW.) Now for the graph of temperature T versus time t:

Process 1 causes a linear increase in temperature for 37 s;

Process 2 causes no change in temperature for 278 s;

Process 3 causes a linear increase in temperature for 349 s;

Process 4 causes no change in temperature for 1880 s.

Graphing Calculators

PROBLEM SOLVING TACTICS

TACTIC 3: *Storing Common Constants*

The Stefan–Boltzmann constant σ is not stored in the built-in list of constants on a TI-85/86. You can add it to a user list of constants. To get ready, press 2nd CONS and choose EDIT. To find the Greek letter to enter as the name of the constant, now press 2nd CHAR, choose GREEK, press MORE twice, and then choose σ. The symbol appears on the screen as the name. Press ENTER, press in 5.6703E-8 (the value of the Stefan–Boltzmann constant), and then press ENTER again. If you want to store another constant, choose NEXT. Otherwise exit the menu. Whenever you insert the symbol σ into a calculation, the calculator will use the value 5.6703E-8.

TACTIC 4: *Converting Negative Temperatures*

On a TI-85/86, you can use the conversion menus to convert temperatures between the Celsius, Kelvin, Fahrenheit, and Rankine scales. (The Rankine scale, like the Kelvin scale, has zero at the lowest possible temperature. However, the Rankine scale is marked off in degrees that are equal to Fahrenheit degrees: ΔT = 1 R° = 1 F°.)

When you convert a negative temperature on one scale to another scale, you *must* enclose the temperature with parentheses. For example, to convert the lowest possible temperature of −273.15 on the Celsius scale to kelvins, go to the temperature conversion menu by pressing 2nd CONV and then choosing TEMP. Then press in (−273.15) using the negation key and the parentheses keys. Next choose °C and then °K (the degree symbol is actually improper on the kelvin symbol). Pressing ENTER then gives the answer of 0, the lowest temperature on the Kelvin scale. However, if you forget the parentheses, you get a very wrong answer of −546.30.

124. Emperor penguins, those large penguins that resemble stuffy English butlers, breed and hatch their young even during severe Antarctic winters. Once an egg is laid, the father balances the egg on his feet to prevent the egg from freezing. He must do this for the full incubation period of 105 to 115 days, during which he cannot eat because his food is in the water. He can survive this long without food only if he can reduce his consumption of his internal energy significantly. If he is alone, he consumes that energy too quickly to stay warm, and will eventually abandon the egg in order to eat. To protect themselves from the cold so as to reduce the consumption of internal energy, the penguin fathers huddle closely together, in groups of perhaps several thousand. In addition to providing other benefits, the huddling reduces the rate at which the penguins radiate thermal energy to their surroundings.

Assume that a penguin father is a cylinder with top surface a, height h, surface temperature T, and emissivity ϵ. (a) Find an expression for the rate P_i at which an individual father would radiate energy to the environment from his top surface and his side surface were he is alone with his egg.

If N identical fathers were well apart from one another, the total rate of energy loss via radiation would be NP_i. Suppose, instead, that they closely huddle to form a *huddled cylinder* with top surface Na and height h. (b) Find an expression for the rate P_h at which energy is radiated by the top surface and the side surface of the huddled cylinder.

(c) Assuming $a = 0.34 \text{ m}^2$ and $h = 1.1$ m and using the expressions for P_i and P_h, graph the ratio P_h/NP_i versus N. Of course, the penguins know nothing about algebra or graphing, but their instinctive huddling reduces this ratio so that more fathers have their eggs survive to the hatching stage. From the graphs (as you will see, you need more than one version), approximate how many penguins must huddle so that P_h/NP_i is reduced to (d) 0.5, (e) 0.4, (f) 0.3, (g) 0.2, and (h) 0.15. (i) For the assumed data, what is the lower limiting value for P_h/NP_i?

Chapter Twenty
The Kinetic Theory of Gases

QUESTIONS

16. In Fig. 20-28*a*, three isothermal processes are shown for the same gas and for the same change in volume (V_i to V_f) but at different temperatures. Rank the processes according to (a) the work done by the gas, (b) the change in the internal energy of the gas, and (c) the heat transferred to the gas, greatest first.

In Fig. 20-28*b*, three isothermal processes are shown along a single isotherm, for the same change ΔV in volume. Rank the processes according to (d) the work done by the gas, (e) the change in the internal energy of the gas, and (f) the heat transferred to the gas, greatest first.

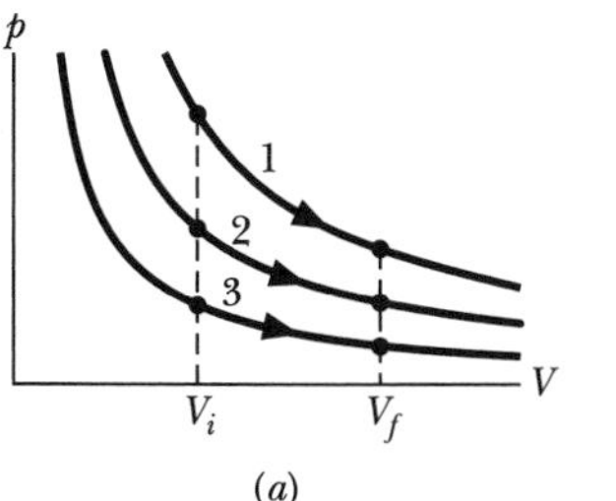

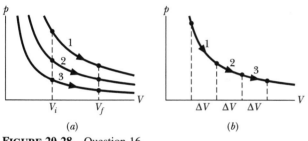

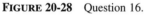

FIGURE 20-28 Question 16.

17. In Figs. 20-29*a* through *d*, ideal gases with identical numbers of moles and identical volumes are confined to identical cylinders. Each cylinder has a movable piston with identical shot on top, as in Fig. 19-12. Rank the gases according to their temperature, greatest first.

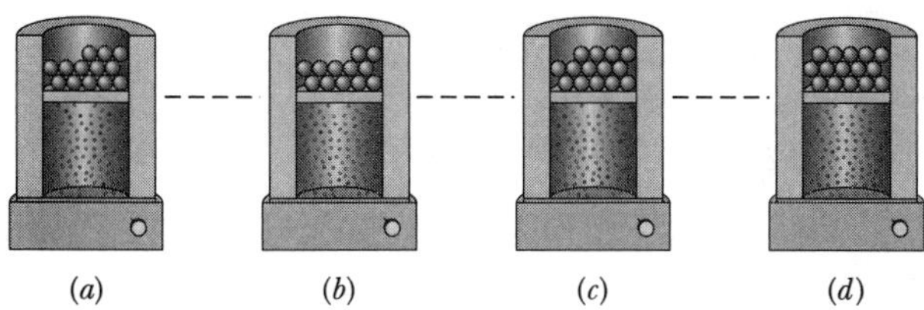

FIGURE 20-29 Question 17.

18. Figure 20-30 shows the initial state of an ideal gas and an isotherm through that state. Which of the paths shown result in (a) an increase in the root-mean-square speed of the gas molecules and (b) a decrease in their average translational kinetic energy?

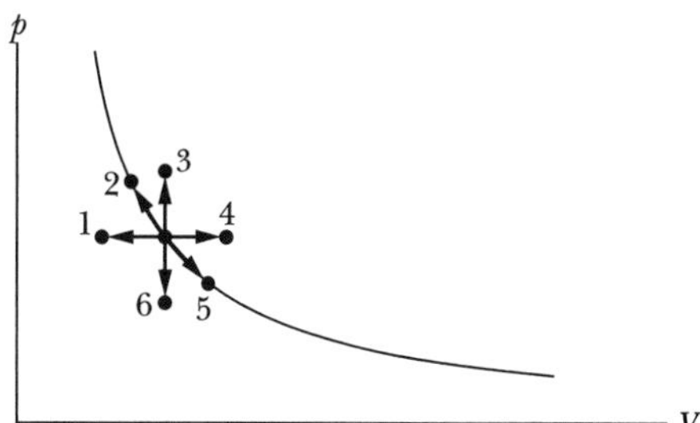

FIGURE 20-30 Question 18.

19. The following table gives the molar masses and the volumes of three ideal gasses, all at the same temperature. Rank the gasses according to (a) their root-mean-square speeds and (b) their average translational kinetic energy, greatest first.

	NITROGEN (N_2)	HYDROGEN (H_2)	OXYGEN (O_2)
Molar Mass	28*M*	2*M*	32*M*
Volume	4*V*	8*V*	6*V*

20. A certain gas is taken to the five states represented by dots in Fig. 20-31; the plotted lines are isotherms. Rank the states according to (a) the most probable speed of the molecules v_p and (b) their average speed v, greatest first.

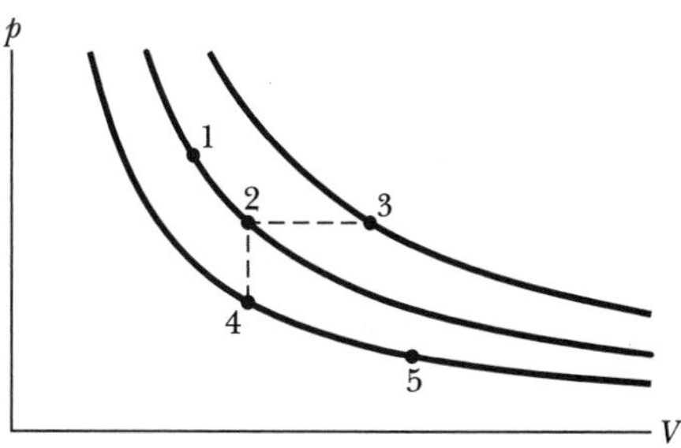

FIGURE 20-31 Question 20.

21. Reconsider Fig. 20-3 and the discussion about it. If we gradually increase the distance between the shaded wall and the wall opposite it while keeping the temperature of the gas constant, determine if the following increase, decrease, or remain the same: (a) the average rate at which momentum is delivered to the shaded wall by a molecule striking it, (b) the pressure on the shaded wall, and (c) the root-mean-square speed of the gas.

22. A one-dimensional gas is a hypothetical gas with molecules that can move along only a single axis. The following table gives, for four situations, the velocities in meters per second of such a gas having four molecules. (The positive and negative signs refer to the direction of the velocity along the axis, as usual.) Without written calculation or a calculator, rank the four situations according to the root-mean-square speed of the molecules, greatest first.

SITUATION	VELOCITIES			
a	−2	+3	−4	+5
b	+2	−3	+4	−6
c	+2	+3	+4	+5
d	+3	+3	−4	−5

23. An instrument that can monitor molecular speeds in a narrow range dv is placed in a gas of molecules. Initially no molecules have speeds in the monitored range; then the number of molecules in the range gradually increases as the temperature of the gas is changed. (a) Is the monitored range initially above or below the root-mean-square speed of the molecules? (b) Is the gas being cooled or warmed during the observations?

24. Suppose that the volume of a certain gas is to be doubled by one of the following processes: (1) isothermal expansion, (2) adiabatic expansion, (3) free expansion, or (4) expansion at constant pressure. Rank those processes according to the change in the average kinetic energy of the gas molecules, most positive change first, most negative last.

25. In each of the following six situations we give the ideal gases 30 J via heating under the circumstances listed. The gases have the same number of moles. Rank the situations according to (a) the increase in the temperature of the gas, (b) the increase in the average kinetic energy of the gas molecules, (c) the work done by the gas, and (d) the increase in the internal energy of the gas, greatest first.

GAS TYPE	CIRCUMSTANCES
1 monatomic	constant volume
2 monatomic	constant pressure
3 diatomic	constant volume, no rotation
4 diatomic	constant pressure, no rotation
5 diatomic	constant volume, with rotation
6 diatomic	constant pressure, with rotation

26. *Organizing question:* The dot in Fig. 20-32*a* represents the initial state of a gas, and the vertical line through the dot divides the p-V diagram into regions 1 and 2. For the following processes, determine whether the work W done by the gas is positive, negative, or zero: (a) the gas moves up along the vertical line, (b) it moves down along the vertical line, (c) it moves to anywhere in region 1, and (d) it moves to anywhere in region 2.

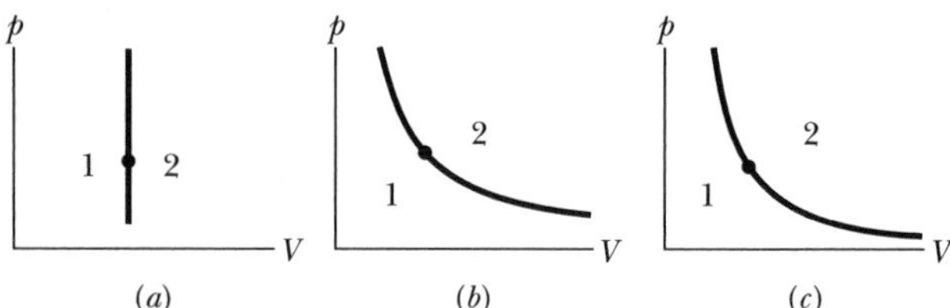

FIGURE 20-32 Questions 26 through 28.

27. *Organizing question:* The dot in Fig. 20-32*b* represents the initial state of a gas, and the isotherm through the dot divides the p-V diagram into regions 1 and 2. For the following processes, determine whether the change ΔE_{int} in the internal energy of the gas is positive, negative, or zero: (a) the gas moves up along the isotherm, (b) it moves down along the isotherm, (c) it moves to anywhere in region 1, and (d) it moves to anywhere in region 2.

Now determine whether the heat Q involved is greater than, less than, or equal to the work W done by the gas for the same processes: (e) the gas moves up along the isotherm, (f) it moves down along the isotherm, (g) it moves to anywhere in region 1, and (h) it moves to anywhere in region 2.

28. *Organizing question:* The dot in Fig. 20-32*c* represents the initial state of a gas, and the adiabat through the dot divides the p-V diagram into regions 1 and 2. For the following processes, determine whether the corresponding heat Q is positive, negative, or zero: (a) the gas moves up along the adiabat, (b) it moves down along the adiabat, (c) it moves to anywhere in region 1, and (d) it moves to anywhere in region 2.

29. Soda pop, beer, and champagne have dissolved carbon dioxide in the fluid and gaseous carbon dioxide in the small space in the container that is not occupied by the fluid. The pressure of the gas exceeds atmospheric pressure; when you open the container, the gas suddenly and adiabatically expands outward from the container against the atmosphere. Do (a) the internal energy and (b) the temperature of the gas increase or decrease? (You can often see evidence of the temperature change; it can cause a slight fog to form at the container's opening.)

30. *Get Back.* An ideal gas undergoes an isothermal expansion and then an adiabatic expansion. By which of the following processes can the gas possibly get back to its initial state: (a) An isothermal compression; (b) a constant-pressure compression and then a constant-volume process; (c) a constant-pressure compression and then an isothermal compression; (d) a constant-pressure compression and then an adiabatic compression; (e) a constant-pressure expansion and then an adiabatic compression; (f) a constant-pressure expansion and then an isothermal compression; (g) a constant-volume process and then an isothermal compression; (h) an isothermal expansion and then an adiabatic compression; (i) an isothermal compression and then an adiabatic compression?

EXERCISES & PROBLEMS

87. An ideal gas initially has a volume of 4.00 m^3, a pressure of 5.67 Pa, and a temperature of -56°C. The gas is then compressed to 7.00 m^3, leaving it with a temperature of 40°C. What then is the pressure?

88. An ideal gas undergoes an isothermal compression from an initial volume of 4.00 m^3 to a final volume of 3.00 m^3. There are 3.5 moles in the gas, and the gas temperature is 10°C. (a) How much work was done by the gas? (b) How much heat was transferred between the gas and its environment?

89. An ideal gas with 3.00 mol is initially at 425 K when it is then compressed until it reaches 350 K, without its pressure changing. The gas loses 4670 J via heat. What is the change in the internal energy of the gas?

90. At what temperature is the average translational kinetic energy of a molecule in a gas equal to 2.50 eV?

91. In an interstellar gas cloud of temperature 50 K, the pressure is 1.00×10^{-8} Pa. Assuming that the molecular diameters of the gas are all 20.0 nm, what is the mean free path?

92. The temperature of 3.00 mol of an ideal diatomic gas is increased by 40.0°C without the pressure of the gas changing. The molecules in the gas rotate but do not oscillate. (a) How much energy is transferred to the gas as heat? (b) What is the change in the internal energy of the gas? (c) How much work is done by the gas? (d) How much does the translational kinetic energy of the gas increase?

93. We give 70 J of heat to a diatomic gas, which then expands at constant pressure. The gas molecules rotate but do not oscillate. By how much did the internal energy of the gas increase?

94. Figure 20-33 shows two paths to be taken by a gas from an initial point i to a final point f. Path 1 consists of an isothermal expansion (work is 50 J in magnitude), an adiabatic expansion (work is 40 J in magnitude), an isothermal compression (work is 30 J in magnitude), and then an adiabatic compression (work is 25 J in magnitude). What is the change in the internal energy of the gas if, instead, the gas goes from point i to point f along path 2?

95. Figure 20-34 shows a cycle consisting of five paths: AB is isothermal at 300 K; BC is adiabatic with work = 5.0 J; CD is at a constant pressure of 5 atm; DE is isothermal; and EA is adiabatic with a change in internal energy of 8.0 J. What is the change in internal energy of the gas along path CD?

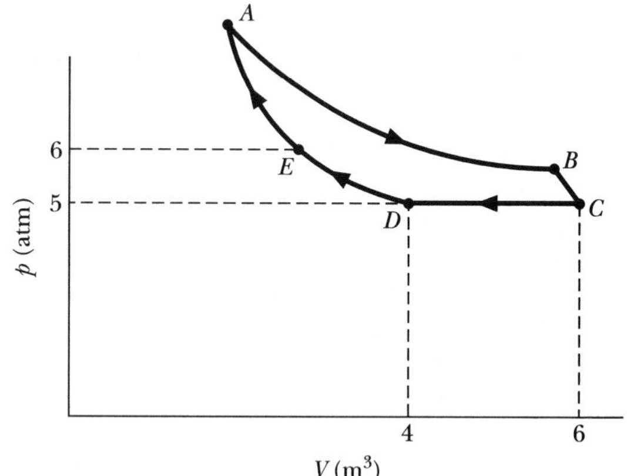

FIGURE 20-34 Problem 95.

96. Figure 20-35 represents an adiabatic compression of an ideal gas from 15 m^3 to 12 m^3, followed by an isothermal compression

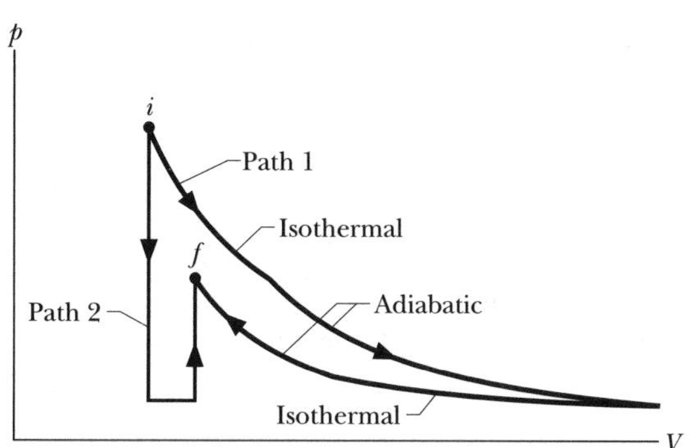

FIGURE 20-33 Problem 94.

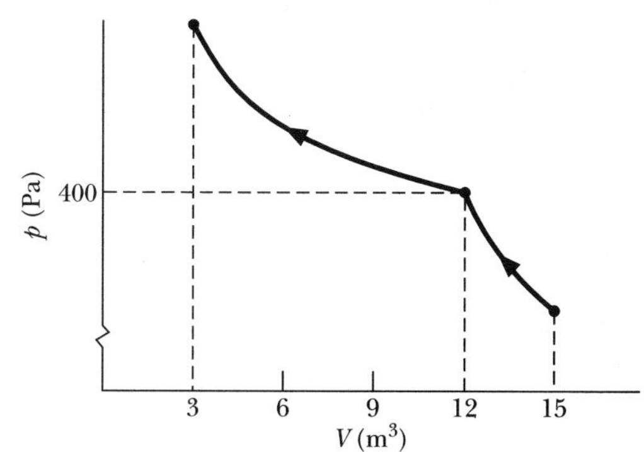

FIGURE 20-35 Problem 96.

at 300 K to a final volume of 3.0 m³. The gas has 2.0 moles. What is the total energy transferred as heat?

97. An ideal gas is suddenly allowed to expand freely so that the ratio of its new volume V_1 to its initial volume V_0 is $V_1/V_0 = 5.00$. The gas is then adiabatically compressed back to its initial volume V_0, leaving it with a pressure p_2 that is $(5.00)^{0.40}$ times its initial pressure p_0. (a) Is the gas monatomic, diatomic with no rotation of the molecules, diatomic with rotating molecules, or polyatomic? In terms of the initial average kinetic K_0 of the molecules, what are the average kinetic energies after (b) the free expansion and (c) the adiabatic compression?

98. *The Hots.* Figure 20-36 shows a temperature scale that is laid out as blocks on a game board. A sample of material begins at a temperature in one of the blocks, and then, in the steps given below, energy is added to or removed from the sample as heat. With each step, you are to determine where the sample is on the game board by giving the block's label. (*Caution:* One wrong move means all further moves are probably wrong.)

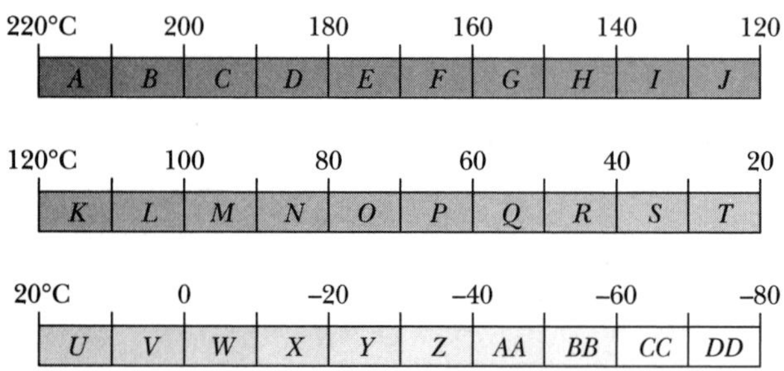

FIGURE 20-36 Problem 98.

The material is fictitious, and its properties, listed in the table, were invented just for this game. In the gaseous state, the material is diatomic and all changes occur at constant volume unless otherwise stated. The initial temperature of the sample is 80.00°C. Remove, in this order, (a) 300.0 kJ, (b) 400.0 kJ, (c) 820.0 kJ, (d) 250.0 kJ, and finally (e) 670.0 kJ. Now add, in this order, (f) 1240.0 kJ, (g) 1280 kJ, (h) 820.0 kJ, (i) 1000 kJ, (j) 583.1 kJ, (k) 166.2 kJ, (l) 277.0 kJ, (m) 581.7 kJ at constant pressure, and finally (n) 249.3 kJ at constant pressure. (*Hint:* For the last step, see Fig. 20-12 and use Eq. 20-38.)

Sample's mass = 4.000 kg
Molar mass = 3.000 g/mol
Freezing point = −15.00°C
Boiling point = 105.0°C
Heat of fusion = 150 kJ/kg
Heat of vaporization = 500 kJ/kg
Molecular rotation: $T \geq 135$°C
Molecular oscillation: $T \geq 185$°C
Specific heat: liquid = 4.00 kJ/kg · K
solid = 2.00 kJ/kg · K

Tutorial Problems

99. An ideal gas of 1.00 mol is initially at a pressure of 1.00×10^5 Pa and a temperature of 47°C. (a) What are the gas's initial temperature in kelvins and its initial volume? Plot graphs of pressure p versus volume V, p versus temperature T, and V versus T for the following processes that the gas in the initial state could undergo: (b) an *isobaric process* (a constant-pressure process) to a temperature of 480 K, (c) a constant-volume process to a temperature of 480 K, and (d) an isothermal process to twice the initial volume. On each graph, mark the initial and final states and give their values on the axes, draw the path accurately, and add an arrowhead to the path to indicate the path's direction.

Answers

(a) The temperature is (47°C + 273 K) = 320 K. From the ideal gas law, the volume is

$$V = \frac{nRT}{p} = \frac{(1.00 \text{ mol})(8.31 \text{ J/K} \cdot \text{mol})(320 \text{ K})}{1.00 \times 10^5 \text{ Pa}}$$
$$= 0.0266 \text{ m}^3.$$

(b) From the ideal gas law, the final volume is 0.0399 m³. For p versus V, plot a horizontal line at the constant pressure, from the initial volume to the final volume. For p versus T, plot a horizontal line at the constant pressure, from the initial temperature to the final temperature. For V versus T, plot a straight line from the initial volume and temperature to the final volume and temperature. This line is straight because here, from the ideal gas law, $V = nRT/p =$ (a constant)T.

(c) From the ideal gas law, the final pressure is 1.50×10^5 Pa. For p versus V, plot a vertical line at the constant volume, from the initial pressure to the final pressure. For p versus T, plot a straight line from the initial pressure and temperature to the final pressure and temperature. This line is straight because here, from the ideal gas law, $p = nRT/V =$ (a constant)T. For V versus T, plot a horizontal line at the constant volume, from the initial temperature to the final temperature.

(d) From the ideal gas law, the final pressure 0.50×10^5 Pa. For p versus V, plot the equation

$$p = \frac{nRT}{V} = \frac{(1.0 \text{ mol})(8.31 \text{ J/K} \cdot \text{mol})(320 \text{ K})}{V}$$
$$= \frac{(2659 \text{ Pa} \cdot \text{m}^3)}{V},$$

from the initial pressure and volume to the final pressure and volume. For p versus T, plot a vertical line at the constant temperature, from the initial pressure to the final pressure. For V versus T, plot a vertical line at the constant temperature, from the initial volume to the final volume.

100. Consider a system containing 1.50 moles of an ideal gas. Initially this system has a pressure of 1.00×10^5 N/m² and a temperature of 27°C = 300 K. Let's suppose that this gas has an internal energy E_{int} that depends on the absolute temperature T as

$E_{int} = 3.5nRT$. For each of the following processes, which all start from the given initial state, determine (1) the final pressure, volume, and temperature of the system. Write out your explanations in complete sentences, making any equations part of the sentence. Don't simply list equations and plug into them. Also, indicate whether you are using a law or definition to determine these quantities. (2) Next, determine the change ΔE_{int} in internal energy of the system, the heat Q added to the system, and the work W done by the system. Explain logically and completely how you determined these quantities, again using complete sentences. Determine ΔE_{int}, Q, and W in a logical order, which might differ from one process to another. (3) Last, determine the path of the process on a p-V diagram. Remember that all graphs have the same initial state. Of course, the graph should be done after you have determined the appropriate numbers.

(a) The gas undergoes a constant-pressure process to a final temperature of 400 K. (b) It undergoes a constant-volume process to a final temperature of 400 K. (c) It undergoes an isothermal process to a final pressure of 1.20×10^5 N/m^2.

Answers

(a) First, let's calculate the initial volume using the ideal gas law:

$$V_i = \frac{nRT_i}{p_i} = \frac{(1.50 \text{ mol})(8.31 \text{ J/mol}\cdot\text{K})(300 \text{ K})}{1.00 \times 10^5 \text{ Pa}} = 0.0374 \text{ m}^3.$$

We know that $p_f = p_i = 1.00 \times 10^5$ Pa and $T_f = 400$ K. So,

$$V_f = \frac{400 \text{ K}}{300 \text{ K}} V_i = 0.0499 \text{ m}^3,$$

which tells us that the gas expanded. During the expansion, the change in the internal energy is

$$\Delta E_{int} = 3.5nR\,\Delta T = (3.5)(1.5 \text{ mol})(8.31 \text{ J/mol}\cdot\text{K})(100 \text{ K}) = 4.36 \text{ kJ}$$

and the work done by the gas is

$$W = p\,\Delta V = (1.00 \times 10^5 \text{ Pa})(0.0499 \text{ m}^3 - 0.0374 \text{ m}^3) = 1.25 \text{ kJ}.$$

Then, by the first law of thermodynamics, we know that the heat is

$$Q = \Delta E_{int} + W = 4.36 \text{ kJ} + 1.25 \text{ kJ} = 5.61 \text{ kJ}.$$

(b) We know that $V_f = V_i = 0.0374$ m^3 and $T_f = 400$ K. From the ideal gas law with constant volume, the pressure is proportional to the absolute temperature, so $p_f = \frac{4}{3}p_i = 1.33 \times 10^5$ N/m^2. The change in the internal energy is 4.36 kJ, the same as in part (a) because the final temperature is the same. And $W = 0$ because the volume does not change. Thus by the first law of thermodynamics,

$$Q = \Delta E_{int} = 4.36 \text{ kJ}.$$

(c) We know that $T_f = T_i = 300$ K, which means that $\Delta E_{int} = 0$. We also know that $p_f = 1.20 \times 10^5$ N/m^2. Then, using the ideal gas law,

$$V_f = \frac{nRT}{p_f} = \frac{V_i}{1.20} = \frac{0.0374 \text{ m}^3}{1.20} = 0.0312 \text{ m}^3.$$

From Eq. 20-10, the work done is

$$W = nRT \ln (V_f/V_i) = (1.5 \text{ mol})(8.31 \text{ J/mol}\cdot\text{K})(300 \text{ K}) \ln (1/1.20) = -0.682 \text{ kJ}.$$

Then, by the first law of thermodynamics, since $\Delta E_{int} = 0$,

$$Q = W = -0.682 \text{ kJ}.$$

Chapter Twenty-One
Entropy and the Second Law of Thermodynamics

QUESTIONS

15. *Organizing question:* Block A and Block B are placed together in an insulated box and allowed to come to thermal equilibrium. The initial temperatures are $T_A = -10°C$ and $T_B = 15°C$. The materials making up the blocks have melting temperatures of $T_{A,F} = 0°C$ and $T_{B,F} = 20°C$. Block A has a mass of 0.20 kg and its material has a specific heat of 4190 J/kg · K and a heat of fusion of 333 kJ/kg. Set up equations, complete with known data, to find the change in the entropy of block A for the block to reach thermal equilibrium if the equilibrium temperature is (a) $-5.0°C$ and (b) 5.0°C.

16. An ideal gas, which is in contact with a controllable thermal reservoir, is taken reversibly from initial state i to final state f via the process plotted in Fig. 21-29. Does the entropy of (a) the gas, (b) the reservoir, and (c) the gas–reservoir system increase, decrease, or remain the same?

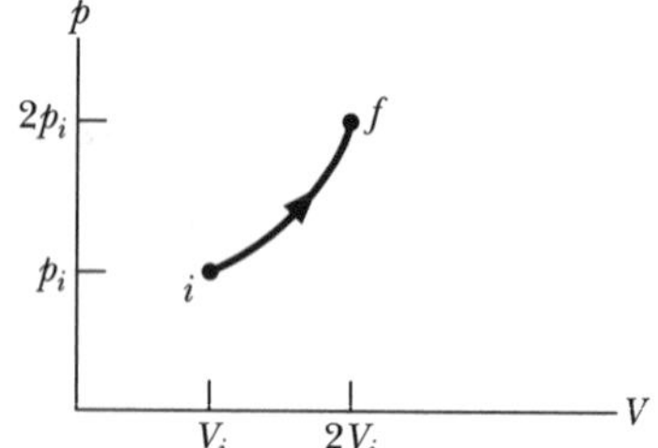

FIGURE 21-29 Question 16.

17. An ideal gas, in contact with a controllable thermal reservoir, can be taken from initial state i to final state f along the four reversible paths in Fig. 21-30. Rank the paths according to the magnitudes of the resulting entropy changes of (a) the gas, (b) the reservoir, and (c) the gas–reservoir system, greatest first.

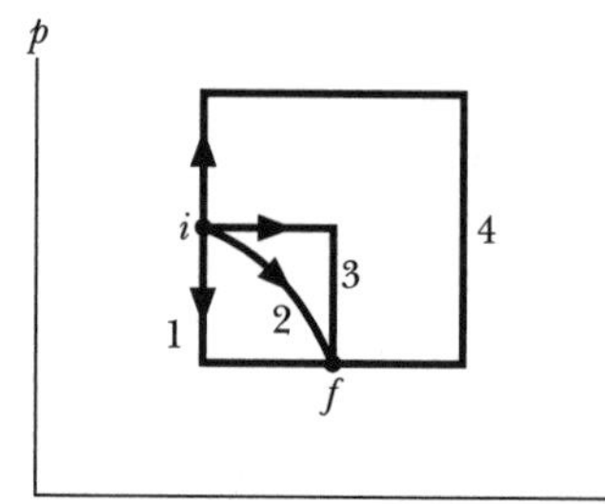

FIGURE 21-30 Question 17.

18. Figure 21-31 shows four reversible paths along which an ideal gas, in contact with a controllable thermal reservoir, can be taken from the common initial state i (lower left corner) to a final state of either f_1 (upper left corner) or f_2 (upper right corner). Rank the paths according to the magnitudes of the resulting entropy changes of (a) the gas, (b) the reservoir, and (c) the gas–reservoir system, greatest first.

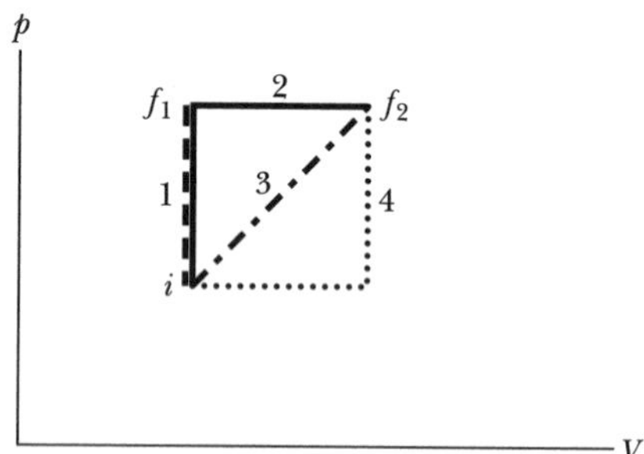

FIGURE 21-31 Question 18.

19. In four experiments, blocks A and B, starting at different initial temperatures, were brought together in an insulating box (as in Sample Problem 21-2) and allowed to reach a common final temperature. The entropy changes for the blocks in the four experiments had the following values (in Joules per kelvin), but not necessarily in the order given. Determine which values for A go with which values for B.

BLOCK	VALUES			
A	8	5	3	9
B	−3	−8	−5	−2

20. A sample of ice at $-10°C$ is placed on a controllable thermal reservoir. The water is then taken through three steps: (1) the ice is warmed to the melting point, (2) the ice is melted, and (3) the liquid is warmed to 10°C. Without written computation or a calculator, rank the three steps according to the resulting change in entropy of the sample, greatest first. (The specific heat of ice is 2220 J/kg · K, the specific heat of liquid water is 4190 J/kg · K, and the heat of fusion of water is 333,000 J/kg.)

21. The following table gives the masses, the heats of fusion, and the melting points (in kelvins) for three samples. The samples are initially frozen and at their melting points; heat is then transferred to them until they are completely unfrozen. Rank the samples according to the entropy change during this melting process, greatest first.

Sample	Mass	Heat of Fusion	Melting Point
1	M	$2L$	T
2	$2M$	$L/2$	$2T$
3	$M/2$	$2L$	$T/2$
4	$2M$	L	$T/2$

22. (a) Which of the curves in Fig. 21-32 gives the limiting efficiency of an engine as a function of the temperature ratio T_H/T_C? (b) Which curve gives the limiting coefficient of performance for a refrigerator?

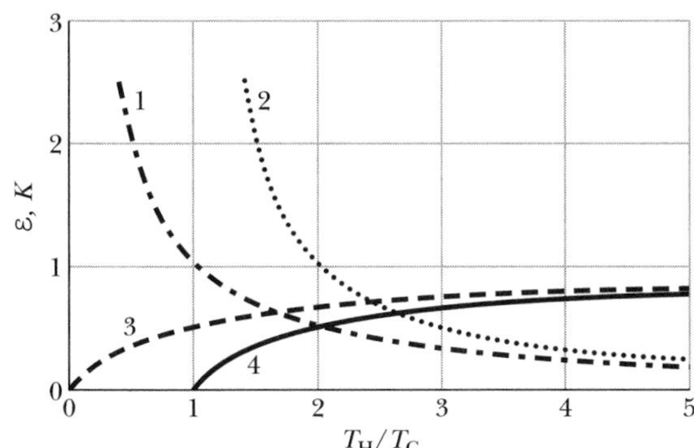

FIGURE 21-32 Question 22.

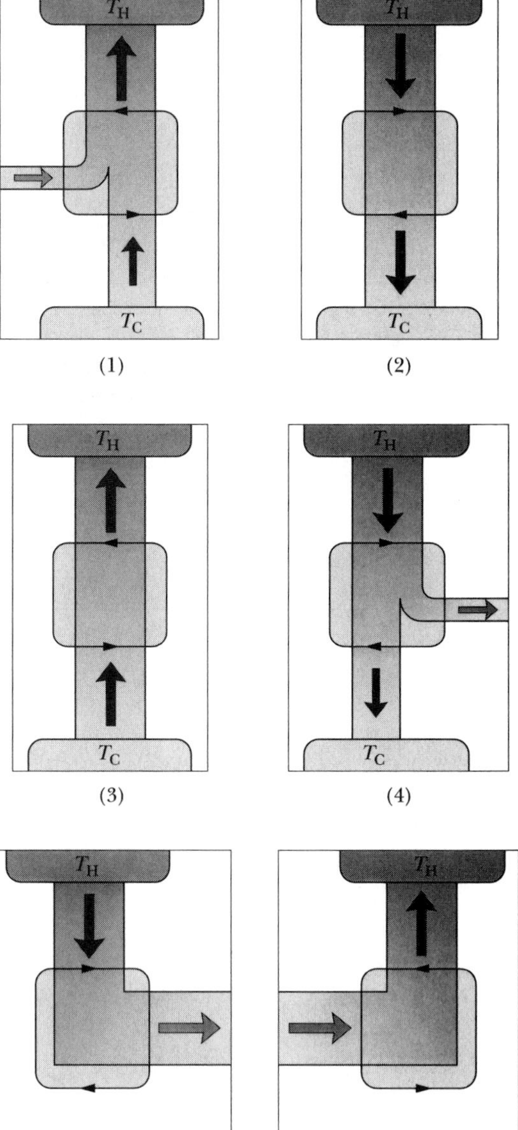

FIGURE 21-33 Question 23.

23. Figure 21-33 shows the elements of refrigerators (heat pumps) or engines in six general arrangements, each drawn in the style of Chapter 21. Which arrangement corresponds to (a) an ideal engine, (b) a real engine, (c) a perfect engine, (d) an ideal refrigerator, (e) a real refrigerator, (f) a perfect refrigerator, (g) an engine that produces only waste heat, and (h) an arrangement in which the work done is transformed entirely to thermal energy by friction. (i) For each of these categories, tell if the net entropy change for the entire closed system during one cycle is positive, negative, or zero. (j) Can these devices be built; which are the theoretical limit of what can be built?

24. Figure 21-34 shows a snapshot at time $t = 0$ of molecules a and b in a box (similar to that of Fig. 21-14). The molecules have the same mass and speed v, and the collisions between the molecules and the walls are elastic. What is the probability that snapshots taken at times (a) $t = 0.10L/v$ and (b) $t = 10L/v$ will show that a is in the left side of the box and b is in the right side of the box? (c) What is the probability that at some later time only half of the kinetic energy of the molecules will be in the right side of the box?

25. (a) Which of the curves of Fig. 21-35 best corresponds to Boltzmann's entropy equation? (Curves 2 and 5 are straight.)

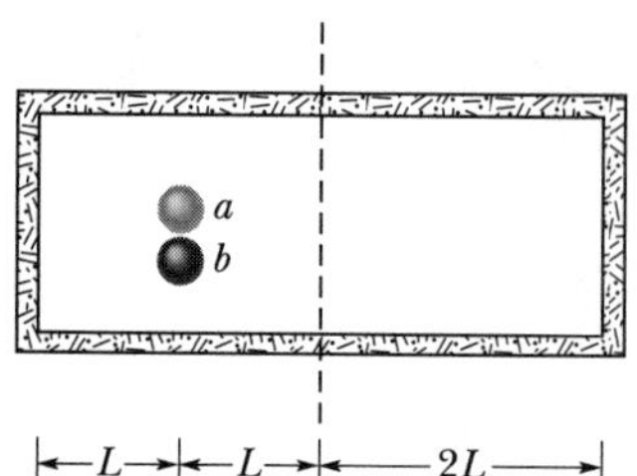

FIGURE 21-34 Question 24.

(b) In a particular system the ratio of the multiplicities of configurations A and B is $W_A/W_B = 2$. Is the corresponding ratio of entropies S_A/S_B equal to 2, greater than 2, or less than 2? That is, if we double the multiplicity, do we double the entropy, more than double the entropy, or less than double the entropy?

26. *Math Tool Time.* Which of the following is equivalent to $\ln(2a/b^3)$? (a) $2 \ln a - 3 \ln b$; (b) $2 \ln a + 3 \ln b$; (c) $\ln 2a - 3 \ln b$; (d) $2 \ln a - \frac{1}{3} \ln b$.

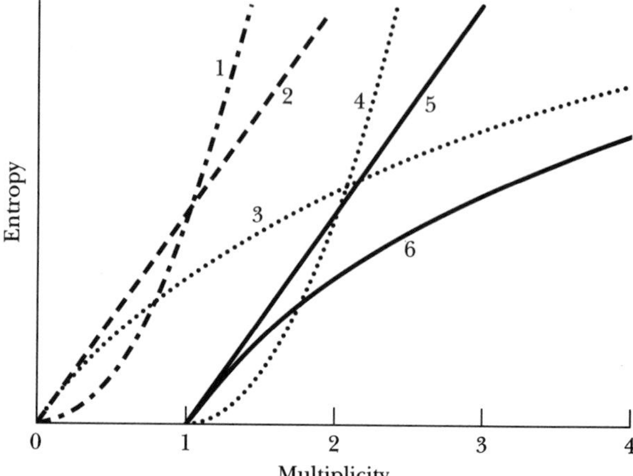

FIGURE 21-35 Question 25.

EXERCISES & PROBLEMS

72. A cylindrical copper rod of length 1.50 m and radius 2.00 cm is insulated to prevent heat loss through its curved surface. One end is attached to a thermal reservoir fixed at 300°C; the other is attached to a thermal reservoir fixed at 30.0°C. What is the rate at which entropy increases for the rod–reservoirs system?

73. A 600 g lump of copper at 80.0°C is placed in 70.0 g of water at 10.0°C in an insulated container. (See Table 19-3 for specific heats.) (a) What is the equilibrium temperature of the copper–water system? What are the entropy changes of (b) the copper, (c) the water, and (d) the copper–water system to reach the equilibrium temperature?

74. A 45.0 g block of tungsten at 30.0°C and a 25.0 g block of silver at −120°C are placed together in an insulated container. (See Table 19-3 for specific heats.) (a) What is the equilibrium temperature? What are the entropy changes of (b) the tungsten, (c) the silver, and (d) the tungsten–silver system to reach the equilibrium temperature?

75. An insulated thermos contains 130 g of water at 80.0°C. You put in a 12.0 g ice cube at 0°C to form a system of *ice + original water.* (a) What is the equilibrium temperature of the system? What are the entropy changes of the water that was originally the ice cube (b) for its melting and then (c) for its warming to the equilibrium temperature? (d) What is the entropy change of the original water to reach the equilibrium temperature? (e) What is the net entropy change of the *ice + original water* system to reach the equilibrium temperature?

76. An ideal gas of 4.00 mol is taken reversibly through a cycle of three processes: (1) an adiabatic expansion that gives the gas 2.00 times its initial volume; (2) a constant-volume process; (3) an isothermal compression back to the initial state of the gas. We do not know if the gas is monatomic or diatomic; if it is diatomic, we do not know if the molecules are rotating or oscillating. What are the entropy changes for (a) the complete cycle, (b) process 1, (c) process 3, and (d) process 2?

77. An ideal engine whose high-temperature reservoir is at 400 K has an efficiency of 0.300. By how much should the temperature of the low-temperature reservoir be changed to increase the efficiency to 0.400?

78. An ideal refrigerator does 200 J of work to remove 600 J of energy from its cold compartment. (a) What is the refrigerator's coefficient of performance? (b) How much heat per cycle is exhausted to the kitchen?

79. A diatomic gas of 2.00 mol is initially at 300 K. Its molecules do not rotate or oscillate both initially and throughout the following complete cycle of three processes: (1) the gas is heated at constant volume to a temperature of 800 K; (2) it is then allowed to expand isothermally to its initial pressure; and (3) it is then compressed at constant pressure to its initial state. (a) During the cycle, how much heat is transferred to the gas? (b) What is the net work done by the gas? (c) What is the efficiency of the cycle?

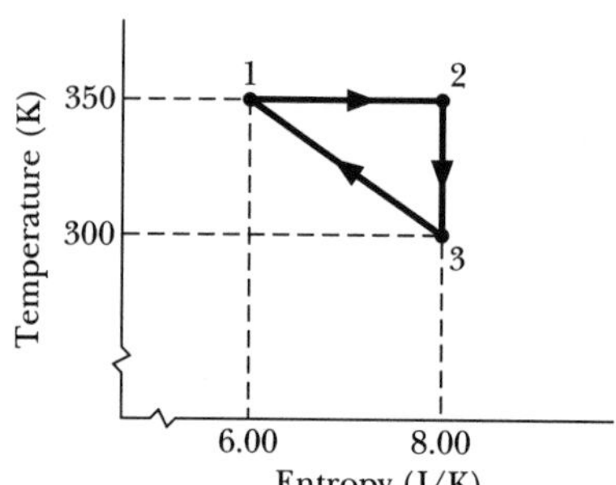

FIGURE 21-36 Problem 80.

80. A diatomic gas of 2 mol is taken reversibly around the cycle shown in the *T-S* diagram of Fig. 21-36. The molecules do not rotate or oscillate. What are the heats Q for (a) the path from point 1 to point 2, (b) the path from point 2 to point 3, and (c) the full cycle? (d) What is the work W for the isothermal process? The volume V_1 at point 1 is 0.200 m^3. What are the volumes at (e) point 2 and (f) point 3?

What are the changes ΔE in internal energy for (g) the path from point 1 to point 2, (h) the path from point 2 to point 3, and (i) the full cycle? (*Hint:* Part (h) can be done with one or two lines of calculation using Section 20-8 or with a page of calculation using Section 20-11.) (j) What is the work W for the adiabatic process?

81. System A of three particles and system B of five particles are in insulated boxes as in the example of Fig. 21-14. What are the least multiplicities W of (a) system A and (b) system B? What are the most multiplicities W of (c) system A and (d) system B? Finally, what are the maximum entropies of (e) system A and (f) system B?

Tutorial Problems

82. An ideal gas (the system) is taken around a cycle back to its initial state through four processes (steps). Initially, the gas has a pressure of 1.00×10^5 Pa, a temperature of 300 K, and a volume of 0.900 m^3. The internal energy is given by $E_{\text{int}} = 2.5nRT$, where T is in kelvins.

The four steps of the cycle follow. For each, find the final pressure, volume, and temperature of the system. Also find the change ΔE_{int} in the system, the heat Q absorbed or lost by the system, and the work W done by or on the system. Finally, graph the path taken by the system on a *p-V* diagram, labeling the path with the step number and indicating its direction with an arrowhead. For notation, use i as a subscript for initial values and the step number as a subscript for the values at the end of each step.

(a) Determine how many moles of gas are in the system. (b) Step 1 of the cyclic process is a constant-volume process with a final pressure of 2.00×10^5 Pa. (c) Step 2 is a constant-pressure process with a final temperature of 800 K. (d) Step 3 is a constant-volume process back to the initial pressure of 1.00×10^5 Pa. (e) Step 4 is a constant-pressure process back to the initial state. (f) For the entire cyclic process, what are the net quantities ΔE_{int}, Q, and W? Explain how you can determine W directly from a *p-V* diagram of the cycle.

The thermal efficiency of a cyclic process is defined as the ratio of the net work done during the cycle to the total positive heat absorbed (counting Q only when $Q > 0$). (g) What is the thermal efficiency of the cycle? Suppose that the cycle were run in reverse; that is, do step 4, then 3, then 2, and then 1, each in reverse. (h) What would be the effect (if any) on the net quantities of part (f)? Imagine a different cyclic process with the same steps 1 and 2 but with steps 3 and 4 replaced by a shortcut directly back to the initial state along a straight line on a *p-V* diagram. (i) Describe in words how the net quantities determined in part (f) would be affected.

Answers

(a) Use the ideal gas law to write

$$n = \frac{p_i V_i}{RT_i} = \frac{(1.00 \times 10^5 \text{ Pa})(0.900 \text{ m}^3)}{(8.31 \text{ J/K} \cdot \text{mol})(300 \text{ K})} = 36.1 \text{ mol}.$$

(b) The volume remains 0.900 m^3. At the pressure of 2.00×10^5 Pa, the new temperature is

$$T_1 = \frac{p_1 V_1}{nR} = \frac{(2.00 \times 10^5 \text{ Pa})(0.900 \text{ m}^3)}{(36.1 \text{ mol})(8.31 \text{ J/K} \cdot \text{mol})} = 600 \text{ K}.$$

The work done by the system is zero because there is no change in volume: $W_1 = 0$. The change in internal energy is

$$\Delta E_{\text{int},1} = 2.5nR\,\Delta T = (2.5)(36.1 \text{ mol})(8.31 \text{ J/K} \cdot \text{mol})(600 \text{ K} - 300 \text{ K}) = 225 \text{ kJ}.$$

The heat is

$$Q_1 = \Delta E_{\text{int},1} + W_1 = 225 \text{ kJ} + 0 = 225 \text{ kJ}.$$

Because the result is positive, the heat is absorbed by the system. On a *p-V* diagram this step forms a vertical line at the constant volume, from pressure 1.00×10^5 Pa to pressure 2.00×10^5 Pa.

(c) Since the pressure remains 2.00×10^5 Pa and the final temperature T_2 is 800 K, the final volume must be

$$V_2 = \frac{nRT_2}{p_2} = \frac{(36.1 \text{ mol})(8.31 \text{ J/K} \cdot \text{mol})(800 \text{ K})}{2.00 \times 10^5 \text{ Pa}} = 1.20 \text{ m}^3.$$

Since the pressure remains constant, the work done by the system is

$$W_2 = p_2\,\Delta V = (2.00 \times 10^5 \text{ Pa})(1.20 \text{ m}^3 - 0.900 \text{ m}^3) = 60 \text{ kJ}.$$

The change in internal energy is

$$\Delta E_{\text{int},2} = 2.5nR\,\Delta T = (2.5)(36.1 \text{ mol})(8.31 \text{ J/K} \cdot \text{mol})(800 \text{ K} - 600 \text{ K}) = 150 \text{ kJ}.$$

Thus the heat is

$$Q_2 = \Delta E_{\text{int},2} + W_2 = 150 \text{ kJ} + 60 \text{ kJ} = 210 \text{ kJ}.$$

Because the result is positive, the heat is absorbed by the system.

On a p-V diagram this step forms a horizontal line at the constant pressure, from volume 0.900 m³ to volume 1.20 m³.

(d) The volume remains 1.20 m³. At a pressure of 2.00 × 10⁵ Pa, the temperature at the end of step 3 is

$$\begin{aligned} T_3 &= \frac{p_3 V_3}{nR} \\ &= \frac{(1.00 \times 10^5 \text{ Pa})(1.20 \text{ m}^3)}{(36.1 \text{ mol})(8.31 \text{ J/K} \cdot \text{mol})} \\ &= 400 \text{ K}. \end{aligned}$$

The work done by the system is zero since there is no change in volume: $W_3 = 0$. The change in internal energy is

$$\begin{aligned} \Delta E_{\text{int},3} &= 2.5nR\,\Delta T \\ &= (2.5)(36.1 \text{ mol})(8.31 \text{ J/K} \cdot \text{mol})(400 \text{ K} - 800 \text{ K}) \\ &= -300 \text{ kJ}. \end{aligned}$$

The heat is

$$Q_3 = \Delta E_{\text{int},3} + W_3 = -300 \text{ kJ} + 0 = -300 \text{ kJ},$$

where the negative sign means that 300 kJ is removed from the system. On a p-V diagram this step forms a vertical line at the constant volume, from pressure 2.00 × 10⁵ Pa to pressure 1.00 × 10⁵ Pa.

(e) The final pressure is $p_f = p_i = 1.00 \times 10^5$ Pa, the final temperature is $T_f = T_i = 300$ K, and the final volume is $V_f = V_i = 0.900$ m³. Since the pressure does not change during this step, the work done by the system is

$$\begin{aligned} W_4 &= p_f\,\Delta V \\ &= (1.00 \times 10^5 \text{ Pa})(0.900 \text{ m}^3 - 1.20 \text{ m}^3) = -30 \text{ kJ}. \end{aligned}$$

The change in internal energy is

$$\begin{aligned} \Delta E_{\text{int},4} &= 2.5nR\,\Delta T \\ &= (2.5)(36.1 \text{ mol})(8.31 \text{ J/K} \cdot \text{mol})(300 \text{ K} - 400 \text{ K}) \\ &= -75 \text{ kJ}. \end{aligned}$$

The heat is then

$$\begin{aligned} Q_4 &= \Delta E_{\text{int},4} + W_4 \\ &= -75 \text{ kJ} - 30 \text{ kJ} = -105 \text{ kJ}. \end{aligned}$$

On a p-V diagram this step forms a horizontal line at the constant pressure, from volume 1.20 m³ to volume 0.900 m³.

(f) The net change in internal energy should be zero. Let's check:

$$\begin{aligned} \Delta E_{\text{int}} &= \Delta E_{\text{int},1} + \Delta E_{\text{int},2} + \Delta E_{\text{int},3} + \Delta E_{\text{int},4} \\ &= 225 \text{ kJ} + 150 \text{ kJ} - 300 \text{ kJ} - 75 \text{ kJ} = 0. \end{aligned}$$

The heat absorbed by the system is

$$\begin{aligned} Q &= Q_1 + Q_2 + Q_3 + Q_4 \\ &= 225 \text{ kJ} + 210 \text{ kJ} - 300 \text{ kJ} - 105 \text{ kJ} = 30 \text{ kJ}. \end{aligned}$$

The work done by the system is

$$\begin{aligned} W &= W_1 + W_2 + W_3 + W_4 \\ &= 0 + 60 \text{ kJ} + 0 - 30 \text{ kJ} = 30 \text{ kJ}. \end{aligned}$$

W is also just the area inside the path of the cyclic process on a p-V diagram:

$$W = (1.00 \times 10^5 \text{ Pa})(0.30 \text{ m}^3) = 30 \text{ kJ}.$$

(g) The net work done is 30 kJ. The heat absorbed during steps 1 and 2, which are the only steps in which $Q > 0$, is 225 kJ + 210 kJ = 435 kJ. The thermal efficiency of this cycle is then (30 kJ)/(435 kJ) = 0.069, or 6.9%.

(h) Running the cyclic process in reverse would reverse all the quantities ΔE_{int}, Q, and W from their values in the corresponding processes. Thus, the net quantities would also be reversed, so we would have $\Delta E_{\text{int}} = 0$, $Q = -30$ kJ, and $W = -30$ kJ.

(i) The shortcut would not affect the fact that $\Delta E_{\text{int}} = 0$, but it would halve the area inside the curve on a p-V diagram. Consequently, the heat absorbed, which equals W for a closed path, would also be halved.

83. This problem examines the heat flow and entropy change involved in processes of cooling and phase changes. Initially steam of mass $m = 200$ g and temperature 100°C is put in contact with a large heat reservoir at −30°C. Eventually, the water is ice at the reservoir's temperature. Let's assume that the process from the initial to this final state is done reversibly. Assume also that specific heats given for ice and liquid water in Table 19-3 hold for the temperature range of this problem.

(a) List the four processes that occur as the steam is converted to ice. For each, list the initial and final temperatures. (b) What does "large" mean in the description of the heat reservoir? (c) Find an expression for the entropy change of the system, with heat of transformation L, when the system undergoes a phase change. What is the corresponding entropy change of the reservoir? (d) Show that the entropy change of the system, with constant specific heat c, when it is cooled from temperature T_i to temperature T_f is given by

$$\Delta S = mc \ln \frac{T_f}{T_i}.$$

What is the corresponding entropy change of the heat reservoir?

(e) Use the expressions of part (c) or part (d) for each of the four processes you listed in part (a) to complete the following table:

PROCESS	1	2	3	4	TOTAL
Water:					
Q (kJ)					
ΔS (kJ/K)					
Reservoir:					
Q_r (kJ)					
ΔS_r (kJ/K)					

Answers

(a) (1) The steam is condensed at 100°C into liquid water. The initial and final temperatures are both 100°C. (2) The water is cooled from its initial temperature of 100°C to its freezing temperature of 0°C. (3) The liquid water freezes and becomes solid water (ice). This takes places at 0°C, which is both the initial temperature and the final temperature. (4) The ice is cooled to −30°C. The initial temperature is 0°C and the final temperature is −30°C.

(b) A "large" heat reservoir is one that can absorb or release relatively large amounts of heat without a significant change in its own temperature. In other words, it has a very large heat capacity. In the context of this problem, it means that all the heat absorbed from the water system does not significantly change the temperature of the reservoir. Maybe it's a big block of ice with a mass of several megagrams.

(c) If a system of mass m undergoes a phase change at a temperature T, it absorbs an amount of heat $Q = mL$, where L is a positive quantity for melting and evaporation and a negative quantity for freezing or condensation. The corresponding change in entropy is $\Delta S = Q/T = mL/T$. If this heat is obtained from a heat reservoir at temperature T_r, the change in entropy of the reservoir is $\Delta S_r = -Q/T_r = -mL/T_r$.

(d) The heat involved is $dQ = mc\,dT$ so

$$dS = \frac{dQ}{T} = \frac{mc\,dT}{T}.$$

Thus

$$\Delta S = S_f - S_i = \int_i^f \frac{mc\,dT}{T} = mc\left[\ln T\right]_i^f = mc\ln\left(\frac{T_f}{T_i}\right).$$

This value is positive if $T_f > T_i$ and negative if $T_f < T_i$.

The heat transferred by the system is

$$Q = mc\,\Delta T = mc(T_f - T_i),$$

so the heat transferred by the heat reservoir is $-Q = -mc(T_f - T_i)$ and the change in entropy of the heat reservoir is $-mc(T_f - T_i)/T_r$.

(e) (1) The water vapor condenses at 100°C = 373 K. For the water,

$$Q = -mL = -(0.200\text{ kg})(2256\text{ kJ/kg}) = -451\text{ kJ}$$

and
$$\Delta S = \frac{Q}{T} = -\frac{mL}{T} = -\frac{(0.200\text{ kg})(2256\text{ kJ/kg})}{373\text{ K}} = -1.21\text{ J/K}.$$

For the reservoir,

$$Q_r = -Q = 451\text{ kJ}$$

and
$$\Delta S = \frac{Q_r}{T_r} = \frac{(451\text{ kJ})}{(243\text{ K})} = 1.86\text{ kJ/K}.$$

(2) Water cools from 100°C = 373 K to 0°C = 273 K. For the water,

$$Q = mc\,\Delta T = (0.200\text{ kg})(4190\text{ J/kg}\cdot\text{K})(0°\text{C} - 100°\text{C}) = -83.8\text{ kJ}$$

and
$$\Delta S = mc\ln\left(\frac{T_f}{T_i}\right) = (0.200\text{ kg})(4190\text{ J/kg}\cdot\text{K})\ln\left(\frac{273\text{ K}}{373\text{ K}}\right) = -0.26\text{ kJ/K}.$$

For the reservoir,

$$Q_r = -Q = 83.8\text{ kJ}$$

and
$$\Delta S = \frac{Q_r}{T_r} = \frac{83.8\text{ kJ}}{243\text{ K}} = 0.35\text{ kJ/K}.$$

(3) Water freezes into ice at 0°C = 273 K. For the water,

$$Q = -mL = -(0.200\text{ kg})(333\text{ kJ/kg}) = -66.6\text{ kJ}$$

and
$$\Delta S = \frac{Q}{T} = -\frac{66.6\text{ kJ}}{273\text{ K}} = -0.24\text{ kJ/K}.$$

For the reservoir,

$$Q_r = -Q = 66.6\text{ kJ}$$

and
$$\Delta S = \frac{Q_r}{T_r} = \frac{66.6\text{ kJ}}{243\text{ K}} = 0.27\text{ kJ/K}.$$

(4) Ice cools from 0°C = 273 K to −30°C = 243 K. For the water,

$$Q = mc\,\Delta T = (0.200\text{ kg})(2220\text{ J/kg}\cdot\text{K})(-30°\text{C} - 0°\text{C}) = -13.3\text{ kJ}$$

and
$$\Delta S = mc\ln\left(\frac{T_f}{T_i}\right) = (0.200\text{ kg})(2220\text{ J/kg}\cdot\text{K})\ln\left(\frac{243\text{ K}}{273\text{ K}}\right) = -0.052\text{ kJ/K}.$$

For the reservoir,

$$Q_r = -Q = 13.3\text{ kJ}$$

and
$$\Delta S = \frac{Q_r}{T_r} = \frac{13.3\text{ kJ}}{243\text{ K}} = 0.055\text{ kJ/K}.$$

The net entropy change of the water plus the reservoir is positive in each process, as we expect from the entropy formulation of the second law of thermodynamics. Overall, the total entropy change is

$$2.53\text{ kJ/K} - 1.76\text{ kJ/K} = +0.77\text{ kJ/K} > 0.$$

Chapter Twenty-Two
Electric Charge

QUESTIONS

19. Figure 22-27 shows three charged particles. Which of the 12 vectors in the figure best gives the force on the particle of charge $+q_2$ due to the presence of (a) the particle of charge $-q_1$ and (b) the particle of charge $+q_3$?

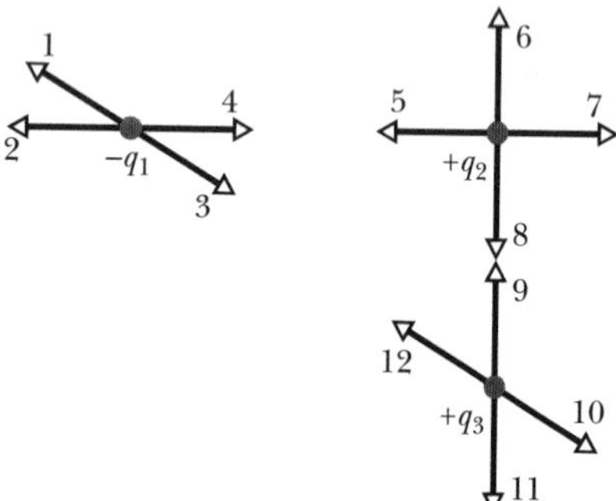

FIGURE 22-27 Question 19.

20. *Organizing question:* Figure 22-28 shows four particles fixed in place in a plane. (a) Set up an equation, complete with known data, to find the x component of the net electrostatic force on the particle at the origin due to the other three particles. Here, $q_1 = 1.6 \times 10^{-19}$ C, $q_2 = 3.2 \times 10^{-19}$ C, $q_3 = 1.6 \times 10^{-19}$ C, $q_4 = -3.2 \times 10^{-19}$ C. (b) Similarly set up an equation to find the corresponding y component.

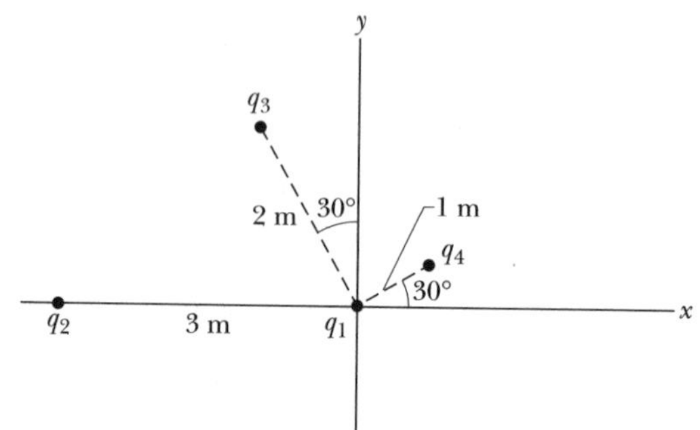

FIGURE 22-28 Question 20.

21. In Fig. 22-29*a*, a particle of charge Q is fixed in place on an x axis. As an electron (unshown) moves slowly along the axis, the electrostatic force F on it due to Q is measured and graphed. When the force is in the positive direction of the axis, a positive value is graphed; when it is in the negative direction, a negative value is graphed. (a) Does Fig. 22-29*b* correspond to the situation where Q is positive or where Q is negative, or is the graph physically impossible? (b) Repeat (a) for Fig. 22-29*c*.

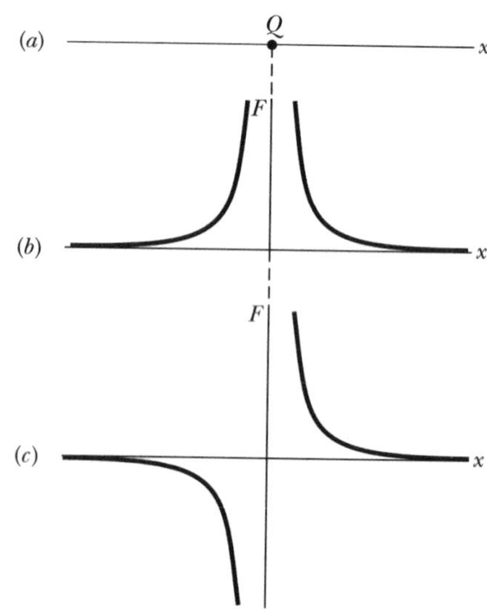

FIGURE 22-29 Question 21.

22. In Fig. 22-30*a*, two particles of charges Q_1 and Q_2 are fixed in place on an x axis. As an electron (unshown) moves slowly along the axis, the net electrostatic force F on it due to Q_1 and Q_2 is measured and graphed. When the force is in the positive direction of the axis, the positive value is graphed; when it is in the negative direction, a negative value is graphed. What are the signs of Q_1 and Q_2 for the graphs of (a) Fig. 22-30*b* and (b) Fig. 22-30*c*?

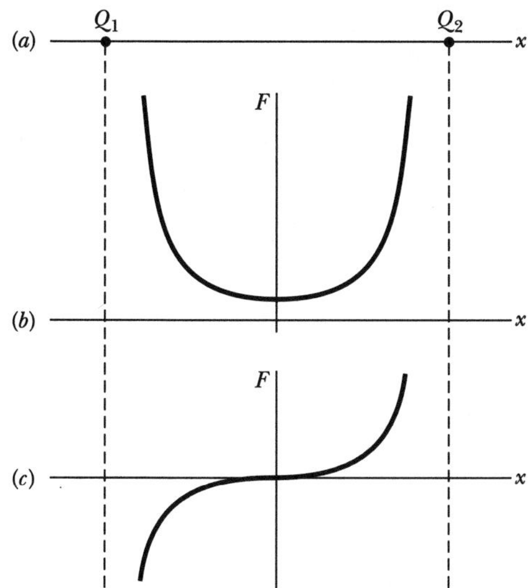

FIGURE 22-30 Question 22.

23. Figure 22-31 shows three situations involving a charged particle and a uniformly charged spherical shell. The charges are given, and the radii of the shells are indicated. Rank the situations according to the magnitude of the force on the particle due to the presence of the shell, greatest first.

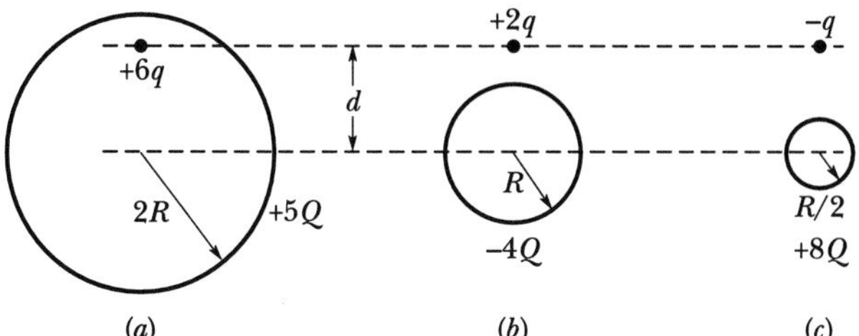

FIGURE 22-31 Question 23.

24. Figure 22-32 shows four situations in which a central proton is partially surrounded by protons or electrons fixed in place along a half circle. The angles θ are identical; the angles ϕ are also. (a) In each situation, what is the direction of the net force on the central proton due to the other particles? (b) Rank the four situations according to the magnitude of that net force on the central proton, greatest first.

25. Figure 22-33 shows three identical conducting bubbles A, B, and C floating in a conducting container that is grounded by a wire. The bubbles initially have the same charge. Bubble A bumps into the container's ceiling and then into bubble B. Then bubble B bumps into bubble C, which then drifts to the container's floor. When bubble C reaches the floor, a charge of $-3e$ is transferred upward through the wire, from the ground to the container, as indicated. (a) What was the initial charge of each bubble? When (b) bubble A and (c) bubble B reach the floor, what is the charge transfer through the wire? (d) During this whole process, what is the total charge transfer through the wire?

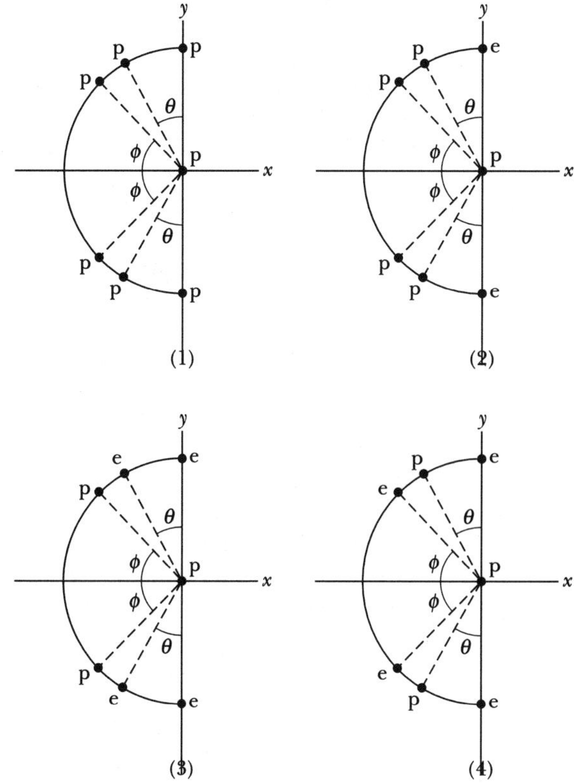

FIGURE 22-32 Question 24.

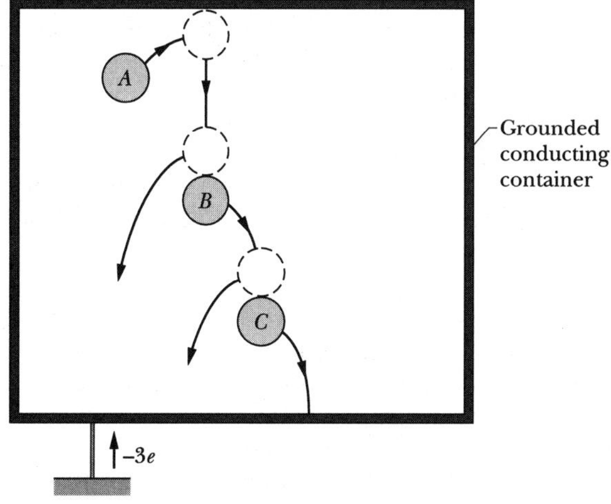

FIGURE 22-33 Question 25.

EXERCISES & PROBLEMS

43. Two point charges of $-80\ \mu C$ and $40\ \mu C$ are held fixed on an x axis at the origin and $x = 20$ cm, respectively. What are the magnitude and direction of the total electrostatic force on a third charge of $20\ \mu C$ placed at (a) $x = 40$ cm and (b) $x = 80$ cm? (c) Where on the x axis can the third charge be placed such that the total electrostatic force on it is zero?

44. A particle of charge Q is fixed at the origin of an xy coordinate system. At $t = 0$ a particle ($m = 0.800$ g, $q = 4.00\ \mu C$) is located on the x axis at $x = 20.0$ cm and is moving with a speed of 50.0 m/s in the positive y direction. For what value of Q will the 0.800 g particle execute circular motion? (Assume that the gravitational force on the particle may be neglected.)

45. Three charged particles form a triangle: particle 1 with charge $Q_1 = 80.0$ nC is at xy coordinates (0, 3.00 mm), particle 2 with charge Q_2 is at (0, -3.00 mm), and particle 3 with charge $q = 18.0$ nC is at (4.00 mm, 0). What are the magnitude and direction of the electrostatic force on particle 3 due to the other two particles if Q_2 is equal to (a) 80.0 nC and (b) -80.0 nC?

46. Two point charges, $40\ \mu C$ and Q, are fixed on an x axis at the points $x = -2.0$ cm and $x = 3.0$ cm, respectively. A third point charge of magnitude $20\ \mu C$ is released from rest at a point on the y axis at $y = 2.0$ cm. (a) What is the value of Q for which the total electrostatic force on this third charge is initially in the positive x direction? (b) What is the value of Q for which the total electrostatic force on this third charge is initially in the positive y direction?

47. Charges of $+6.0\ \mu C$ and $-4.0\ \mu C$ are placed on an x axis at $x = 8.0$ m and $x = 16$ m, respectively. What charge must be placed at $x = 24$ m so that any charge placed at the origin would experience no electrostatic force?

48. Two point charges of 30 nC and -40 nC are held fixed on an x axis at the origin and $x = 72$ cm, respectively. A third point charge of $42\ \mu C$ is released from rest at $x = 28$ cm. If the initial acceleration of the particle has a magnitude of 100 km/s^2, what is its mass?

49. A nonconducting spherical shell, with an inner radius of 4.0 cm and an outer radius of 6.0 cm, has charge spread through its volume (between inner and outer surfaces). *Volume charge density* ρ is the charge per unit volume, with the unit of coulomb per cubic meter. For the shell it is given by $\rho = b/r$. Here r is the distance in meters from the center of the shell and $b = 3.0\ \mu C/m^2$. What is the net charge within the shell? (*Hint:* Because ρ is not uniform over the shell, you must integrate.)

50. A charged nonconducting rod, with a length of 2.00 m and a cross-sectional area of 4.00 cm^2, lies along the positive side of an x axis with one end at the origin. *Volume charge density* ρ is the charge per unit volume, with the unit of coulomb per cubic meter. How many excess electrons are on the rod if the rod's volume charge density is (a) uniform, with a value of $-4.00\ \mu C/m^3$, and (b) nonuniform, with a value given by $\rho = bx^2$, where $b = -2.00\ \mu C/m^5$?

51. In Fig. 22-34 the particles with charges q_1 and q_2 are fixed in place but the third particle is free to move. If the net electrostatic force on that third particle due to the other particles is zero, what is q_1 in terms of q_2?

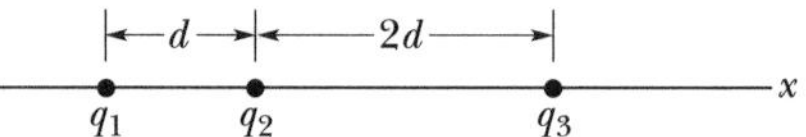

FIGURE 22-34 Problem 51.

52. In Fig. 22-35, how far from the charged particle on the right and in what direction is there a point where a third charged particle will be in balance?

FIGURE 22-35 Problem 52.

53. In Fig. 22-36, if $Q = +3.20 \times 10^{-19}$ C, $q = 1.60 \times 10^{-19}$ C, and $a = 2.00$ cm, what is the net electrostatic force on the particle at the origin due to the other charged particles?

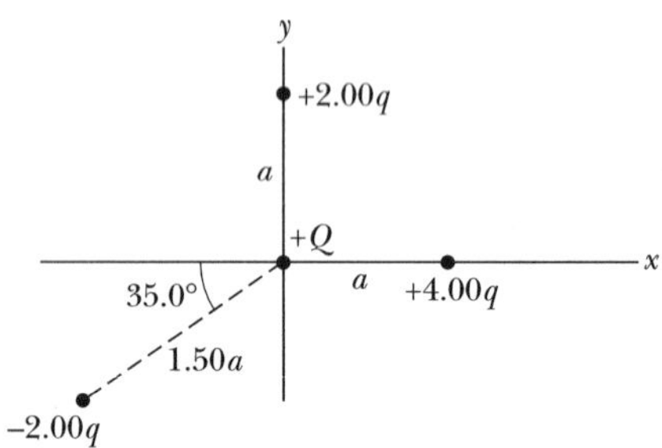

FIGURE 22-36 Problem 53.

54. A charge of $6.0\ \mu C$ is to be split into two parts that are then separated by 3.0 mm. What is the maximum possible magnitude of the electrostatic force between those two parts?

55. In Fig. 22-37, what is q in terms of Q if the net electrostatic force on the charged particle at the upper left corner of the square array is to be zero?

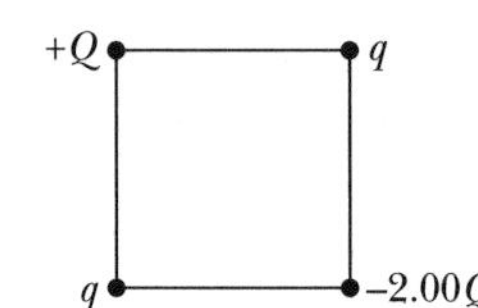

FIGURE 22-37 Problem 55.

56. The initial charges on the three identical metal spheres in Fig. 22-38 are the following: sphere A, Q; sphere B, $-Q/4$, and sphere C, $Q/2$, where $Q = 2.00 \times 10^{-14}$ C. Spheres A and B are fixed in place, with a center-to-center separation of $d = 1.20$ m, which is much larger than the spheres. Sphere C is touched first to sphere A and then to sphere B and is then removed. What then is the magnitude of the electrostatic force between spheres A and B?

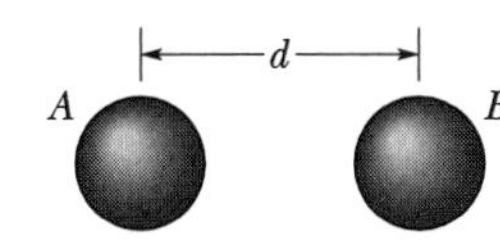

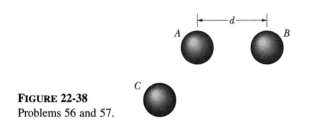

FIGURE 22-38 Problems 56 and 57.

57. In Fig. 22-38, three identical conducting spheres initially have the following charges: sphere A, $4Q$; sphere B, $-6Q$, and sphere C, 0. Spheres A and B are fixed in place, with a center-to-center separation that is much larger than the spheres. Two experiments are conducted. In experiment 1, sphere C is touched to sphere A and then (separately) to sphere B, and then it is removed. In experiment 2, starting with the same initial states, the procedure is reversed: sphere C is touched to sphere B and then (separately) to sphere A, and then it is removed. What is the ratio of the electrostatic force between A and B at the end of experiment 2 to that at the end of experiment 1?

Tutorial Problem

58. Let's look at the electrostatic forces in a system of four charged particles that form a square with sides of length L (Fig. 22-39). The variable Q is a positive charge of unspecified magnitude. Use the following notation: $\mathbf{F}_{12}$, $\mathbf{F}_{13}$, and $\mathbf{F}_{14}$ are the forces on particle 1 due to particles 2, 3, and 4, respectively.

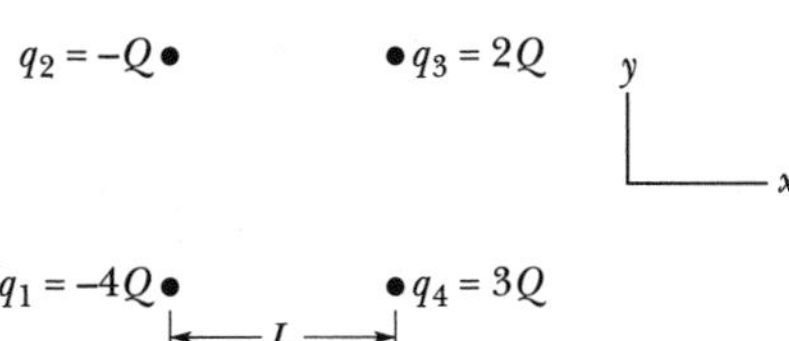

FIGURE 22-39 Problem 58.

(a) Show $\mathbf{F}_{12}$, $\mathbf{F}_{13}$, and $\mathbf{F}_{14}$ on a figure similar to Fig. 22-39. (b) Express each of these forces in the form $\mathbf{F} = F\hat{F}$, where F is a magnitude and $\hat{F}$ is a unit vector in the direction of $\mathbf{F}$. Each $\hat{F}$ should be expressed in $\hat{\imath}$-$\hat{\jmath}$ notation. (c) Name the principle of physics that allows us to find the total force $\mathbf{F}_1$ from the individual forces $\mathbf{F}_{12}$, $\mathbf{F}_{13}$, and $\mathbf{F}_{14}$. Express $\mathbf{F}_1$ in (d) $\hat{\imath}$-$\hat{\jmath}$ notation and (e) as a magnitude and direction. (f) Show $\mathbf{F}_1$ on a figure similar to Fig. 22-39.

(g) Now assume that only charges q_3 and q_4 are present ($q_1 = q_2 = 0$) and that a positive charge q is placed on the line between charges q_3 and q_4. At what point on the line could the charge be located so that the net electrostatic force on q is zero? (h) Is the point you found one of stable, unstable, or neutral equilibrium? If the charge q were negative instead of positive, would the answers to this question and part (g) be the same or different? Explain.

Answers

(b) $\mathbf{F}_{14}$ is an attractive force because the two charges have opposite signs, so the force is directed from q_1 toward q_4, that is, in the direction $+\hat{\imath}$. The magnitude of the force is

$$F = \frac{k|q|\,|q|}{L^2} = \frac{k(3Q)(4Q)}{L^2} = \frac{12kQ^2}{L^2},$$

where we have used k from Eq. 22-5. So,

$$\mathbf{F}_{14} = \frac{12kQ^2}{L^2}\,\hat{\imath}.$$

$\mathbf{F}_{13}$ is an attractive force because the two charges have opposite signs, so the force is directed from q_1 toward q_3, that is, at an angle of 45° from the x axis. Thus the force is in the direction

$$\hat{F} = (\cos 45°)\hat{\imath} + (\sin 45°)\hat{\jmath} = 0.707\hat{\imath} + 0.707\hat{\jmath}.$$

The magnitude of the force is

$$F = \frac{k|q_3|\,|q_1|}{L^2 + L^2} = \frac{4kQ^2}{L^2}.$$

Thus
$$\mathbf{F}_{13} = \frac{2.828kQ^2}{L^2}(\hat{\imath} + \hat{\jmath}).$$

This can be written in the form $F\hat{F}$ as

$$\mathbf{F}_{13} = \frac{4kQ^2}{L^2}\,\frac{(\hat{\imath} + \hat{\jmath})}{\sqrt{2}}.$$

$\mathbf{F}_{12}$ is a repulsive force, directed away from q_2, or in the $-\hat{\jmath}$ direction. So $\hat{F} = -\hat{\jmath}$. The magnitude of the force is

$$F_{12} = \frac{k|q_2|\,|q_1|}{L^2} = \frac{4kQ^2}{L^2},$$

so
$$\mathbf{F}_{12} = -\frac{4kQ^2}{L^2}\,\hat{\jmath}.$$

(c) To determine the total force on q_1 we use the principle of superposition: the total force on q_1 is the vector sum of all the individual forces acting on q_1 due to the other particles.

(d) In the notation using the unit vectors along the axes,

$$\begin{aligned}\mathbf{F}_1 &= \mathbf{F}_{12} + \mathbf{F}_{13} + \mathbf{F}_{14}\\ &= -\frac{4kQ^2}{L^2}\,\hat{\jmath} + \frac{2.828kQ^2}{L^2}(\hat{\imath} + \hat{\jmath}) + \frac{12kQ^2}{L^2}\,\hat{\imath}\\ &= \frac{kQ^2}{L^2}(14.828\hat{\imath} - 1.172\hat{\jmath}).\end{aligned}$$

(e) The magnitude of $\mathbf{F}_1$ is

$$F_1 = \left(\frac{kQ^2}{L^2}\right)\sqrt{(14.828)^2 + (1.172)^2} = 14.9\left(\frac{kQ^2}{L^2}\right).$$

The direction has angle (in the fourth quadrant)

$$\theta = \tan^{-1}\left(\frac{-1.172}{14.828}\right) = -4.5° \text{ or } 355.5°.$$

So we can say the total force is $14.9(kQ^2/L^2)$ in a direction at an angle 355.5° from the positive direction of the x axis (corresponding to 4.5° below the positive direction of the x axis).

(g) Suppose the charge q is a distance y from q_4, and thus a distance $L - y$ from q_3. For a positive charge q the force on q due to q_4 would be down and the force on q due to q_3 would be up. The magnitudes of these forces would be proportional to $|q_4|/y^2$ or $3/y^2$ and to $|q_3|/(L - y)^2$ or $2/(L - y)^2$, respectively. So the condition for the net force to be zero is

$$\frac{3}{y^2} = \frac{2}{(L - y)^2},$$

which leads to

$$y^2 - 6Ly + 3L^2 = 0.$$

This is a quadratic equation with two roots: $y = 3L \pm \sqrt{6}L$. One of these is $y = 0.55L$, which is the root of interest, and the other is $y = 5.45L$, which is outside the range of interest. So the point where the net force on q is zero is about 55% of the way from q_4 to q_3. Reasonably, the equilibrium point is closer to q_3, which has the smaller magnitude of charge.

(h) The point is a point of unstable equilibrium, because if the charge q were displaced from the equilibrium point the net force on it would not be back toward the equilibrium point. This may be seen by imagining a displacement along the line between q_3 and q_4; a displacement toward either of these charges would result in a net force toward that charge and not back to the equilibrium point. If the charge q were negative instead of positive, the net force would be zero at the same point. The equilibrium point would be unstable in this case, too, as may be seen by considering a displacement perpendicular to the line between q_3 and q_4; the charge would be repelled away from the line.

Graphing Calculators

59. Figure 22-40 shows two electrons (charge $-e$) on an x axis and two charged ions of identical charges $-q$ and identical angles θ. The central electron is free to move; the other particles are fixed in place and are intended to hold the free electron in place. (a) Plot the required magnitude of q versus angle θ if this is to happen. (b) From the plot, determine which values of θ will be needed for physically possible values of $q \leq 5e$.

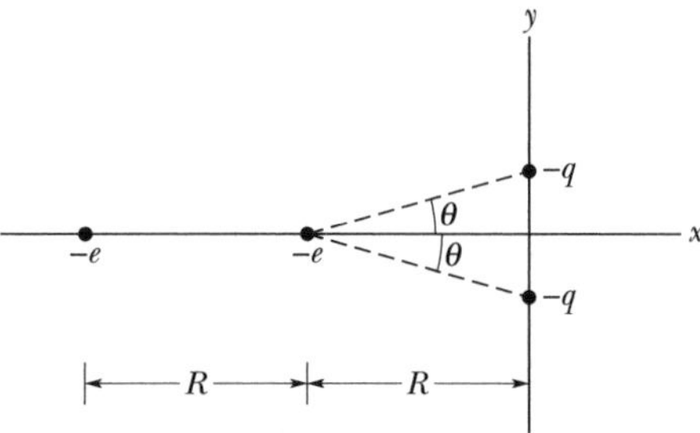

FIGURE 22-40 Problem 59.

SAMPLE PROBLEM 22-6

Solving Multiple-Force Problems Quickly. In Sample Problem 22-2 we found the net electrostatic force $\mathbf{F}_1$ on q_1 due to the other five particles by taking a vector sum of the five electrostatic forces they produce. The solution given there involves arguments of symmetry to simplify the calculation.

Instead, we could resolve the five forces into x and y components, sum the x components and then the y components, and then finally find the magnitude of the net force (using the Pythagorean theorem with the net x component and the net y component) and the angle of the net force (using an inverse tangent of the net y component divided by the net x component). This procedure could be called the long way because it is painfully long, with many chances for error.

Now, we are going to remove particles 4 and 5 and then solve for the net force on q_1. Because the symmetry is no longer present, our only resort appears to be the long way. However, there is an easier way. We can use the vector capabilities of a graphing calculator. There is even a shortcut in that use.

SOLUTION: Recall that on a TI-85/86, a vector can be entered in the form

[magnitude ∠ angle]

as examined in the Graphing Calculators section of Chapter 3 in this booklet. For example, from Sample Problem 22-2 we know that force $\mathbf{F}_{12}$ has a magnitude of $kq_1q_2/(2a)^2$ and an angle of 180° (the factor k comes from Eq. 22-5). So, using the given data for the charges and the distance a and putting the calculator in degree mode, we could enter that force as

[8.99E9*3E−6²/ (2*.02)² ∠ 180]

on the calculator.

Similarly, we could enter force $\mathbf{F}_{13}$, with magnitude kq_1q_3/a^2 and angle 90° + 30°, as

[8.99E9*3E−6²/ .02² ∠ 90+30]

and force $\mathbf{F}_{16}$, with magnitude kq_1q_6/a^2 and angle 0°, as

[8.99E9*3E−6²/ .02² ∠ 0]

which would involve a lot of keystrokes. However, rather than entering all three vectors in this way to sum them, we can use a shortcut.

The magnitudes of all the vectors include the expression 8.99E9*3E−6²/ .02². So we can pull that common expression out front of the sum and enter the sum as

8.99E9*3E−6²/ .02²([1/2² ∠ 180]
+ [1 ∠ 90+30] + [1 ∠ 0])

in which the calculator will multiply the expression at the left of the parentheses with only the magnitude portion of each vector, not the angle portion. Of course, we could save a few keystrokes by mentally performing some of the easy mathematics (such as 2^2 or 90 + 30).

We can save a few more keystrokes and avoid looking in the book for the value of k by using the value stored in the calculator under the symbol Cc (for Coulomb's constant). To get it, press 2nd CONS, choose BLTIN (for built-in constants), and then choose Cc. Or, if you happen to remember that Cc is the symbol needed here, press it in using the ALPHA key. Then complete the line so that

Cc*3E−6²/ .02² ([1/2² ∠ 180]
+ [1 ∠ 90+30] + [1 ∠ 0])

is our calculation.

If the calculator is in SphereV mode (check the mode menu), pressing ENTER gives us the sum as

[1.82E2 ∠ 73.9]

which tells us that the net force $\mathbf{F}_1$ on particle 1 has

magnitude = 182 N

and angle = 73.9°

counterclockwise from the positive x direction. (Answer)

Chapter Twenty-Three
Electric Fields

QUESTIONS

15. Figure 23-47 shows two charged particles. Which of the figure's 12 vectors best depicts the electric field at point P due to (a) the particle of charge $+q$ and (b) the particle of charge $-q$?

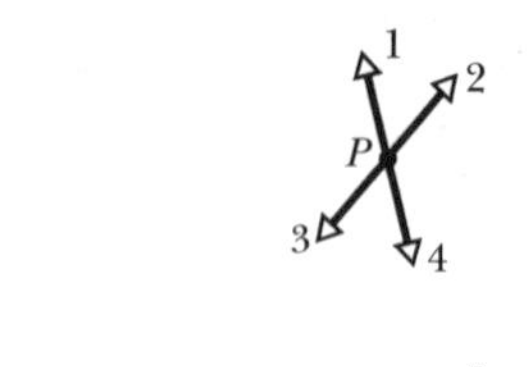

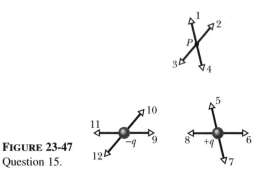

FIGURE 23-47 Question 15.

16. Dust devils, vortexes of swirling air and dust that can extend upward by a considerable distance, are electrified by the transfer of charge between dust grains and between the grains and the ground. The center of negative charge and the center of positive charge in a dust devil can then separate vertically to form an electric dipole.

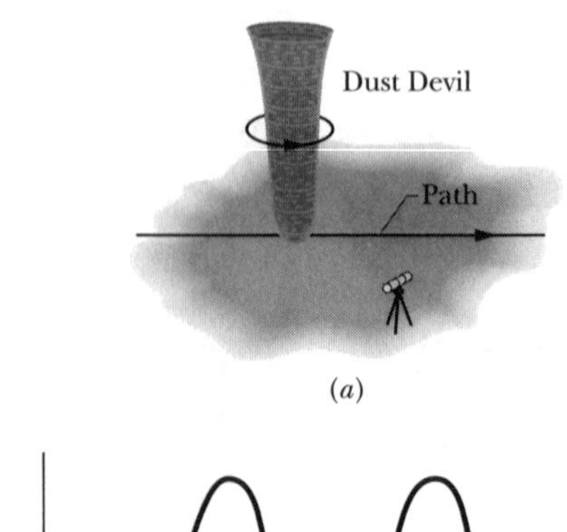

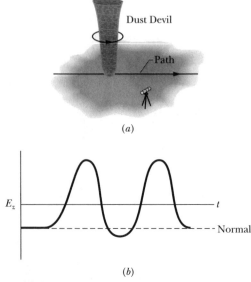

FIGURE 23-48 Question 16.

Figure 23-48a depicts an experiment that revealed the dipole in a dust devil. The vertical component E_z of the atmosphere's electric field was measured as the dust devil moved past the measuring equipment at approximately constant velocity. (In clear weather, E_z is normally downward.) Figure 23-48b gives E_z versus time t during the experiment. (a) Which was higher in the dust devil, the center of negative charge or the center of positive charge? (b) Was the electric dipole moment directed upward or downward? (c) Was the lower center of charge near ground level or well above ground level?

17. *Organizing question:* Figure 23-49 shows a rod of uniform positive charge density λ lying along an x axis; it also shows a differential charge element dq located at distance x from one end of the rod. The net electric field $\mathbf{E}$ at point P due to the rod is to be evaluated. (a) What is the direction of the differential electric field $d\mathbf{E}$ at P due to element dq? In the expression

$$dE = \frac{1}{4\pi\epsilon_0}\frac{dq}{r^2},$$

what should be substituted for (b) dq and (c) r^2 to ready for the required integration? (d) What are the limits of the integration? (e) If $\mathbf{E}$ is to be expressed in terms of the charge Q on the rod, what should be substituted for the charge density?

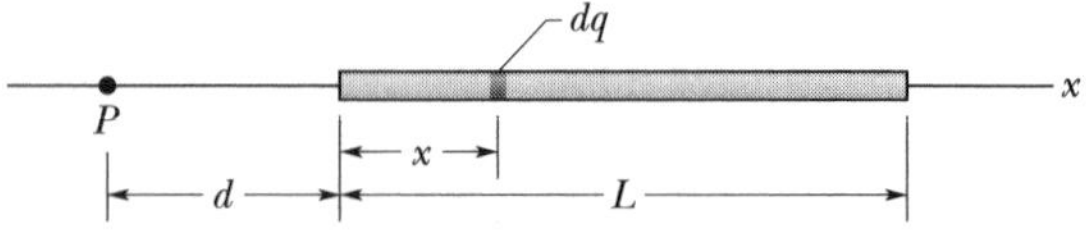

FIGURE 23-49 Question 17.

18. Figure 23-50 shows three rods, each with the same charge Q spread uniformly along its length. Rods a (of length L) and b (of length $L/2$) are straight, and points P are aligned with their midpoints. Rod c (of length $L/2$) forms a complete circle about point P. Rank the rods according to the magnitude of the electric field they create at points P, greatest first.

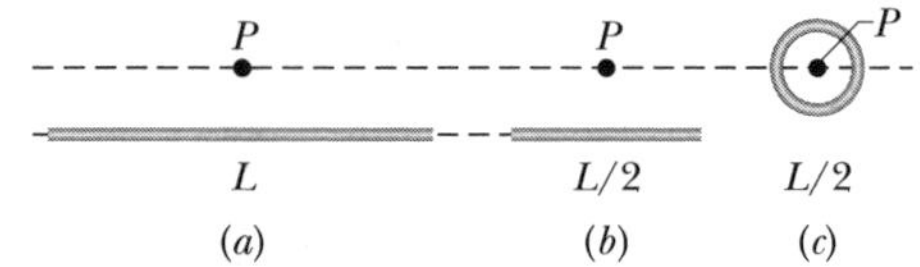

FIGURE 23-50 Question 18.

19. Figure 23-51 shows two disks and a flat ring, each with the same uniformly spread charge Q. Rank the objects according to the magnitude of the electric field they create at points P (which are at the same vertical heights), greatest first.

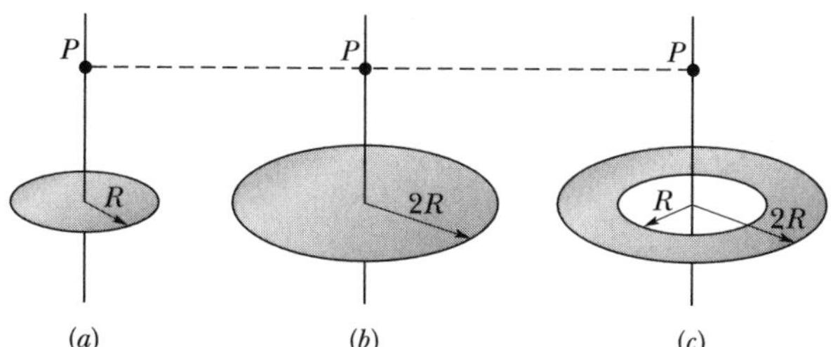

FIGURE 23-51 Question 19.

20. An electron is released in a uniform electric field given by $\mathbf{E} = 4\mathbf{k}$, where $\mathbf{k}$ is the usual unit vector and $\mathbf{E}$ is in newtons per coulomb. What is the direction of (a) the force on the electron, (b) the acceleration of the electron, and (c) the rate of change $d\mathbf{p}/dt$ of the electron's momentum?

21. Figure 23-52 shows five protons that are launched in a uniform electric field $\mathbf{E}$; the magnitude and direction of the launch velocities are indicated. Rank the protons according to the magnitude of their accelerations due to the field, greatest magnitude first.

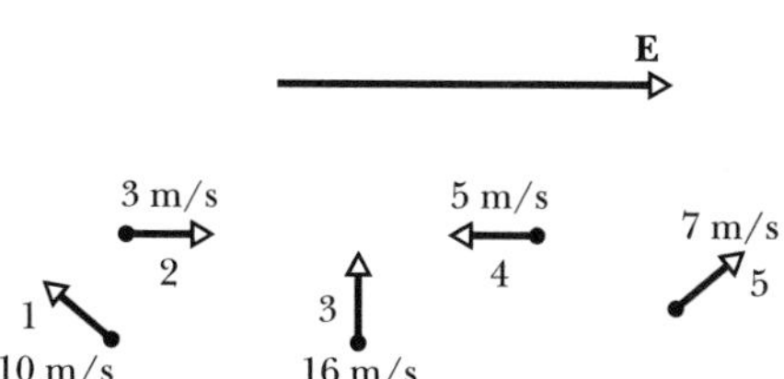

FIGURE 23-52 Question 21.

22. In Fig. 23-53 electron e_1 is launched leftward between two charged parallel plates where the electric field is uniform. The field causes the electron to double back toward the right (the path is displaced upward in the figure only for clarity). (a) What is the direction of the electric field? (b) Rank positions A, B, and C of electron e_1 according to the magnitude of the acceleration of the electron there, greatest first. (c) Rank those three positions according to the kinetic energy of electron e_1 there, greatest first. (This question continues in the next column.)

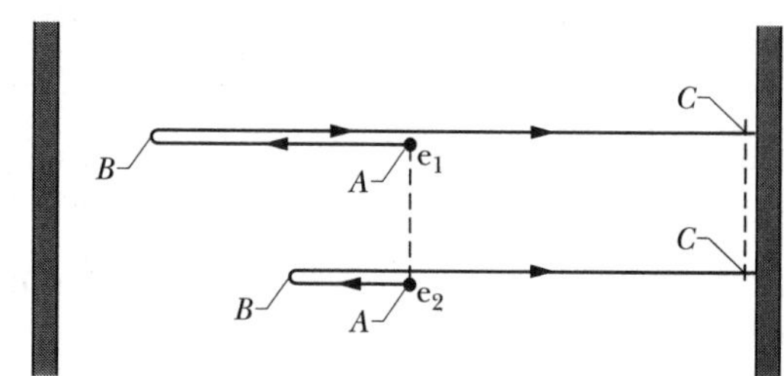

FIGURE 23-53 Question 22.

Electron e_2 is also launched to the left and doubles back to the right. (d) Is the magnitude of the acceleration of electron e_2 at its point A greater than, less than, or equal to that of electron e_1 at its point A? (e) Is the kinetic energy of electron e_2 at its point B greater than, less than, or equal to that of electron e_1 at its point B? (f) Repeat (e) for points C.

23. When two electric dipoles are near each other, they each experience the electric field of the other, and the two-dipole system has a certain potential energy. (a) If the electric dipoles are arranged side by side, does the system have greater potential energy when the dipole moments are parallel (Fig. 23-54*a*) or antiparallel (Fig. 23-54*b*)? (b) If, instead, the electric dipoles are arranged on the same axis, does the system have greater potential energy when the dipole moments are parallel (Fig. 23-54*c*) or antiparallel (Fig. 23-54*d*)?

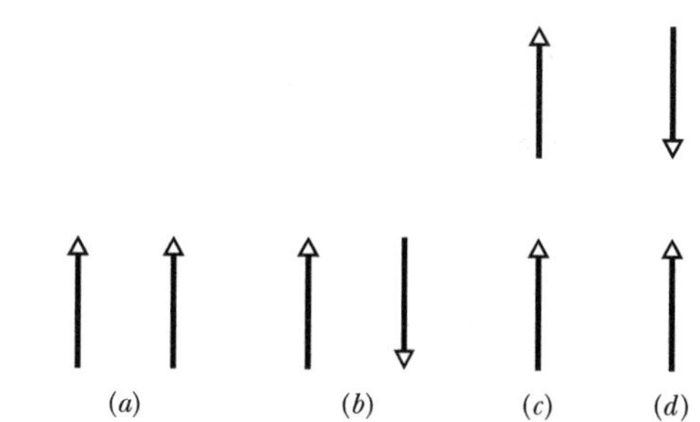

FIGURE 23-54 Question 23.

24. When three electric dipoles are near each other, they each experience the electric field of the other two, and the three-dipole system has a certain potential energy. Figure 23-55 shows two arrangements in which three electric dipoles are side by side. Each dipole has the same magnitude of electric dipole moment, and the spacings between adjacent dipoles are identical. In which arrangement is the potential energy of the three-dipole system greater?

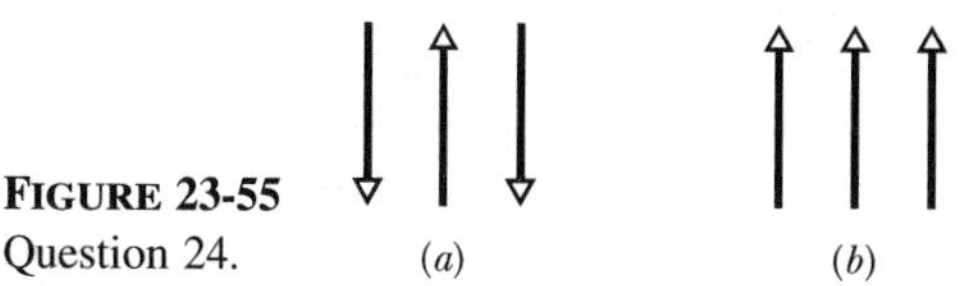

FIGURE 23-55 Question 24.

EXERCISES & PROBLEMS

65. Two particles, each with a charge of magnitude 12 nC, are placed at two of the vertices of an equilateral triangle. The length of each side of the triangle is 2.0 m. What is the magnitude of the electric field at the third vertex of the triangle if (a) both of the charges are positive and (b) one of the charges is positive and the other is negative?

66. A particle of charge Q lies at the origin of an xy coordinate system. A particle of charge q lies at coordinates $(4a, 0)$. We are concerned about the electric field these particles produce at point P at coordinates $(4a, 3a)$. What is Q in terms of q if at P (a) the x and y components of the electric field are equal and (b) the electric field has no y component?

67. Three particles, each with positive charge Q, form an equilateral triangle, with each side of length d. What is the magnitude of the electric field produced by the particles at the midpoint of any side?

68. The electric field in the xy plane produced by a positively charged particle is $7.2(4\mathbf{i} + 3\mathbf{j})$ N/C at the point (3.0, 3.0) cm and $100\mathbf{i}$ N/C at the point (2.0, 0) cm. What are (a) the x and y coordinates and (b) the charge of the particle?

69. An electron enters a region of uniform electric field (E = 50 N/C) with an initial velocity of 40 km/s in the same direction as the electric field. (a) What is the speed of the electron 1.5 ns after entering this region? (b) How far does the electron travel during the 1.5 ns interval?

70. Figure 23-56 shows the deflection-plate system of a conventional TV tube. The length of the plates is 3.0 cm and the electric field between the two plates is 10^6 N/C (vertically up). If the electron enters the plates with a horizontal velocity of 5.9×10^7 m/s, what is the vertical deflection Δy at the end of the plates?

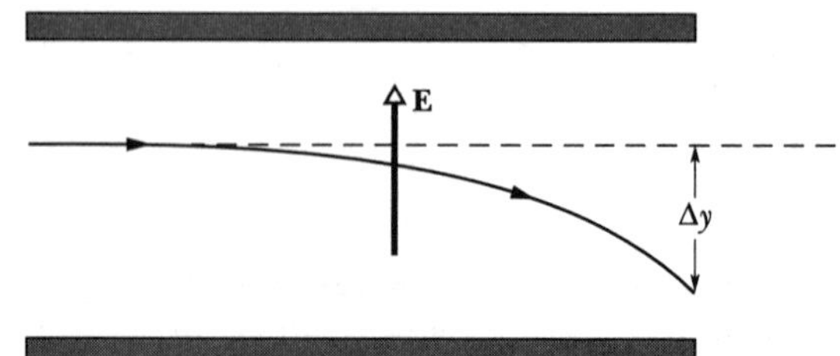

FIGURE 23-56 Problem 70.

71. A charge of 20 nC is uniformly distributed along a straight rod of length 4.0 m that is bent into a circular arc with a radius of 2.0 m. What is the magnitude of the electric field at the center of curvature of the arc?

72. A charge (uniform linear density = 9.0 nC/m) is distributed along an x axis from x = 0 to x = 3.0 m. Determine the magnitude of the electric field at x = 4.0 m on the x axis.

73. In Fig. 23-57, eight charged particles form a square array; charge $q = e$ and distance d = 2.0 cm. What are the magnitude and direction of the net electric field at the center?

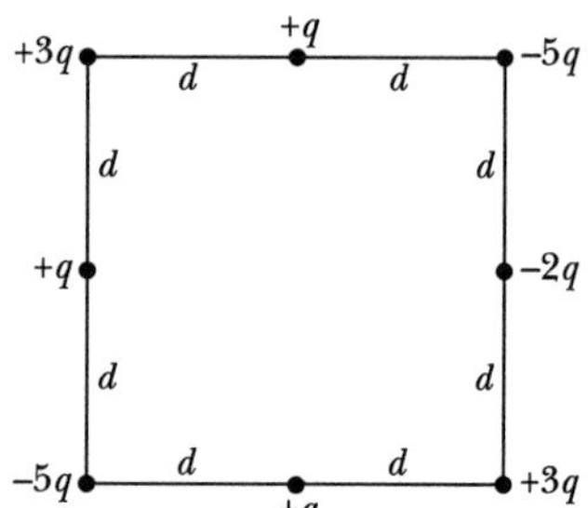

FIGURE 23-57 Problem 73.

74. Two charged particles are fixed in place in Fig. 23-58. What is the x coordinate of the point at which the net electric field is zero?

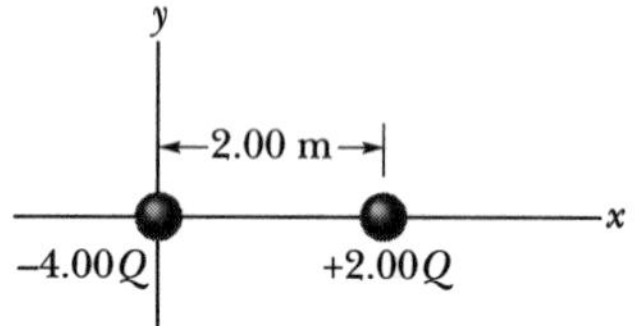

FIGURE 23-58 Problem 74.

75. What are the magnitude and direction of the net electric field at point P in Fig. 23-59?

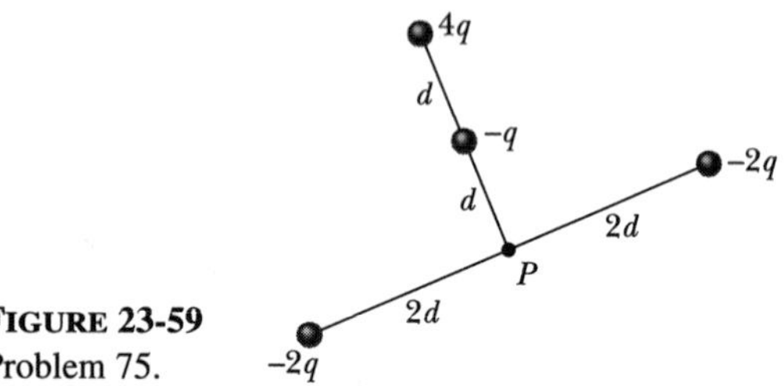

FIGURE 23-59 Problem 75.

76. Figure 23-60 shows two charged particles on an x axis: $-q = -3.20 \times 10^{-19}$ C at $x = -3.00$ m and $q = 3.20 \times 10^{-19}$ C at $x = +3.00$ m. What are the magnitude and direction of the net electric field they produce at point P at y = 4.00 m?

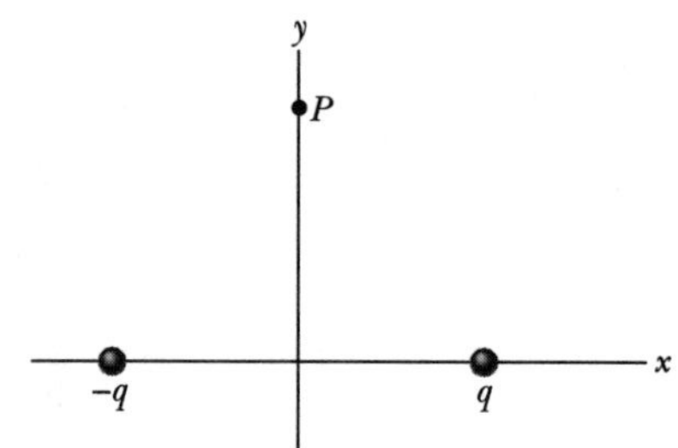

FIGURE 23-60 Problem 76.

77. Figure 23-61 shows an uneven arrangement of electrons (e) and protons (p) on a circular arc of radius $r = 2.00$ cm. What are the magnitude and direction of the net electric field produced by the particles at the center of the arc?

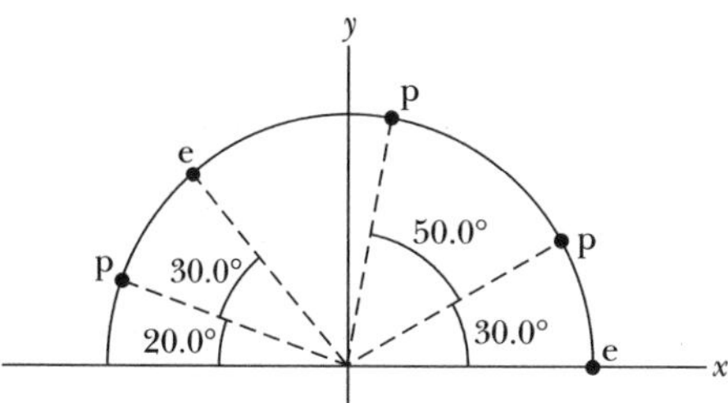

FIGURE 23-61 Problem 77.

78. How much energy is needed to flip an electric dipole from being lined up with a uniform external electric field to being lined up opposite the field? The dipole consists of an electron and a proton at a separation of 2.00 nm, and it is in a uniform field of magnitude 3.00×10^6 N/C.

79. In Fig. 23-62, an electric dipole swings from an initial orientation i to a final orientation f in a uniform external electric field. The electric dipole moment is 1.60×10^{-27} C · m; the field is 3.00×10^6 N/C. What is the change in the dipole's potential energy?

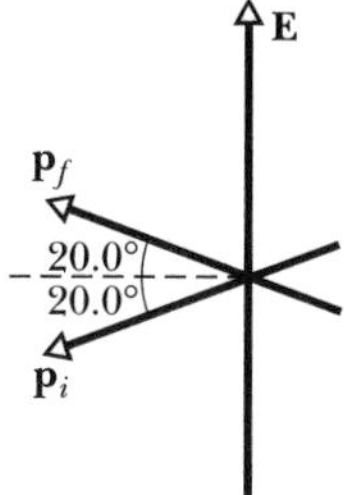

FIGURE 23-62 Problem 79.

Tutorial Problem

80. Let's look at a system of four charges arranged in a square with sides of length L, with charges $q_1 = Q$, $q_2 = -2Q$, $q_3 = 3Q$, and $q_4 = -4Q$, as shown in Fig. 23-63. We are interested in determining the electric field at the center of the square. Then we will determine how this electric field would affect a charged particle placed at the center.

(a) On a diagram similar to Fig. 23-63, sketch the contributions to the electric field vector from each of the four charges by drawing the vectors in the correct direction and approximately to scale. Label these vectors, using a fitting notation, such as $\mathbf{E}_1$ for the contribution to the electric field from charge q_1, $\mathbf{E}_2$ for charge q_2, and so on, reserving $\mathbf{E}$ for the total (net) electric field.

(b) Determine the contributions of each charge to the electric field; do this algebraically (with symbols, such as Q, L, etc.), using the standard unit-vector notation. (c) Name the principle of physics that is used to determine the electric field from the individual contributions found in part (b). Determine the total electric field at the center of the square. Show this vector on your diagram. (d) Now assume that $Q = 0.150\ \mu$C and $L = 60.0$ cm. What is the numerical value of the electric field at the center of the square?

(e) Suppose an electron were placed at the center of the square. What would be the force on the electron? In which direction would the electron accelerate? What would be its initial acceleration? (f) Suppose, instead, a proton were placed at the center of the square. What would be the force on it? In which direction would it accelerate? What would be its initial acceleration? Compare the force and the acceleration of the proton with those of the electron. (g) At a great distance from this system of charges, what does it appear to be?

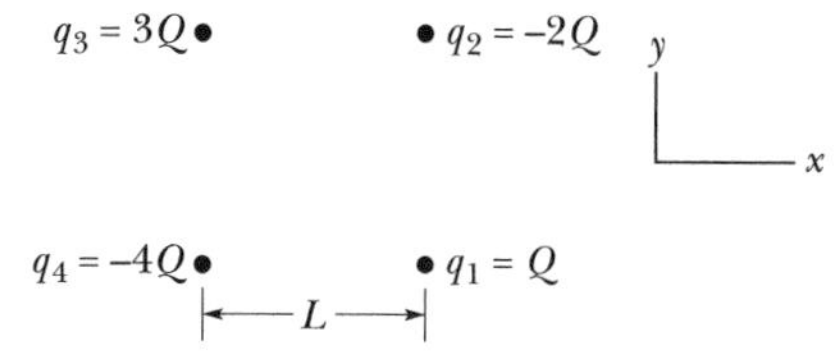

$q_3 = 3Q$ $q_2 = -2Q$ $q_4 = -4Q$ $q_1 = Q$ L x y

FIGURE 23-63 Problem 80.

Answers

(b) The individual charges q_i contribute $\mathbf{E}_i = -(kq_i/r_i^2)\hat{r}_i$ to the electric field at the center, which we can take as the origin, $\hat{r}_i$ being the unit vector toward the charge; the negative sign indicates that the electric field points away from a positive charge q_i. To determine these contributions, we can make a table listing all the terms that show up in $\mathbf{E}_i$.

i	q_i	r_i	$\hat{r}_i$	$\mathbf{E}_i = -(kq_i/r_i^2)\hat{r}_i$
1	Q	$L/\sqrt{2}$	$(+\hat{\imath} - \hat{\jmath})/\sqrt{2}$	$kQ(\sqrt{2}/L^2)(-\hat{\imath} + \hat{\jmath})$
2	$-2Q$	$L/\sqrt{2}$	$(+\hat{\imath} + \hat{\jmath})/\sqrt{2}$	$2kQ(\sqrt{2}/L^2)(+\hat{\imath} + \hat{\jmath})$
3	$3Q$	$L/\sqrt{2}$	$(-\hat{\imath} + \hat{\jmath})/\sqrt{2}$	$3kQ(\sqrt{2}/L^2)(+\hat{\imath} - \hat{\jmath})$
4	$-4Q$	$L/\sqrt{2}$	$(-\hat{\imath} - \hat{\jmath})/\sqrt{2}$	$4kQ(\sqrt{2}/L^2)(-\hat{\imath} - \hat{\jmath})$

(c) The principle of linear superposition can be used to find the total electric field at the center of the square from the individual contributions found in part (b). The sum of the four electric fields is

$$\begin{aligned}\mathbf{E} &= \mathbf{E}_1 + \mathbf{E}_2 + \mathbf{E}_3 + \mathbf{E}_4\\ &= \frac{\sqrt{2}kQ}{L^2}[(-1 + 2 + 3 - 4)\hat{\imath} + (1 + 2 - 3 - 4)\hat{\jmath}]\\ &= -\frac{4\sqrt{2}kQ}{L^2}\hat{\jmath}.\end{aligned}$$

Note that the x components of the electric field cancel one another.

(d) Substituting the given values into the result of part (c), we find

$$\mathbf{E} = -\frac{4\sqrt{2}kQ}{L^2}\hat{\jmath}$$
$$= -\frac{4\sqrt{2}(8.99 \times 10^9\ \mathrm{N \cdot m^2/C^2})(0.150 \times 10^{-6}\ \mathrm{C})}{(0.600\ \mathrm{m})^2}\hat{\jmath}$$
$$= -(2.12 \times 10^4\ \mathrm{N/C})\hat{\jmath}.$$

(e) The force on a charge q at a point where the electric field is $\mathbf{E}$ is $\mathbf{F} = q\mathbf{E}$. For an electron, $q = -e = -1.602 \times 10^{-19}$ C, so the force on the electron at the center of the square would be

$$\mathbf{F} = -e\mathbf{E} = -(1.602 \times 10^{-19}\ \mathrm{C})(-2.12 \times 10^4\ \mathrm{N/C})\hat{\jmath}$$
$$= +(3.40 \times 10^{-15}\ \mathrm{N})\hat{\jmath}.$$

This force is in the $+\hat{\jmath}$ direction, so the electron, which is attracted to the positive charges and repelled by the negative charges, is accelerating toward the midpoint of the top part of the square. The initial acceleration of the electron is found from Newton's second law:

$$\mathbf{a} = \frac{\mathbf{F}}{m} = \frac{(3.40 \times 10^{-15}\ \mathrm{N})\hat{\jmath}}{9.11 \times 10^{-31}\ \mathrm{kg}} = (3.73 \times 10^{15}\ \mathrm{m/s^2})\hat{\jmath}.$$

(f) The force on a charge q at a point where the electric field is $\mathbf{E}$ is $\mathbf{F} = q\mathbf{E}$. For a proton, $q = +e = +1.602 \times 10^{-19}$ C, so the force on the proton at the center of the square would be

$$\mathbf{F} = +e\mathbf{E} = +(1.602 \times 10^{-19}\ \mathrm{C})(-2.48 \times 10^4\ \mathrm{N/C})\hat{\jmath}$$
$$= -(3.40 \times 10^{-15}\ \mathrm{N})\hat{\jmath}.$$

The initial acceleration of the proton is found from Newton's second law:

$$\mathbf{a} = \frac{\mathbf{F}}{m} = \frac{-(3.40 \times 10^{-15}\ \mathrm{N})\hat{\jmath}}{1.67 \times 10^{-27}\ \mathrm{kg}}$$
$$= -(2.04 \times 10^{12}\ \mathrm{m/s^2})\hat{\jmath}.$$

The force on the proton is equal in magnitude but opposite in direction to the force experienced by the electron in part (e). The initial acceleration of the proton is smaller in magnitude and opposite in direction to the acceleration of the electron in part (e).

(g) At a great distance from this system of charges, the system will appear to be a single charge equal to the net charge: $(+Q - 2Q + 3Q - 4Q) = -2Q$. In other words, at a great distance from this system of charges, the electric field will be almost exactly the same as the electric field of a single charge $-2Q$; the difference would be difficult to detect.

Chapter Twenty-Four
Gauss' Law

QUESTIONS

16. Figure 24-43 shows, in cross section, two Gaussian spheres and two Gaussian cubes that are centered on a positively charged particle. (a) Rank the net flux through the four Gaussian surfaces, greatest first. (b) Rank the magnitudes of the electric field on the surfaces, greatest first, and indicate if the magnitude is uniform or variable along the surface.

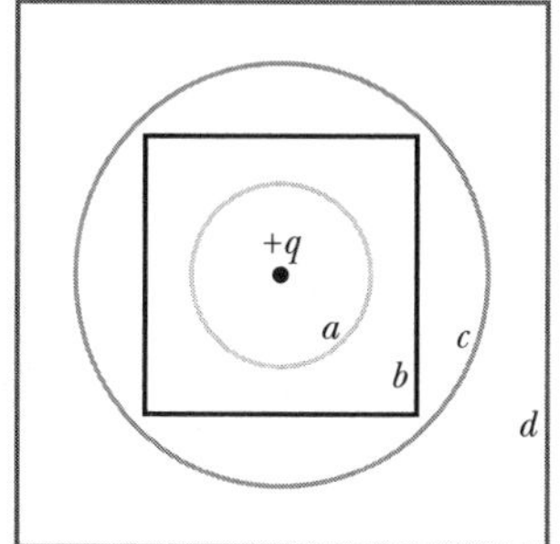

FIGURE 24-43 Question 16.

17. Figure 24-44 shows four spheres, each with the same charge Q uniformly distributed through its volume. (a) Rank the spheres according to their volume charge density, greatest first. The figure also shows a point P for each sphere, at the same distance from the center of the sphere. (b) Rank the spheres according to the magnitude of the electric field they produce at points P, greatest first.

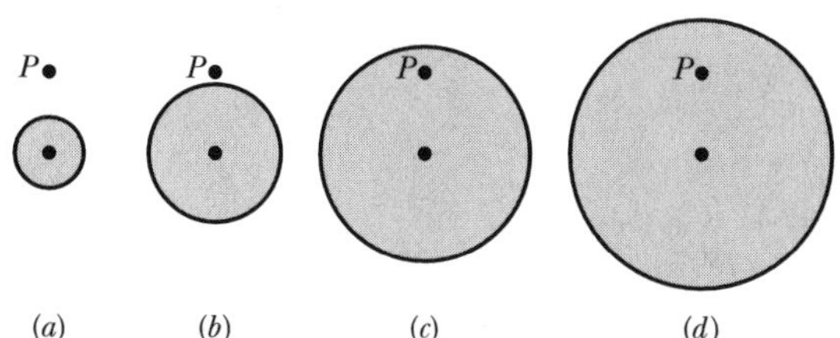

FIGURE 24-44 Question 17.

18. A hollow conducting sphere with inner radius R_i and outer radius R_o initially has a net charge of $+Q$. If we double the net charge to $+2Q$, what happens to the magnitude of the electric field at a point located (a) at radius $R_i/2$, (b) between radius R_i and radius R_o, and (c) at radius $2R_o$?

19. Figure 24-45 shows three hollow conducting spheres of the same size; the net charge of each sphere is given. Rank the spheres according to the magnitude of the electric field they produce at (a) points P_1, which are at the same radial distance within the hollows, (b) points P_2, which are at the same radial distance within the material of the spheres, and (c) points P_3, which are at the same radial distance outside the spheres.

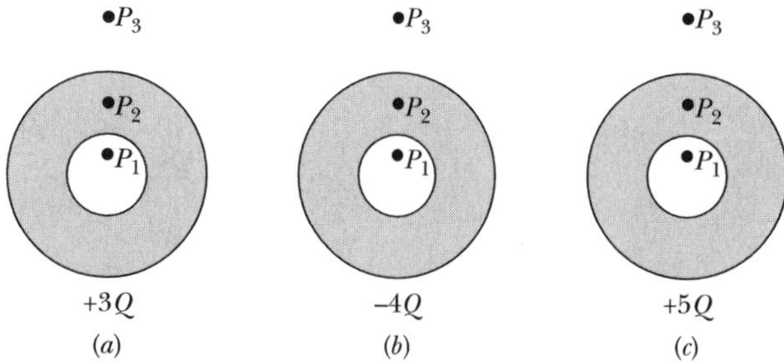

FIGURE 24-45 Question 19.

20. In Fig. 24-46 an electron is released between two infinite nonconducting sheets that are horizontal and have uniform surface charge densities $\sigma_{(+)}$ and $\sigma_{(-)}$, as indicated. Rank the magnitude of the electron's acceleration for the surface charge densities and sheet separations in the following three situations, greatest first:

SITUATION	$\sigma_{(+)}$	$\sigma_{(-)}$	SEPARATION
1	$+4\sigma$	-4σ	d
2	$+7\sigma$	$-\sigma$	$4d$
3	$+3\sigma$	-5σ	$9d$

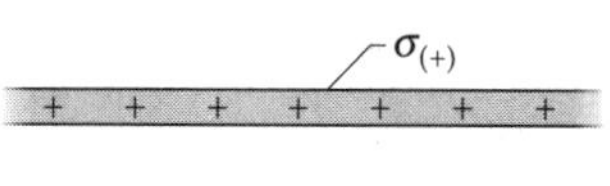

e⁻

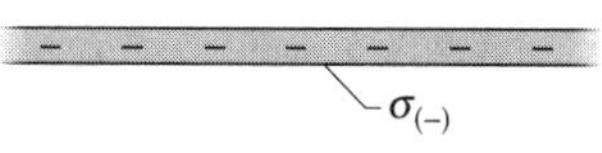

FIGURE 24-46 Question 20.

21. Figure 24-47a shows four electrons launched at the same distance from an infinite nonconducting sheet of uniform positive surface charge density; each electron reaches the sheet. The launch velocities of the electrons are

Electron	A	B	C	D
Velocity (m/s)	$-2.0\mathbf{i}$	$+2.0\mathbf{j}$	0	$+2.0\mathbf{i}$

(a) Rank the electrons according to the time they take to reach the sheet, greatest first. (b) Rank them according to the y component of their velocity as they reach the sheet, greatest first. Figure 24-47b shows six plots of vertical velocity v_y versus time t; plots 2 and 4 are straight, the others are curved either up or down. (c) Which plot best corresponds to which particle?

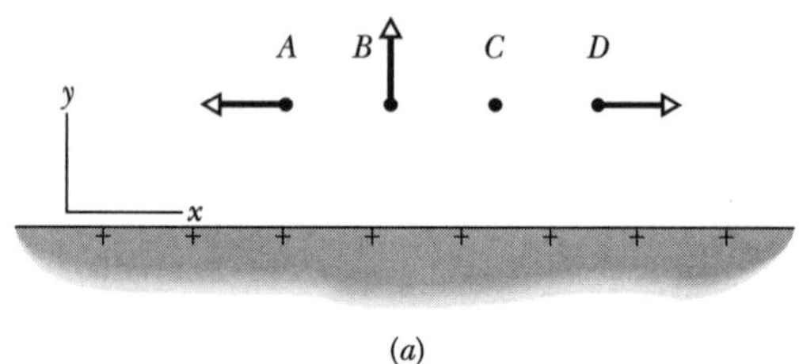

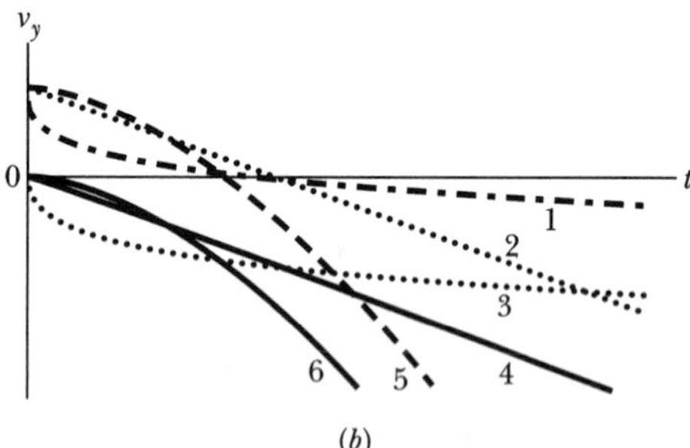

FIGURE 24-47 Question 21.

22. In Fig. 24-48a, an electron e$^-$ is released from rest at time $t = 0$ within a sphere that has negative charge $-Q$ uniformly distributed within its volume. Figure 24-48b is identical except that the sphere has positive charge $+Q$. Assume that the electron can move through the sphere without collisions.

Figure 24-48c gives several plots of radial velocity v versus time (positive corresponds to radially outward, negative to radially inward); plots 2 and 7 are straight, plots 4 and 5 are sinusoidal, and the other four are curved either up or down. Which of the eight plots best corresponds to the velocity of the electron as it moves inside the sphere of (a) Fig. 24-48a and (b) Fig. 24-48b?

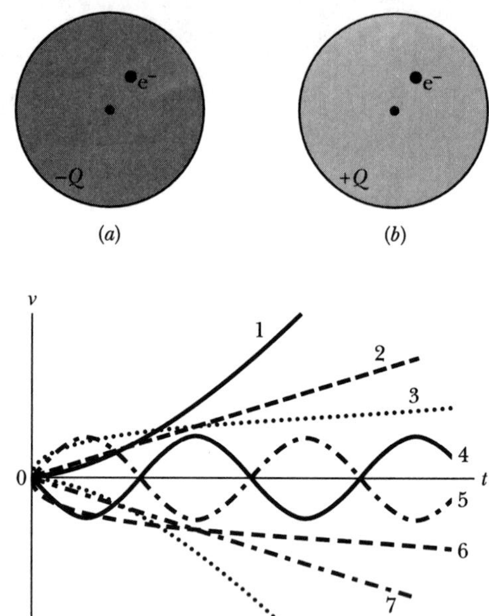

FIGURE 24-48 Question 22.

23. *Organizing question:* Figure 24-49 shows a section of a very long cylindrical rod of nonconducting material in which the volume charge density ρ is uniform. It also shows two cylindrical Gaussian surfaces A and B that are coaxial with the rod (they have the same central axis as the rod). Surface A is inside the rod; surface B is outside. The radius of the rod is R, that of surface A is r_A, and that of surface B is r_B. Let L be a length along the axis of the rod, and let E_A and E_B be the magnitudes of the electric field on surfaces A and B, respectively.

In terms of these symbols and ϵ_0, set up the left and right sides of Gauss' law (Eq. 24-7) for (a) surface A and (b) surface B. (For the left side of Eq. 24-7 you need to replace the integral with its equivalent in terms of the given symbols; for the right side, you need to replace q_{enc} with its equivalent in terms of the given symbols.)

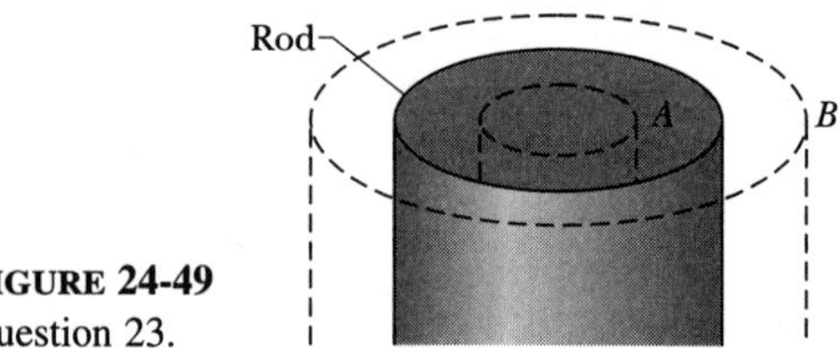

FIGURE 24-49 Question 23.

24. *Organizing question:* Figure 24-50 shows a ball of nonconducting material in which the volume charge density ρ is uniform. It also shows, in cross section, two spherical Gaussian surfaces A and B centered on the ball. The radius of the ball is R, that of surface A is r_A, and that of surface B is r_B. Let E_A and E_B be the magnitudes of the electric field on surfaces A and B, respectively.

In terms of these symbols and ϵ_0, set up the left and right sides of Gauss' law (Eq. 24-7) for (a) surface A and (b) surface B. (For the left side of Eq. 24-7 you need to replace the integral with its equivalent in terms of the given symbols; for the right side, you need to replace q_{enc} with its equivalent in terms of the given symbols.)

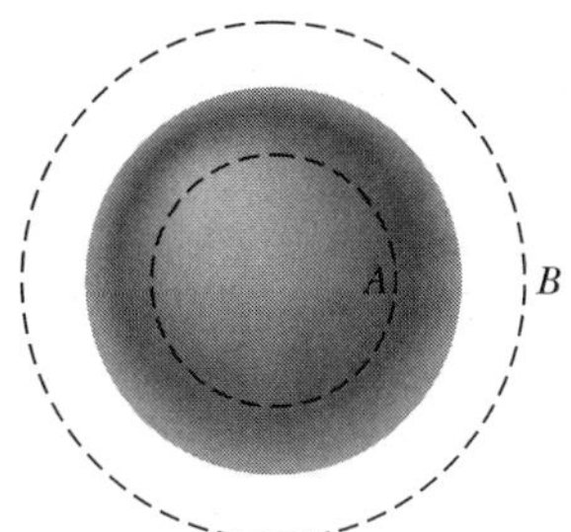

FIGURE 24-50 Question 24.

25. Figure 24-51 shows, in cross section, a large horizontal slab of nonconducting material in which the volume charge density ρ is uniform. The slab occupies the region between $-y_s$ and y_s. The figure also shows two cylindrical Gaussian surfaces A and B as seen from the side; they are centered on a central x axis through the slab. Both cylinders have radius R; cylinder A extends from $-y_A$ to y_A; cylinder B extends from $-y_B$ to y_B. E_A is the magnitude of the electric field at $-y_A$ and y_A. E_B is the magnitude of the electric field at $-y_B$ and y_B.

In terms of these symbols and ϵ_0, set up the left and right sides of Gauss' law (Eq. 24-7) for (a) surface A and (b) surface B. (For the left side of Eq. 24-7 you need to replace the integral with its equivalent in terms of the given symbols; for the right side, you need to replace q_{enc} with its equivalent in terms of the given symbols.)

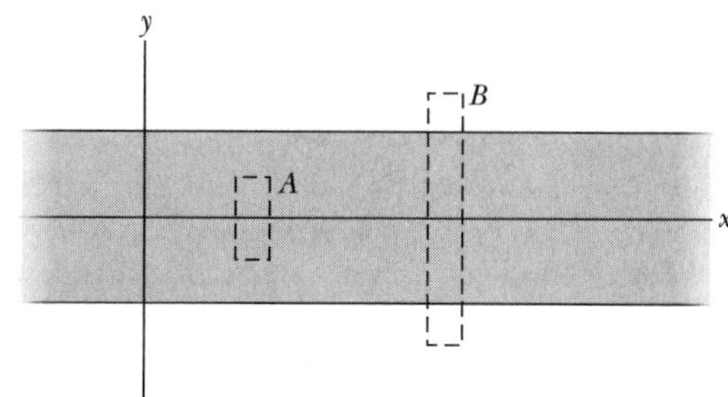

FIGURE 24-51 Question 25.

EXERCISES & PROBLEMS

61. When a shower is turned on in a closed bathroom, the splashing of the water on the bare tub can fill the room's air with negatively charged ions and produce an electric field of as much as 1000 N/C. Consider a bathroom with dimensions of 2.5 m $\times$ 3.0 m $\times$ 2.0 m. Along the ceiling, floor, and four walls, approximate the electric field as being directed perpendicular to the surface and as having a uniform magnitude of 600 N/C. (a) Treat the bathroom as a Gaussian container and find the volume charge density ρ in the room's air. (b) What is the number of electronic charges e per cubic meter?

62. A uniform surface charge of 8.0 nC/m^2 is distributed over the entire xy plane. Consider a spherical Gaussian surface centered on the origin and having a radius of 5.0 cm. Determine the electric flux for this surface.

63. At each point on the surface of the cube shown in Fig. 24-52, the electric field is in the y direction. The length of each edge of the cube is 3.0 m. On the top surface of the cube $\mathbf{E} = -34\mathbf{j}$ N/C, and on the bottom face of the cube $\mathbf{E} = +20\mathbf{j}$ N/C. Determine the net charge contained within the cube.

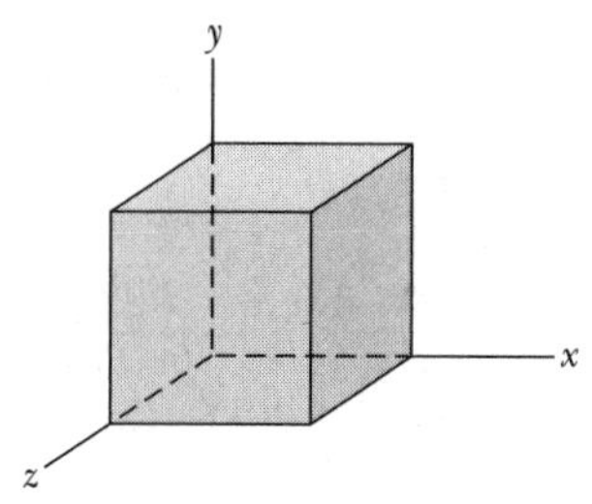

FIGURE 24-52 Problem 63.

64. The electric field throughout all space is $\mathbf{E} = (x + 2)\mathbf{i}$ N/C, with x measured in meters. Consider a cylinder (radius = 20 cm) that is coaxial with the x axis. One end of the cylinder is positioned at $x = 0$ and the other end at $x = 2.0$ m. (a) What is the magnitude of the electric flux through the end of the cylinder at $x = 2.0$ m? (b) What amount of charge is enclosed within the cylinder?

65. A uniform charge density of 500 nC/m^3 is distributed throughout a spherical volume of radius 6.0 cm. Consider a cube with its center at the center of the sphere. What is the electric flux through this cubical surface if the edge length is (a) 4.0 cm and (b) 8.0 cm?

66. A long nonconducting cylinder of radius 4.0 cm has a nonuniform charge density that is a function of the radial distance r from the axis of the cylinder. This charge distribution is such that at any point within the cylinder $\rho = Ar^2$, with $A = 2.5\ \mu$C/m^5. What are the magnitudes of the electric field at radial distances of (a) 3.0 cm and (b) 5.0 cm from the axis of the cylinder?

67. A charge of uniform linear density 2.0 nC/m is distributed along a long, thin, nonconducting rod. The rod is coaxial with a long, hollow, conducting cylinder (inner radius = 5.0 cm, outer radius = 10 cm). The net charge on the conductor is zero. (a) What is the magnitude of the electric field 15 cm from the axis of the cylinder? What are the surface charge densities on (b) the inner surface and (c) the outer surface of the conductor?

68. Charge of uniform density 8.0 nC/m^2 is distributed over the entire xy plane; charge of uniform density 3.0 nC/m^2 is distributed over the parallel plane defined by $z = 2.0$ m. Determine the magnitudes of the electric field for any point with (a) $z = 1.0$ m and (b) $z = 3.0$ m.

69. Charge of uniform density $\rho = 1.2$ nC/m^3 fills an infinite slab between $x = -5.0$ cm and $x = +5.0$ cm. What are the magnitudes of the electric field at any point for which (a) $x = 4.0$ cm and (b) $x = 6.0$ cm?

70. Charge of uniform density $\rho = 3.2\ \mu$C/m^3 fills a nonconducting sphere of radius 5.0 cm. What are the magnitudes of the electric field (a) 3.5 cm and (b) 8.0 cm from the center of the sphere?

71. The electric field just outside the outer surface of a hollow spherical conductor of inner radius = 10 cm and outer radius = 20 cm has a magnitude of 450 N/C and is directed outward. When an unknown point charge Q is introduced into the center of the sphere, the electric field is still directed outward but has decreased to 180 N/C. (a) What is the net charge enclosed within the outer spherical surface of the conductor before Q is introduced? (b) What is the charge Q? After Q is introduced, what are the charges on (c) the inner surface and (d) the outer surface of the conductor?

72. Figure 24-53a shows a closed Gaussian surface in the shape of a cube of edge length 2.00 m. It lies in a region where the electric field is given by

$$\mathbf{E} = (3.00x + 4.00)\mathbf{i} + 6.00\mathbf{j} + 7.00\mathbf{k},$$

with $\mathbf{E}$ in newtons per coulomb and x in meters. What is the net charge contained by the cube?

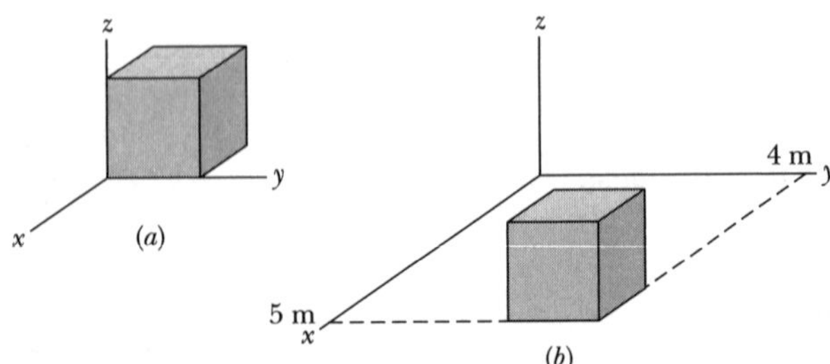

FIGURE 24-53 Problems 72 and 73.

73. Figure 24-53b shows a closed Gaussian surface in the shape of a cube of edge length 2.00 m. It lies in a region where the electric field is given by

$$\mathbf{E} = -3.00\mathbf{i} - 4.00y^2\mathbf{j} + 3.00\mathbf{k},$$

with $\mathbf{E}$ in newtons per coulomb and y in meters. What is the net charge contained by the cube?

74. A spherical conducting shell has a charge of $-14\ \mu$C on its outer surface and a charged particle in its hollow. If the net charge on the shell is $-10\ \mu$C, what are the charges (a) on the inner surface of the shell and (b) of the particle?

75. A spherical conducting shell with a net charge of $-3.00Q$ contains a particle of charge $+5.00Q$ in its hollow. The shell has an inner radius of 0.80 m and an outer radius of 1.4 m. What are the magnitudes and directions of the electric fields at (a) point A at radius 0.500 m, (b) point B at radius 1.00 m, and (c) point C at radius 2.00 m?

76. Point P is between two infinite, parallel, nonconducting sheets of charge, at distance d from one sheet with surface charge density $-3.00\sigma_1$ and at distance $3.00d$ from the other sheet with surface charge density $+\sigma_1$. What are the magnitude and direction of the electric field at P due to the two sheets?

77. Figure 24-54 shows, in cross section, three infinitely large nonconducting sheets on which charge is uniformly spread. The surface charge densities are $\sigma_1 = +2.0\ \mu$C/m^2, $\sigma_2 = +4.0\ \mu$C/m^2, and $\sigma_3 = -5.0\ \mu$C/m^2, and distance $L = 1.50$ cm. What are the magnitude and direction of the net electric field at point P?

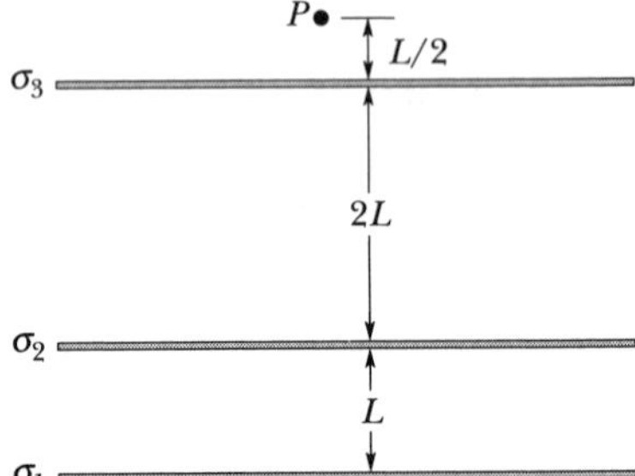

FIGURE 24-54 Problem 77.

78. Charge Q is uniformly distributed in a sphere of radius R. (a) How much charge is contained within radius $r = R/2$? (b) What is the ratio of the electric field magnitude at $r = R/2$ to that on the surface of the sphere?

79. A charge of 6.00 pc is spread uniformly throughout the volume of a sphere of radius $r = 4.00$ cm. What are the magnitude and direction of the electric fields at the radial distances of (a) 6.00 cm and (b) 3.00 cm?

Tutorial Problems

80. A solid metal sphere of radius a is concentric with, and inside, a metal shell of inner radius b_1 and outer radius b_2. The sphere has a net positive charge Q, and the shell has a net negative charge $-2Q$. Let's determine the electric fields in this problem by making use of Gauss' law.

(a) First, sketch this physical situation. (b) What is the type of symmetry exhibited in this problem? What does the symmetry tell you about the direction and magnitude of the electric field in this problem? How is this symmetry used in the application of Gauss' law? What type of Gaussian surface is appropriate to this situation?

(c) What is the electric field for $r < a$? (d) What is the electric field for $a < r < b_1$? (e) What is the electric field for $b_1 < r < b_2$? What does this imply about the total charge on the inside surface of the metal shell? (f) What is the electric field for $r > b_2$? (g) Plot the electric field component as a function of r on a graph. Mark a, b_1, and b_2 for $b_1 \approx b_2 \approx 2a$, and plot the field approximately to scale. (h) Determine the electric charge densities on each of the three metal surfaces and show that they lead to the correct electric field magnitudes just outside those surfaces.

Answers

(b) This problem exhibits spherical symmetry: everything is symmetric about the center of the sphere. This symmetry implies that the direction of the electric field must be in the radial direction, that is, along the line from the center of the sphere to the point at which the field is being calculated. The direction of the electric field at a point $\mathbf{r}$ must be either $\hat{r}$ or $-\hat{r}$; the field cannot have a component in any other direction. Also, the electric field strength must be a function of just r. Thus all points with the same value of r must have the same magnitude of the electric field. We'll write the electric field as $E_r(r)\hat{r}$.

Applying Gauss' law in this situation involves using spherical Gaussian surfaces centered at the center of the sphere. The electric flux through a Gaussian surface of radius r will be $4\pi r^2 E_r(r)$.

(c) There is no electric field inside the sphere because that region is a conducting material. Recall that in an electrostatic situation (no moving charges) the electric field must be zero in a conductor.

(d) The total charge inside a Gaussian surface of radius r with $a < r < b_1$ is just Q, the net charge on the metal sphere. Gauss' law in this region gives

$$4\pi r^2 E_r(r) = \frac{Q}{\epsilon_0},$$

so

$$E_r(r) = \frac{Q}{4\pi\epsilon_0 r^2}.$$

Thus the electric field is

$$E_r(r)\,\hat{r} = \frac{Q}{4\pi\epsilon_0 r^2}\,\hat{r}.$$

(e) The region $b_1 < r < b_2$ is in the metal shell (a conducting material) and must have zero electric field. Gauss' law then implies that the total charge inside a Gaussian surface through this region must be zero. Since the metal sphere has a charge of Q, the inside surface (at $r = b_1$) of the metal shell must have a charge of $-Q$.

(f) The total charge of the metal sphere and the metal shell is $Q - 2Q = -Q$, so Gauss' law in the region outside the metal shell gives

$$4\pi r^2 E_r(r) = -\frac{Q}{\epsilon_0},$$

so

$$E_r(r) = -\frac{Q}{4\pi\epsilon_0 r^2}.$$

(g) Plot $E = 0$ for $0 < r < a$; $E > 0$ for $a < r < b_1$; $E = 0$ for $b_1 < r < b_2$; and $E < 0$ for $b_2 < r$. Set $E(b_1) \approx \frac{1}{4}E(a)$ and $E(b_2) \approx -E(b_1)$.

(h) On the surface at $r = a$, the total charge is Q. Since the total surface area of the sphere is $4\pi a^2$, the charge density at that surface is $\sigma = Q/4\pi a^2$. That means the electric field just outside the sphere is $\sigma/\epsilon_0 = Q/4\pi\epsilon_0 a^2$, which agrees with the result of part (d) for $r = a$.

On the surface at $r = b_1$, the total charge is $-Q$ and the surface area is $4\pi b_1^2$. So $\sigma = -Q/4\pi b_1^2$, which leads to $\sigma/\epsilon_0 = -Q/4\pi\epsilon_0 b_1^2$. This agrees with the result of part (d), since at $r = b_1$ the field E_r is then $-\sigma/\epsilon_0 = +Q/4\pi\epsilon_0 b_1^2$.

On the surface at $r = b_2$, the surface density is $\sigma = -Q/4\pi b_2^2$ and the electric field component is $\sigma/\epsilon_0 = -Q/4\pi\epsilon_0 b_2^2$. This matches the value of E_r at $r = b_2$ we found in part (f).

81. Let's consider an infinitely long, solid, insulating cylinder with an outer radius a and a uniform charge density ρ. (a) First, sketch this physical situation. (b) What type of symmetry is exhibited in this problem? What does the symmetry tell you about the direction and magnitude of the electric field in this problem? (c) How can the symmetry of this problem be used in the application of Gauss' law? What type of Gaussian surface is appropriate to this situation?

(d) What is the electric field for $r < a$? (e) What is the electric field for $r > a$? (f) Plot the magnitude of the electric field as a function of r on a graph. Mark a and $2a$, and plot the field approximately to scale. (g) Suppose that the cylinder had been a solid metal cylinder instead of an insulating cylinder, but had the same total electric charge per unit length. Where would the charge be found? Plot the electric field magnitude as a function of r for this situation.

Answers

(b) This problem exhibits cylindrical symmetry: everything is symmetric about the axis of the cylinder. This symmetry implies that the direction of the electric field must be in the radial direction, that is, along the line from the axis of the cylinder to the point at which the field is being calculated. In terms of the cylindrical coordinates r, θ, and z, where the z axis is along the axis of the cylinder, the direction of the electric field at a point $\mathbf{r}$ must be either $\hat{r}$ or $-\hat{r}$; the field cannot have a component in any other direction. Also, the electric field strength must be a function of just r. Thus all points with the same value of r must have the same magnitude of the electric field, since there is no difference in the physical situation for different values of θ. We'll write the electric field as $E_r(r)\,\hat{r}$. Note that $\hat{r}$ is not a unit vector from the origin, but a unit vector perpendicular to the axis.

(c) Applying Gauss' law in this situation involves using cylindrical Gaussian surfaces centered on the axis of the cylinder. For such a Gaussian surface of radius r and length L, the electric flux through the Gaussian surface is $2\pi r L E_r(r)$. All this flux occurs through the curved side of the Gaussian surface. There is no flux through the base and top of the cylinder because a normal to the surface there is perpendicular to the electric field.

(d) To apply Gauss' law to the Gaussian surface described in part (b), we need to determine the total charge inside the Gaussian surface. Since the whole region is charged, with a uniform charge density ρ, the total charge is just ρ times the enclosed volume. The volume of a cylinder of radius r and length L is $\pi r^2 L$. Gauss' law thus becomes

$$2\pi r L E_r(r) = \frac{\rho \pi r^2 L}{\epsilon_0},$$

so

$$E_r(r) = \frac{\rho r}{2\epsilon_0}.$$

The electric field is then

$$E_r(r)\,\hat{r} = \frac{\rho r}{2\epsilon_0}\,\hat{r}.$$

(e) We can use arguments similar to those for part (c), except that the charge exists only in the insulating cylinder, in the region $r < a$. The total charge inside the cylinder, and inside the Gaussian surface of radius r and length L, is then

$$Q = \rho(\pi a^2)L.$$

Gauss' law thus becomes

$$2\pi r L E_r(r) = \frac{\rho \pi a^2 L}{\epsilon_0},$$

so

$$E_r(r) = \frac{\rho a^2}{2 r \epsilon_0}.$$

The electric field is then $E_r(r)\,\hat{r} = (\rho a^2/2r\epsilon_0)\,\hat{r}$.

(g) The electric charge would be located on the outer surface of the cylinder; none would be found inside the cylinder. The electric field would be zero inside the cylinder instead of what we found in (d), but outside it would be indistinguishable from that of the insulating cylinder.

Chapter Twenty-Five
Electric Potential

QUESTIONS

25. A proton is released at rest or launched in a uniform electric field. Which of the graphs in Fig. 25-54 could possibly show how the kinetic energy of the proton changes during the proton's motion?

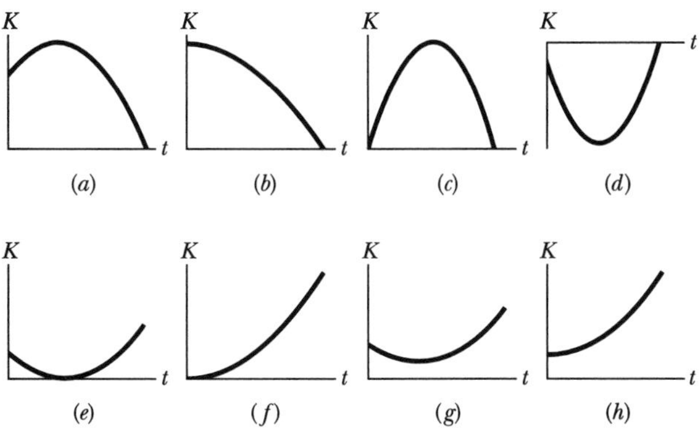

FIGURE 25-54 Question 25.

26. In Fig. 25-55, a particle is to be released at rest at point A and then is to be accelerated directly through point B by an electric field. The potential difference between points A and B is 100 V. Which point should be at higher electric potential if the particle is (a) an electron, (b) a positron (the positively charged antiparticle of the electron), (c) a proton, and (d) an alpha particle (a nucleus of two protons and two neutrons)? (e) For each situation and without a calculator or written calculation, determine the particle's kinetic energy in electron-volts as the particle reaches point B.

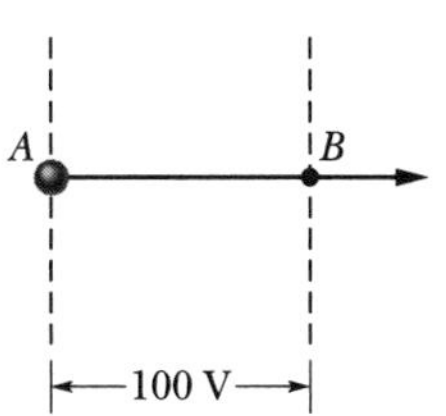

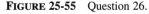

FIGURE 25-55 Question 26.

27. Figure 25-56 shows a family of equipotential lines with values -10 V, -20 V, -30 V, . . . , -70 V (the last three lines are not labeled). Two electrons are released from rest at the point indicated, with electron A slightly to the left of electron B. Electron B travels through point c, at distance d from the release point. Electron A travels through point b, at the same distance d from the release point, and then through point a. (a) Which electron takes more time to travel the distance d, or do the electrons take the same time? (b) Is the speed of electron A at point a greater than, less than, or equal to the speed of electron B at point c?

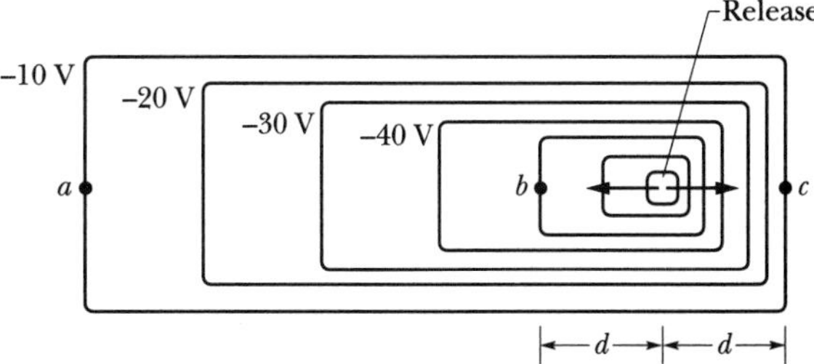

FIGURE 25-56 Question 27.

28. Figure 25-57 show a thin, uniformly charged rod and three points at the same distance d from the rod. Without calculation, rank the magnitude of the electric potential the rod produces at those three points, greatest first.

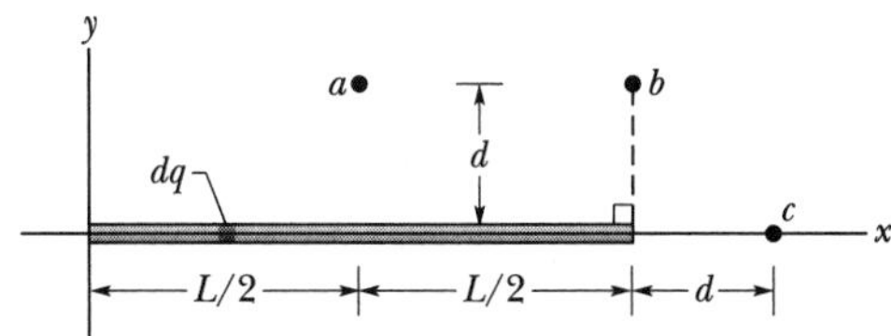

FIGURE 25-57 Questions 28 and 29.

29. *Organizing question:* Figure 25-57 shows a rod of uniform positive charge density λ lying along an x axis; it also shows a differential charge element dq. The net electric potential V at point

c due to the rod is to be evaluated, with the potential taken to be zero at infinite distance. In the expression

$$dV = \frac{1}{4\pi\epsilon_0}\frac{dq}{r},$$

what should be substituted for (a) dq and (b) r to ready for the required integration? (c) What are the limits of the integration? (d) If V is to be expressed in terms of the charge Q on the rod, what should be substituted for the charge density?

30. Figure 25-58 shows a loop of nine path segments along which we move a positively charged particle clockwise in a uniform electric field. The segments are of length L or $L/2$, as drawn. Segments 1 and 3 are symmetric; so are segments 5 and 6. Rank the segments according to (a) the magnitude of the change ΔV of the electric potential along them and (b) the magnitude of the work we do along them, greatest first. (c) Along which segments do we do positive work? (d) Is the total (or net) work we do for the complete loop positive, negative, or zero?

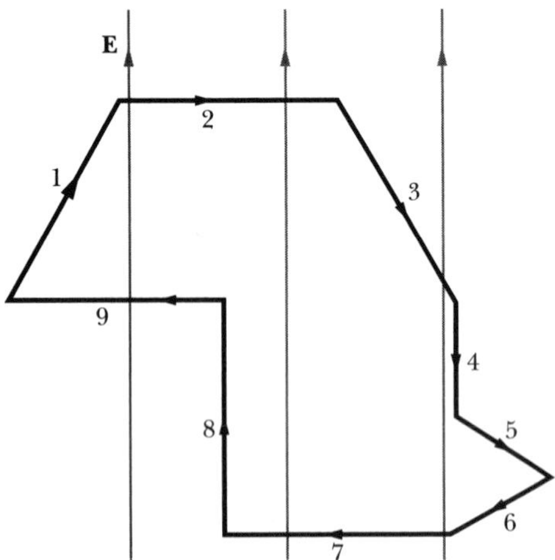

FIGURE 25-58 Question 30.

31. Figure 25-59 shows three parallel paths (A, B, and C) of equal lengths along which we move electrons in a nonuniform electric field. (The field lines converge.) Rank the paths according to (a) the magnitude of the change ΔV of the electric potential along them and (b) the magnitude of the work we do along them, greatest first. (*Hint:* Of the several ways the answer can be reasoned, using Section 25-9 is probably easiest.) (c) Is the work we do positive or negative along the paths?

32. Figures 25-60a through c are graphs of the electric potential V versus x and y in a certain region for three situations. (In Fig. 25-60a, the plots are identical.) Figure 25-60d gives 16 choices of electric field in the region. In choices 3, 7, 11, and 15, the electric field makes an angle of 45° with the x axis. Which of the 16 choices best gives the direction of the electric field corresponding to (a) Fig. 25-60a, (b) Fig. 25-60b, and (c) Fig. 25-60c?

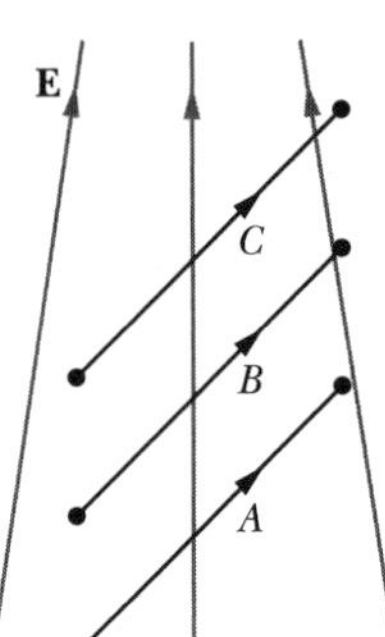

FIGURE 25-59 Question 31.

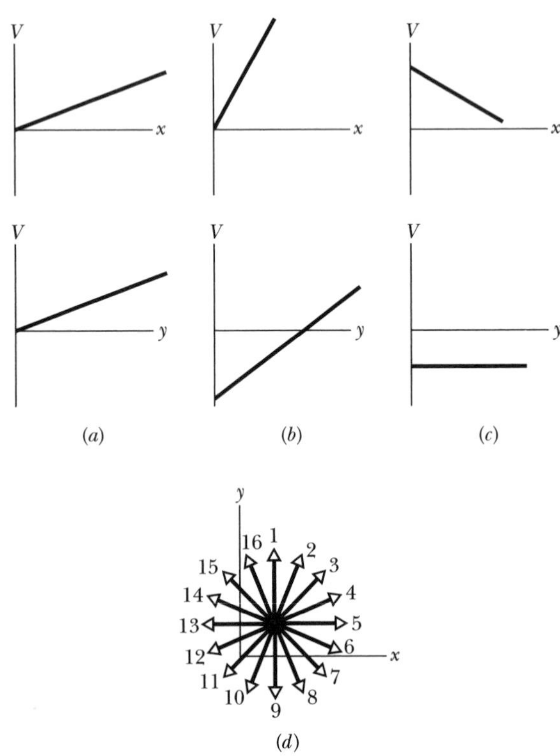

FIGURE 25-60 Question 32.

33. Figure 25-61 shows an electric dipole that is fixed in place and three points at which we can place an electron. Points a and c are on the dipole axis; point b is on the perpendicular bisector to the dipole. The points are at the same distance from the center of the dipole. (a) Rank the points according to the electric potential energy that the dipole–electron system would have if we placed the electron there, most positive first. (b) For each point, what is the value of θ in the equation

$$V = \frac{1}{4\pi\epsilon_0}\frac{p\cos\theta}{r^2}?$$

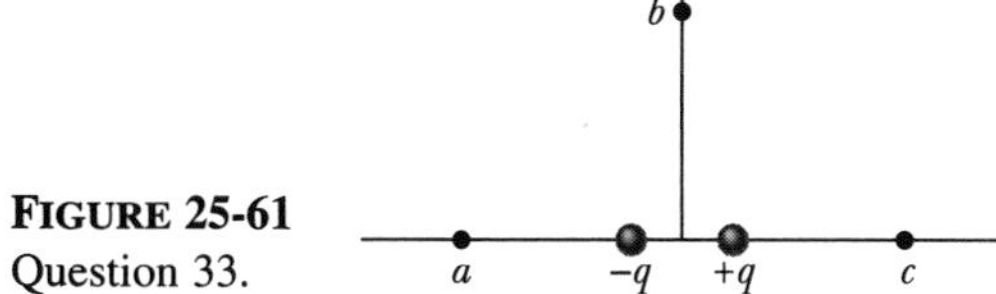

FIGURE 25-61 Question 33.

34. In Fig. 25-62, electron 1 enters the region between two parallel plates having different electric potentials, with uniform electric field between them. Electron 2 enters the same region through a small hole in the top plate. As each electron begins to travel through this region, do its (a) kinetic energy and (b) momentum increase, decrease, or remain the same? (The gravitational forces on the electrons can be neglected.)

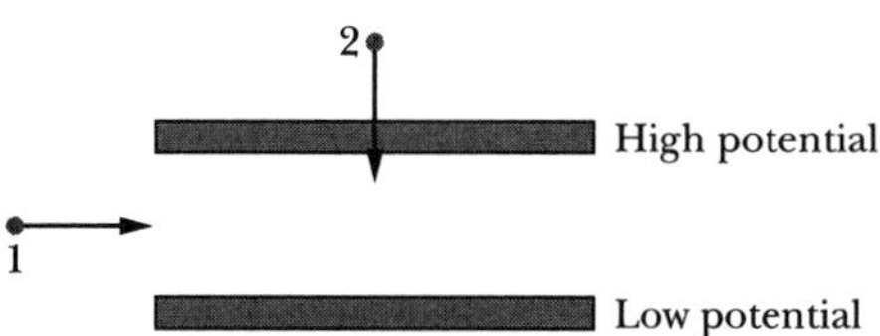

FIGURE 25-62 Question 34.

35. Figure 25-63 shows three spherical shells in separate situations, with each shell having the same positive net charge. Points 1, 4, and 7 are at the same radial distances from the center of their respective shells; so are points 2, 5, and 8; and so are points 3, 6, and 9. With the electric potential equal to zero at an infinite distance, rank the points according to the electric potential at them, greatest first.

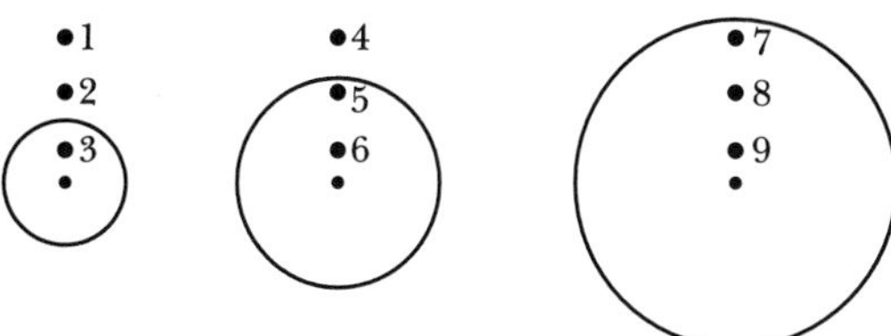

FIGURE 25-63 Question 35.

36. Figure 25-64 shows the electric potential that is associated with an electric field directed along an x axis. At which of the lettered points is the magnitude of the electric field (a) maximum with the field pointing in the positive direction of the x axis, (b) maximum with the field pointing in the negative direction of the x axis, and (c) zero?

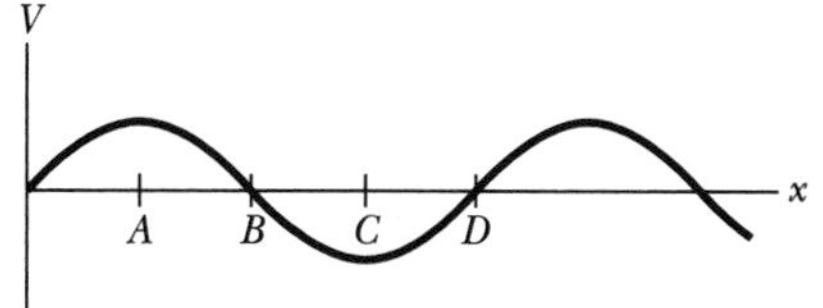

FIGURE 25-64 Question 36.

37. Figure 25-65 shows four situations in which a proton moves through point a and then through point b, with either protons (p) or electrons (e) fixed in place as indicated. In each situation, the moving proton has the same velocity $\mathbf{v}_i$ when it passes through a, and points a and b are equal distances from the x axis. (a) Rank the situations according to the time the proton requires to move from a to b, greatest first. (b) Now rank them according to the proton's speed at b, greatest first.

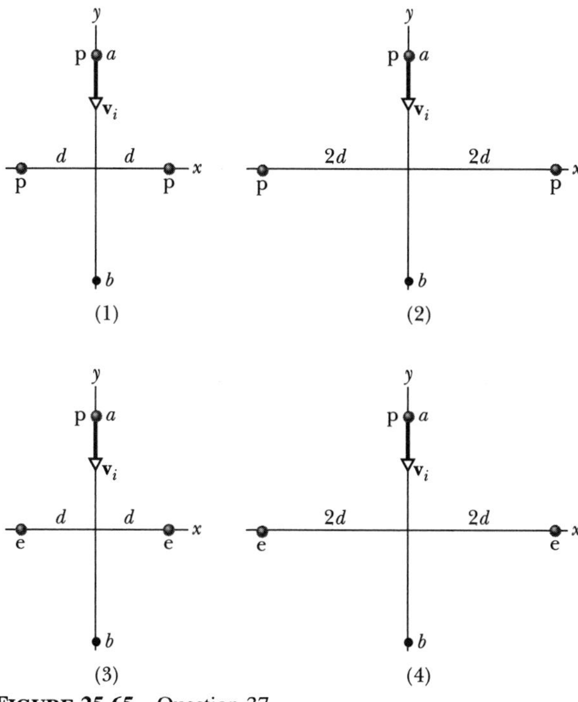

FIGURE 25-65 Question 37.

EXERCISES & PROBLEMS

85. The electric field in a region of space has the components $E_y = E_z = 0$ and $E_x = (4.00\text{ N/C})x$. Point A is on the y axis at $y = 3.0$ m, and point B is on the x axis at $x = 4.0$ m. What is the potential difference $V_B - V_A$?

86. Two uniformly charged, infinite, nonconducting planes are parallel to the yz plane and positioned at $x = -50$ cm and $x = +50$ cm. The charge densities on the planes are -50 nC/m^2 and $+25$ nC/m^2, respectively. What is the magnitude of the potential difference between the origin and the point on the x axis at $x = +80$ cm? (*Hint:* Use Gauss' law for planar symmetry to determine the electric field in each region of space; see Section 24-8.)

87. A graph of the x component of the electric field as a function of x in a region of space is shown in Fig. 25-66. Both the y and z components of the electric field are zero in this region. If the electric potential at the origin is 10 V, (a) what is the electric potential at $x = 2.0$ m, (b) what is the greatest positive value of the electric potential for points on the x axis for which $0 \leq x \leq 6.0$ m, and (c) for what value of x is the electric potential zero?

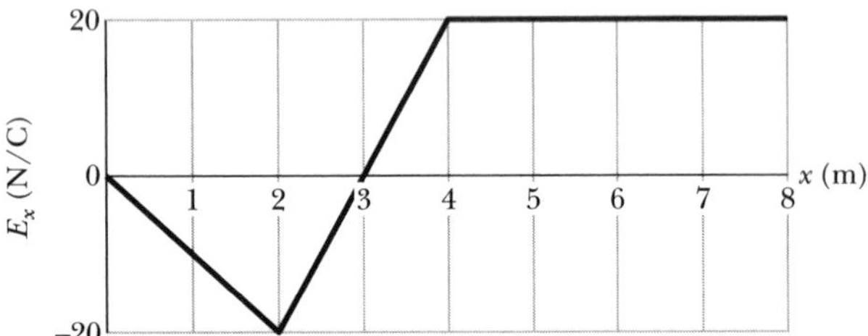

FIGURE 25-66 Problem 87.

88. Point charges of equal magnitudes (25 nC) and opposite signs are placed on diagonally opposite corners of a 60 cm × 80 cm rectangle. Point A is the corner of this rectangle nearest the positive charge, and point B is the corner of this rectangle nearest the negative charge. Determine the potential difference $V_B - V_A$.

89. A nonuniform linear charge distribution given by $\lambda = bx$, where b is a constant, is distributed along an x axis from $x = 0$ to $x = 0.20$ m. If $b = 20$ nC/m^2, what is the electric potential (relative to a potential of zero at infinity) at (a) the origin and (b) the point $y = 0.15$ m on the y axis?

90. A net charge of $+16$ μC is uniformly distributed on a circular ring that lies in an xy plane with its center at the origin. The radius of the ring is 3.0 cm. If point A is at the origin and point B is on the z axis at $z = 4.0$ cm, what is $V_B - V_A$?

91. What is the magnitude of the electric field at the point $(3\mathbf{i} - 2\mathbf{j} + 4\mathbf{k})$ m if the electric potential is given by $V = 2xyz^2$, where V is in volts and x, y, and z are in meters?

92. Three point charges $q_1 = +10$ μC, $q_2 = -20$ μC, and $q_3 = +30$ μC are positioned at the vertices of an isosceles triangle as shown in Fig. 25-67. If $a = 10$ cm and $b = 6.0$ cm, how much work must an external agent do to exchange the positions of (a) q_1 and q_3 and, instead, (b) q_1 and q_2?

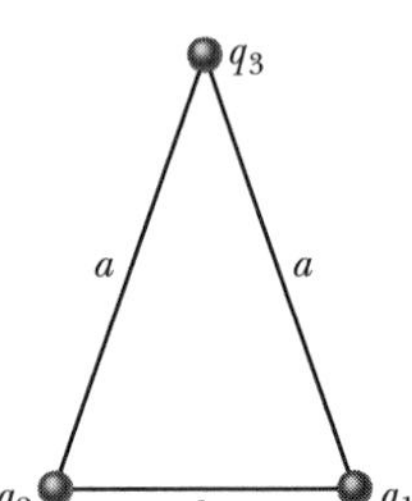

FIGURE 25-67 Problem 92.

93. Identical 50 μC charges are fixed on an x axis at $x = \pm 3.0$ m. A particle of charge $q = -15$ μC is then released from rest at a point on the positive part of the y axis. Due to the symmetry of the situation, the particle moves along the y axis and has a kinetic energy of 1.2 J as it passes through the point $x = 0$, $y = 4.0$ m. (a) What is the kinetic energy of the particle as it passes through the origin? (b) At what negative value of y will the particle momentarily stop?

94. A particle of charge $= 7.5$ μC is released from rest at a point on an x axis at $x = 60$ cm. It begins to move due to the presence of a charge Q that remains fixed at the origin. What is the kinetic energy of the particle at the instant it has moved 40 cm if (a) $Q = +20$ μC and (b) $Q = -20$ μC?

95. A solid conducting sphere of radius 3.0 cm has a charge of 30 nC distributed over its surface. Let A be a point 1.0 cm from the center of the sphere, S be a point on the surface of the sphere, and B be a point 5.0 cm from the center of the sphere. What are the electric potential differences (a) $V_S - V_B$ and (b) $V_A - V_B$?

96. A long, conducting, solid cylinder has a radius of 2.0 cm. The electric field at the surface of the cylinder is 160 N/C, directed radially outward. Let A, B, and C be points that are 1.0 cm, 2.0 cm, and 5.0 cm, respectively, from the central axis of the cylinder. (a) What is the electric field at point C? (b) What are the electric potential differences (b) $V_B - V_C$ and (c) $V_A - V_B$?

97. In Fig. 25-68 two charged particles are fixed in place on an x axis. If $d = 1.00$ m, what is the electric potential difference $V_B - V_A$ between points B and A?

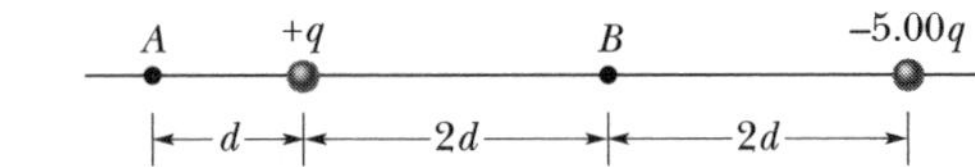

FIGURE 25-68 Problem 97.

98. Figure 25-69 shows a rectangular array of charged particles fixed in place. What is the net electric potential at the center of the array? (*Hint:* First consider just the corner particles.)

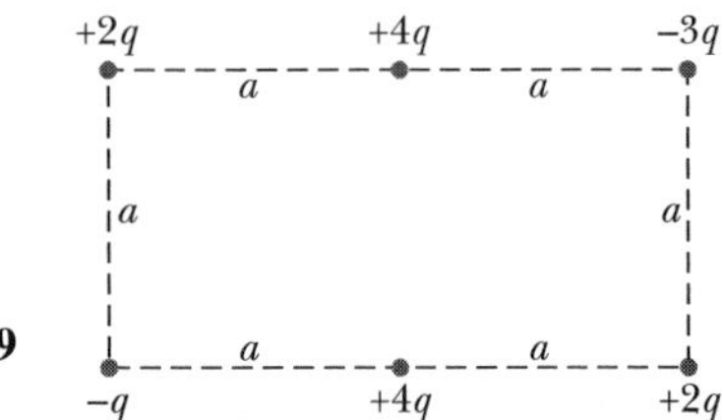

FIGURE 25-69 Problem 98.

99. Figure 25-70 shows three circular arcs, of radius R and total charge as indicated. What is the net electric potential at the center of curvature?

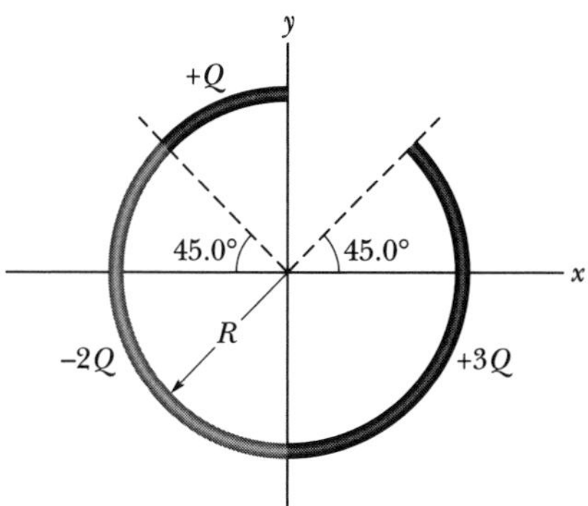

FIGURE 25-70 Problem 99.

100. In Fig. 25-71, what is the net electric potential at the origin due to the circular arc of charge $+Q$ and the two particles of charges $+4Q$ and $-2Q$? Take the radius of curvature as $R =$ 2.00 m.

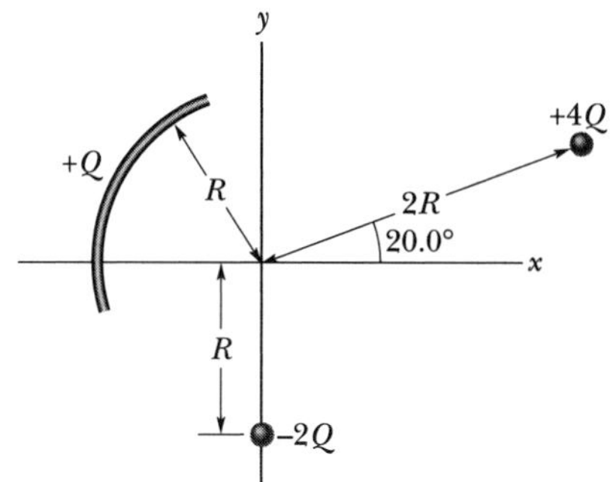

FIGURE 25-71 Problem 100.

101. In Fig. 25-72, a particle with charge $+5q$ is brought in from infinity to the position shown; the other two particles in the figure are fixed in place. What is the ratio of the potential energy of this three-particle system to that of the original two-particle system?

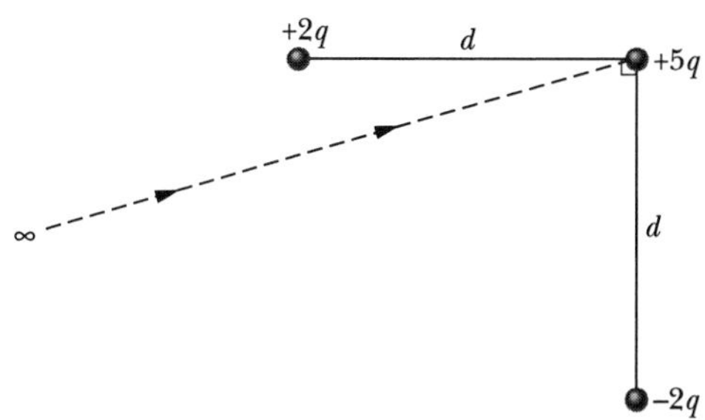

FIGURE 25-72 Problem 101.

102. (a) In Fig. 25-73, what is the net potential at point P due to the two charged particles, which are fixed in place? We bring a particle of charge $+2e$ from infinity to point P. (b) How much work do we do? (c) What is the potential energy of the three-particle system once the third particle is in place?

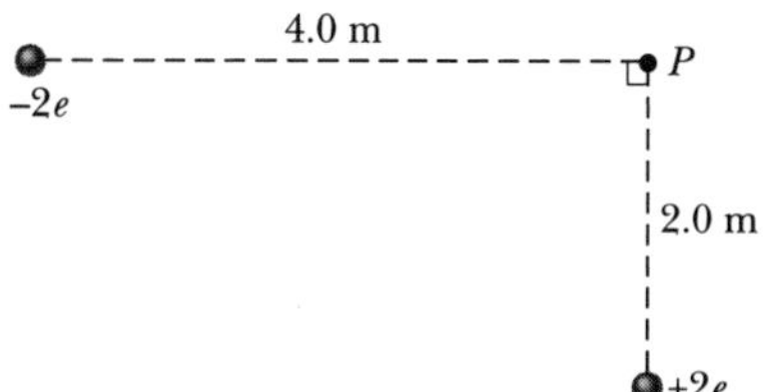

FIGURE 25-73 Problem 102.

103. Initially two electrons are fixed in place with a separation of 2.00 μm. How much work must we do to bring a third electron in from infinity to complete an equilateral triangle?

104. In Fig. 25-74, we move a particle of charge $+2e$ in from infinity to the x axis. How much work do we do? Distance D is 4.00 m.

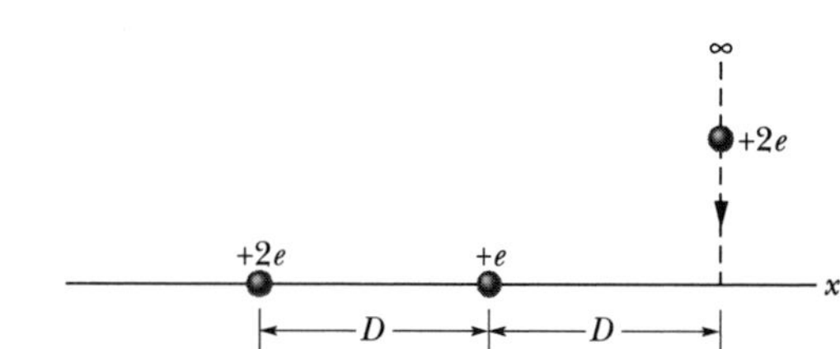

FIGURE 25-74 Problem 104.

105. In Fig. 25-75, a charged particle (which is either an electron or a proton) is moving rightward between two parallel charged plates separated by distance $d = 2.00$ mm. The particle is slowing from an initial speed of 90.0 km/s at the left plate. (a) Is the particle an electron or a proton? (b) What is its speed just as it reaches the plate at the right?

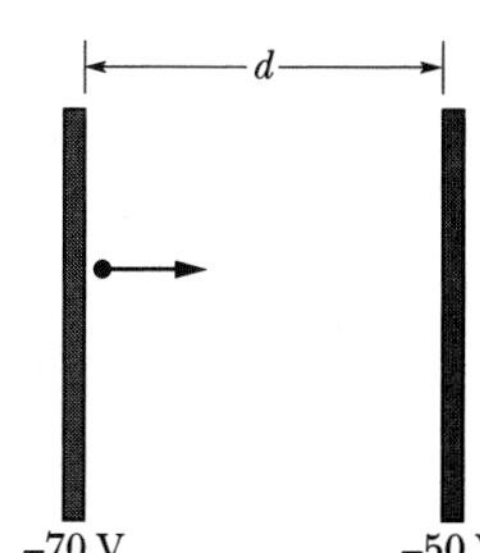

FIGURE 25-75 Problem 105.

106. An alpha particle (which has two protons) is sent directly toward a target nucleus with 92 protons. The alpha particle has an initial kinetic energy of 0.48 pJ. What is the least center-to-center distance that the alpha particle will be from the target nucleus, assuming that the nucleus does not move?

107. The magnitude E of an electric field depends on the radial distance r according to $E = A/r^4$, where A is a constant with the unit volt-cubic meter. What is the magnitude of the electric potential difference between $r = 2.00$ m and $r = 3.00$ m?

Clustered Problems

Cluster 1

Problems in this cluster involve the electric field and the electric potential set up by nested spherical shells *that are thin, charged, conducting, and concentric; they are shown in cross section in Fig. 25-76. Take the electric potential to be zero at infinite radius from the shells. Also take it to be zero on any shell that is grounded (that is, connected to ground), as indicated in Figs. 25-76b through d with the standard symbol for grounded objects.*

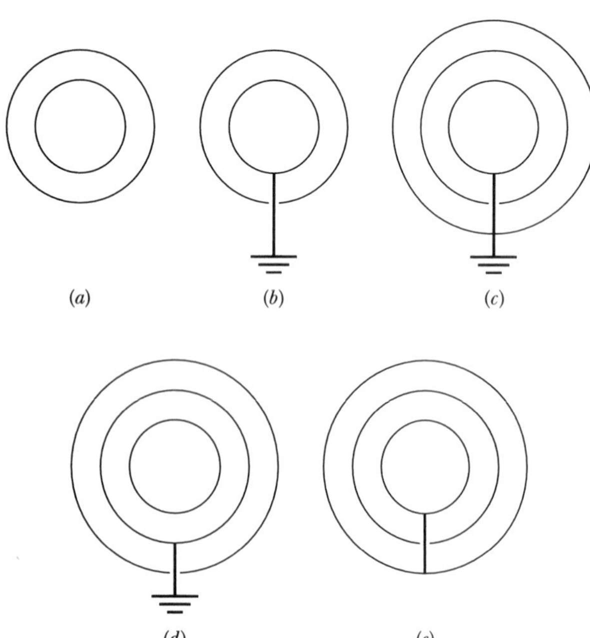

FIGURE 25-76 Problems 108 through 110.

For a point outside the nested shells, expressions for the electric potential can be found in the usual way. For a point on or inside a shell (closer to the center than the shell), the electric potential has the form

$$V = \frac{kQ_{\text{enc}}}{r} + c,$$

where $k = 1/4\pi\epsilon_0$, r is the radial distance of the point, Q_{enc} is the net charge enclosed by an imaginary sphere through that point, and c is a constant. The values of Q_{enc} and c for points inside a shell differ from those for points outside a shell. However, the inside and outside functions for $V(r)$ must give the same value at the shell itself.

108. See this cluster's setup. In Fig. 25-76*a*, the smaller shell has radius R_1 and charge Q_1, and the larger shell has radius R_2 and charge Q_2. For all radii, find expressions in terms of given symbols for (a) the electric field and (b) the electric potential set up by the shells. (c) Taking $R_1 = 10.0$ cm, $R_2 = 20.0$ cm, $Q_1 = 1.00$ nC, and $Q_2 = 3.00$ nC, graph the electric potential for $4R_2 > r > 0$.

A positron (which has charge $+e$ and the mass of an electron) is released at rest on the outer surface of the smaller shell. The positron then accelerates radially outward, passing through a microscopic hole in the larger shell. What are the speeds of the positron (d) as it passes through that hole and (e) when it is very far ($r \gg R_2$) from the shells?

109. See this cluster's setup. In Fig. 25-76*b*, the smaller shell has radius R_1 and is grounded (the wire passes through a small insulated hole in the larger shell). The larger shell has radius R_2 and charge Q. In terms of given symbols, find expressions for (a) the charge on the smaller shell and then, for all radii, (b) the electric field and (c) the electric potential set up by the shells.

Next, a grounded third spherical shell of radius R_3 is placed around the first two shells (Fig. 25-76*c*). It, too, is thin, conducting, and concentric. What are the charges (d) on the smallest shell and (e) on the third shell? (f) Show that the answer to (d) becomes that of (a) in the limit $R_3 \to \infty$.

110. See this cluster's setup. In Fig. 25-76*d*, the smallest shell has radius R_1 and charge $-Q$; the middle shell has radius R_2 and is grounded; and the largest shell has radius R_3 and charge Q. (The grounding wire passes through a small insulated hole in the largest shell.) In terms of given symbols, find expressions for (a) the charge on the middle shell and then, for all radii, (b) the electric field and (c) the electric potential that are set up by the shells.

The grounding wire is now removed, and a wire is run from the smallest shell, through a small insulated hole in the middle shell, to the largest shell, so that the smallest and largest shells are then electrically connected (Fig. 25-76*e*). This is done without changing the charge on the middle shell from the value found in (a). (d) Is the electric potential on the smallest shell greater than, less than, or equal to that on the largest shell? Find the charge on (e) the largest shell and (f) the smallest shell. Also find the electric potential on (g) the largest shell and (h) the middle shell.

Cluster 2

111. In Fig. 25-77*a*, a particle of charge q and mass m is released from rest at a point midway between two large conducting plates that are vertical and separated by distance D. As the particle falls due to its weight, it moves rightward because of a potential difference ΔV between the plates. Just as the particle runs into a plate, (a) what distance d has it traveled downward and (b) what

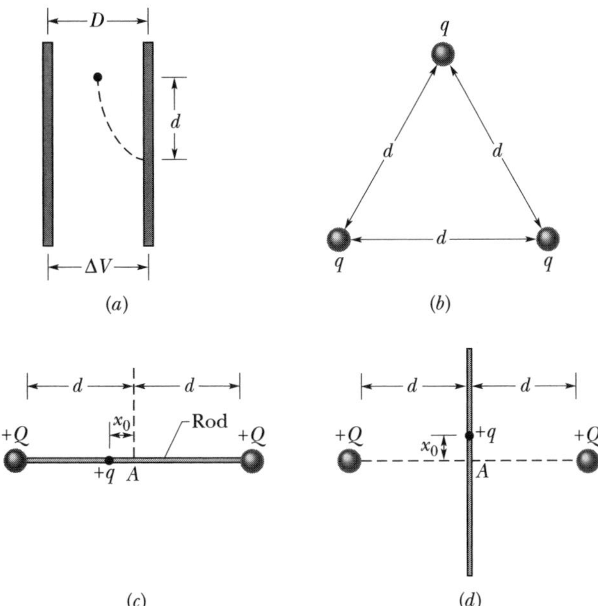

FIGURE 25-77 Problems 111, 112, and 114.

is its speed v? Evaluate (c) distance d and (d) speed v using the values $q = 0.10$ nC, $m = 1.0\ \mu$g, $D = 10$ cm, and $\Delta V = 10$ V.

112. In Fig. 25-77*b*, three particles with the same charge q and same mass m are initially fixed in place to form an equilateral triangle with edge lengths d. (a) If the particles are released simultaneously, what are their speeds when they have traveled a large distance (effectively an infinite distance) from each other? (Measure the speeds in the original rest frame of the particles.)

Suppose, instead, the particles are released one at a time: the first one is released, and then, when that first one is at a large distance, a second one is released, and then, when that second one is at a large distance, the last one is released. What then are the final speeds of (b) the first particle, (c) the second particle, and (d) the last particle?

113. Two small spheres of radius r and mass m are released from rest when their centers are separated by distance $D \gg r$. Sphere 1 has charge Q; sphere 2 has charge $2Q$. What are the final speeds of (a) sphere 1 and (b) sphere 2? (Let speeds be measured in the original rest frame of the particles.) Suppose, instead, that sphere 2 has charge $-2Q$. What then are the speeds of (c) sphere 1 and (d) sphere 2 just as the spheres meet? Next suppose that the collision is elastic. What then are the speeds of (e) sphere 1 and (f) sphere 2 when the separation between the spheres is again D?

114. In Fig. 25-77*c*, two particles of identical charge $+Q$ are fixed in place at opposite ends of an insulated rod of length $2d$. A third particle, of charge $+q$ and mass m, can slide along the rod. The particle is displaced from the rod's midpoint (point A) by distance x_0 and then released from rest. (a) Find an expression for the velocity of the particle as it passes back through point A. (b) Simplify the expression using the assumption that $x_0 \ll d$. (c) Then evaluate that simplified expression using the values $Q = 1.0\ \mu$C, $m = 1.0\ \mu$g, $q = 1.0$ nC, $x_0 = 0.10$ mm, and $d = 1.0$ cm.

The particles of charge $+Q$ are kept in place while the rod is rotated by 90° about its midpoint (Fig. 25-77*d*). Again the particle of charge $+q$ is displaced from point A by distance x_0 and then released from rest. (d) Repeat parts (a) and (b) for this arrangement.

Chapter Twenty-Six
Capacitance

QUESTIONS

17. Figure 26-40 shows an open switch, a battery of potential difference V, a current-measuring meter A, and three identical uncharged capacitors of capacitance C. When the switch is closed and the circuit reaches equilibrium, what are (a) the potential difference across each capacitor and (b) the charge on the left-hand plate of each capacitor? (c) During the charging process, what was the net charge that passed through the meter?

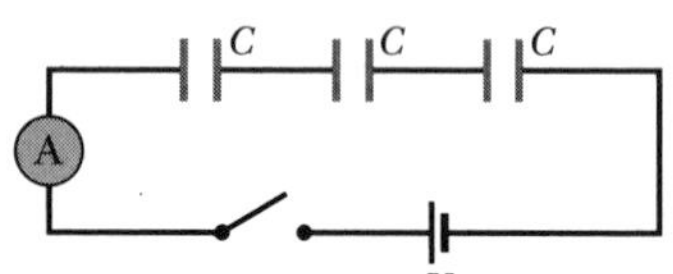

FIGURE 26-40 Question 17.

18. Figure 26-41 shows a circuit containing a battery and three parallel plate capacitors with identical plate separations (filled with air). The capacitors lie along an x axis. The electric potential V along that axis is indicated. (a) Which point at the battery, A or B, is at lower electric potential? (b) Is the plate at the right of capacitor C_3 positively or negatively charged? Rank the capacitors according to (c) the charge on them, (d) their capacitance, (e) their plate area, and (f) the magnitude of the electric field between their plates, greatest first. (g) What is the direction of that electric field on the x axis?

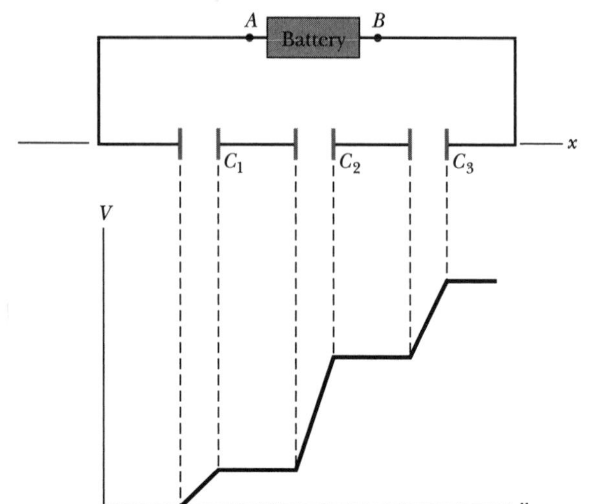

FIGURE 26-41 Question 18.

19. Which of the following are true of the capacitances and the equivalent capacitances (labeled with multiple subscripts) in the circuit of Fig. 26-42?

(a) C_1 is in parallel with C_3 (b) C_{12} is in parallel with C_3
(c) C_{12} is in series with C_4 (d) C_{123} is in series with C_4
(e) C_{123} is in series with C_{45} (f) C_{12345} is in parallel with C_6

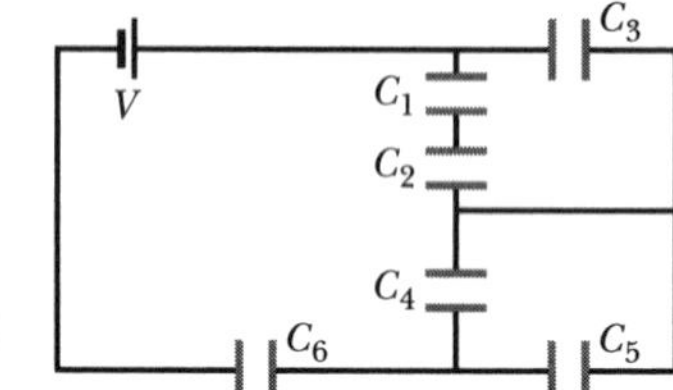

FIGURE 26-42 Question 19.

20. *Organizing question:* Figure 26-43*a* shows a circuit with three capacitors and Fig. 26-43*b* shows the circuit with their equivalent capacitor C_{eq}, which has a charge of 60 μC. Without a calculator (and, if possible, without written calculation), find (a) the charge on and (b) the voltage across capacitor C_3 and then (c) the charge on and (d) the voltage across capacitor C_1.

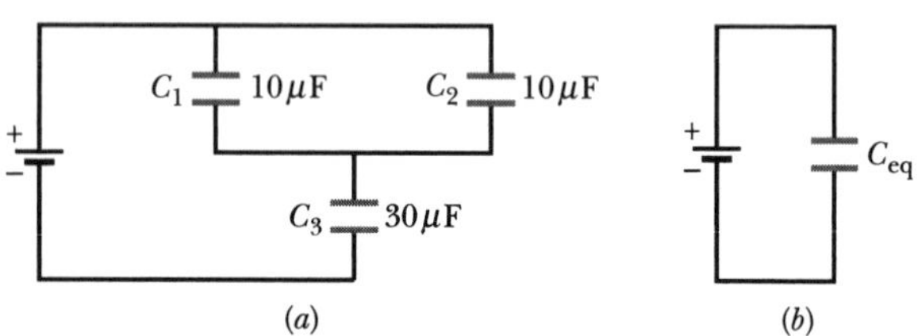

FIGURE 26-43 Question 20.

21. In Fig. 26-44, the capacitors are initially uncharged. When the switch is thrown to position A, does the battery (of potential difference V) then begin to charge (a) capacitor C_1 and (b) capacitor C_2? When equilibrium is reached, what is the potential across (c) C_1 and (d) C_2? (e) When, later, the switch is thrown to position B, is C_2 then charged by the battery or by capacitor C_1? When equilibrium is again reached, (f) is the potential across C_1

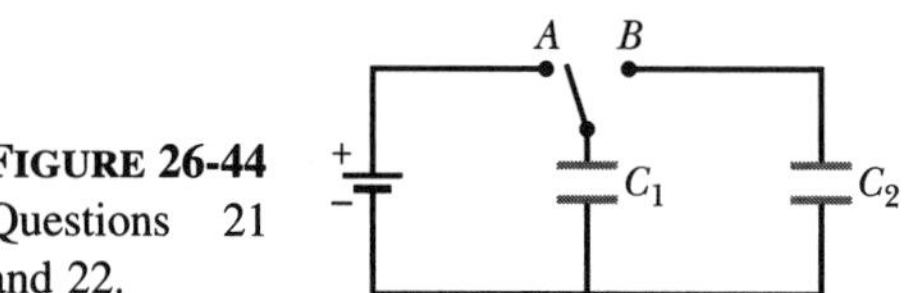

FIGURE 26-44 Questions 21 and 22.

greater than, less than, or equal to that across C_2 and (g) is it greater than, less than, or equal to the battery's potential difference?

22. *Organizing question:* In Question 21 and Fig. 26-44, take $C_1 = 10\ \mu F$ and $C_2 = 20\ \mu F$. (a) Using this data, set up an equation that gives q_1 in terms of q_2, where q_1 is the charge on capacitor 1 and q_2 is the charge on capacitor 2 when the capacitors reach equilibrium. (b) Then set up an equation that gives q_2 in terms of the initial charge q_0 stored on capacitor 1 by the battery.

23. A parallel plate capacitor is connected to a battery of electric potential difference V. If the plate separation is decreased, do the following quantities of the capacitor increase, decrease, or remain the same: (a) capacitance, (b) potential difference across the capacitor, (c) charge on the capacitor, (d) energy stored by the capacitor, (e) magnitude of the electric field between the plates, and (f) energy density of that electric field?

24. A parallel plate capacitor is first charged by a battery and then disconnected. If the capacitor plates of the isolated capacitor are then decreased, do the following quantities of the capacitor increase, decrease, or remain the same: (a) capacitance, (b) potential difference across the capacitor, (c) charge on the capacitor, (d) energy stored by the capacitor, (e) magnitude of the electric field between the plates, and (f) energy density of that electric field?

25. Figure 26-45 shows three circuits with identical capacitors. The capacitors at the left in the circuits initially have identical charges q_0; the capacitors at the right are initially uncharged. As in Sample Problem 26-4, we transfer charge from the charged capacitor by closing the switch. Without written calculation, rank the three circuits according to (a) the equivalent capacitance of the capacitors at the right and (b) the final (equilibrium) charge on the initially charged capacitor, greatest first.

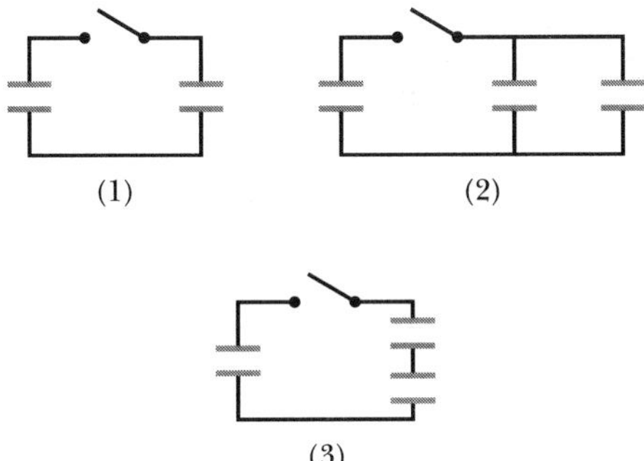

FIGURE 26-45 Question 25.

26. Figure 26-46 shows three identical spherical metal shells of inner radius R_i and outer radius R_o. The hollow of shell 1 is empty, but those of shells 2 and 3 contain a particle of charge $+5q$ at the center of the hollow. The charges on the shells themselves (alone) are shell 1, $+5q$; shell 2, zero; and shell 3, $-5q$. Rank the shells according to the energy density of the electric field, greatest first, at the following radii from the center of hollows: (a) $R_i/2$; (b) between R_i and R_o; (c) $2R_o$.

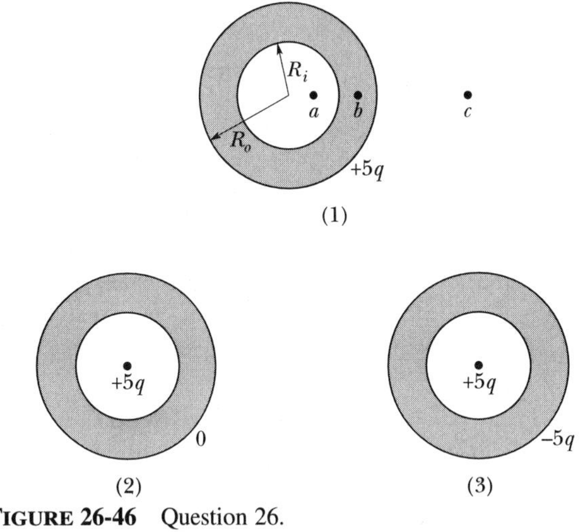

FIGURE 26-46 Question 26.

EXERCISES & PROBLEMS

73. The capacitances of the four capacitors shown in Fig. 26-47 are given in terms of a basic quantity C. (a) If $C = 50\ \mu F$, what is the equivalent capacitance of the four-capacitor segment between points A and B? (*Hint:* First imagine that a battery is connected between those two points; then reduce the circuit to an equivalent capacitance.) (b) Repeat the question for points A and D.

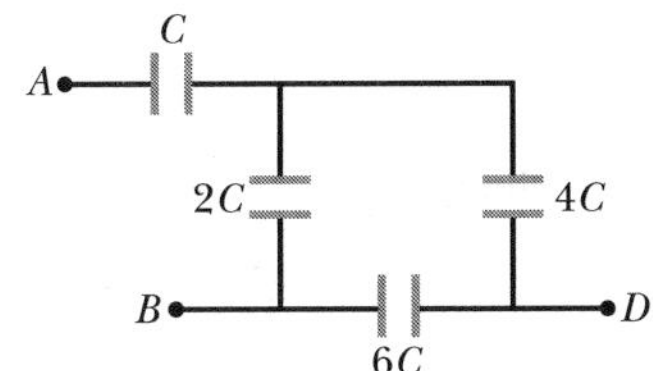

FIGURE 26-47 Problem 73.

74. Figure 26-48 shows a four-capacitor arrangement that is connected to a larger circuit at points A and B. The capacitances are $C_1 = 10\ \mu F$ and $C_2 = C_3 = C_4 = 20\ \mu F$. The charge on the capacitor with capacitance C_1 is $30\ \mu C$. What is the magnitude of the potential difference $V_A - V_B$?

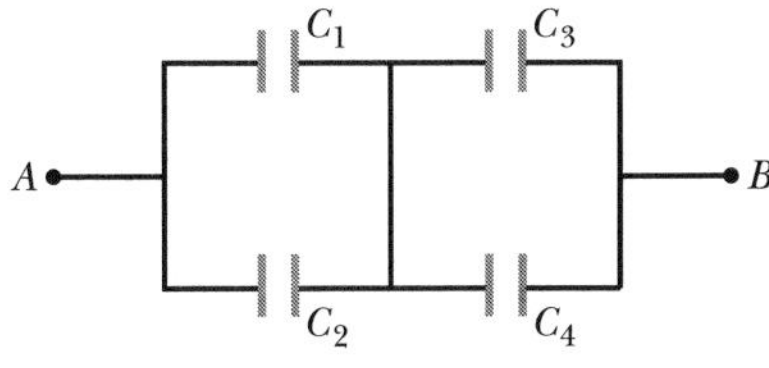

FIGURE 26-48 Problem 74.

75. A capacitor of unknown capacitance C is charged to 100 V and connected across an initially uncharged 60 μF capacitor. If the final potential difference across the 60 μF capacitor is 40 V, determine C.

76. Three capacitors are initially uncharged and attached as shown in Fig. 26-49. If no capacitor can withstand a potential difference of more than 100 V without failure, (a) what is the magnitude of the maximum potential difference that can exist between points A and B and (b) what is the maximum energy that can be stored in the three-capacitor arrangement?

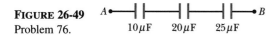

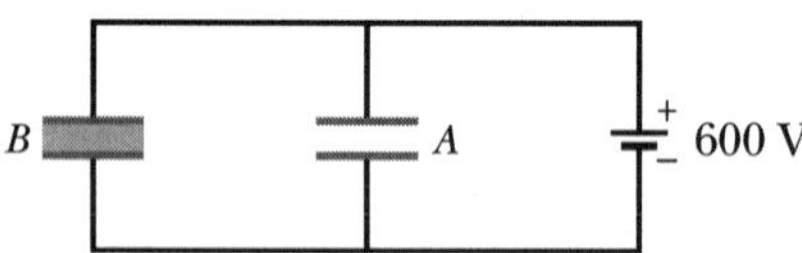

FIGURE 26-49 Problem 76.

77. A certain parallel plate capacitor is filled with a dielectric for which $\kappa = 5.5$. The area of each plate is 0.034 m^2, and the plates are separated by 2.0 mm. The capacitor will fail (short out and burn up) if the electric field between the plates exceeds 200 kN/C. What is the maximum energy that can be stored in the capacitor?

78. In Fig. 26-50, two parallel plate capacitors A and B are connected in parallel across a 600 V battery. Each plate has area 80.0 cm^2; the plate separations are 3.00 mm. Capacitor A is filled with air; capacitor B is filled with a dielectric of dielectric constant $\kappa = 2.60$. Find the electric field within (a) the dielectric of capacitor B and (b) the air of capacitor A. What are the free charge densities σ on the higher potential plate of (c) capacitor A and (d) capacitor B? (e) What is the induced charge density σ' on the top surface of the dielectric?

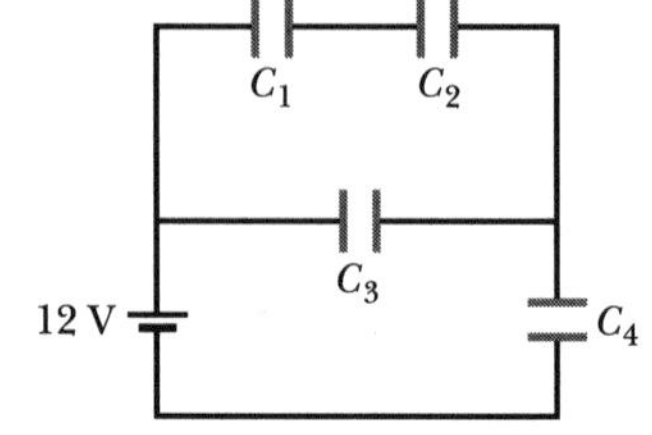

FIGURE 26-50 Problem 78.

79. In Fig. 26-51, $C_1 = C_2 = C_4 = 8.0$ μF and $C_3 = 4.0$ μF. What is the potential across C_4?

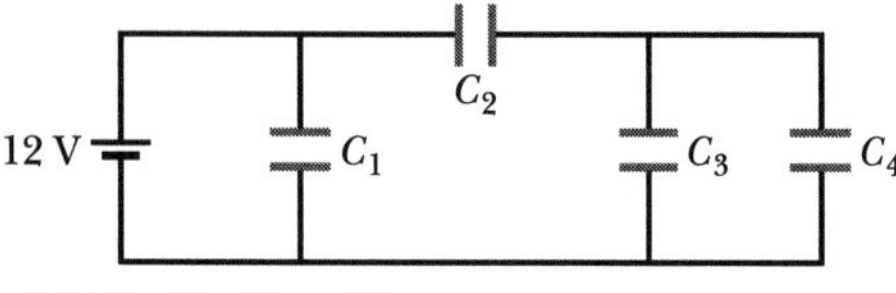

FIGURE 26-51 Problem 79.

80. In Fig. 26-52, $C_1 = C_2 = C_3 = C_4 = 2.00$ μF. (a) What is the charge on C_1? (b) What is the voltage across C_4?

FIGURE 26-52 Problem 80.

81. In Fig. 26-53, (a) what is the equivalent capacitance of the capacitors and (b) what is the net charge stored on them?

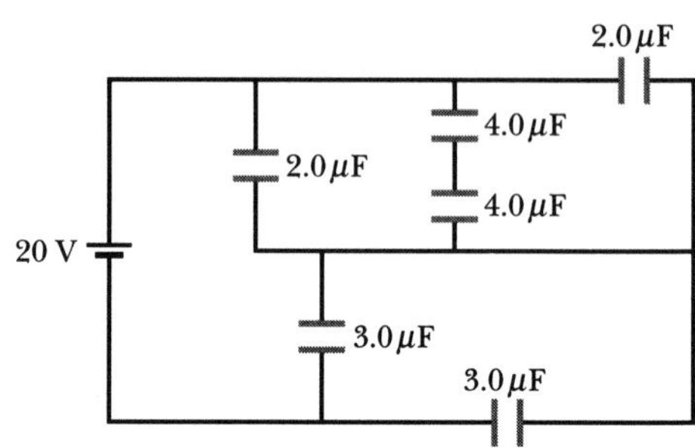

FIGURE 26-53 Problem 81.

82. In Fig. 26-54, what are (a) the charge on the bottom capacitor and (b) the potential across it?

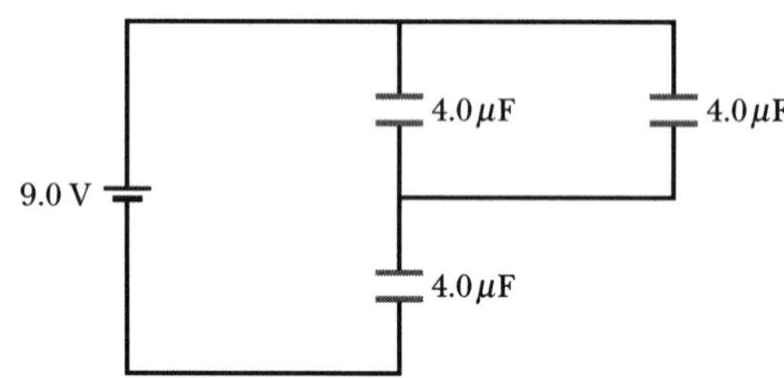

FIGURE 26-54 Problem 82.

83. In Fig. 26-55, $C_1 = C_4 = 2.0$ μF, $C_2 = 4.0$ μF, and $C_3 = 1.0$ μF. What is the charge on capacitor C_4?

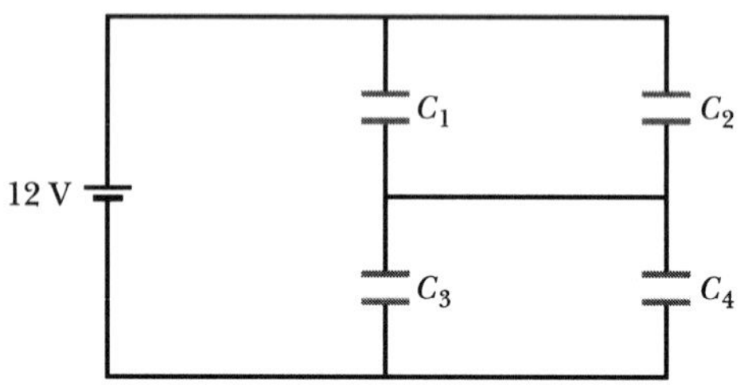

FIGURE 26-55 Problem 83.

84. The capacitors in Fig. 26-56 each have capacitance 10 μF. What are the charges on (a) C_1 and (b) C_2?

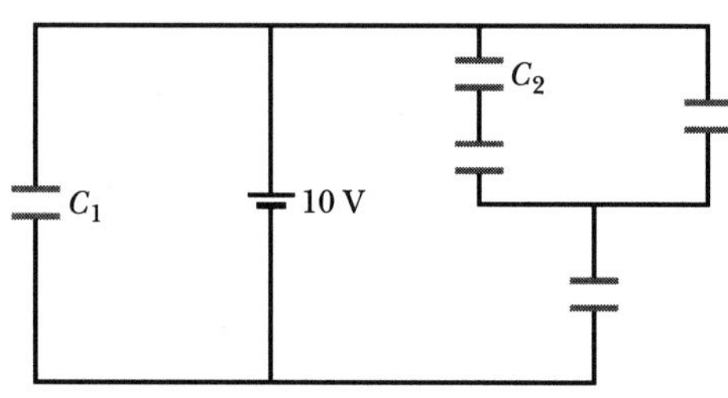

FIGURE 26-56 Problem 84.

85. In Fig. 26-57, $C_1 = C_2 = 30$ μF and $C_3 = C_4 = 15$ μF. What is the charge on C_4?

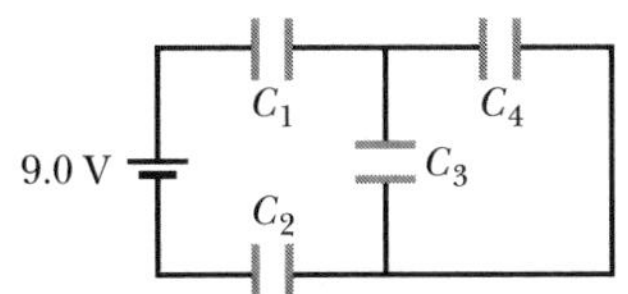

FIGURE 26-57 Problem 85.

86. In Fig. 26-58, $C_1 = C_5 = C_6 = 6.0\ \mu$F and $C_2 = C_3 = C_4 = 4.0\ \mu$F. What are (a) the net charge stored on the capacitors and (b) the charge on C_4?

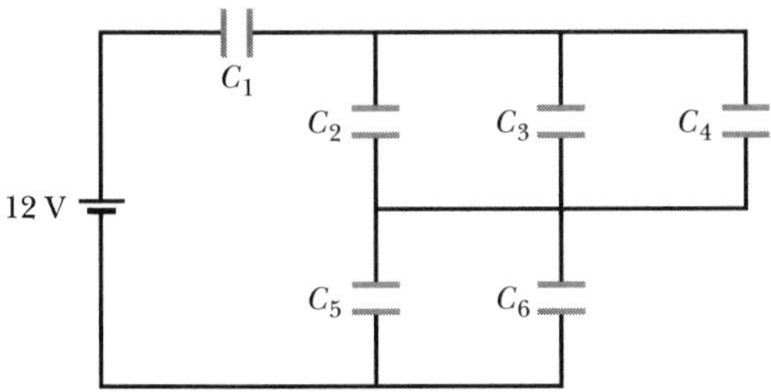

FIGURE 26-58 Problem 86.

87. In Fig. 26-59, $C_1 = C_2 = C_3 = 2.00\ \mu$F. Switch S is first thrown to the left to charge C_1 and C_2 to 48 μC. Then the switch is thrown to the right. When equilibrium is reached, how much charge is on C_3?

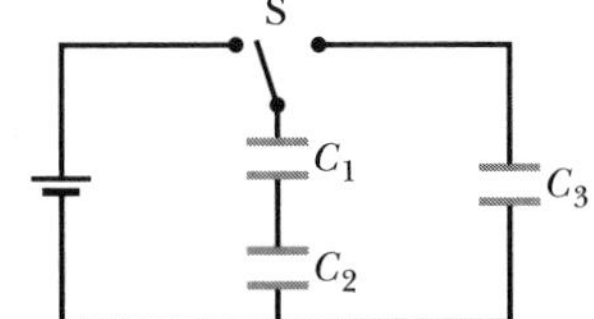

FIGURE 26-59 Problem 87.

88. In Fig. 26-60, $C_1 = C_2 = 2.00\ \mu$F and $C_3 = 4.00\ \mu$F. The switch is first thrown leftward until C_1 reaches equilibrium. Then it is thrown rightward. When equilibrium is again reached, what is the charge on C_3?

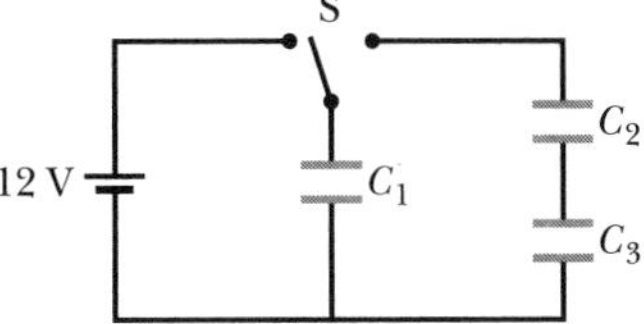

FIGURE 26-60 Problem 88.

89. In Fig. 26-61, $C_1 = 10\ \mu$F and $C_2 = C_3 = 20\ \mu$F. Switch S is first thrown to the left until C_1 reaches equilibrium. Then the switch is thrown to the right. When equilibrium is again reached, how much charge is on C_1?

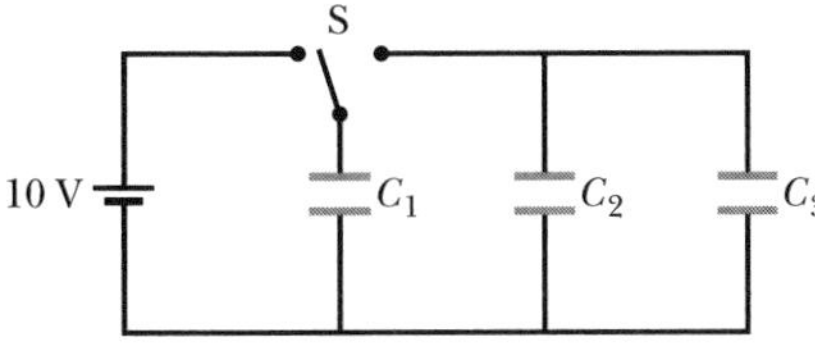

FIGURE 26-61 Problem 89.

90. In Fig. 26-62, $C_1 = 2.0\ \mu$F, $C_2 = 16\ \mu$F, and $C_3 = C_4 = 8.0\ \mu$F. Switch S is first thrown to the left until C_1 reaches equilibrium. Then the switch is thrown to the right. When equilibrium is again reached, (a) how much charge is on C_2 and (b) what is the potential across C_2?

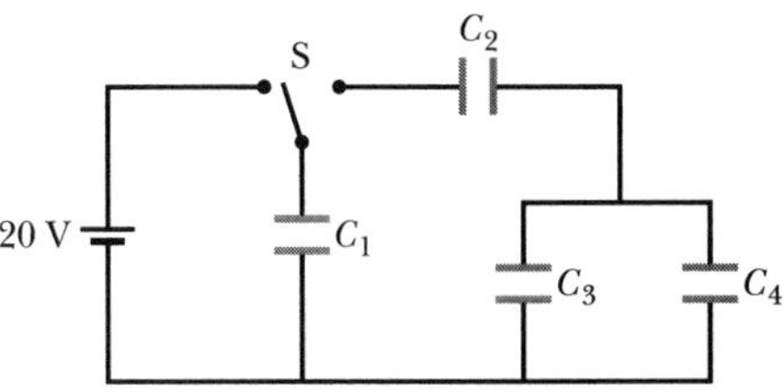

FIGURE 26-62 Problem 90.

91. In Fig. 26-63, how much charge is stored on the parallel plate capacitors? One is filled with air, and the other has a dielectric with $\kappa = 3.00$; both have a plate area of 5.00×10^{-3} m^2 and a plate separation of 2.00 mm.

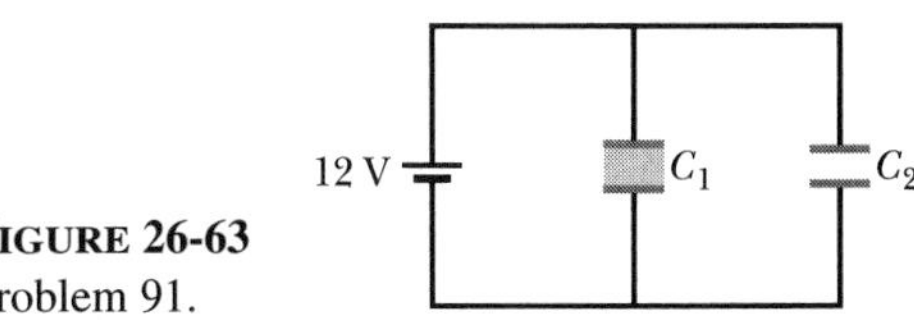

FIGURE 26-63 Problem 91.

92. In Fig. 26-64, the parallel plate capacitor of plate area 2.00×10^{-2} m^2 is filled with two dielectric slabs, each with a thickness of 2.00 mm. One slab has dielectric constant of 3.00; the other has dielectric constant of 4.00. How much charge is on the capacitor?

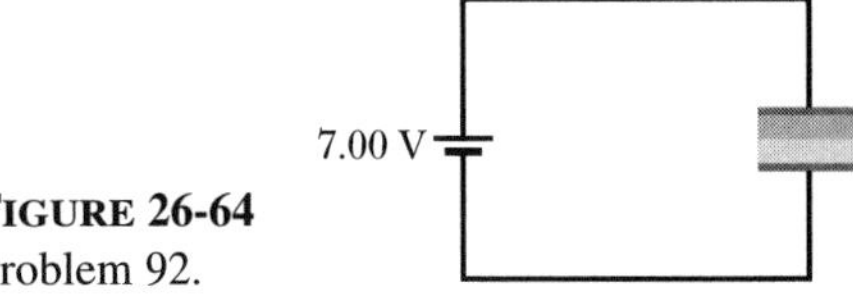

FIGURE 26-64 Problem 92.

Clustered Problems

Cluster 1

93. A 40.0 μF capacitor C_1 is charged to a potential difference of 10.0 V by a battery. Then the battery is removed, C_1 is connected to the uncharged 30.0 μF capacitor C_2 in Fig. 26-65*a*, and switch S is closed. When the system comes to equilibrium, what are (a) the charge on C_1, (b) the charge on C_2, (c) the potential difference across C_1, and (d) the potential difference across C_2? (e) What is the change in the potential energy of the two-capacitor system between the time the switch is closed and the time equilibrium is reached?

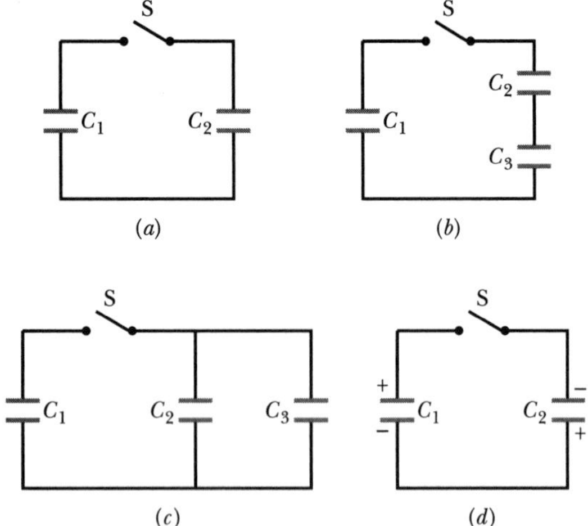

FIGURE 26-65 Problems 93 through 97.

94. Repeat Problem 93 with the following change: The initial potential difference of C_2 is 20.0 V instead of 0 V.

95. A 40.0 μF capacitor C_1 is charged to a potential difference of 10.0 V by a battery. Then the battery is removed, C_1 is connected to the uncharged 15.0 μF capacitor C_2 and the uncharged 30.0 μF capacitor C_3 in Fig. 26-65*b*, and switch S is closed. When the system comes to equilibrium, what are the charges on (a) C_1, (b) C_2, and (c) C_3? Also, what are the potential differences across (d) C_1, (e) C_2, and (f) C_3?

96. A 40.0 μF capacitor C_1 is charged to a potential difference of 10.0 V by a battery. Then the battery is removed, C_1 is connected to the uncharged 30.0 μF capacitor C_2 and the uncharged 15.0 μF capacitor C_3 in Fig. 26-65*c*, and switch S is closed. When the system comes to equilibrium, what are the charges on (a) C_1, (b) C_2, and (c) C_3? Also, what are the potential differences across (d) C_1, (e) C_2, and (f) C_3?

97. A 40.0 μF capacitor C_1 and a 15.0 μF capacitor C_2 are connected in parallel across a 10.0 V battery until equilibrium is established. Then the capacitors are disconnected from the battery and their connections with each other are reversed, with a switch S between them on one side (Fig. 26-65*d*). The switch is then closed. When the capacitors reach equilibrium, (a) how much charge has flowed through the switch and (b) what is the potential difference across either capacitor?

Cluster 2

The problems in this cluster involve a variable capacitor, *which is a capacitor whose capacitance can be varied by, say, changing the plate separation or inserting a dielectric material.*

98. See the setup for this cluster. A 100 μF capacitor C_1 and a variable capacitor C_2 set at 10.0 μF are connected in parallel across a 100 V battery. When equilibrium is reached, the battery is removed. What then are the potential energies of (a) C_1, (b) C_2, and (c) the two-capacitor system? Next, the capacitance of C_2 is reduced to 2.00 μF. (d) To do this, do we increase or decrease the plate separation? What then are the potential energies of (e) C_1, (f) C_2, and (g) the two-capacitor system? What are the changes in the potential energies of (h) C_1 and (i) C_2? (j) How much work do we do in changing the plate separation?

99. See the setup for this cluster. We connect a variable capacitor C that is set at 10.0 μF across a 100 V battery and allow the system to come to equilibrium. Then, with the battery still connected, we reduce the capacitance of C_2 to 2.00 μF. As the system again comes to equilibrium, what are the changes in (a) the potential energy and (b) the charge of C_2? (c) Using the answer to (b), find the change in the energy of the battery. (d) How much work do we do in reducing the capacitance of C_2?

100. Capacitor C_1 has a capacitance of 1.00 nF when the space between its plates is empty. We want to increase the capacitance of 0.3 μF by filling that space with a dielectric wafer. (a) What must be the dielectric constant of the wafer? Our aim is then to charge an initially uncharged second capacitor C_2 with the following steps:

1. Insert the wafer into C_1.
2. Connect C_1 across a 1.00 V battery to charge it.
3. Disconnect the battery.
4. Remove the wafer from C_1.
5. Connect C_1 to C_2.

(b) What is the potential difference across C_1 at the end of step 4? (c) What is the potential difference across C_2 at the end of step 5?

We next disconnect C_1 from C_2, leaving C_2 with whatever charge it has. Then we repeat steps 1 through 5, connecting C_1 to C_2 with the same orientation as previously. (d) What now is the potential difference across C_2 at the end of step 5?

(e) If this procedure is repeated many times, what is the approximate limiting potential difference across C_2? (f) If, after reaching this limiting value, C_2 is discharged in 1 ms, what is the average rate at which it provides its energy? This procedure of charging a capacitor with repeated steps and then suddenly discharging it has many applications, such as in a strobe-flash unit on a camera.

Chapter Twenty-Seven
Current and Resistance

QUESTIONS

14. Figure 27-27 gives, for three wires of radius R, the current density $J(r)$ versus radius r, as measured from the center of a circular cross section through the wire. The wires are made from the same material. Rank the wires according to the magnitude of the electric field at (a) the center, (b) halfway to the surface, and (c) the surface, greatest first.

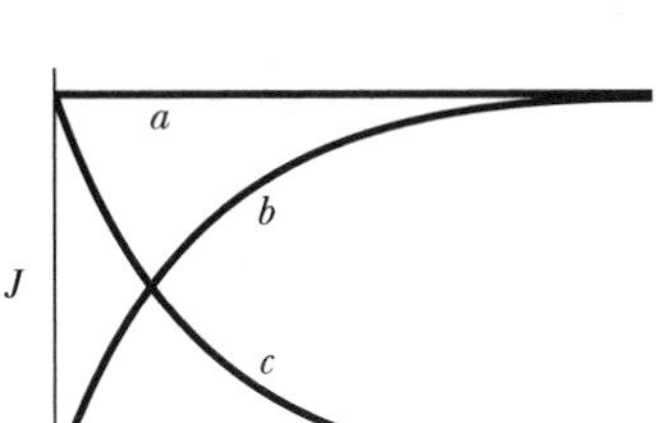

FIGURE 27-27 Question 14.

15. Figure 27-28 displays the circular cross section of a wire and three sections in the form of thin rings concentric with the wire. The rings have the same radial width dr, and the wire has a uniform current density. Rank the rings according to the amount of current passing through their widths, greatest first.

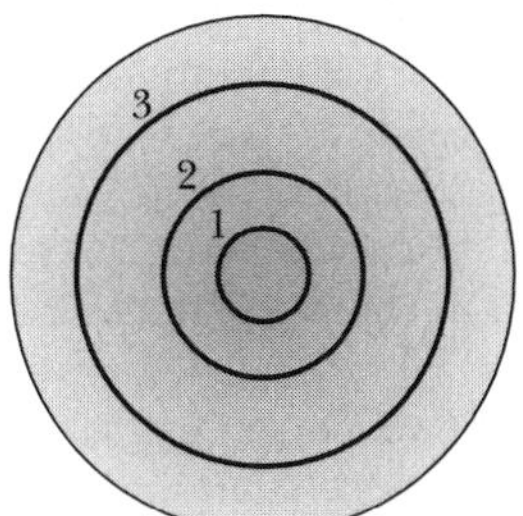

FIGURE 27-28 Question 15.

16. Figure 27-29 shows, as a function of time, the energy dissipated by current through a resistor. Rank the three lettered time periods according to (a) the current through the resistor and (b) the power of the dissipation in the resistor, greatest first.

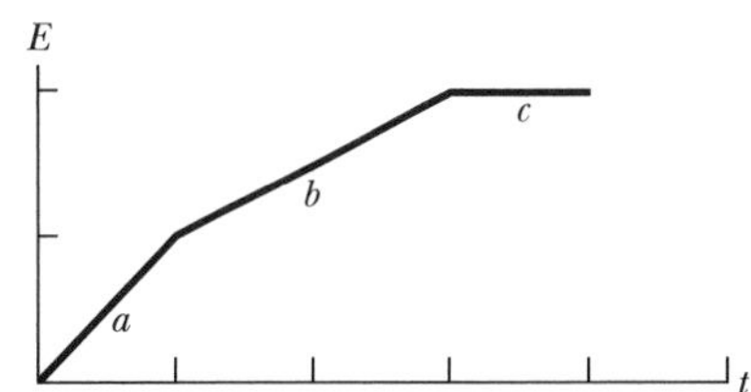

FIGURE 27-29 Question 16.

17. Figure 27-30 gives, for four experiments with the same wire, the charge Q that passes through the wire versus the time t required for the passage. Rank the experiments according to (a) the current in the wire, (b) the power of the energy dissipation in the wire, and (c) the energy dissipated, greatest first.

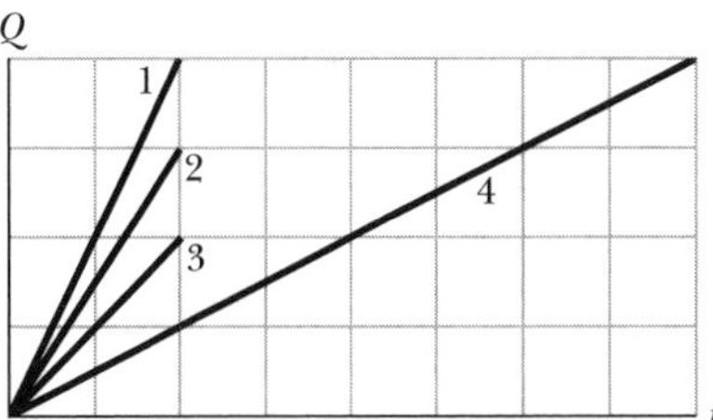

FIGURE 27-30 Question 17.

18. *Organizing question:* The following table gives the current density $J(r)$ in four wires, each with a radius R. J is in amperes per square meter and radial distance r is in meters. For each wire, (a) give the units of the constant (a, b, c, or d) in the expression for $J(r)$ and (b) set up an equation to find the current in the wire between $r = 0$ and $r = R/2$.

WIRE	CURRENT DENSITY
A	$J = ar$
B	$J = br^2$
C	$J = cr^3$
D	$J = d$

19. *Organizing question:* The following list gives data for three situations in which a current is in a wire. For each situation, set up an equation, complete with known data, to find the energy dissipated in the wire in 2 min.

Situation 1: potential difference = 12 V, resistance = 2.0 Ω

Situation 2: current = 4.0 A, resistance = 3.0 Ω

Situation 3: potential difference = 12 V, current = 4.0 A

EXERCISES & PROBLEMS

64. Aluminum wire has been used to replace copper wire during times of high copper prices. (See Table 27-1 for the resistivities of aluminum and copper.) (a) If 20-gauge copper wire has a resistance of 33 Ω per kilometer of length, what is its diameter? (b) What is the diameter of an aluminum wire with the same resistance per unit length as 20-gauge copper wire?

65. A straight conducting wire (diameter = 1.0 mm) carries a current of 2.0 A that is produced by a uniform electric field of 5.3 V/m inside the wire. What is the resistivity of the wire's material?

66. A 200 m-long copper wire connects points A and B. The electric potential at point B is 50 V less than that at point A. If the resistivity of copper is $1.7 \times 10^{-8}\ \Omega\cdot\text{m}$, what are the magnitude and direction of the current density $\mathbf{J}$ in the wire?

67. A certain brand of hot-dog cooker works by applying a potential difference of 120 V across opposite ends of the hot dog and allowing it to cook by means of the thermal energy produced. If the current is 10 A and 60 kJ is required to cook a hot dog, how long will cooking three hot dogs take simultaneously?

68. A 2.0 kW heater element from a dryer has a length of 80 cm. If a 10 cm section is removed, what power is used by the now shortened element at 120 V?

69. The current density J in a certain wire with a circular cross section of radius $R = 2.00$ mm is given by $J = (3.0 \times 10^8)r$, with J in amperes per square meter and radial distance r in meters. What is the current through the outer section bounded by $r = R/2$ and $r = R$?

70. The current density in a wire is uniform and equal to $2.0 \times 10^6\ \text{A/m}^2$; the wire's length is 5.0 m; and the density of conduction electrons is $8.5 \times 10^{28}/\text{m}^3$. How long does an electron take (on the average) to travel the length of the wire?

71. A cylindrical rod is reformed so that its length is 4.00 times its original length (with no change in its volume). What is the ratio of its resistance (end to end) to its original resistance?

72. A resistor, with a potential difference of 200 V across it, transfers electrical energy to thermal energy at the rate of 3000 W. What is the resistance of the resistor?

73. An 18.0 W device has 9.00 V across it. How much charge goes through the device in 4.00 h?

74. How much energy is consumed in 2.0 h by an electrical resistance of 400 Ω when the potential applied across it is 90.0 V?

75. A potential difference of 12 V is applied to (circular) copper wire that has a length of 45 m and a radius of 2.0 mm. How much thermal energy is produced in the wire by the current in 40 s?

76. A copper wire of cross-sectional area $2.0 \times 10^{-6}\ \text{m}^2$ and length 4.0 m has a current of 2.0 A uniformly distributed across that area. (a) What is the magnitude of the electric field along the wire? (b) How much energy is converted to thermal energy in 30 min?

77. The current density in a certain circular wire of radius 3.00 mm is given by $J = (2.75 \times 10^{10}\ \text{A/m}^4)r^2$, where r is the radial distance. The potential applied to the wire (end to end) is 60.0 V. How much energy is converted to thermal energy within the wire in 1.00 h?

Clustered Problems

Cluster 1

78. Figure 27-31a shows a resistive device consisting of n wires in an arrangement resembling a ladder. Each wire has length L, resistivity ρ, and cross-sectional area A. The wires are strung between conducting strips A and B, which have negligible resistance and are held in place by insulating rods. A potential difference V is set up between the two strips. Find expressions for (a) the current through any one wire and (b) the total current through all the wires. (c) Comparing the latter with Eq. 27-8, write an expression for the resistance R of the device as a whole.

79. Figure 27-31b shows a resistive device consisting of n wires of the same resistivity ρ and cross-sectional area A and with lengths in the sequence $L, L/2, L/3, \ldots, L/n$. They are strung between conducting strips that have negligible resistance and that are held in place by insulating rods. A potential difference V is set up between the two strips. Find expressions for the currents through the wires of (a) length L, (b) length $L/2$, (c) length $L/3$, and so on. (d) Sum the currents and then simplify the series in the sum to find an expression for the total current through the wires. (e) Comparing that expression with Eq. 27-8, write an expression for the resistance R of the device as a whole.

80. Figure 27-31c shows a solid resistive wedge formed from a truncated isosceles triangle. It has resistivity ρ, thickness t, left-right width b at the top, left-right width a at the bottom, and height L. (a) If the front and back surfaces are coated with a conducting film having negligible resistance and an electric potential is set up between those two sides, show that the resistance of the wedge is given by

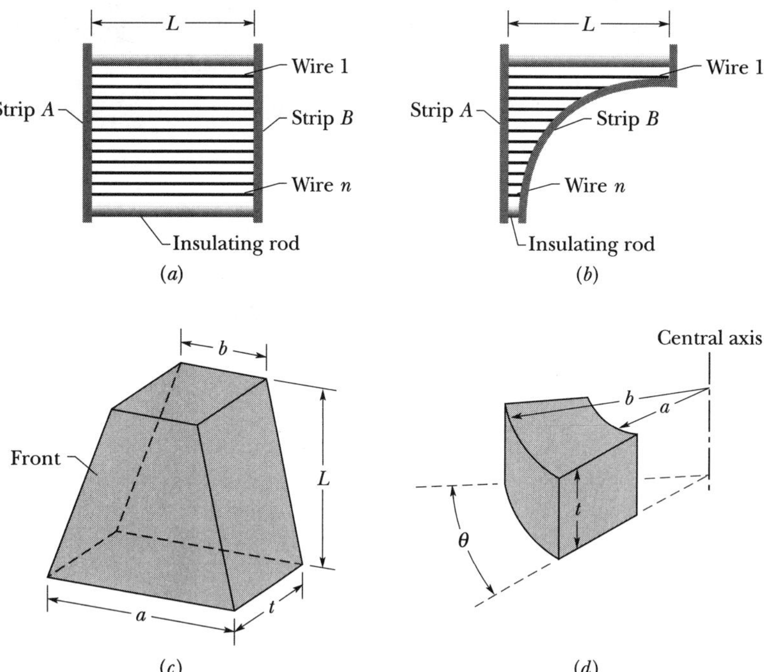

FIGURE 27-31 Problems 78 through 81.

$$R = \frac{2\rho t}{L(a + b)}.$$

(b) If, instead, the top and bottom are the coated sides and an electric potential is set up between them, show that the resistance of the wedge is given by

$$R = \frac{\rho L}{t(a - b)} \ln \frac{a}{b}.$$

(Assume that the electric field lines extend radially from the vertex of the truncated triangle.) (c) What does this latter expression for R reduce to in the limit that a approaches b?

81. Figure 27-31*d* shows an angular section that has been sliced from what was originally a flat doughnut of material. The section has uniform thickness t, outer radius b, inner radius a, and angular width θ, and the material has resistivity ρ. (a) If the top and bottom surfaces are coated with a conducting film having negligible resistance and an electric potential is set up between those two surfaces, what is the resistance of the section? (Assume that the electric field lines are perpendicular to the top and bottom surfaces.)

(b) If, instead, the curved outer and inner surfaces are coated with the film and an electric potential is set up between them, what is the resistance of the section? (Assume that the electric field lines extend radially outward from the central axis of the original doughnut.)

(c) If, instead, the remaining two sides are coated with the film and an electric potential is set up between them, what is the resistance of the section? (Assume that the electric field lines follow circles that are concentric with the central axis of the original doughnut.)

Cluster 2

82. Wire A and wire B are made from the same material. Wire A has twice the diameter and half the length of wire B and a resistance of 8.0 Ω. (a) What is the resistance of wire B? (b) If the same currents are in the two wires, what is the ratio J_A/J_B of their current densities?

83. Wire C and wire D are made from different materials. The resistivity and diameter of wire C are $2.0 \times 10^{-6}\ \Omega\cdot\text{m}$ and 1.00 mm, and those of wire D are $1.0 \times 10^{-6}\ \Omega\cdot\text{m}$ and 0.50 mm. The wires are joined as shown in Fig. 27-32 and a current of 2.0 A is set up in them. What are the electric potential differences between (a) points 1 and 2 and (b) points 2 and 3? What are the rates at which energy is dissipated between (c) points 1 and 2 and (d) points 2 and 3?

84. A certain length of wire C in Problem 83 has a resistance of 8.0 Ω at 300 K. At 400 K it has a resistance of 10 Ω. (a) What is its resistance at 600 K? The material composing wire D in Problem 83 has a temperature coefficient of resistivity that is twice that of the material in wire C. (b) What is the resistance of wire D at 600 K?

85. The tungsten filament in a certain light bulb has a resistance of 2.0 Ω when the filament is at room temperature (300 K) with no current in it. When 12 V is applied to the filament, the current in the filament dissipates energy at the rate of 10 W. What then is the approximate temperature of the filament?

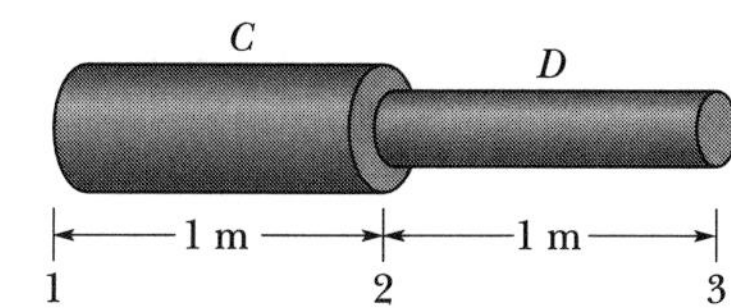

FIGURE 27-32 Problem 83.

Chapter Twenty-Eight
Circuits

QUESTIONS

21. In Fig. 28-60, the four graphs are intended to give the electric potential around a closed, one-loop circuit containing an ideal battery and one or more resistors. (a) Which, if any, are physically possible? What is wrong with any wrong graph? (b) For any physically possible graph, which part of the graph corresponds to the battery and how many resistors are in the circuit?

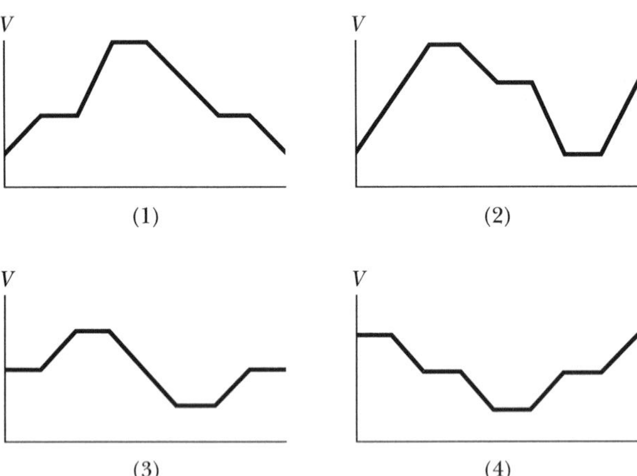

FIGURE 28-60 Question 21.

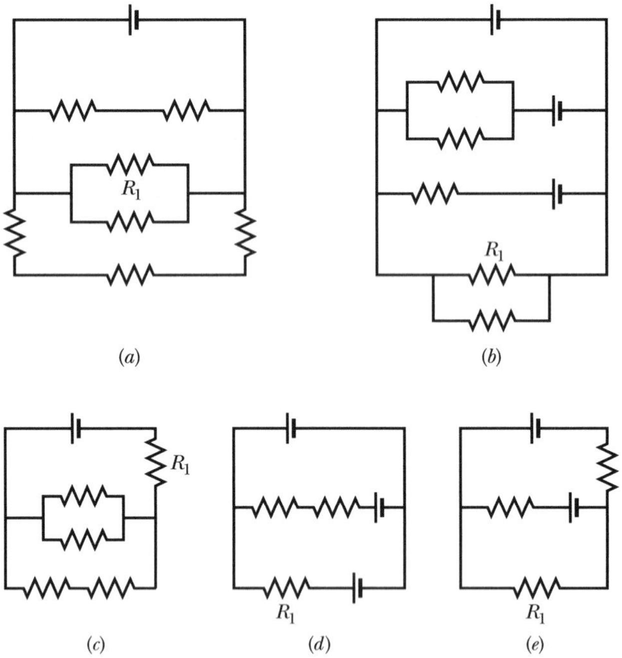

FIGURE 28-61 Question 22.

22. *Organizing question:* In this chapter we have seen three basic techniques for solving for the current through a resistor located in a circuit of batteries and resistors: (1) solving a single-loop equation, (2) simplifying the circuit by finding the equivalent resistance of resistors in parallel or in series, and (3) solving two-loop equations simultaneously. Which of these techniques is best for finding the current through resistor R_1 in the five circuits of Fig. 28-61?

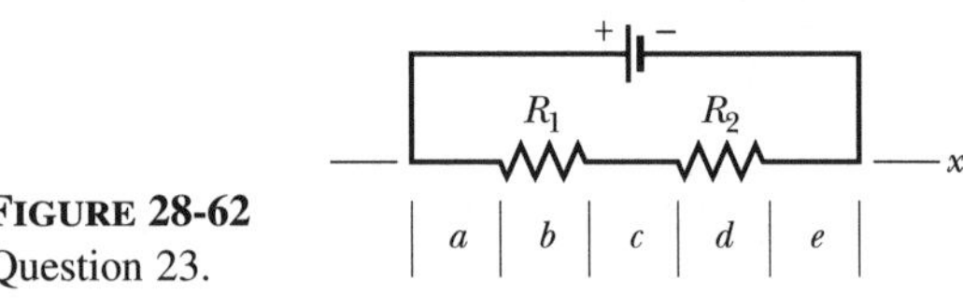

FIGURE 28-62 Question 23.

23. In Fig. 28-62, a circuit consists of a battery and two uniform resistors, and the section lying along an x axis is divided into five segments of equal lengths. (a) Assume that $R_1 = R_2$ and rank the segments according to the magnitude of the average electric field in them, greatest first. (b) Now assume that $R_1 > R_2$ and then again rank the segments. (c) What is the direction of the electric field along the x axis?

24. In Fig. 28-63, a circuit initially consists of a resistance R and two parallel branches that contain identical ideal batteries and identical resistances. If a third branch with the same battery and resistance is added in parallel, as suggested in the figure, does the current through R increase, decrease, or remain the same?

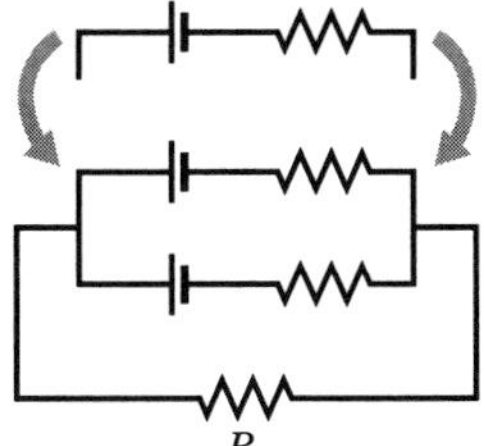

FIGURE 28-63 Question 24.

25. Figure 28-64 shows four circuits with identical resistors and identical ideal batteries. In Fig. 28-64*c*, a broken wire is indicated along one branch (the resistance of that branch is taken to be infinite); in Fig. 28-64*d*, a *short* is indicated along one branch (the resistance of that branch is taken to be zero). Rank the circuits according to (a) the equivalent resistance of the resistors, (b) the current through the battery, and (c) the rate at which the battery supplies energy to the resistors, greatest first.

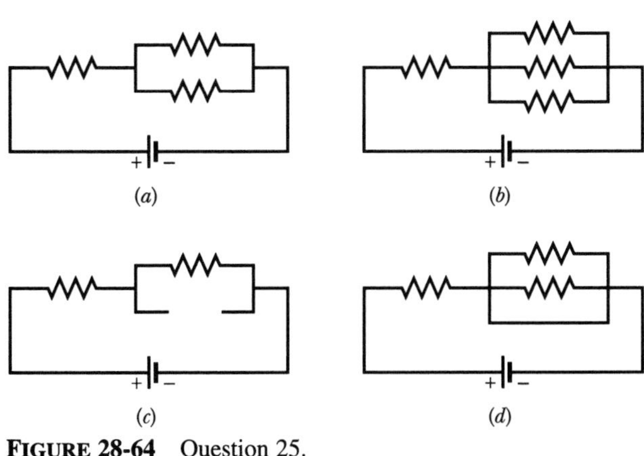

FIGURE 28-64 Question 25.

26. *Organizing question:* The switch in the circuit of Fig. 28-65 is closed at time $t = 0$. Set up a loop equation, complete with known values, to find the charge q on the capacitor when the current in the resistor has dropped to 25 mA.

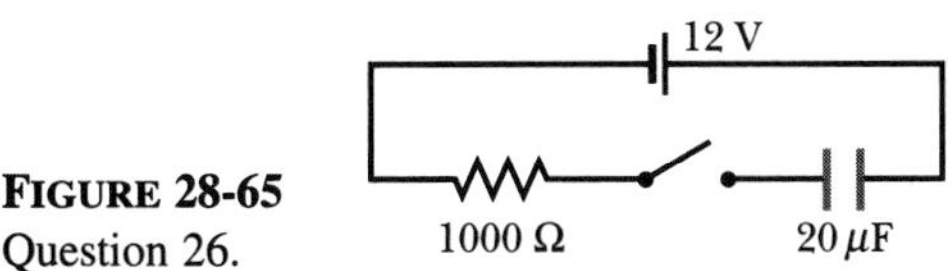

FIGURE 28-65 Question 26.

27. After the switch in Fig. 28-13 is closed on point *a*, there is current *i* through resistance *R*. Figure 28-66 gives that current for four sets of values of *R* and capacitance *C*: (1) R_0 and C_0; (2) $2R_0$ and C_0; (3) R_0 and $2C_0$; (4) $2R_0$ and $2C_0$. Which set goes with which curve?

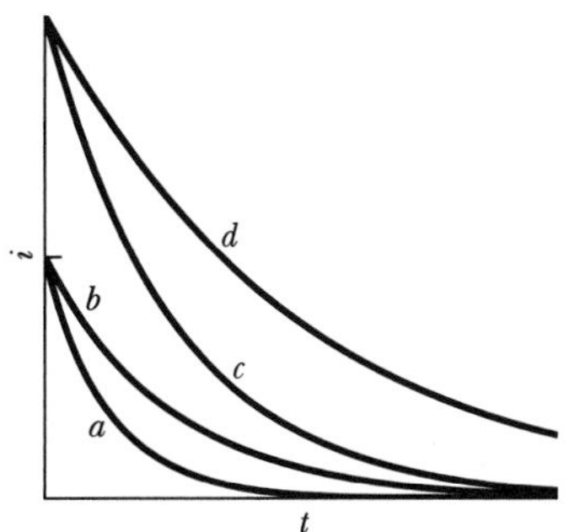

FIGURE 28-66 Question 27.

28. When the switch in the circuit of Fig. 28-67*a* is closed, which of the curves in Fig. 28-67*b* gives the potential difference across (a) the capacitor, (b) the resistor, and (c) the battery, each versus time?

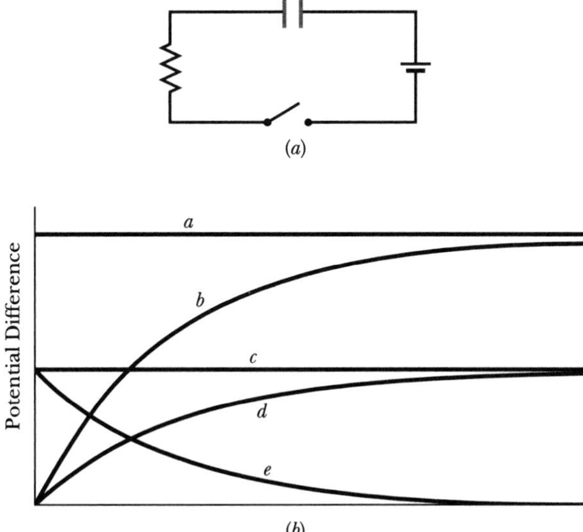

FIGURE 28-67 Question 28.

29. Figure 28-68 shows three circuits consisting of identical resistors, identical capacitors, and identical ideal batteries. In each circuit, the capacitor charges when the switch is closed. Which of the curves in the graph of charge Q versus time t best represents the charging for which of the circuits?

30. *Organizing question:* The switch in Fig. 28-69 is closed at time $t = 0$. (a) Thereafter, what is the current through R_3 in terms of the currents through R_1 and R_2? (b) What is the relation between the charge q on the capacitor and the current through R_2? Using those relations and data given in the figure, set up loop equations for (c) the big loop and (d) the left-hand loop. (e) To produce a single differential equation for charge q, what step should now be taken? When the circuit finally reaches equilibrium, what are (f) the current through R_2 and (g) the potential difference across the capacitor? (*Hint:* No calculator or even writ-

ten calculation is needed for (f) and (g).)

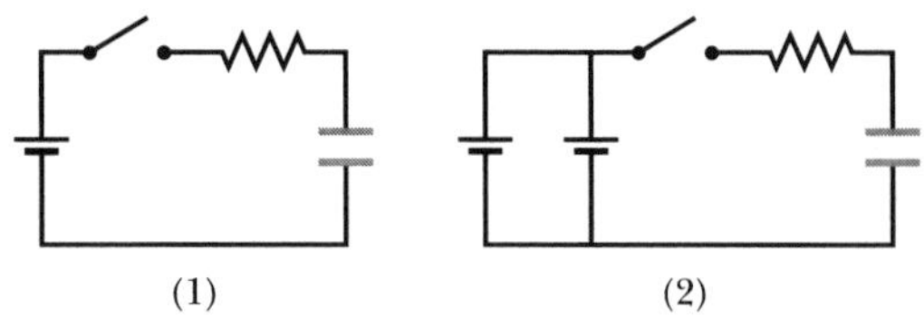

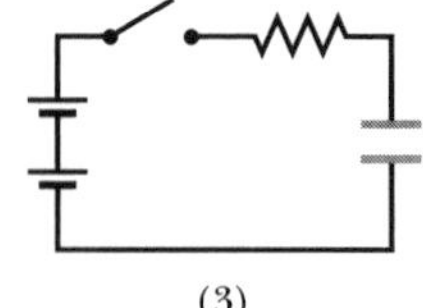

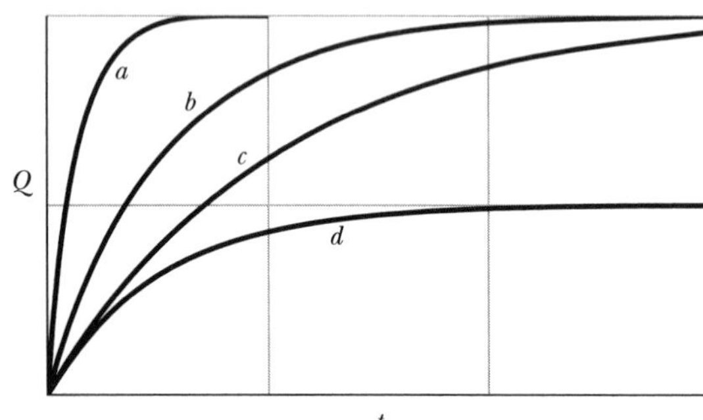

FIGURE 28-68 Question 29.

The switch is reopened at $t = 0$ (we reset our reference clock). (h) What now is the relation between the charge q on the capacitor and the current through R_2? (i) What is the relation between the currents through R_2 and R_3? (j) Using these relations and known data, set up a loop equation for the discharge of the capacitor.

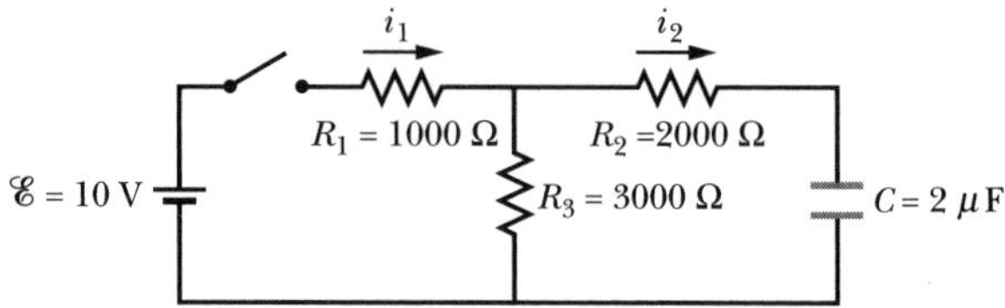

FIGURE 28-69 Question 30.

31. Without using a graphing calculator, determine which of the following functions $f(t)$ best correspond to which of the plots of Fig. 28-70.

(1) $f(t) = 1 - e^{-2t}$ (2) $f(t) = 1 - e^{t/1.75}$

(3) $f(t) = e^{t/3}$ (4) $f(t) = e^{-t}$

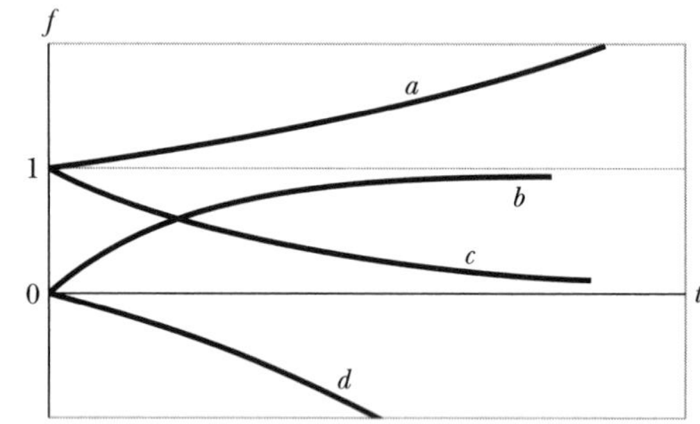

FIGURE 28-70 Question 31.

EXERCISES & PROBLEMS

83. Figure 28-71 shows a section of a circuit. The electric potential difference between points A and B that connect the section to the rest of the circuit is $V_A - V_B = 78$ V, and the current through the 6.0 Ω resistor is 6.0 A. Is the device represented by "Box" absorbing or providing energy to the circuit and at what rate?

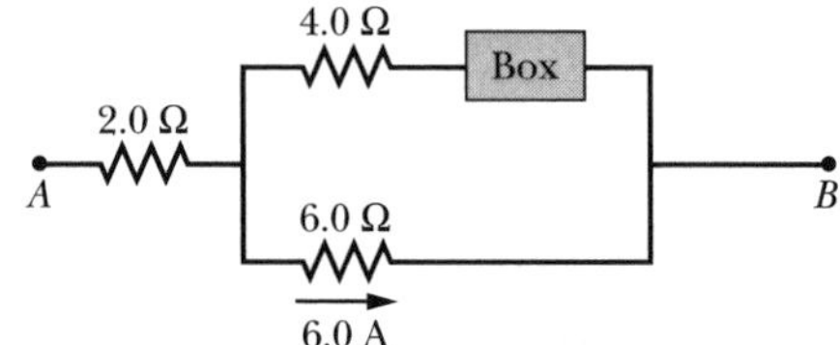

FIGURE 28-71 Problem 83.

84. In Fig. 28-72, the current through the 4.0 Ω resistor is indicated. What is the emf of the (ideal) battery?

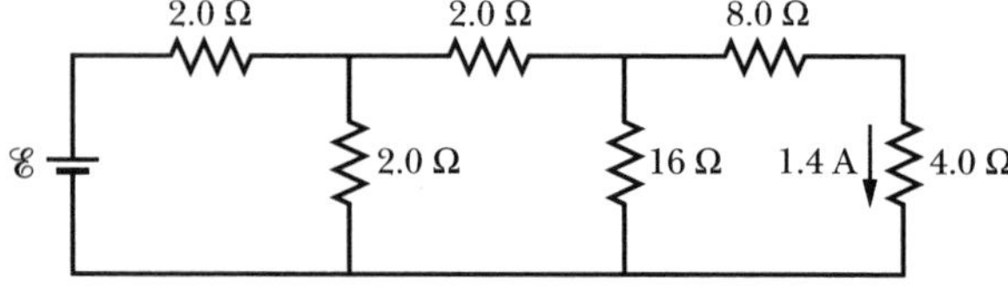

FIGURE 28-72 Problem 84.

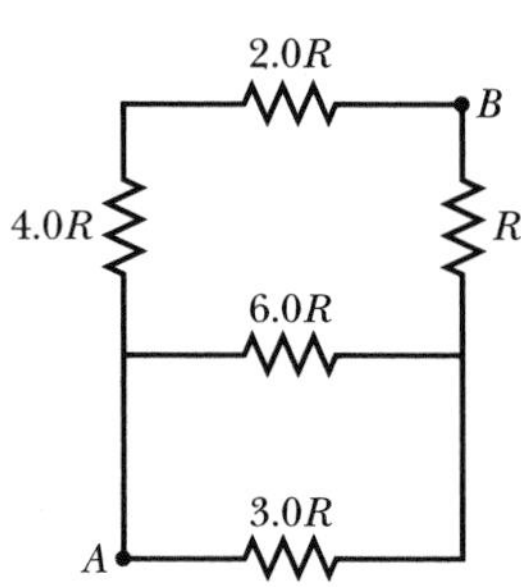

FIGURE 28-73 Problem 85.

85. In Fig. 28-73, $R = 10\ \Omega$. What is the equivalent resistance between points A and B? (*Hint:* This circuit section might look simpler if you first assume that points A and B are connected to a battery.)

86. In the circuit of Fig. 28-74, what value of R will result in no current through the 20.0 V battery?

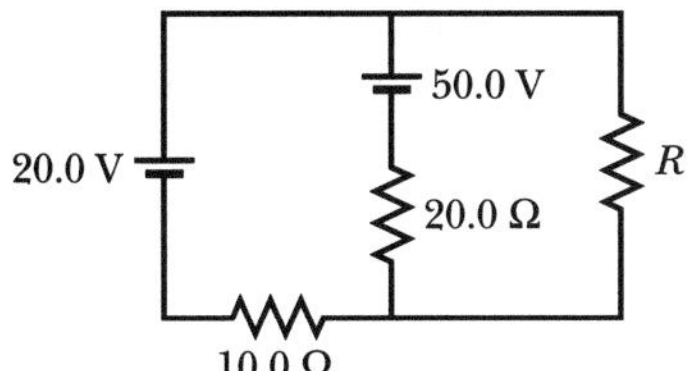

FIGURE 28-74 Problem 86.

87. When steady-state conditions are reached in the circuit of Fig. 28-75, what is the total energy stored in the two capacitors?

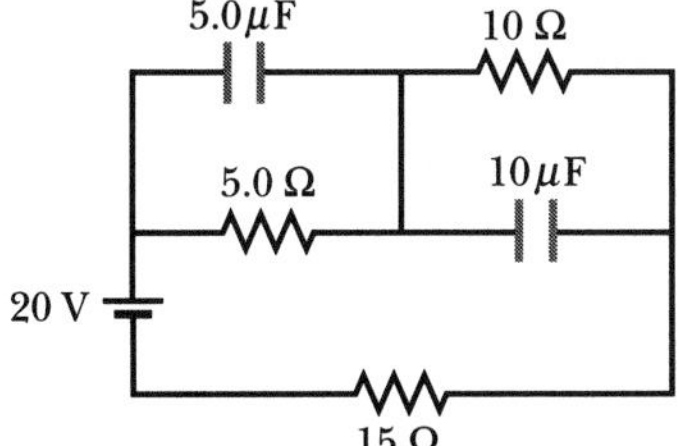

FIGURE 28-75 Problem 87.

88. The switch in Fig. 28-76 is left closed for a long time so that the steady state is reached. Then at time $t = 0$ the switch is opened. What is the current through the 15 kΩ resistor at $t =$ 4.00 ms?

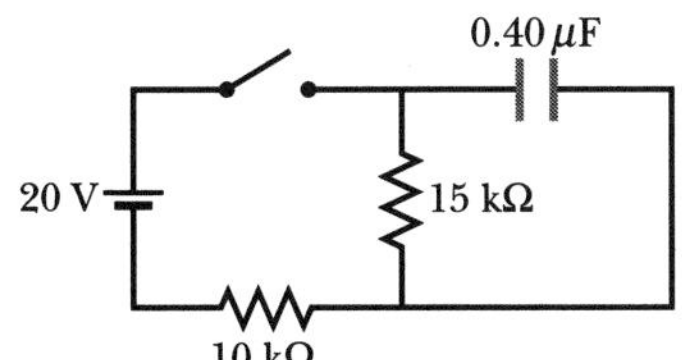

FIGURE 28-76 Problem 88.

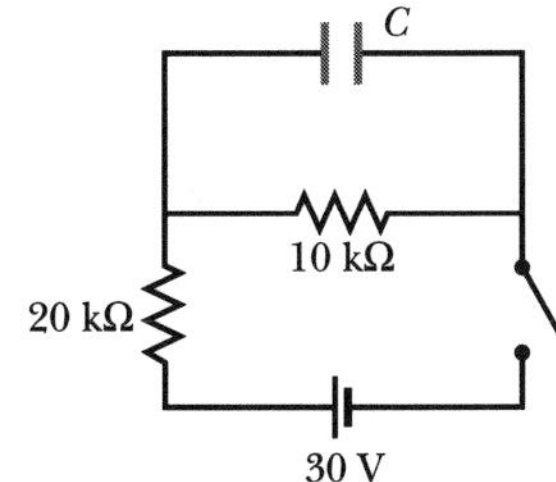

FIGURE 28-77 Problem 89.

89. The capacitor in Fig. 28-77 is uncharged when the switch is closed. What are the initial currents through (a) the 10 kΩ resistor and (b) the 20 kΩ resistor? (c) What is the current through the 10 kΩ resistor a very long time after the switch is closed?

90. (a) What are the size and direction of current i_1 in Fig. 28-78? (Can you answer this making only mental calculations?) (b) At what rate is the battery supplying energy?

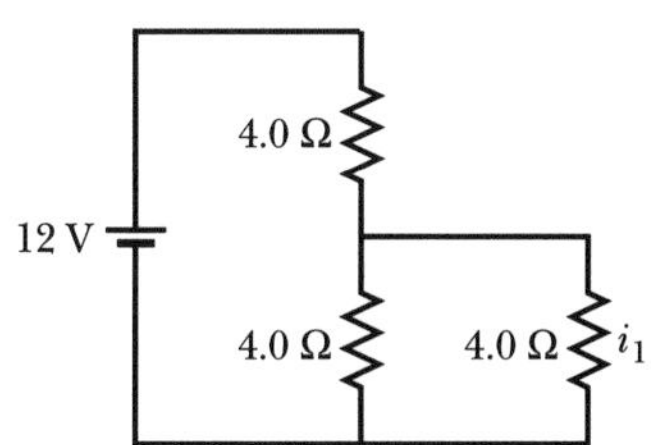

FIGURE 28-78 Problem 90.

91. (a) What are the size and direction of current i_1 in Fig. 28-79, where each resistance is 2.0 Ω? What are the powers of (b) the 20 V battery, (c) the 10 V battery, and (d) the 5.0 V battery, and for each, is energy being supplied or absorbed?

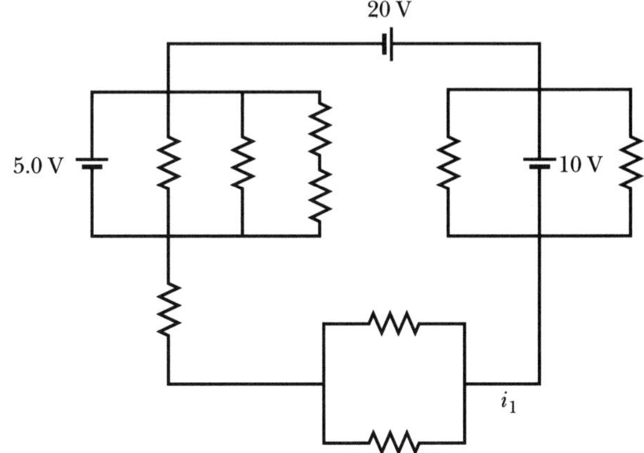

FIGURE 28-79 Problem 91.

92. What are the sizes and directions of (a) current i_1 and (b) current i_2 in Fig. 28-80, where each resistance is 2.00 Ω? (Can you answer this making only mental calculations?) (c) At what rate is energy being transferred in the 5.00 V battery at the left, and is the energy being supplied or absorbed by the battery?

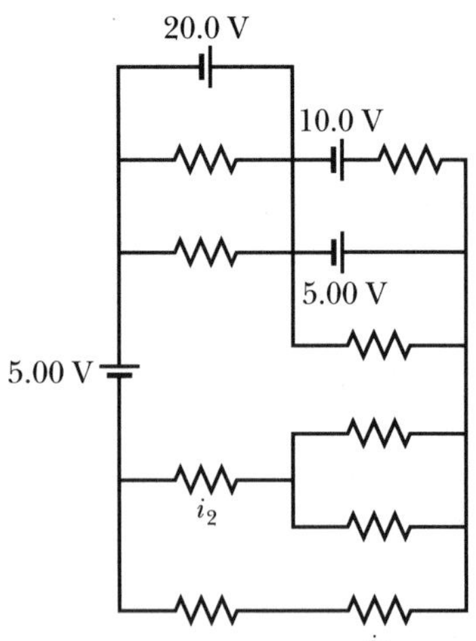

FIGURE 28-80 Problem 92.

93. What are the sizes and directions of currents (a) i_1 and (b) i_2 in Fig. 28-81? (Can you answer this using only mental calculation?) At what rates is energy being transferred at (c) the 16 V battery and (d) the 8.0 V battery, and for each, is energy being supplied or absorbed?

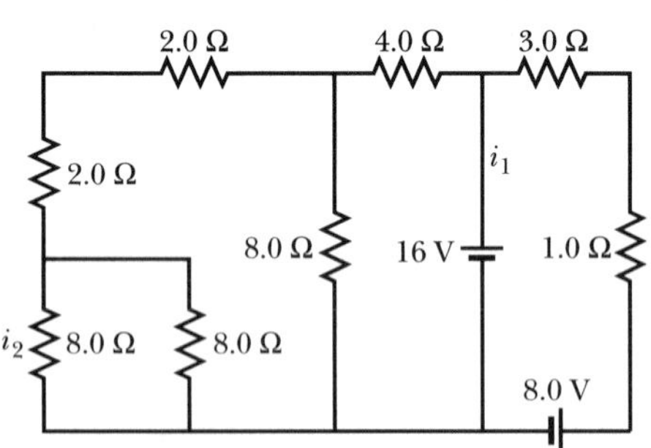

FIGURE 28-81 Problem 93.

94. Figure 28-82 shows a portion of a circuit. What is the size of current i_1?

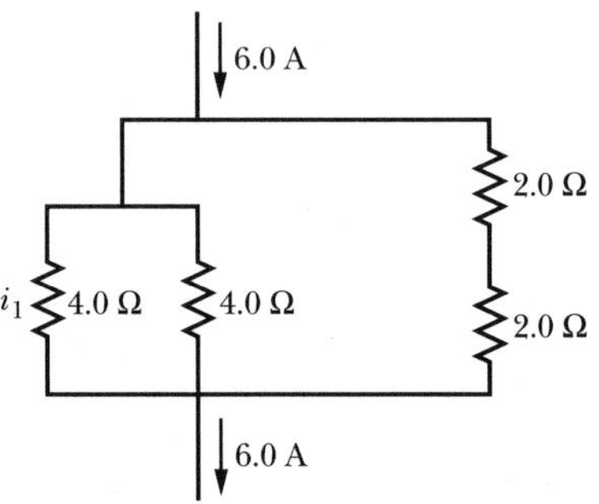

FIGURE 28-82 Problem 94.

95. What are the size and direction of current i_1 in Fig. 28-83? (*Hint:* This can be answered by using only mental calculation.)

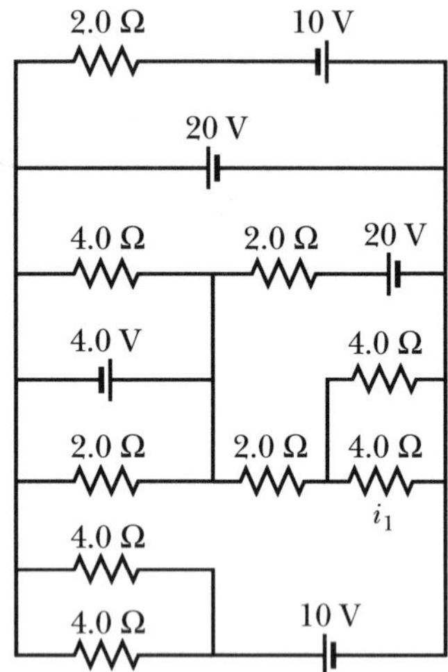

FIGURE 28-83 Problem 95.

96. In Fig. 28-84, what are the currents through (a) the 6.0 V battery and (b) the 12 V battery? What are the rates at which energy is being transferred in (c) the 6.0 V battery and (d) the 12 V battery, and for each, is energy being supplied or absorbed?

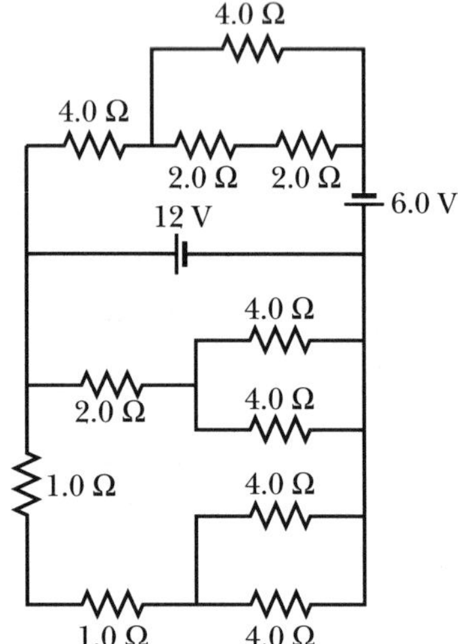

FIGURE 28-84 Problem 96.

97. What are the potentials (a) V_1 and (b) V_2 at the points indicated in Fig. 28-85, where each resistance is 2.0 Ω? (The symbol at the upper right indicates that the circuit is grounded there; the potential is defined to be zero at that point.)

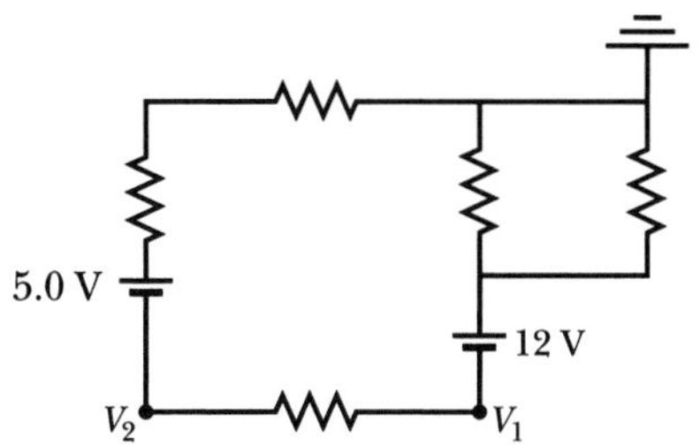

FIGURE 28-85 Problem 97.

98. (a) What are the size and direction of current i_1 in Fig. 28-86? (b) How much energy is dissipated by all four resistors in 1.0 min?

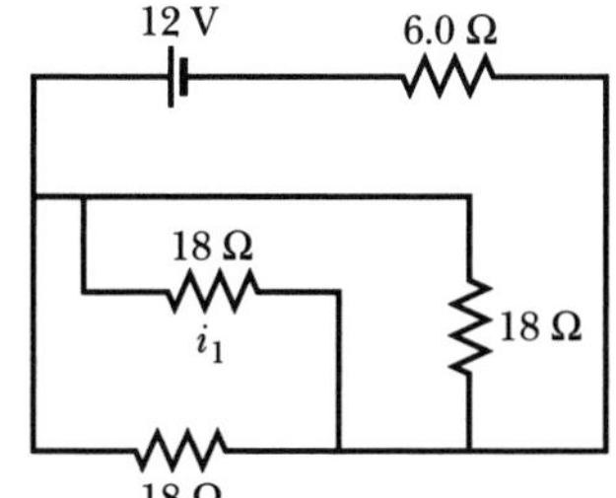

FIGURE 28-86 Problem 98.

99. What are the sizes and directions of the currents through resistors (a) R_2 and (b) R_3 in Fig. 28-87, where each resistance is 4.0 Ω?

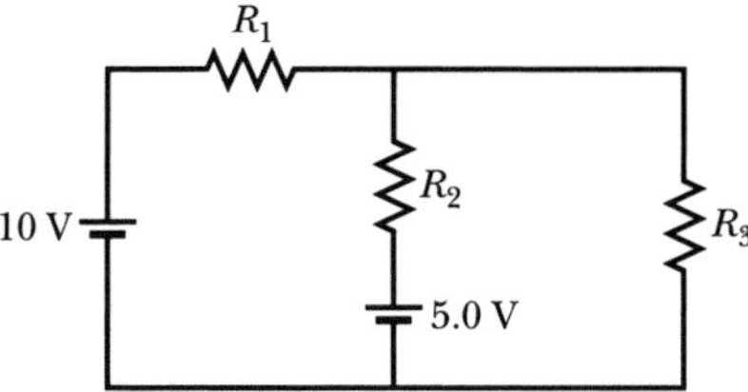

FIGURE 28-87 Problem 99.

100. In Fig. 28-88, where each resistance is 4.00 Ω, what are the sizes and directions of currents (a) i_1 and (b) i_2? At what rates is energy being transferred at (c) the 4.00 V battery and (d) the 12.0 V battery, and for each, is the battery supplying or absorbing energy?

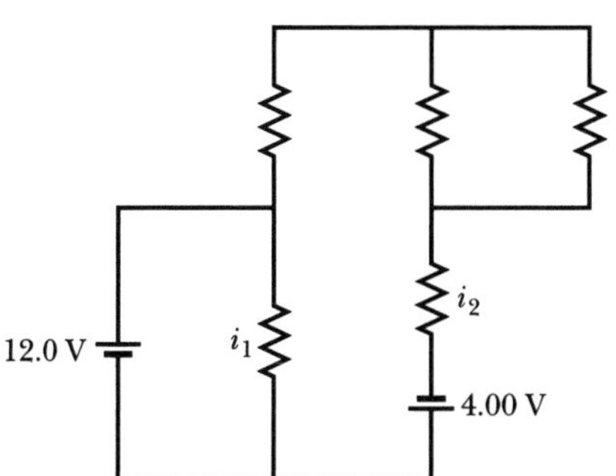

FIGURE 28-88 Problem 100.

101. Considering all possible values of R in Fig. 28-89, find the maximum possible rate at which energy can be supplied by the battery.

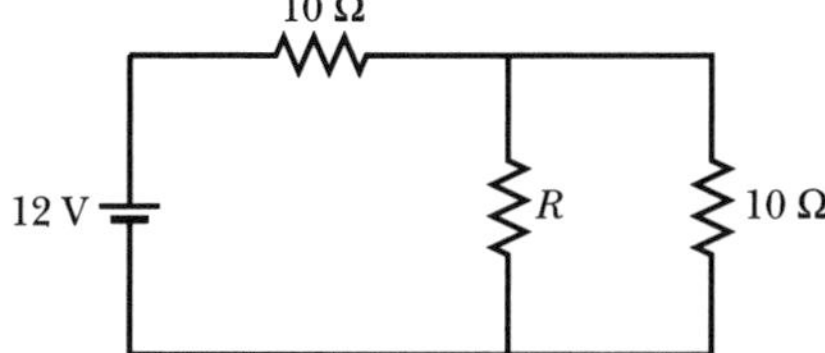

FIGURE 28-89 Problem 101.

102. The six real batteries in Fig. 28-90 all have an emf of 20 V and a resistance of 4.0 Ω. (a) What is the current through the 4.0 Ω resistor? (b) What is the potential difference across each battery? (c) What is the power of each battery? (d) At what rate does each battery transfer energy to internal thermal energy?

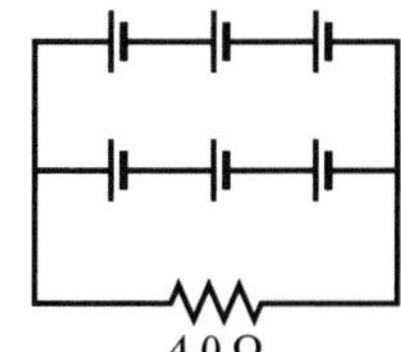

FIGURE 28-90 Problem 102.

103. In Fig. 28-91, what are currents (a) i_2, (b) i_4, (c) i_1, (d) i_3, and (e) i_5?

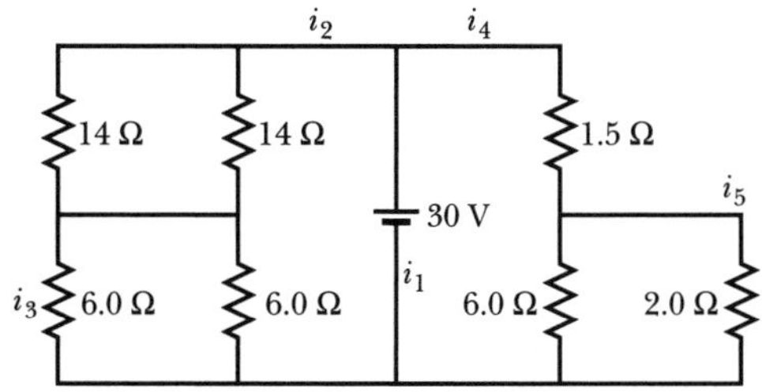

FIGURE 28-91 Problem 103.

104. What are the size and direction of the current indicated in Fig. 28-92, where all resistances are 4.0 Ω and batteries have an emf of 10 V? (*Hint:* This can be answered using only mental calculation.)

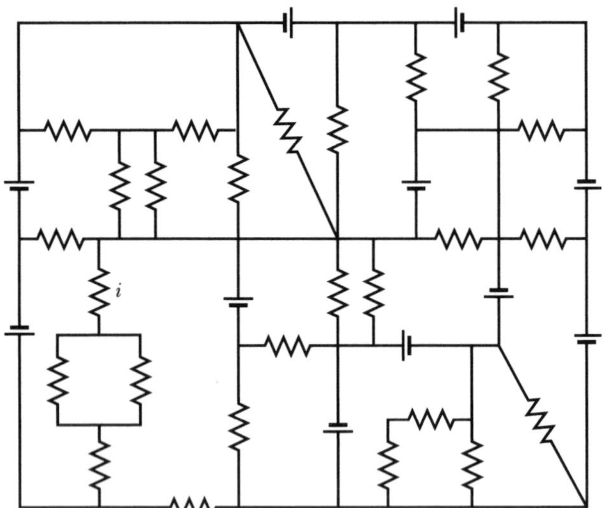

FIGURE 28-92 Problem 104.

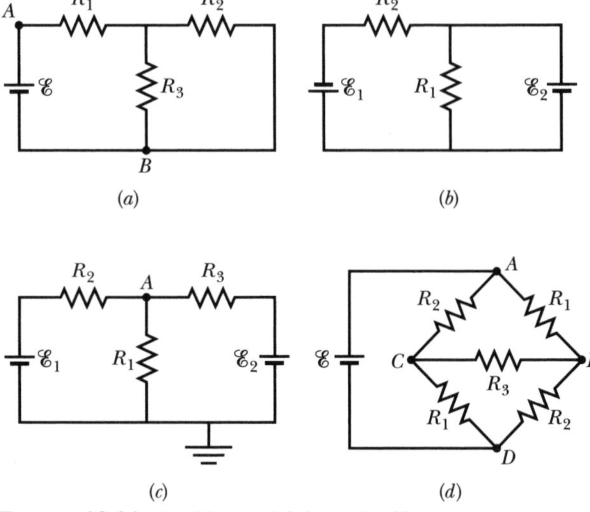

FIGURE 28-94 Problems 106 through 109.

105. In Fig. 28-93, the symbol at the left indicates that the circuit is grounded there, which means that the potential V is defined to be zero there. What are the potentials (a) V_1, (b) V_2, and (c) V_3 at the points indicated? (*Hint:* The whole circuit need not be solved. Only two independent loop equations need to be solved.)

Clustered Problems

Cluster 1

106. In Fig. 28-94*a*, $\mathscr{E} = 6.00$ V, $R_1 = 100\ \Omega$, $R_2 = 300\ \Omega$, and $R_3 = 600\ \Omega$. (a) What is the equivalent resistance between points A and B? (b) What is the electric potential across R_1? (c) What is the current through R_3?

107. In Fig. 28-94*b*, $\mathscr{E}_1 = 6.00$ V (note battery orientation), $\mathscr{E}_2 = 12.0$ V, $R_1 = 200\ \Omega$, and $R_2 = 100\ \Omega$. What is the current (magnitude and direction) through (a) R_1, (b) R_2, and (c) the 12 V battery?

108. In Fig. 28-94*c*, $\mathscr{E}_1 = 6.00$ V, $\mathscr{E}_2 = 12.0$ V, $R_1 = 100\ \Omega$, $R_2 = 200\ \Omega$, and $R_3 = 300\ \Omega$. One point of the circuit is grounded, as indicated by the standard symbol for grounding; the electric potential of that point is defined to be zero. Grounding a point in a circuit does not alter the electric potential differences around the circuit and thus does not alter the currents in the circuit. What is the current (magnitude and direction) through (a) R_1, (b) R_2, and (c) R_3? (d) What is the electric potential at point A?

109. In Fig. 28-94*d*, $\mathscr{E} = 12.0$ V, $R_1 = 2000\ \Omega$, $R_2 = 3000\ \Omega$,

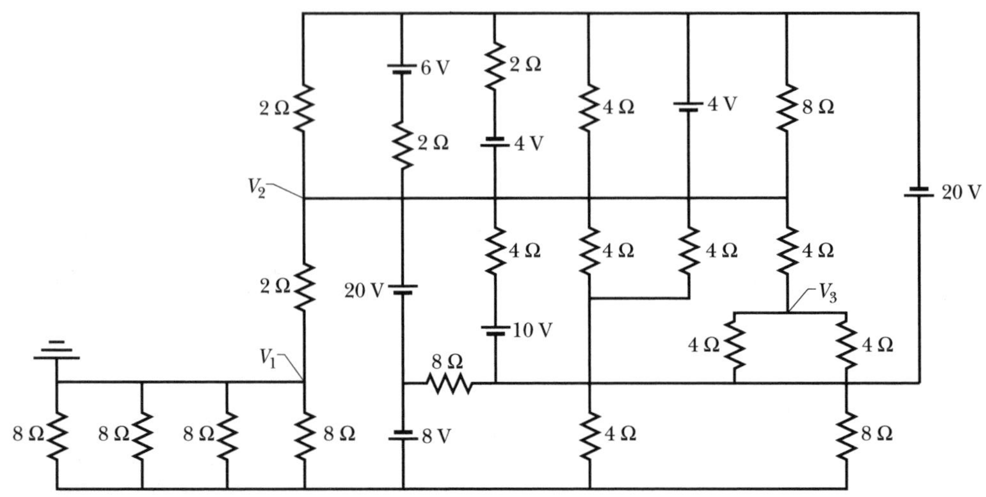

FIGURE 28-93 Problem 105.

and $R_3 = 4000\ \Omega$. What are the potential differences (a) $V_A - V_B$, (b) $V_B - V_C$, (c) $V_C - V_D$, and (d) $V_A - V_C$?

Cluster 2

110. In the circuit of Fig. 28-95a, $\mathscr{E} = 12.0$ V, $R_1 = 5.00$ kΩ, $R_2 = 10.0$ kΩ, and $C = 20.0\ \mu$F. The capacitor is initially uncharged when the switch is closed at time $t = 0$. Just then, what are (a) the current through R_1, (b) the current through R_2, and (c) the rate at which the capacitor is being charged? After the switch has been closed for a long time (that is, many capacitive time constants), what are (d) the current through R_1, (e) the current through R_2, and (f) the rate at which the capacitor is being charged?

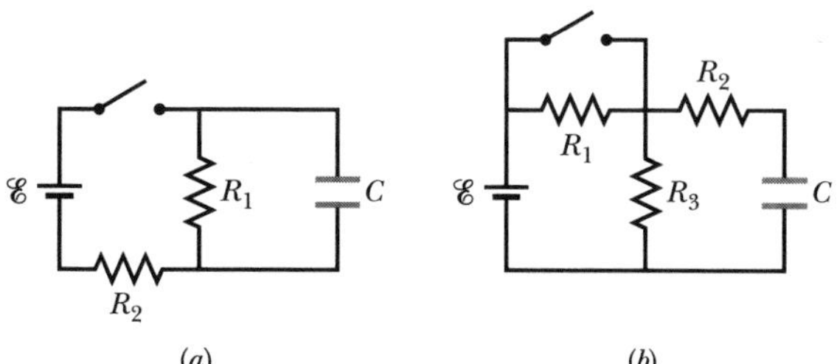

FIGURE 28-95 Problems 110 and 111.

During the charging process, the charge q on the capacitor in *this* circuit can be expressed as

$$q = Q_f(1 - e^{-t/\tau}),$$

where Q_f is the charge at the end of the charging process and τ is the capacitive time constant of the process. (g) Find Q_f from the result of (d) (resistor R_1 and the capacitor have the same potential difference). (h) Then find τ by differentiating the equation just given for q and evaluating the expression for dq/dt at time $t = 0$, using the result of (c). (i) What is the electric potential difference across the capacitor at time $t = \tau$?

After the switch has been closed for a long time, it is reopened. (j) Find an expression for the potential difference across the capacitor as the capacitor discharges through R_1.

111. In the circuit of Fig. 28-95b, $\mathscr{E} = 10.0$ V, $R_1 = 6.00$ kΩ, $R_2 = 2.00$ kΩ, $R_3 = 4.00$ kΩ, and $C = 100\ \mu$F. The switch has been open for a long time (the circuit is in equilibrium). What are (a) the potential difference across the capacitor and (b) the charge on the capacitor?

The switch is closed at time $t = 0$. Just then, what are (c) the charge on the capacitor and (d) the rate at which that charge is changing? (Be careful on (d); from (a) we know that the initial potential difference across the capacitor is not zero, as in the standard problem.)

After the switch is closed for a long time, what are (e) the potential difference across the capacitor and (f) the charge on the capacitor? Before those values are reached (as the charge on the capacitor changes), the potential difference V across the capacitor can be expressed as

$$V = V_0 - V_1 e^{-t/\tau}.$$

Find (g) V_0, (h) V_1, and (i) τ. (*Hint:* From (a) we know V at $t = 0$; substitute those values into the equation. From (e) we know V at $t = \infty$; substitute those values into the equation. To find τ, use the equation for $V(t)$ to write an equation for $q(t)$. Next, differentiate $q(t)$ to write an equation for dq/dt. Then, using the result of (d), evaluate this last equation for time $t = 0$.)

After equilibrium is reached with the switch closed, it is reopened at time $t = 0$ (we reset our reference clock just then). (j) Find an expression for the potential difference $V(t)$ across the capacitor as the charge on the capacitor again changes. (*Hint:* Assume that $V(t)$ has the form

$$V = V_2 + V_3 e^{-t/\tau'}$$

and use the techniques of the previous hint. This time, however, the current through R_2 is $-dq/dt$ of the capacitor.)

Cluster 3

112. In the circuit of Fig. 28-96a, with the switch open, capacitor C_1 is charged to a charge Q while capacitor C_2 is left uncharged. Then at time $t = 0$, the switch is closed. Just then, charge begins to flow through resistor R. Also, the electric potentials across C_1, C_2, and R are immediately equalized, as required by the loop rule (immediate because no resistance slows the process). This means that, immediately, C_1 loses a certain amount of charge to C_2. We take this state of equal potentials as the initial state of the system. The system then moves toward a final equilibrium state. Let's call C_1, C_2, Q, and R the *given symbols.*

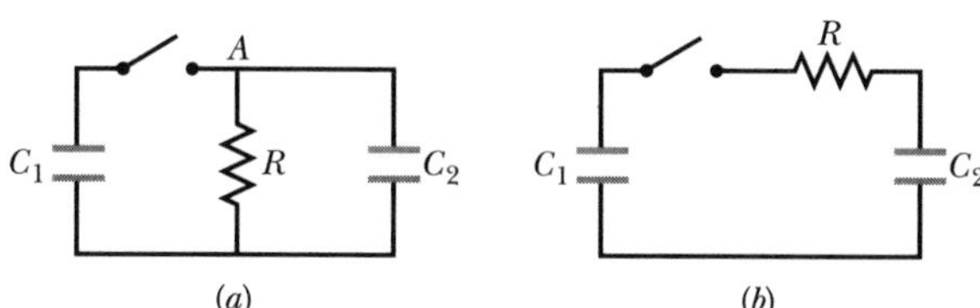

FIGURE 28-96 Problems 112 and 113.

At $t = 0$, in terms of the given symbols, find (a) the initial charge on capacitor C_1, (b) the initial charge on capacitor C_2, (c) the initial potential across C_1, C_2, and R, and (d) the initial current through R.

Let $q_1(t)$ represent the charge lost by C_1 by time t and $q_2(t)$ represent the charge gained by C_2 by that same time. (e) Using the given symbols and also q_1 and q_2, write an expression that relates the potential across C_1 to that across C_2. (f) Differentiate the expression to relate dq_1/dt to dq_2/dt. (g) Using the given symbols and q_1, q_2, and the current i through R, apply the junction rule to junction A. (h) Relate current i to charge $q_2(t)$.

(i) Now, combine the expressions in (f), (g), and (h) to eliminate i and dq_1/dt. What you then have is a differential equation involving $q_2(t)$ and fixed quantities; it resembles Eq. 28-35 and has a solution in the form of Eq. 28-36. (j) Write the solution, using the result of (b) as the initial charge.

(k) Next write an expression for the potential V_2 across C_2. Recall that the potential across C_1 and the potential across R must both equal V_2.

(l) Now find the total energy that is dissipated in the resistor until the system reaches equilibrium by integrating the power of the dissipation from $t = 0$ to $t = \infty$. (m) What is the initial total potential energy U_i of the capacitors? (n) How much energy do the capacitors have when equilibrium is reached?

113. In the circuit of Fig. 28-96*b*, with the switch open, capacitor C_1 is charged to a potential difference V_0 while capacitor C_2 is left uncharged. Then at time $t = 0$, the switch is closed and charge begins to flow through resistor R. Let's call C_1, C_2, V_0, and R the given symbols.

For any time t, let $q(t)$ represent the charge on C_2 and $i(t)$ represent the current in resistor R. (a) Using a loop equation, find an expression for $i(t)$ in terms of q and the given symbols. For a time long after $t = 0$, find the charges on (b) C_1 and (c) C_2 in terms of the given symbols.

(d) Next replace i in the equation of (a) with dq/dt. What you then have is a differential equation involving $q(t)$ and fixed quantities; it resembles Eq. 28-38 and has a solution in the form of Eq. 28-39. (e) Using the steps following Eq. 28-39, solve the differential equation here.

Next find (f) the potential difference $V_2(t)$ across C_2 and (g) the potential difference $V_1(t)$ across C_1 in terms of the given symbols. (h) Find $i(t)$ in terms of only the given symbols (don't include symbol q this time).

(i) Now find the total energy that is dissipated in the resistor until the system reaches equilibrium by integrating the power of the dissipation from $t = 0$ to $t = \infty$. What are (j) the initial potential energy U_i of the system as the switch is closed, (k) the final potential energy U_f of the system as equilibrium is reached, and (l) the change $U_f - U_i$ in the potential energy of the system?

Graphing Calculators

SAMPLE PROBLEM 28-8

Simultaneous Linear Equations. In Sample Problem 28-4 we needed to solve two simultaneous linear equations, Eqs. 28-24 and 28-26:

$$i_1(4.0\ \Omega) - i_2(4.0\ \Omega) = 3.0\ \text{V}$$

and

$$i_1(4.0\ \Omega) + i_2(8.0\ \Omega) = 0.$$

Although these equations are easily solved by "hand," now we will solve them with a graphing calculator.

SOLUTION: On a TI-85/86, press 2nd SIMULT. For the Number, press 2 (because we have two simultaneous equations) and then ENTER. A template for the first equation appears at the top of the screen.

a1,1x1 + a1,2x2 = b1

Comparing this template with our first equation, we see that coefficient a1,1 is 4.0, coefficient a1,2 is −4.0, and term b1 is 3.0. Press in this data, each datum followed by ENTER. The calculator then shows the template for the second equation; similarly press in the corresponding data. Next choose the SOLVE option in the menu. The calculator then shows the solution (depending on the mode settings for the style of displayed results) as

x1 = .50

x2 = −.25

which we identify as

$$i_1 = 0.50\ \text{A}$$

and

$$i_2 = -0.25\ \text{A}. \qquad \text{(Answer)}$$

You can then use F1 to alter the "a" coefficients and "b" terms (to repeat the problem) or use F2 to store the "a" coefficients, F3 to store the "b" terms, and F4 to store the "x" results. Each storing process requires that you press in a name. Later, to view the stored data from the home screen, use 2nd RCL and then the corresponding name.

SAMPLE PROBLEM 28-9

Differential Equations for *RC* Circuits. In Fig. 28-97*a*, $\mathscr{E} = 10.0$ V, $R_1 = 1000\ \Omega$, $R_2 = 2000\ \Omega$, and $C = 2.00\ \mu$F. The circuit is initially in equilibrium (steady state) when switch S is closed at time $t = 0$ (Fig. 28-97*b*).

(*a*) Find the charge $q(t)$ on the capacitor for $t \geq 0$.

SOLUTION: We can find $q(t)$ by going through three steps: (1) Find the initial charge q_i on the capacitor when the switch

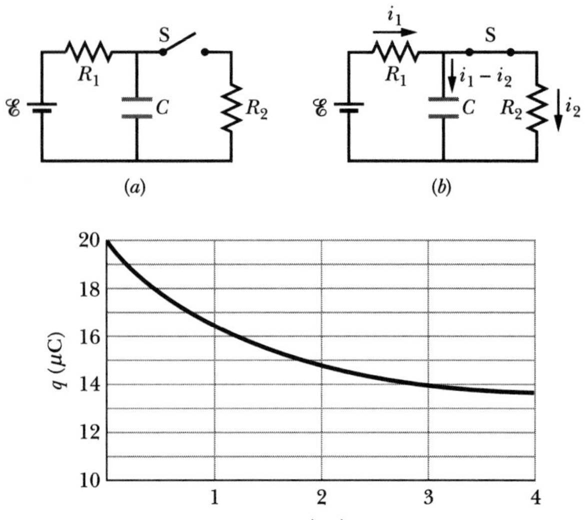

FIGURE 28-97 Sample Problem 28-9 and Problem 114.

is closed. (2) Apply the loop and junction rules to the circuit to generate an equation where the only variable is q; this equation is a differential equation. (3) Solve the differential equation, either by hand or with a short program on a calculator. In spite of what you might think, step 3 is actually easier than steps 1 and 2.

STEP 1: find q_i. The initial charge q_i that is on the capacitor in Fig. 28-97*b* is equal to the charge that is on the capacitor in Fig. 28-97*a* as the switch is closed. Because the circuit in Fig. 28-97*a* is in equilibrium, the charge on the capacitor is not changing in that circuit and the current through R_1 must be zero. Writing a loop equation for that circuit (moving clockwise), we then find

$$\mathscr{E} - i_1 R_1 - \frac{q_i}{C} = 0 \qquad (28\text{-}44)$$

and
$$10\text{ V} - (0)(1000\ \Omega) - \frac{q_i}{2.00 \times 10^{-6}\text{ F}} = 0,$$

which gives us

$$q_i = 20.0 \times 10^{-6}\text{ C} = 20.0\ \mu\text{C}. \qquad (28\text{-}45)$$

STEP 2: generate a differential equation in terms of q. We can solve the circuit of Fig. 28-97*b* to get an equation in terms of q in several ways. Here is one way. Writing a loop equation for the left-hand loop as we did in step 1 (but with variable q instead of value q_i), we find

$$i_1 = \frac{\mathscr{E} - q/C}{R_1}. \qquad (28\text{-}46)$$

Similarly, writing a loop equation for the right-hand loop gives us

$$-i_2 R_2 + \frac{q_i}{C} = 0$$

and
$$i_2 = \frac{q/C}{R_2}. \qquad (28\text{-}47)$$

From Fig. 28-97*b*, we see that the current that changes the charge on the capacitor can be written as $i_1 - i_2$, that is,

$$\frac{dq}{dt} = i_1 - i_2. \qquad (28\text{-}48)$$

(Note that we do not yet know if the capacitor is being charged or discharged, that is, whether dq/dt is positive or negative. In fact, we do not yet need to know, but we must be consistent with the directions of the currents in our equations.)

Substituting into Eq. 28-48 for i_1 and i_2 from Eqs. 28-46 and 28-47, we have

$$\frac{dq}{dt} = \frac{\mathscr{E} - q/C}{R_1} - \frac{q/C}{R_2}, \qquad (28\text{-}49)$$

which is a differential equation written in terms of known quantities and a single variable q, as we wanted.

STEP 3: solve the differential equation. To solve Eq. 28-49, we first rearrange it to be in the general form for *RC* circuits:

$$\frac{dq}{dt} + Aq = B \quad \text{(general } RC \text{ differential equation).} \qquad (28\text{-}50)$$

Doing so, Eq. 28-49 becomes

$$\frac{dq}{dt} + \frac{1}{C}\left(\frac{1}{R_1} + \frac{1}{R_2}\right)q = \frac{\mathscr{E}}{R_1}. \qquad (28\text{-}51)$$

Comparing Eqs. 28-50 and 28-51, we see that for the circuit of Fig. 28-97*b*,

$$A = \frac{1}{C}\left(\frac{1}{R_1} + \frac{1}{R_2}\right) \qquad (28\text{-}52)$$

and
$$B = \frac{\mathscr{E}}{R_1}. \qquad (28\text{-}53)$$

Substituting known data then gives us

$$A = \frac{1}{2.00 \times 10^{-6}\text{ F}}\left(\frac{1}{1000\ \Omega} + \frac{1}{2000\ \Omega}\right) = 750\text{ s}^{-1} \qquad (28\text{-}54)$$

and
$$B = \frac{10\text{ V}}{1000\ \Omega} = 0.010\text{ C/s}. \qquad (28\text{-}55)$$

The general solution to Eq. 28-50 is

$$q = q_p + Ke^{-At} = q_p + Ke^{-t/\tau} \quad \text{(general solution).} \qquad (28\text{-}56)$$

Here q_p is called a *particular solution,* K is a constant, and τ is the capacitive time constant for the change of charge on the capacitor. We see immediately that

$$\tau = \frac{1}{A}. \qquad (28\text{-}57)$$

Substituting for A gives us

$$\tau = \frac{1}{750\text{ s}^{-1}} = 1.333 \times 10^{-3}\text{ s} \approx 1.33\text{ ms} \qquad (28\text{-}58)$$

To find q_p and K, we evaluate Eqs. 28-50 and 28-56 for the initial or final conditions of the capacitor. First we get q_p from a final condition: When the circuit again reaches equilibrium, the capacitor's charge will no longer change. That is, at $t = \infty$ (which means "a long time later"), the charging $dq/dt = 0$ and the capacitor has its final charge $q = q_f$. Making these substitutions in Eq. 28-50, we have

$$0 + Aq_f = B,$$

or
$$q_f = \frac{B}{A}. \qquad (28\text{-}59)$$

But from Eq. 28-56 we also know that at $t = \infty$,

$$q_f = q_p + Ke^{-\infty/\tau} = q_p + 0 = q_p. \qquad (28\text{-}60)$$

Comparing Eqs. 28-59 and 28-60, we see that the particular solution q_p is the final charge q_f and that

$$q_p = \frac{B}{A}. \qquad (28\text{-}61)$$

This is a general result, true for any capacitor charging or discharging in an RC circuit.

Next we get K from an initial condition: At time $t = 0$, the charge is $q = q_i$. Making these substitutions in Eq. 28-56, we obtain

$$q_i = q_p + Ke^{-0/\tau} = q_p + K. \qquad (28\text{-}62)$$

Thus,

$$K = q_i - q_p. \qquad (28\text{-}63)$$

This, too, is a general result, true for any capacitor charging or discharging in an RC circuit.

Substituting the values for A and B that we found earlier, Eq. 28-61 gives us the final charge as

$$q_p = \frac{B}{A} = \frac{0.0100\ \text{C/s}}{750\ \text{s}^{-1}}$$
$$= 1.333 \times 10^{-6}\ \text{C} \approx 13.3\ \mu\text{C}. \qquad (28\text{-}64)$$

Next, substituting this and the initial charge $q_i = 20.0\ \mu$C, found in step 1, Eq. 28-63 yields

$$K = 20.0\ \mu\text{C} - 13.3\ \mu\text{C} = 6.67\ \mu\text{C}. \qquad (28\text{-}65)$$

Putting these results for q_p, K, and τ into Eq. 28-56, we find that when the switch is closed at $t = 0$, the charge on the capacitor varies as

$$q = 13.3\ \mu\text{C} + (6.67\ \mu\text{C})e^{-t/(1.33\ \text{ms})}. \qquad \text{(Answer)}$$

(*b*) When the switch is closed at $t = 0$, does the capacitor charge or discharge as the circuit moves toward equilibrium? Graph q versus t.

SOLUTION: Comparing the final charge $q_p = 13.3\ \mu$C with the initial charge $q_i = 20.0\ \mu$C, we see that the capacitor partially discharges to reach its new equilibrium state. Figure 28-97*c* is a graph for the first few time constants of the discharge.

(*c*) Write a program to solve the differential equation for q in an RC circuit where a capacitor is being charged or discharged.

SOLUTION: Table 28-2 gives such a program for a TI-85/86 (if you don't know how to program such a calculator, read Sample Problem 2-14 in this Supplement). The program displays the general differential equation (Eq. 28-50) as a reminder. It then requests values of A, B, and q_i for the circuit you are solving. Next it displays the general solution to the differential equation (Eq. 28-56) as a reminder and the values of the final charge (qf in the program is the particular solution q_p) in coulombs, the constant K in coulombs, and the time constant τ in seconds.

You can run the program from the home screen by pressing in its name RC and then pressing ENTER. Be careful about the powers of ten in the data: 20.0 μC should be entered as 20 E-6. The program has a pause so that the general solution and the values of qf, K, and τ are simultaneously on the screen. When you have copied the results, press ENTER to continue and thus also stop the program. Press ENTER again to rerun the program.

Using the program might save you a little time in solving problems about RC circuits. However, the challenge lies in steps 1 and 2, which require practice if you are to carry them out smoothly on a test.

TABLE 28-2 SAMPLE PROBLEM 28-9

```
PROGRAM:RC
:Disp "dq/dt+Aq=B"
:Prompt A
:Prompt B
:Prompt qi
:B/A→qf
:Disp "q=qf+Ke^(-t/τ)"
:Disp "qf",qf
:qi-qf→K
:Disp "K",K
:1/A→τ
:Disp "τ",τ
:Pause
:Stop
```

114. Switch S in Fig. 28-97*b* has been closed a long time (the circuit has reached equilibrium) when it is opened at $t = 0$. Find the charge on the capacitor for $t \geq 0$, and graph the results.

115. In Fig. 28-98, $\mathscr{E} = 10.0$ V, $R_1 = 1000\ \Omega$, $R_2 = 2000\ \Omega$, and $C = 2.00\ \mu$F. The circuit is initially in equilibrium (steady state) when switch S is closed at time $t = 0$. Find the charge on the capacitor for $t \geq 0$.

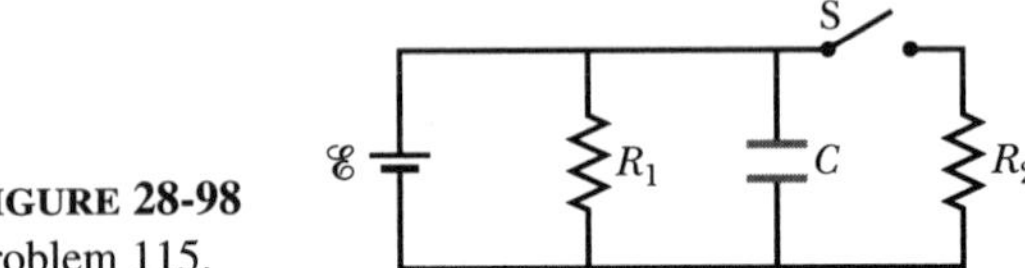

FIGURE 28-98 Problem 115.

116. In Fig. 28-99, $\mathscr{E} = 10.0$ V, $R_1 = 1000\ \Omega$, $R_2 = R_3 = 4000\ \Omega$, and $C = 2.00\ \mu$F. The circuit is initially in equilibrium (steady state) when switch S is closed at time $t = 0$. For $t \geq 0$, find (a) the charge on the capacitor, (b) the rate at which that charge changes, and (c) the electric potential across R_1. (d) Graph each of those three quantities. (e) Is the capacitor charging or discharging? When equilibrium is reached, the switch is opened at $t = 0$ (reset the reference clock then). (f) Again find the charge on the capacitor for $t \geq 0$.

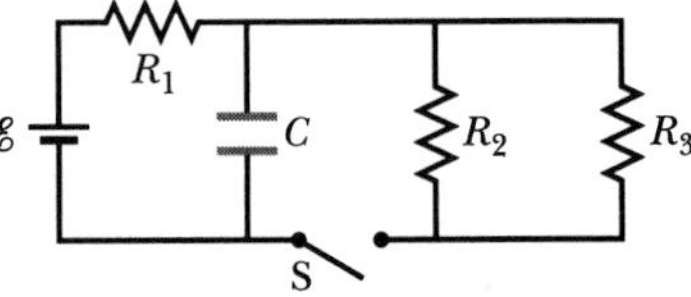

FIGURE 28-99 Problem 116.

117. In Fig. 28-100, $\mathscr{E} = 10.0$ V, $R_1 = R_2 = R_3 = 1000\ \Omega$, and $C = 2.00\ \mu$F. The circuit is initially in equilibrium (steady state) when switch S is closed at time $t = 0$. For $t \geq 0$, find (a) the charge on the capacitor, (b) the rate at which that charge changes, (c) the current through R_2 and R_3, and (d) the current through R_1. (e) Graph those four quantities. When equilibrium is reached, the switch is opened at $t = 0$ (reset the reference clock then). (f) Again find the charge on the capacitor for $t \geq 0$.

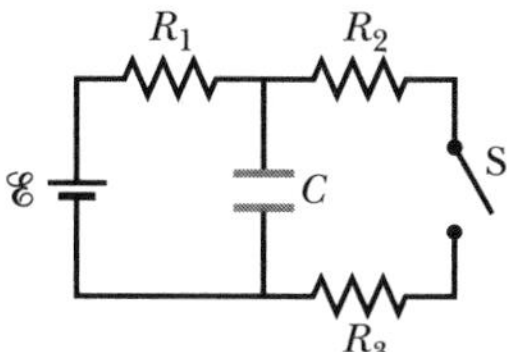

FIGURE 28-100
Problem 117.

118. In Fig. 28-101, $\mathscr{E} = 20.0$ V, $R_1 = 6000\ \Omega$, $R_2 = R_3 = 4000\ \Omega$, and $C = 8.00\ \mu$F. The circuit is initially in equilibrium (steady state) when switch S is closed at time $t = 0$. For $t \geq 0$, find (a) the charge on the capacitor and (b) the rate at which energy is dissipated in R_1. (c) Graph both quantities. When equilibrium is reached, the switch is opened at $t = 0$ (reset the reference clock then). (d) Again find the charge on the capacitor for $t \geq 0$.

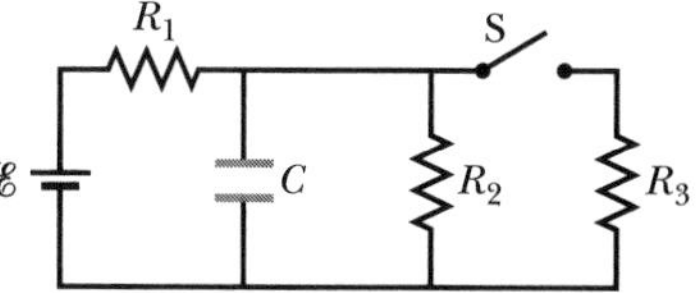

FIGURE 28-101
Problem 118.

Chapter Twenty-Nine
Magnetic Fields

QUESTIONS

18. Figure 29-51 gives snapshots for three situations in which a positively charged particle passes through a uniform magnetic field **B**. The velocities **v** of the particle differ in orientation in the three snapshots but not in magnitude. Rank the situations according to (a) the period, (b) the frequency, and (c) the pitch of the particle's motion, greatest first.

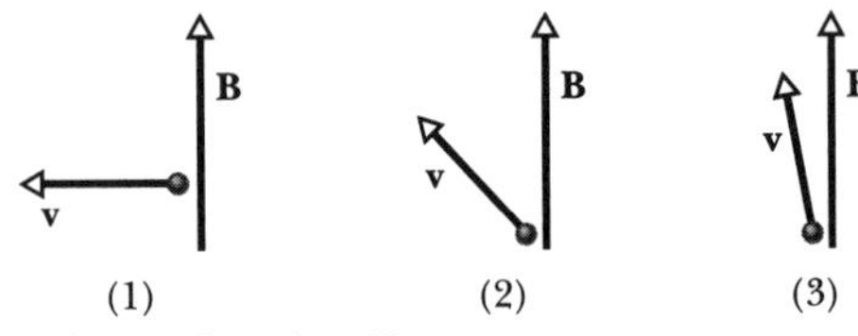

FIGURE 29-51 Question 18.

19. Figure 29-52 gives snapshots for three situations in which a positively charged particle moves through uniform magnetic fields **B** that differ in magnitude but not in direction. The particle has the same velocity **v** in each snapshot. Rank the situations according to (a) the period, (b) the frequency, and (c) the pitch of the particle's motion, greatest first.

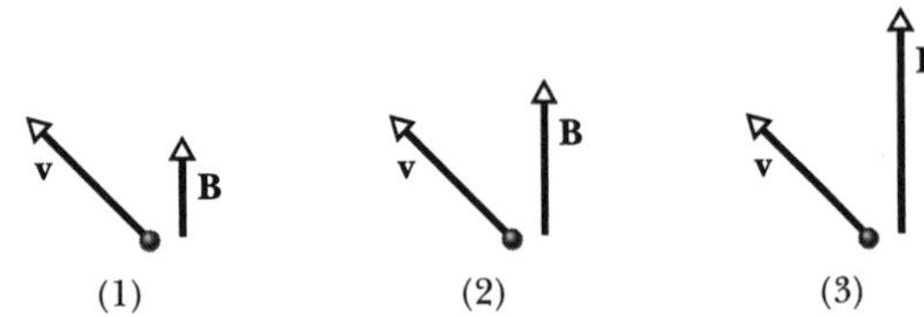

FIGURE 29-52 Question 19.

20. For three situations in Fig. 29-53, wires with identical currents lie along an x axis in nonuniform magnetic fields. In each situation, the field is drawn to the same scale for five points on the axis: $x = 0, 1, 2, 3, 4$ m. (a) Which of the following functions $\mathbf{B}(x)$ best corresponds to the fields? (b) For each situation, rank the five points according to the magnitude of the differential magnetic force $d\mathbf{F}_B$ on a differential length ds centered at them, greatest first.

$$\mathbf{B}_1 = (-0.5x\mathbf{i} + 0.5x\mathbf{j})\ \text{mT}$$
$$\mathbf{B}_2 = (0.5x\mathbf{i} + 2\mathbf{j})\ \text{mT}$$
$$\mathbf{B}_3 = [(2 - x)\mathbf{i} + 2\mathbf{j}]\ \text{mT}$$
$$\mathbf{B}_4 = [(2 - x)\mathbf{i} + (3 - x)\mathbf{j}]\ \text{mT}$$

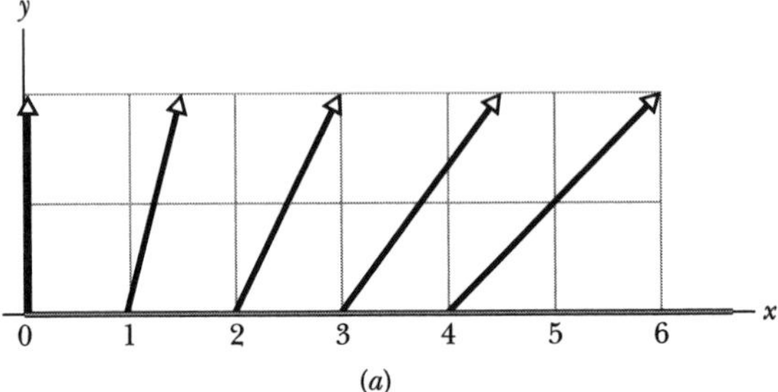

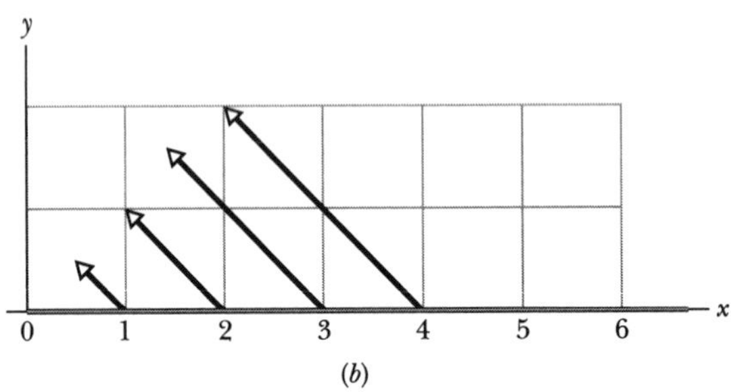

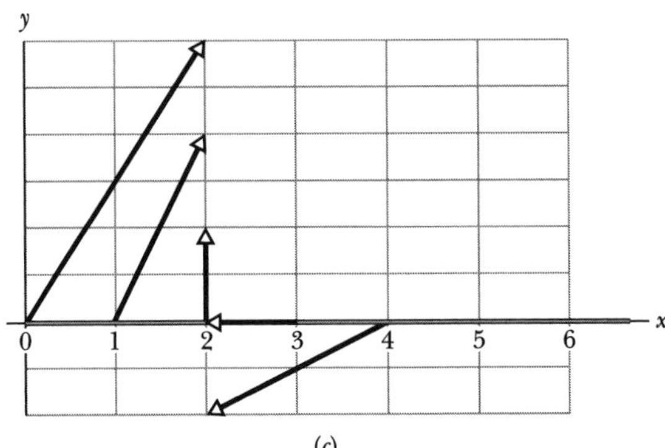

FIGURE 29-53 Question 20.

21. *Organizing question:* Figure 29-54 shows a wire of length 2 m carrying a 4 A current along a y axis through a magnetic field. The magnetic field is given in the table for six situations (angles are measured counterclockwise from the positive direction of the x axis). For each situation, set up an equation, complete with known data, to find the magnitude of the net magnetic force on the 2.0 m length of wire. (Here, x and y are in meters.)

SITUATION	MAGNETIC FIELD
a	$B = 3\ \mu\text{T}$ at 60°
b	$B = 3y\ \mu\text{T}$ at 30°
c	$\mathbf{B} = (2\mathbf{i} + 3y\mathbf{j})\ \mu\text{T}$
d	$\mathbf{B} = (2\mathbf{i} + 3x\mathbf{j})\ \mu\text{T}$
e	$\mathbf{B} = (2y\mathbf{i} + 3\mathbf{j})\ \mu\text{T}$
f	$\mathbf{B} = (2x^2\mathbf{i} + 3y^2\mathbf{j})\ \mu\text{T}$

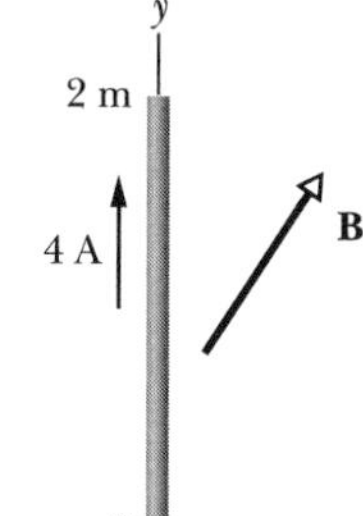

FIGURE 29-54 Question 21.

22. Figure 29-55 shows six wires that carry identical currents i rightward through the same uniform magnetic field (directed into the page) in six separate experiments. Each wire extends from $x = 0$ to $x = d$. Rank the wires according to the magnitude of the net magnetic force on them, greatest first.

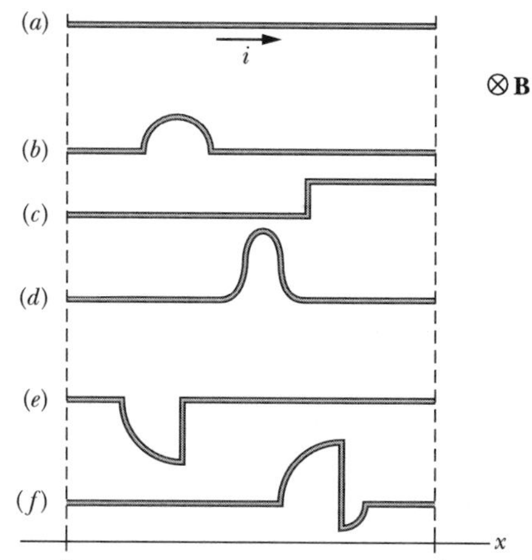

FIGURE 29-55 Question 22.

23. Figure 29-56 shows eight wires that carry identical currents through the same uniform magnetic field (directed into the page) in eight separate experiments. Each wire consists of two straight sections (each of length L and either parallel or perpendicular to the x and y axes shown) and one curved section (with radius of curvature R). The directions of the currents are indicated by the arrows. (a) Give the direction of the net magnetic force on each wire in terms of an angle measured counterclockwise from the positive direction of the x axis. (b) Rank wires 1 through 4 according to the magnitude of the net magnetic force on them, greatest first. (c) Do the same for wires 5 through 8.

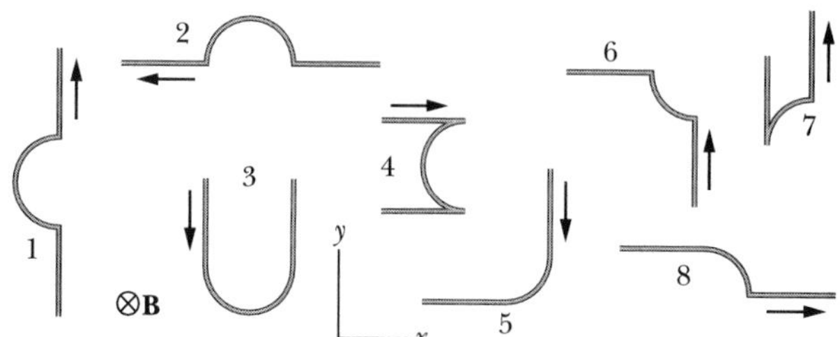

FIGURE 29-56 Question 23.

24. The potential energies associated with four orientations of a magnetic dipole in a magnetic field are (1) $-6U_0$, (2) $8U_0$, (3) $6U_0$, and (4) $-7U_0$, where U_0 is positive. Rank the orientations according to (a) the angle between the directions of the magnetic dipole moment $\boldsymbol{\mu}$ and the magnetic field $\mathbf{B}$ and (b) the magnitude of the torque on the magnetic dipole, greatest first.

25. In Fig. 29-57, three circular (single-turn) loops are in the uniform magnetic field $\mathbf{B}$ of a large magnet. The loops are identical in size and carry identical currents in the directions indicated. Magnetic field $\mathbf{B}$ is directed perpendicular to the planes of loops 1 and 2; it is directed along the plane of loop 3, which is seen on edge in the figure. Rank the loops according to (a) the potential energy associated with the orientation of their magnetic dipole moments and (b) the torque on the loop, greatest first.

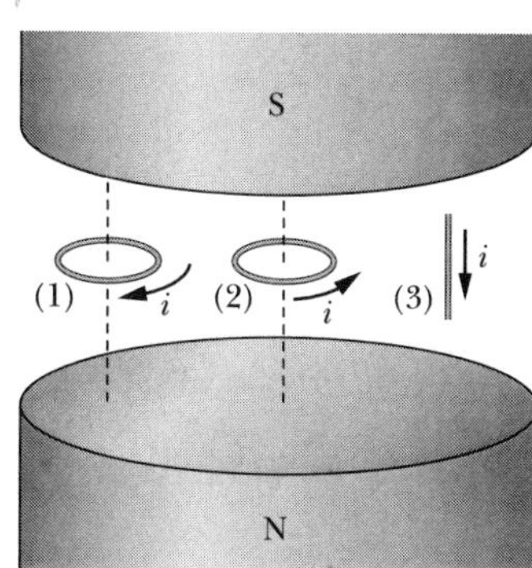

FIGURE 29-57 Question 25.

26. In Fig. 29-58 there are three pairs of concentric circular loops of radii R or $2R$, carrying currents of i or $4i$ as indicated. Rank the pairs according to the magnitude of the net magnetic moment of each pair, greatest first.

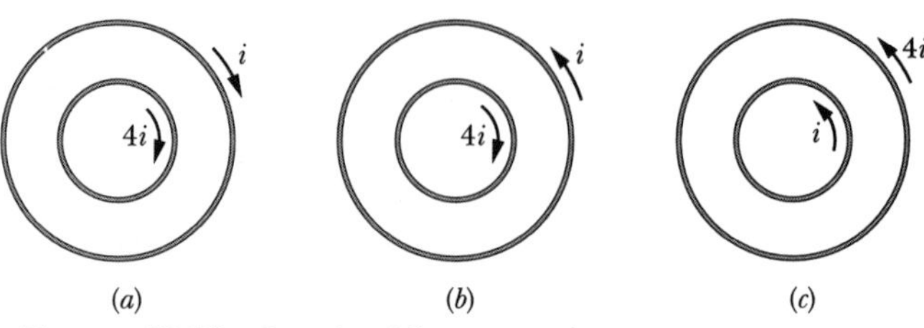

FIGURE 29-58 Question 26.

EXERCISES & PROBLEMS

72. A particle of mass 6.0 g moves at 4.0 km/s along an xy plane in a region with a uniform magnetic field given by 5.0**i** mT. At one instant, when the particle's velocity is directed 37° counterclockwise from the positive direction of the x axis, the magnetic force on the particle is 0.48**k** N. What is the particle's charge?

73. A particle with charge 2.0 C moves through a uniform magnetic field. At one instant the particle's velocity is (2.0**i** + 4.0**j** + 6.0**k**) m/s and the magnetic force on the particle is (4.0**i** − 20**j** + 12**k**) N. The x and y components of the magnetic field are equal. What is the magnetic field?

74. A particle of mass 10 g and charge 80 μC moves through a uniform magnetic field, in a region where the free-fall acceleration is −9.8**j** m/s². The velocity of the particle is a constant 20**i** km/s. What, then, is the magnetic field?

75. A 5.0 μC particle moves through a region containing a magnetic field of −20**i** mT and an electric field of 300**j** V/m. At one instant the velocity of the particle is (17**i** − 11**j** + 7**k**) km/s. At that instant, what is the net electromagnetic force (the sum of the electric and magnetic forces) on the particle?

76. An electron moves through a region of uniform magnetic field of magnitude 60 μT, directed along the positive direction of an x axis. As the electron enters the field, it has a velocity of (32**i** + 40**j**) km/s. What are (a) the radius of the helical path taken by the electron and (b) the pitch of that path? (c) To an observer looking into the magnetic field region from the entrance point of the electron, does the electron spiral clockwise or counterclockwise as it moves deeper into the region?

77. The bent wire shown in Fig. 29-59 lies in a uniform magnetic field. The two straight sections of the wire each have length 2.0 m, and the wire carries a current of 2.0 A. What is the net magnetic force on the wire in unit-vector notation if the magnetic field is given by (a) 4.0**k** T and (b) 4.0**i** T?

78. In Fig. 29-60, a particle moves along a circle in a region of uniform magnetic field of magnitude **B** = 4.00 mT. The particle is either a proton or an electron (you must decide which). It experiences a magnetic force of magnitude 3.20×10^{-15} N. What are (a) the particle's speed, (b) the radius of the circle, and (c) the period of the motion?

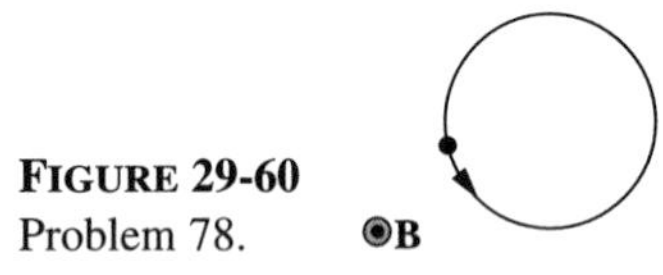

FIGURE 29-60 Problem 78.

79. At one instant a proton has velocity **v** = (−2.00**i** + 4.00**j** − 6.00**k**) m/s in a uniform magnetic field **B** = (2.00**i** − 4.00**j** + 8.00**k**) mT. At that instant, what are (a) the magnetic force **F** on the proton, in unit-vector notation, (b) the angle between **v** and **F**, and (c) the angle between **v** and **B**?

80. In Fig. 29-61, a conducting rectangular solid moves at constant velocity **v** = (20.0 m/s)**i** through a uniform magnetic field **B** = (30.0 mT)**j**. What are (a) the induced electric field, in unit-vector notation, and (b) the induced potential difference across the solid?

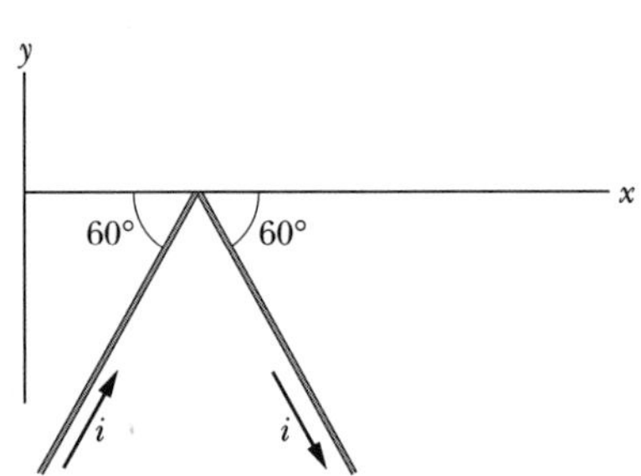

FIGURE 29-59 Problem 77.

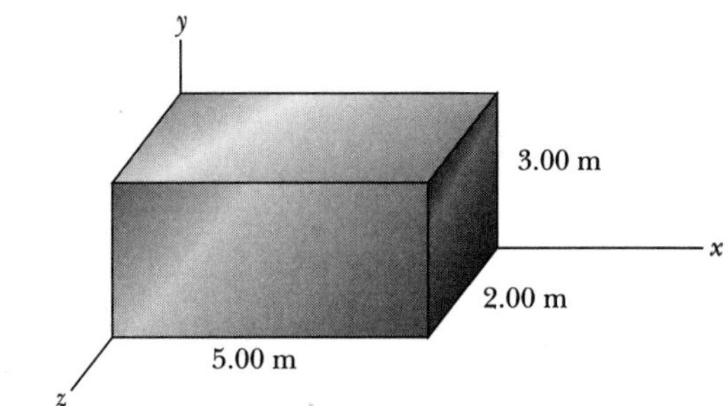

FIGURE 29-61 Problem 80.

81. In Fig. 29-62, an electron moves at speed $v = 100$ m/s along a straight line toward the right through uniform electric and magnetic fields. The magnetic field **B** is directed into the page and has magnitude 5.00 T. What are the magnitude and direction of the electric field?

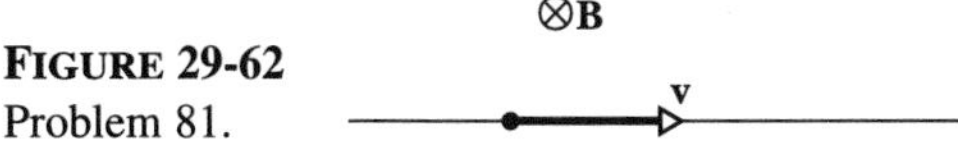

FIGURE 29-62 Problem 81.

82. In Fig. 29-63, a charged particle moves into a region of uniform magnetic field **B**, goes through half a circle, and then exits that region. The particle is either a proton or an electron (you must decide which). It spends 130 ns within the region. (a) What is the magnitude of **B**? (b) If the particle is sent back through the magnetic field (along the same initial path) but with 2.00 times its previous kinetic energy, how much time does it spend within the field?

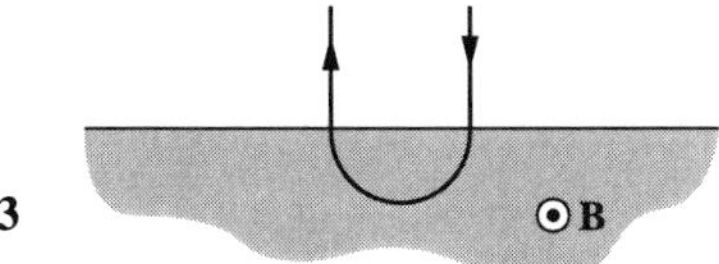

FIGURE 29-63 Problem 82.

83. A wire lying along an x axis from $x = 0$ to $x = 1.00$ m carries a current of 3.00 A. It is immersed in a magnetic field given by

$$\mathbf{B} = (4.00\ \text{T/m}^2)x^2\mathbf{i} - (0.600\ \text{T/m}^2)x^2\mathbf{j}.$$

In unit-vector notation, what is the magnetic force on the wire?

84. The coil in Fig. 29-64 carries a current of 2.00 A in the direction indicated, is parallel to an xz plane, has 3.00 turns and an area of 4.00×10^{-3} m², and lies within a uniform magnetic field $\mathbf{B} = (2.00\mathbf{i} - 3.00\mathbf{j} - 4.00\mathbf{k})$ mT. What are (a) the magnetic potential energy of the coil–magnetic field system and (b) the magnetic torque (in unit-vector notation) on the coil?

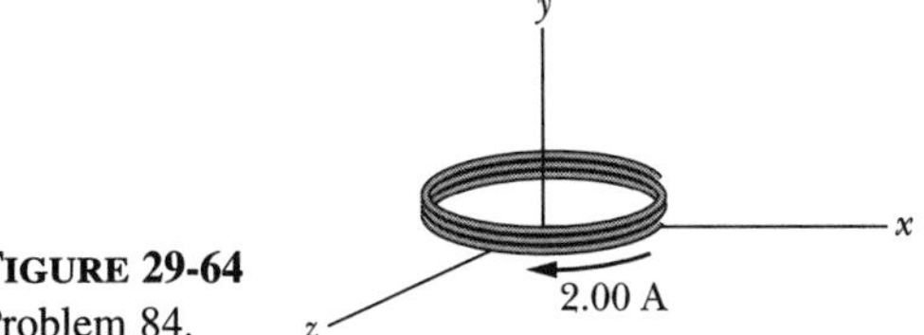

FIGURE 29-64 Problem 84.

Tutorial Problems

85. In the first parts of this problem, we try to determine the electric and magnetic fields at a point by observing the net force on a charged particle at that point. The particle has a mass of 1.80×10^{-25} kg and an electric charge of $+4e$. We use a Cartesian coordinate system chosen with the y axis directed upward and the origin on the ground. (a) If the particle is placed at rest at point A for which $\mathbf{r} = (2.45\ \text{m})\mathbf{j}$, what is the gravitational force on the particle? Express the force in unit-vector notation.

(b) Suppose that in addition to the gravitational force there is a force $+(7.31 \times 10^{-14}\ \text{N})\mathbf{i}$ on the particle when it is at rest. Compare the magnitude of this force with that of the gravitational force in part (a). What fundamental interaction is responsible for this force? Name the field providing the force and determine its value in unit-vector notation.

(c) Now suppose that the same particle is sent through point A with velocity $\mathbf{v} = (4.12 \times 10^6\ \text{m/s})\mathbf{i}$ and is found to experience an additional force $-(5.43 \times 10^{-14}\ \text{N})\mathbf{k}$. What can you say about the magnetic field at point A? Explain your reasoning in complete sentences. (d) Devise an experiment that might enable you, with the information already gathered, to determine the complete magnetic field vector. Also, explain the experiment using complete sentences.

(e) Now consider a new physical situation. If a proton moves through a uniform magnetic field **B** with a velocity of

$$\mathbf{v} = (3.0 \times 10^6\ \text{m/s})\mathbf{i} - (4.0 \times 10^6\ \text{m/s})\mathbf{j},$$

it experiences a magnetic force $\mathbf{F}_{mag} = (3.2 \times 10^{-15}\ \text{N})\mathbf{k}$. If, instead, it moves through the field with a velocity along **k**, the magnetic force on it is parallel to **j**. Determine **B**, explaining your solution logically in full sentences (not with only a string of formulas and a few words).

Answers

(a) The gravitational force is

$$\mathbf{F}_{grav} = -mg\mathbf{j} = -(1.80 \times 10^{-25}\ \text{kg})(9.80\ \text{m/s}^2)\mathbf{j}$$
$$= -(1.76 \times 10^{-24}\ \text{N})\mathbf{j}.$$

(b) This force has a magnitude greater than 10^{10} times that of the gravitational force. Presumably an electromagnetic interaction is responsible for this force. Since the particle is at rest, it must be the electric field that produces this force. From the relation $\mathbf{F} = q\mathbf{E}$ we can determine the electric field:

$$\mathbf{E} = \frac{\mathbf{F}}{q} = \frac{(7.31 \times 10^{-14}\ \text{N})\mathbf{j}}{(4)(1.602 \times 10^{-19}\ \text{C})}$$
$$= (1.14 \times 10^5\ \text{N/C})\mathbf{j}.$$

The electric field is directed upward from Earth's surface, in the same direction as the force on the particle because the particle has a positive electric charge.

(c) First, since the velocity is along **i** and the force is along **k**, there must clearly be a component of the magnetic field along **j**. From the magnetic force law,

$$F_z = q(v_x B_y - v_y B_x) = qv_x B_y,$$

because $v_y = 0$. So we see that

$$B_y = +\left(\frac{F_z}{qv_x}\right)$$
$$= \frac{-5.43 \times 10^{-14}\ \text{N}}{(4)(1.602 \times 10^{-19}\ \text{C})(4.12 \times 10^6\ \text{m/s})}$$
$$= -0.0206\text{T}.$$

We also know that there cannot be a component of **B** along **k**, for it would have shown up as a component of force along **j**. However, we can't learn anything about the component of the magnetic field along **i**, because that is also the direction of **v** and thus such a component wouldn't make any contribution to the magnetic force.

(d) We can determine the component of **B** along **i** if we send the particle through point A with a velocity along the **j** direction (or the **k** direction or any direction other than exactly along **i**). We then find a component of the magnetic force along **k** that depends on B_x, and that allows us to determine B_x.

(e) The magnetic forces in the two situations must both be perpendicular to the magnetic field, since $\mathbf{F}_{mag} = q\mathbf{v} \times \mathbf{B}$. In one case the force is parallel to **k** and in the other case the force is parallel to **j**, so **B** must be perpendicular to both. That means **B** is along $\pm\mathbf{i}$. Knowing this, we can make use of the numerical information. In doing so, we note that v_x has no effect on the magnetic force since it is the component of the velocity parallel to the magnetic field; so only v_y needs to be taken into account.

We know that $\mathbf{F}_{mag} = (3.2 \times 10^{-15}\ \text{N})\mathbf{k}$ in the first case. We also see that

$$\mathbf{F}_{mag} = e(\mathbf{v} \times B_x\mathbf{i})$$
$$= (1.60 \times 10^{-19}\ \text{C})(-4.0 \times 10^6\ \text{m/s})B_x(\mathbf{j} \times \mathbf{i})$$
$$= (6.4 \times 10^{-13}\ \text{C}\cdot\text{m/s})B_x\mathbf{k}.$$

So

$$B_x = \frac{3.2 \times 10^{-15}\ \text{N}}{6.4 \times 10^{-13}\ \text{C}\cdot\text{m/s}} = 5.0 \times 10^{-3}\ \text{T} = 5.0\ \text{mT}$$

and $$\mathbf{B} = (5.0\ \text{mT})\mathbf{i}.$$

86. Consider a current loop that is a plane square with sides of length L, lying in an xy plane with sides parallel to the axes and one corner of the loop at the origin. Suppose the loop carries current i in a counterclockwise direction as viewed by us on the positive side of the z axis. Let's number the sides of the square 1, 2, 3, 4 in counterclockwise order beginning with the bottom side. Suppose also that the loop is in a uniform magnetic field of magnitude B, directed in the positive z direction.

(a) Sketch this physical situation, labeling the relevant physical quantities (such as **B** and i). (b) Determine the force vector on each side of the current loop, writing it in unit-vector notation. Give the force on side 1 the symbol $\mathbf{F}_1$, and use similar notation for the other three sides. Add these forces to the sketch in part (a). (c) How is the net force—let's call it $\mathbf{F}_{net}$—on the current loop related to the quantities you calculated in part (b)? Determine the net force. Explain, using complete sentences, what this tells you about the motion of the center of mass of the current loop.

(d) Determine the magnetic dipole moment of the current loop, writing it in unit-vector notation. (e) What is the net torque on the current loop? Describe, in complete sentences and from our viewpoint, how the current loop will start to rotate. (For example, you might say that the left edge of the loop will tend to rise up off the xy plane.) Through what angle would the current loop have to rotate before the torque on it would be zero?

(f) What would be the torque on the current loop if the magnetic field had magnitude B and was directed in the positive y direction? Describe, again, in complete sentences, how the current loop would start to rotate. Through what angle would the current loop have to rotate before the torque on it would be zero? (g) Now suppose that the magnetic field has magnitude B and is directed in the positive x direction. Describe in words how the current loop would start to rotate. In what direction will the torque have to be? Calculate the torque to check your answer.

Answers

(b) On side 1, we have

$$\mathbf{F}_1 = i\mathbf{L} \times \mathbf{B} = iL\mathbf{i} \times B\mathbf{k} = -iLB\mathbf{j}.$$

On side 2, we have

$$\mathbf{F}_2 = i\mathbf{L} \times \mathbf{B} = iL\mathbf{j} \times B\mathbf{k} = +iLB\mathbf{i}.$$

On side 3, we have

$$\mathbf{F}_3 = i\mathbf{L} \times \mathbf{B} = (-iL\mathbf{i}) \times B\mathbf{k} = +iLB\mathbf{j}.$$

On side 4, we have

$$\mathbf{F}_4 = i\mathbf{L} \times \mathbf{B} = (-iL\mathbf{j}) \times B\mathbf{k} = -iLB\mathbf{i}.$$

(c) The net force on the loop is the sum of the four forces calculated in part (b):

$$\mathbf{F} = \mathbf{F}_1 + \mathbf{F}_2 + \mathbf{F}_3 + \mathbf{F}_4 = -iLB\mathbf{j} + iLB\mathbf{i} + iLB\mathbf{j} - iLB\mathbf{i} = 0.$$

This means there is no acceleration of the center of mass of the current loop; thus the center of mass is either at rest or moving with a constant velocity.

(d) The magnetic dipole moment is $\boldsymbol{\mu} = NiA\mathbf{n} = iL^2\mathbf{k}$, since the normal **n** to the plane of the current loop is in the positive z direction. This direction is found by the right-hand rule, in which we wrap the fingers of the right hand in the direction of the current so that the thumb is pointing along the direction of the magnetic dipole moment.

(e) The torque on the current loop is $\boldsymbol{\tau} = \boldsymbol{\mu} \times \mathbf{B} = (iL^2\mathbf{k}) \times (B\mathbf{k}) = 0$. Since the torque is zero, the current loop will not rotate but will remain in its present orientation, which is an equilibrium orientation.

(f) The torque on the current loop in this case would be

$$\boldsymbol{\tau} = \boldsymbol{\mu} \times \mathbf{B} = (iA\mathbf{k}) \times (B\mathbf{j}) = -iL^2B\mathbf{i}.$$

The magnetic moment of the loop always rotates toward the direction of the magnetic field. In this case the moment is up toward us, in the positive z direction, so the bottom edge of the loop will rise above the xy plane and the top edge will move down below that plane. The total rotation will be 90°.

How does this jibe with the fact that the torque is in the negative x direction? Point the thumb of your right hand in the direction of the torque, and your fingers will wrap around in the direction the loop rotates. *Voilà!*

(g) This time we first figure out how the loop will rotate and then figure out the direction of the torque. The loop will rotate so that its magnetic moment (up, toward us) moves toward the magnetic field (to our right). Sweep the fingers of your right hand from up toward you to your right and your thumb will point in the positive y direction—at least it will if your right hand has an oriention like most people's! So the torque will be in the positive y direction, which means it is along $\mathbf{j}$.

Let's check this out. The torque on the current loop is

$$\boldsymbol{\tau} = \boldsymbol{\mu} \times \mathbf{B} = (iA\mathbf{k}) \times (B\mathbf{i}) = +iL^2B\mathbf{j}.$$

Yes! It checks out.

Chapter Thirty
Magnetic Fields Due to Currents

QUESTIONS

19. Figure 30-71 shows a wire with current i, a differential length element ds of the wire (at the origin), and five numbered points. The points are in the xy plane, at equal distances from element ds. A straight line through the element connects points 2 and 4; another straight line through the element connects points 3 and 5. (a) Rank the points according to the magnitude of the magnetic field $d\mathbf{B}$ produced there by the current in element ds, greatest first. (b) What is the direction of that field at each point?

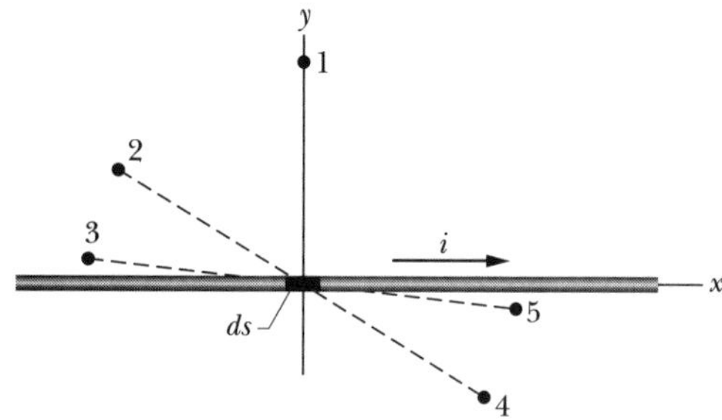

FIGURE 30-71 Question 19.

20. In Fig. 30-72, a wire in the shape of a semicircle of radius R carries current i. Three length elements with identical lengths ds are indicated; element 3 spans the central section of the semicircle. (a) Rank the elements according to the magnitudes of the magnetic field they produce at point P_1 (at the center of curvature), greatest first. (b) Repeat the ranking for the field magnitudes at point P_2, which is at distance R from the semicircle. (c) Is the magnitude of the net magnetic field at P_1 greater than, less than, or equal to that at P_2?

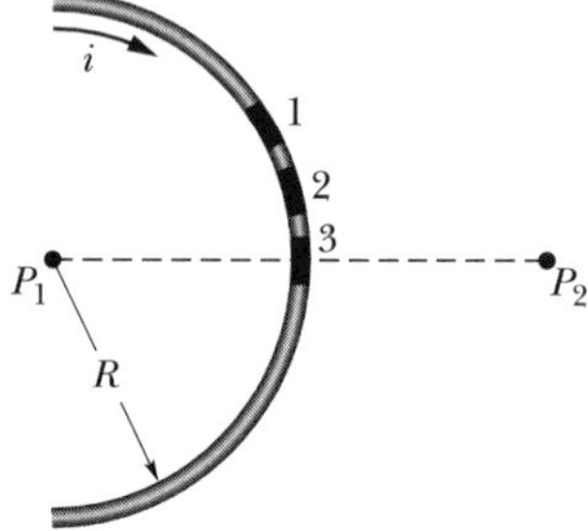

FIGURE 30-72 Question 20.

21. *Organizing question:* Figure 30-73 shows five situations in which a wire's current i produces a magnetic field $\mathbf{B}$ at a point P. For each situation, we have seen how the magnitude of $\mathbf{B}$ can be determined immediately or by means of a formula derived from the Biot–Savart law. Give that immediate result or the formula. The situations are (a) wire in circular arc, P at center of curvature; (b) straight wire of finite length, P along an extension of the wire; (c) straight wire of semi-infinite length, P along an extension of the wire; (d) straight wire of infinite length, P off the wire; and (e) straight wire of semi-infinite length, P aligned with the end but off the wire.

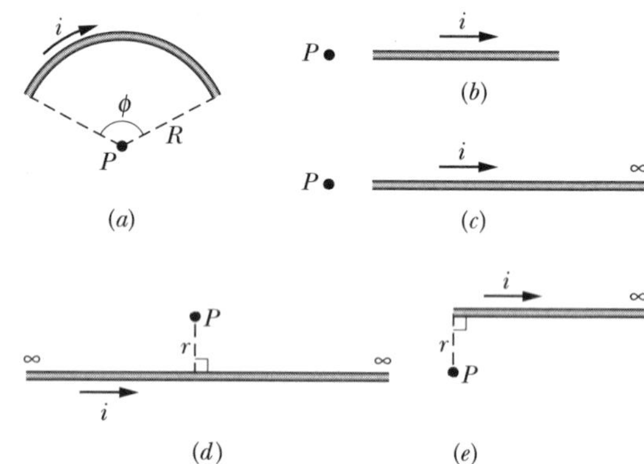

FIGURE 30-73 Question 21.

22. Figure 30-74 gives three situations in which identical currents are in wires consisting of three straight sections: a short section of length either d or $2d$ and two very long sections. The latter are

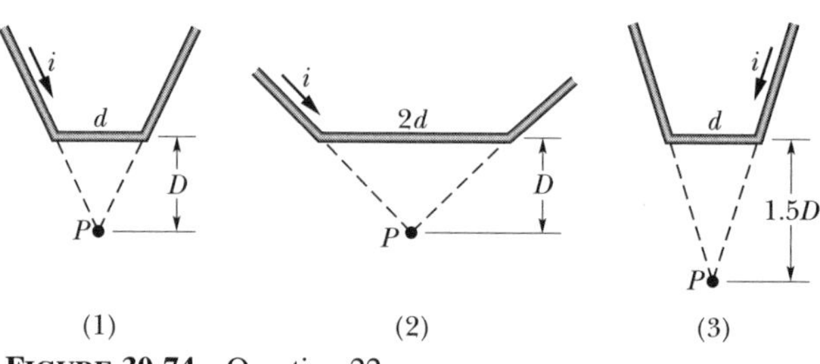

FIGURE 30-74 Question 22.

radial to a point P, which is at a perpendicular distance of either D or $1.5D$ from the short section. (a) Rank the situations according to the magnitude of the net magnetic field the current produces at point P, greatest first. (b) Give the direction of that field in each situation.

23. In Fig. 30-75*a*, an infinite wire carries a current past a point at perpendicular distance d from the wire. A mark splits the wire into two parts, a semi-infinite section extending to the left and a semi-infinite section extending to the right. The other three parts of the figure contain the same two semi-infinite sections and also a curved section connecting them. (a) Rank the four parts of the figure according to the magnitude of the net magnetic field produced by the two semi-infinite sections at the point at distance d, greatest first. (Do not include the magnetic field produced by the curved sections.) (b) In each part, what is the direction of that net magnetic field?

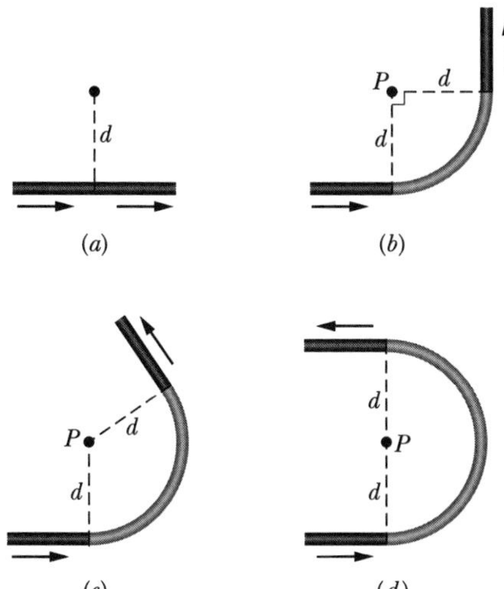

FIGURE 30-75 Question 23.

24. Three pairs of parallel currents in long straight wires are represented in Fig. 30-76. The values and directions of the currents and the distance between the wires (either d or $2d$) are indicated for each pair. (a) Rank the pairs according to the magnetic force per unit length the wires exert on each other, greatest first. (b) In which situation do the wires attract each other?

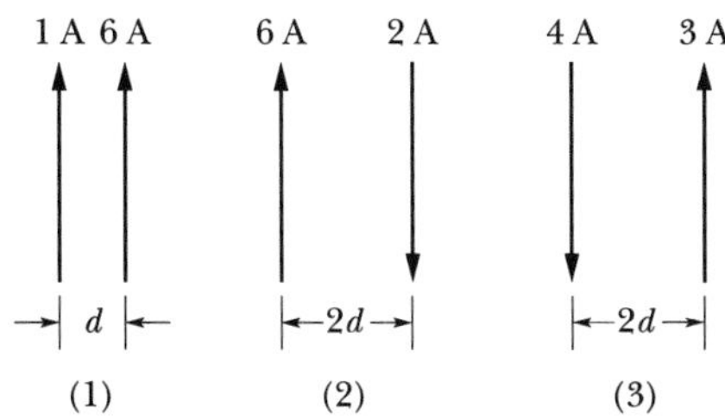

FIGURE 30-76 Question 24.

25. In Fig. 30-77, three long wires, with identical currents either directly into or directly out of the page, form three partial squares. Rank the squares according to the magnitude of the net magnetic field produced by the currents at the (empty) upper right corner of the square, greatest first.

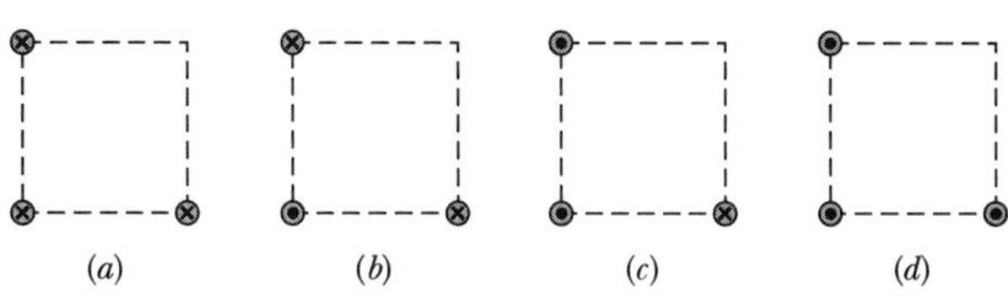

FIGURE 30-77 Question 25.

26. In each part of Fig. 30-78 two long straight wires (shown in cross section) carry identical currents either directly into or directly out of the page and are at equal distances from a y axis, and in each part, the net magnetic field at a point P on the y axis due to the pair of currents is indicated by the vector. (a) For each part, what are the directions of the currents in the left-hand wire and the right-hand wire, respectively? In part 1, if the current in the left-hand wire is increased somewhat, what happens to (b) the magnitude and (c) the direction of the net magnetic field at point P on the y axis?

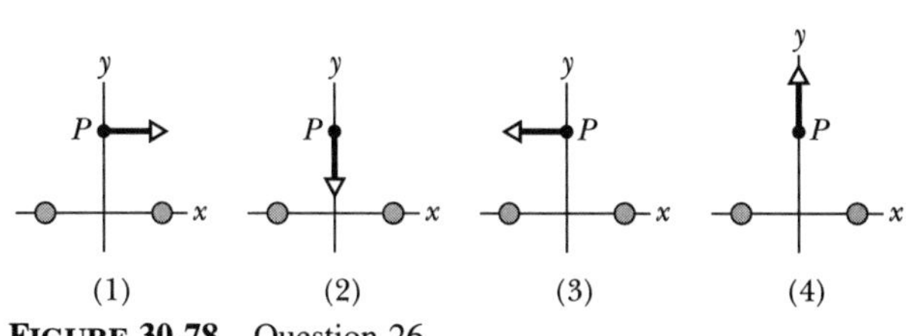

FIGURE 30-78 Question 26.

27. *Organizing question:* In Fig. 30-79, two long straight wires (shown in cross section) carry currents i_1 and i_2 either directly into or directly out of the page and are at equal distances from a y axis. The currents set up a net magnetic field $\mathbf{B}$ at a point P on the y axis; an example of a possible field is shown. Directions around P are divided into four regions: K, L, M, and N. The example of $\mathbf{B}$ is directed into angular region K. What directions (into the page or out of the page) are required of the currents if field $\mathbf{B}$ is to be in angular region (a) K, (b) L, (c) M, and (d) N?

Suppose that the currents can range between 0 and 5 A. (e) For each of the four borders between angular regions, give the currents in the wires as being 0 or 5 A. (f) For each angular region, tell if we should increase or decrease a current (consider i, and then i_2) to cause a $\mathbf{B}$ vector in that region to rotate counterclockwise by a given angle.

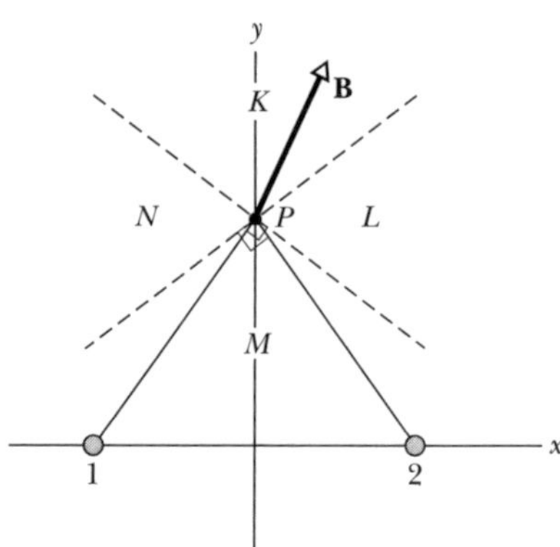

FIGURE 30-79 Question 27.

28. Figure 30-80 represents a snapshot of the velocity vectors of four electrons near a wire with current i. The four velocities have the same magnitude; velocity $\mathbf{v}_2$ is directed into the page. Particles 1 and 2 are at the same distance from the wire, as are particles 3 and 4. Rank the particles according to the magnitude of the magnetic force on them due to current i, greatest first.

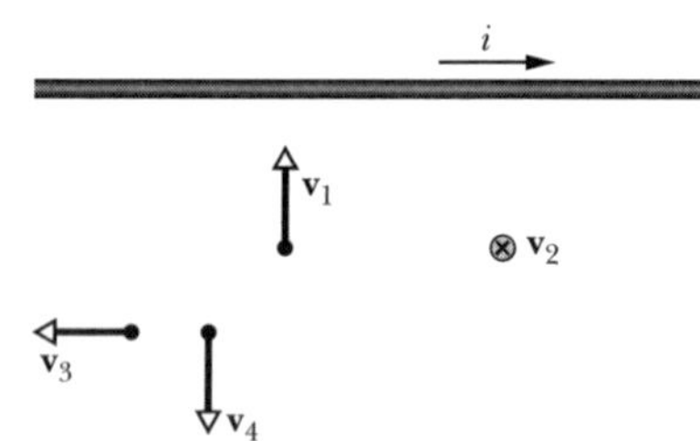

FIGURE 30-80 Question 28.

29. *Organizing question:* Figure 30-81 shows a circular region of radius R in which a current i is directed out of the page. It also shows two Amperian loops (of radii r_1 and r_2) that form circles concentric with the circular region. Set up the right side of Ampere's law (Eq. 30-16) for loops 1 and 2 for the following situations, where either the total current i, the current density J, or the current i_{enc} encircled by a loop at radius r is given. (Worry about worrisome subscripts.)

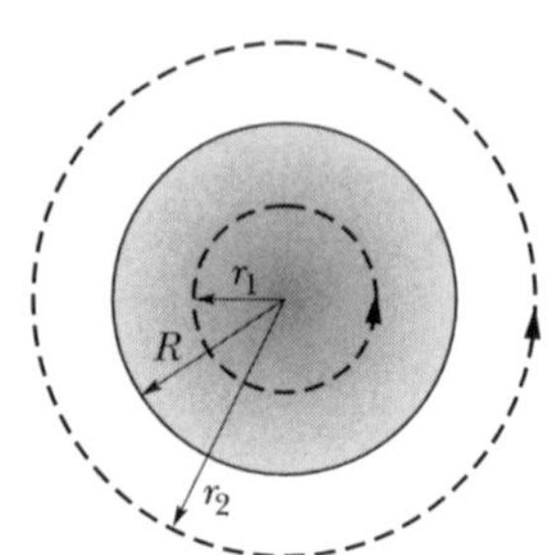

FIGURE 30-81 Question 29.

SITUATION	CURRENT OR CURRENT DENSITY
a	$i = 2$ A, uniform
b	$J = 3$ A/m^2, uniform
c	$i_{enc} = (4\text{ A})\left(\dfrac{r}{R}\right)$
d	$i_{enc} = (5\text{ A})\left(\dfrac{r}{R}\right)^2$
e	$J = (6\text{ A/m}^2)\left(1 - \dfrac{r}{R}\right)$

30. Figure 30-82 gives the magnitude of the magnetic field inside and outside four wires (a, b, c, and d) carrying currents that are uniformly distributed across the cross sections of the wires. Overlapping portions of the plots are indicated by double labels. Rank the wires according to (a) their radii, (b) the magnitudes of the magnetic field on their surface, and (c) the values of their currents, greatest first. (d) Is the current density in wire a greater than, less than, or equal to that in wire c?

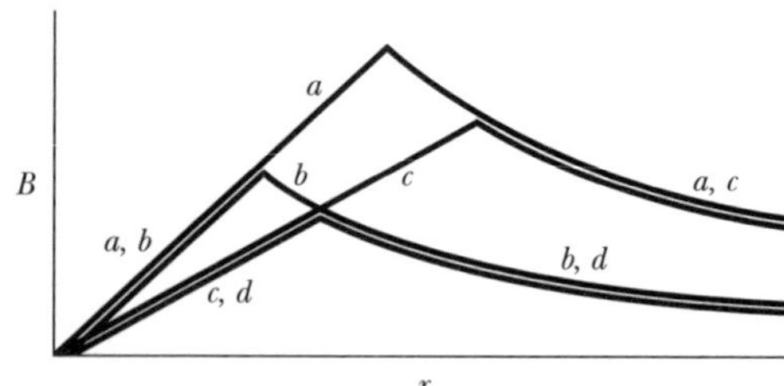

FIGURE 30-82 Question 30.

EXERCISES & PROBLEMS

76. As shown in Fig. 30-83, an arc of radius 4.0 cm subtends an angle of 120° and has a common center at point P with a semicircle of radius 5.0 cm. (a) If $I = 0.40$ A, what is the net magnetic field at point P due to these current segments? (b) If the direction of the current in the semicircle is reversed, what now is the net magnetic field at point P?

77. One long wire lies along an entire x axis and carries a current of 30 A in the positive x direction. A second long wire is perpendicular to the xy plane, passes through the point (0, 4 m, 0), and carries a current of 40 A in the positive z direction. What is the magnitude of the resulting magnetic field at the point y = 2.0 m on the y axis?

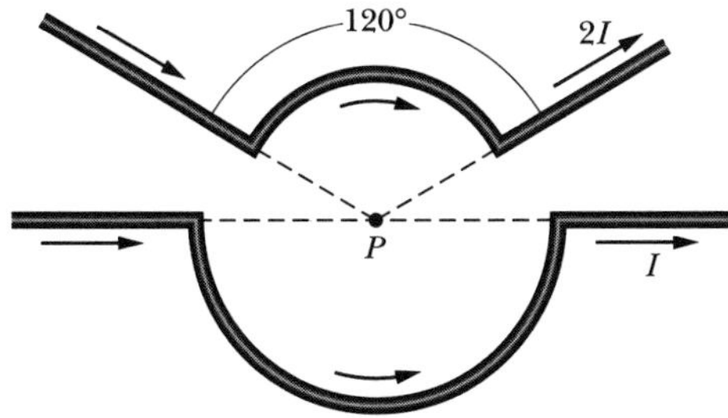

FIGURE 30-83 Problem 76.

78. Two long parallel wires lie in an xy plane. One wire lies along the line $y = 10.0$ cm and carries a current of 6.00 A in the positive x direction. The other wire lies along the line $y = 5.00$ cm and carries a current of 10.0 A in the positive x direction. (a) What is the resulting magnetic field at the origin? (b) For what value of y is the resulting magnetic field zero? (c) If the 6.00 A current is reversed so that it is now in the negative x direction, for what value of y is the resulting magnetic field now zero?

79. Three long wires all lie in an xy plane parallel to the x axis. They are spaced equally, 10 cm apart. The two outer wires each carry a current of 5.0 A in the positive x direction. What is the magnitude of the force on a 3.0 m section of either of the outer wires if the current in the center wire is (a) 3.2 A in the positive x direction and (b) 3.2 A in the negative x direction?

80. Three long wires are parallel to a z axis and each carries a current of 10 A in the positive z direction. Their points of intersection with the xy plane form an equilateral triangle with sides of 50 cm as shown in Fig. 30-84. A fourth wire (wire b) passes through the midpoint of the base of the triangle and is parallel to the other three wires. What must be the current in wire b for the net magnetic force on wire a to be zero?

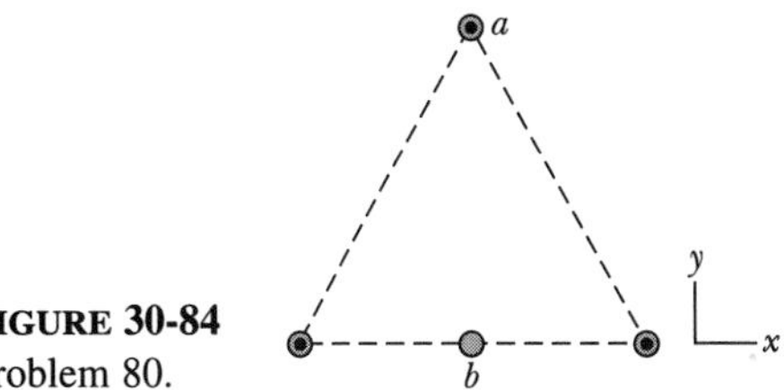

FIGURE 30-84 Problem 80.

81. A long straight wire (radius = 3.0 mm) carries a constant current distributed uniformly over a cross section perpendicular to the axis of the wire. If the current density is 100 A/m^2, what are the magnitudes of the magnetic fields (a) 2.0 mm from the axis of the wire and (b) 4.0 mm from the axis of the wire?

82. A long, hollow, cylindrical conductor (inner radius = 2.0 mm, outer radius = 4.0 mm) carries a current of 24 A distributed uniformly across its cross section. A long thin wire that is coaxial with the cylinder carries a current of 24 A in the opposite direction. What are the magnitudes of the magnetic fields (a) 1.0 mm, (b) 3.0 mm, and (c) 5.0 mm from the central axis of the wire and cylinder?

83. A long thin wire carries an unknown current. Coaxial with the wire is a long, thin, cylindrical conducting surface that carries a current of 30 mA. The cylindrical surface has a radius of 3.0 mm. If the magnitude of the magnetic field at a point 5.0 mm from the wire is 1.0 μT, what is the current in the wire?

84. A long wire is known to have a radius greater than 4.0 mm and to carry a current uniformly distributed over its cross section. The magnitude of the magnetic field is 0.28 mT at a point 4.0 mm from the axis of the wire and 0.20 mT at a point 10 mm from the axis of the wire. What is the radius of the wire?

85. Figure 30-85 shows, in cross section, two long straight wires; the 3.0 A current in the right-hand wire is out of the page. What are the size and direction of the current in the left-hand wire if the net magnetic field at point P is to be zero?

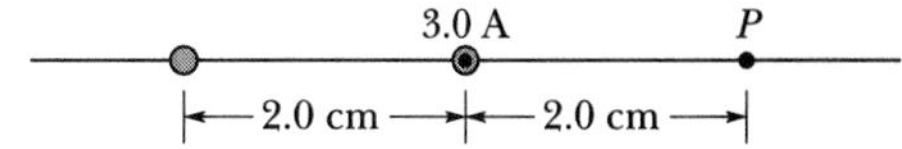

FIGURE 30-85 Problem 85.

86. Figure 30-86 shows two very long straight wires (in cross section) that carry currents of 4.00 A directly out of the page. Distance $d_1 = 6.00$ m and distance $d_2 = 4.00$ m. What is the magnitude of the net magnetic field at point P, which lies on a perpendicular bisector to the wires?

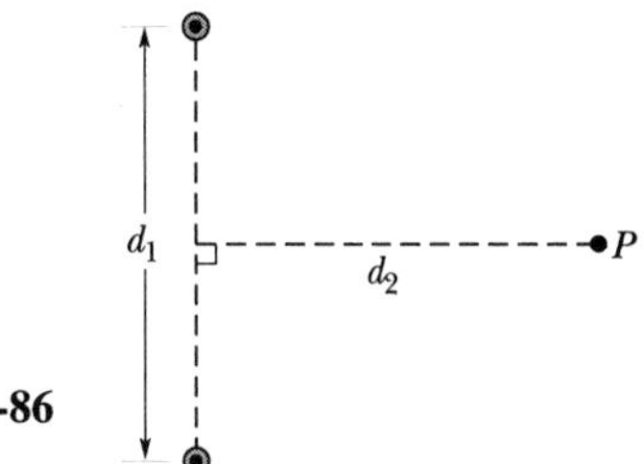

FIGURE 30-86 Problem 86.

87. Figure 30-87 shows a closed loop carrying a current of 2.00 A. The loop consists of a half circle of radius 4.00 m, two quarter circles of radii 2.00 m, and three radial straight wires. What is the magnitude of the net magnetic field at the common center of the circular sections?

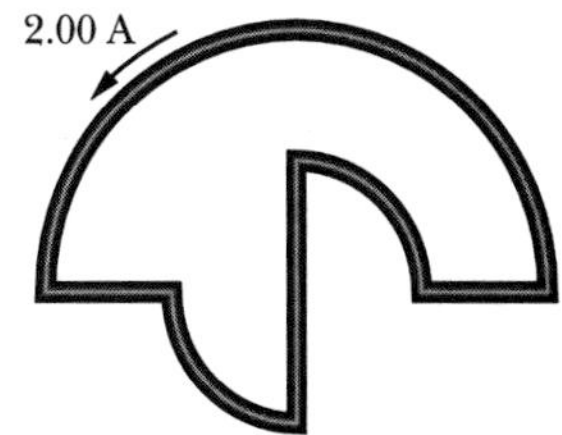

FIGURE 30-87 Problem 87.

88. In Fig. 30-88, a closed loop carries a current of 200 mA. The loop consists of two radial straight wires and two concentric circular arcs of radii 2.00 m and 4.00 m. The angle θ is $\pi/4$ rad. What are the magnitude and direction of the net magnetic field at point P, which is at the center of curvature?

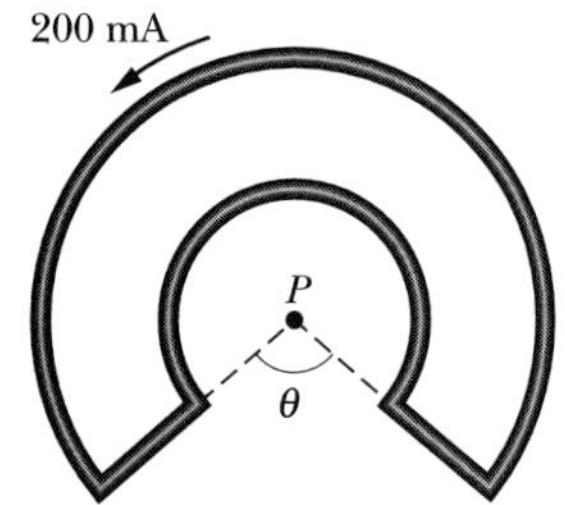

FIGURE 30-88 Problem 88.

89. Figure 30-89 shows five very long wires (in cross section) that carry currents directly into or out of the page, as indicated. The wires are uniformly spaced by 0.50 m; the currents are $i_1 = 2.00$ A, $i_2 = 4.00$ A, $i_3 = 0.25$ A, $i_4 = 4.00$ A, and $i_5 = 2.00$ A. What is the net force per unit length acting on the central wire due to the currents in the other wires?

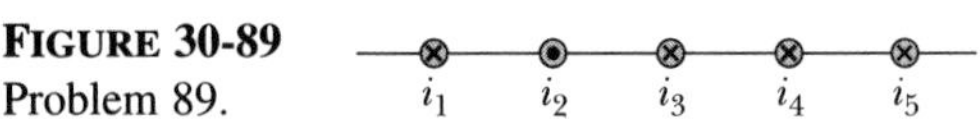

FIGURE 30-89 Problem 89.

90. A cylindrical cable, with radius 8.00 mm, carries a current of 25.0 A, uniformly spread over its cross-sectional area. At what distance from the center of the wire is there a point within the wire where the magnetic field is 0.100 mT?

Tutorial Problem

91. Let's consider an infinitely long wire that is a solid cylinder of radius a. The wire carries a current i; we will make the simplifying assumption that this current is uniformly distributed throughout the interior of the wire. (a) Make two cross-sectional sketches of the wire, one along the axis and one perpendicular to the axis. Mark the direction of the current on the diagram. (b) What is the magnitude of the current density (current per unit area) in the wire?

(c) For a problem with cylindrical symmetry, it is most appropriate to write the magnetic field as a function of cylindrical coordinates r, θ, and z, where r and θ are polar coordinates in the xy plane and z is the coordinate along the cylinder axis. Explain in complete sentences why the magnetic field would not be expected to depend on θ or z, but might depend on r. (d) What do you expect the magnetic field lines to look like inside and outside the wire? Sketch them on your diagram of part (a), showing the correct direction of the field. (e) The appropriate unit vectors to use in a problem with cylindrical symmetry are $\hat{r}$ (lying in the xy plane, directed straight out from the z axis), $\hat{\theta}$ (lying in the xy plane, perpendicular to $\hat{r}$, and pointing in the counterclockwise direction of increasing θ), and $\hat{z}$ (along the z axis; $\hat{z}$ is equivalent to $\mathbf{k}$). In terms of these unit vectors, what is the direction of the current in the wire?

(f) Name and describe in a sentence the fundamental principle of physics that can be used to determine the magnitude of the magnetic field. (g) Use this principle to determine the magnetic field in the regions $r < a$ (inside the wire) and $r > a$ (outside the wire). Explain clearly any path or surface you use. (h) Graph the magnitude of $\mathbf{B}$ as a function of r.

Answers

(b) The wire has radius a and cross-sectional area πa^2, so its current density $\mathbf{J}$ has magnitude $|\mathbf{J}| = i/\pi a^2$.

(c) All values of z are equivalent because the wire is infinitely long, so the magnetic field should not depend on z. Also, all directions from the wire (all values of θ) are equivalent, so the magnetic field should not depend on θ. However, clearly, all values of r are not equivalent, so we expect that the magnetic field will depend on r.

(d) The magnetic field lines should be circles centered on the axis of the wire. From the right-hand rule, if the current in the wire is directed up out of the plane of the paper, the field lines will be counterclockwise. That the field lines are circles follows from the cylindrical symmetry of the wire.

(e) The current in the inner wire is along $\hat{z}$, by our choice of direction for $\hat{z}$.

(f) This is the highly symmetric type of situation in which Ampere's law can be used. That law says that the line integral of the magnetic field vector around a closed path equals $\mu_0 i_{enc}$, where i_{enc} is the net current through the path. Mathematically,

$$\oint \mathbf{B} \cdot d\mathbf{s} = \mu_0 i_{enc}.$$

(g) The appropriate Amperian path to use is a circle of radius r lying in the xy plane and concentric with the wire. Along this path the magnetic field is parallel to $d\mathbf{s}$. The current is simple outside the wire, but not inside. For $r < a$,

$$\oint \mathbf{B}(r) \cdot d\mathbf{s} = \mu_0 i_{enc}$$

becomes

$$2\pi r B = \mu_0 \left(\frac{i}{\pi a^2}\right) \pi r^2.$$

So

$$B(r) = \frac{\mu_0 i r}{2\pi a^2}.$$

For $a < r$,

$$\oint \mathbf{B}(r) \cdot d\mathbf{s} = \mu_0 i_{enc}$$

becomes

$$2\pi r B = \mu_0 i.$$

So

$$B(r) = \frac{\mu_0 i}{2\pi r}.$$

Chapter Thirty-One
Induction and Inductance

QUESTIONS

19. *Organizing question:* In Fig. 31-75, a wire loop forms a rectangle of height H and width W. The loop lies in a magnetic field that does not vary with time, is directed out of the page, and has a magnitude B. We wish to find the magnetic flux Φ_B through the loop for three choices of $B(x, y)$: (a) $B = ax$, (b) $B = by$, and (c) $B = cxy$, where a, b, and c are constants. For each choice, which of the following expressions can be used to find Φ_B?

(1) BHW (2) $\int BH\,dx$
(3) $\int BW\,dx$ (4) $\int BH\,dy$
(5) $\int BW\,dy$ (6) $\int B\,dx\,dy$

(d) For each choice, is the emf induced around the loop clockwise, counterclockwise, or nonexistent?

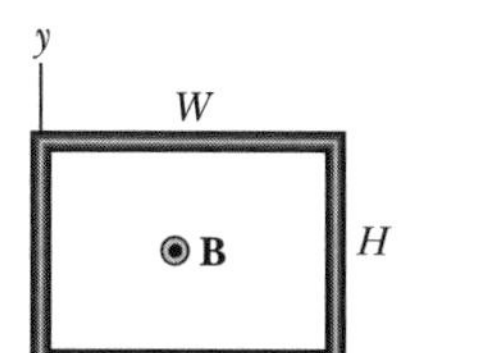

FIGURE 31-75 Question 19.

20. In Fig. 31-76, for three situations, magnetic fields are directed through rectangular wire loops that have areas of either A_0 or $A_0/2$. In each situation, the field is perpendicular to the plane of the loop and is increasing in magnitude at the rate of 5 mT/s. At the instant shown, the magnitudes of the fields are either B_0, $2B_0$, or $3B_0$. (a) At that instant, rank the loops according to the magnitude of the emf induced in them, greatest first. (b) For each situation, what is the direction of the emf induced in the loop?

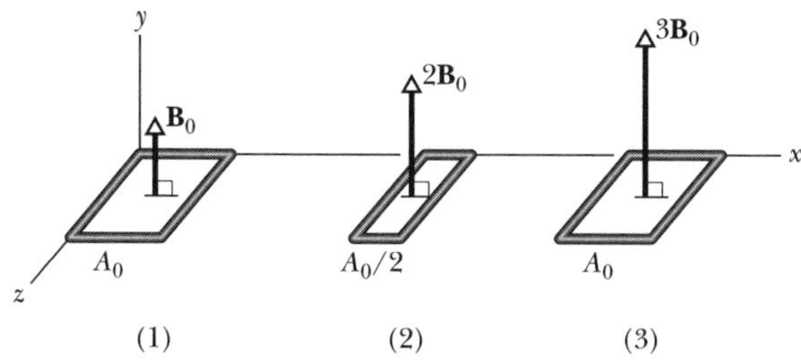

FIGURE 31-76 Question 20.

21. Figure 31-77*a* is a plot of the magnitude of a magnetic field that is directed perpendicularly through the plane of a wire loop. Which of the plots in Fig. 31-77*b* best gives the current induced in that loop?

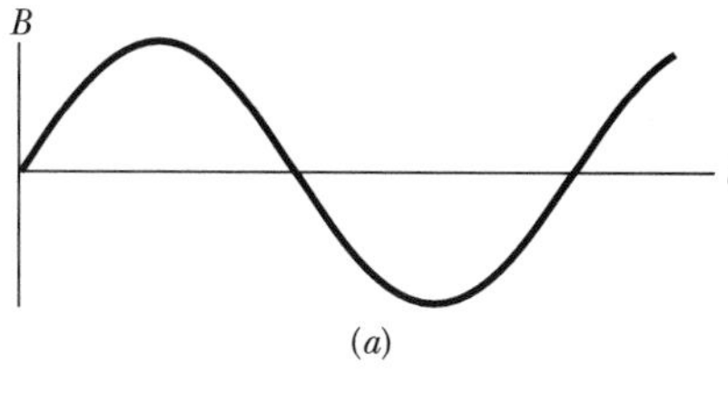

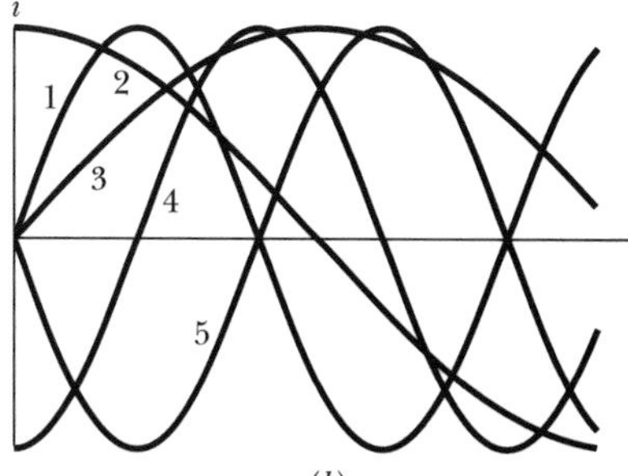

FIGURE 31-77 Question 21.

22. Figure 31-78 shows three situations in which a wire loop lies partially in a magnetic field. The magnitude of the field is either

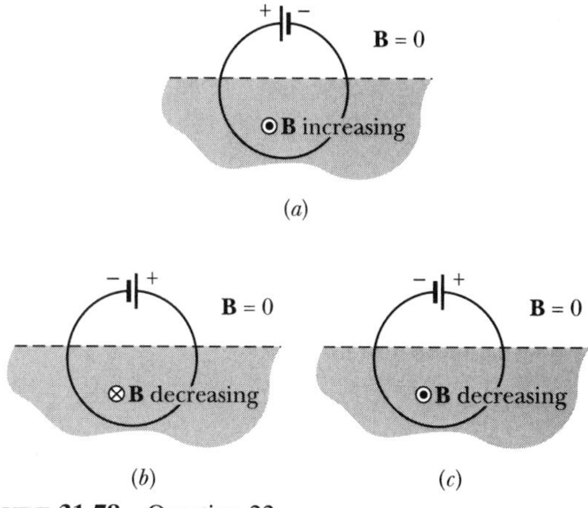

FIGURE 31-78 Question 22.

increasing or decreasing, as indicated. In each situation, a battery is part of the loop. In which situations are the induced emf and the battery emf in the same direction along the loop?

23. In Fig. 31-79, a wire loop has been bent so that it has three segments: segment *ab* (a quarter circle), *bc* (a square corner), and *ca* (straight). Here are three choices for a magnetic field through the loop:

(1) $\mathbf{B}_1 = 3\mathbf{i} + 7\mathbf{j} - 5t\mathbf{k}$,
(2) $\mathbf{B}_2 = 5t\mathbf{i} - 4\mathbf{j} - 15\mathbf{k}$,
(3) $\mathbf{B}_3 = 2\mathbf{i} - 5t\mathbf{j} - 12\mathbf{k}$,

where **B** is in milliteslas and t is in seconds. Without written calculation, rank the choices according to (a) the work done per unit charge on setting up the induced current and (b) that induced current, greatest first. (c) For each choice, what is the direction of the induced current in the figure?

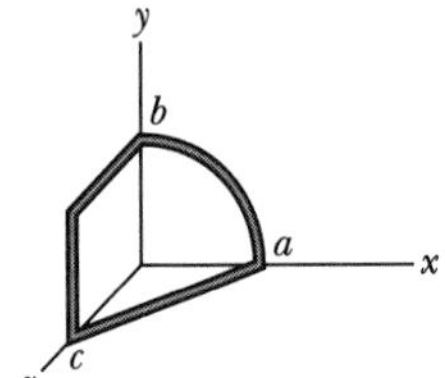

FIGURE 31-79 Question 23.

24. Figure 31-80 gives four situations (similar to that in Fig. 31-10) in which we pull rectangular wire loops out of identical magnetic fields (directed into the page) at the same constant speed. The loops have edge lengths of either L or $2L$, as drawn. Rank the situations according to (a) the magnitude of the force required of us and (b) the rate at which energy is transferred from us to thermal energy of the loop, greatest first.

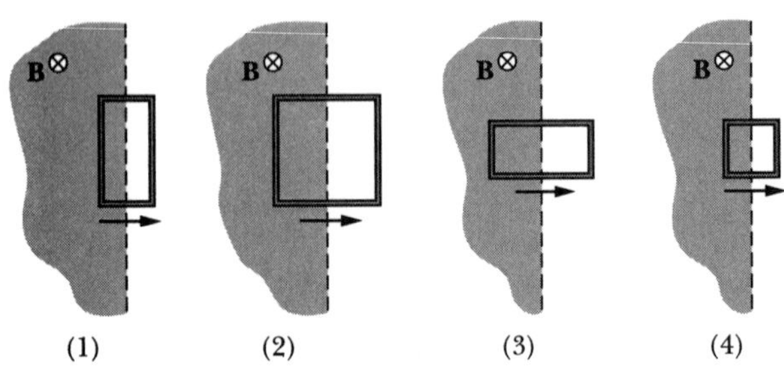

FIGURE 31-80 Question 24.

25. In Fig. 31-81*a*, a wire loop is pulled into, through, and then out of four regions in which uniform magnetic fields are either directly into or out of the page, as indicated by the encircled × and the encircled dot. (Compare with Fig. 31-13.) Figure 31-81*b* gives the current induced in the loop as a function of the position of the right edge of the loop during the motion. (Counterclockwise current is taken to be negative.) Rank the four regions according to the magnitudes of their magnetic fields, greatest first.

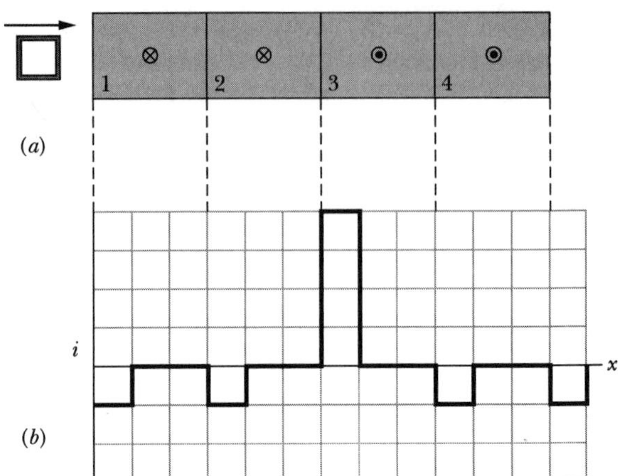

FIGURE 31-81 Question 25.

26. In Fig. 31-12, is the direction of the eddy currents in the plate clockwise or counterclockwise when the plate is (a) entering the magnetic field from the left (as shown), (b) leaving the field toward the right, (c) entering the field from the right, and (d) leaving the field toward the left?

27. Figure 31-82 gives the magnitude of the electric field inside and outside four circular regions (*a*, *b*, *c*, and *d*) in which the magnetic field is changing, as in Fig. 31-14. The plots are versus the radial distance from the center of the circular regions (as in Fig. 31-15), and overlapping portions of the plots are indicated by double labels. Rank the regions according to (a) the radii of their borders, (b) the magnitudes of the electric field along their borders, and (c) the rate dB/dt at which their magnetic fields are changing, greatest first.

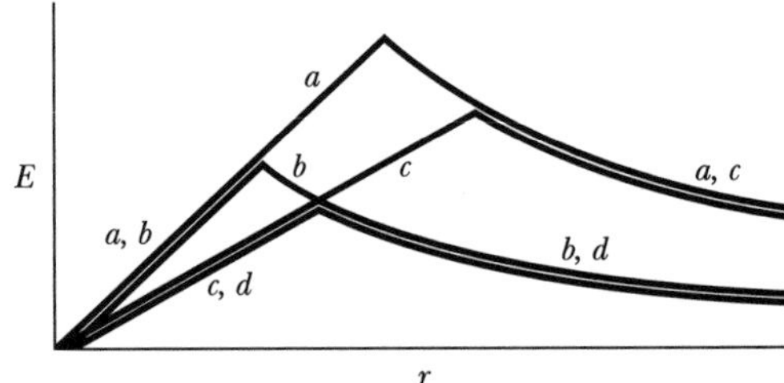

FIGURE 31-82 Question 27.

28. *Organizing question:* Figure 31-83 shows a circular region of radius R in which a magnetic flux Φ_B is directed out of the page. It also shows two integration paths (of radii r_1 and r_2) that form circles concentric with the circular region. Set up the right side of Faraday's law (Eq. 31-22) for paths 1 and 2 for the following situations, where either the total flux Φ_B, the flux $\Phi_{B,\text{enc}}$ encircled by a path of radius r, or the magnitude B of the associated magnetic field is given. Neglect the negative sign in Eq. 31-22. Perform any differentiation that may be required, but not any integration (and be warned of menacing subscripts).

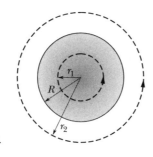

FIGURE 31-83 Question 28.

SITUATION	FLUX OR FIELD
a	$\Phi_B = (4\ \text{T}\cdot\text{m}^2/\text{s})t$, uniform
b	$B = (2\ \text{T/s})t$, uniform
c	$B = (3\ \text{T})\left(\frac{r}{R}\right)$,
d	$\Phi_{B,\text{enc}} = (5\ \text{T}\cdot\text{m}^2/\text{s})\left(\frac{r}{R}\right)t$
e	$B = (4\ \text{T/s})\left(1 - \frac{r}{R}\right)t$

29. In Fig. 31-19, assume that the (ideal) battery has an emf of 5 V. After the switch is thrown to position *a*, consider the inductor at the stages when the potential difference across the resistor is (1) 1 V, (2) 2 V, and (3) 3 V. Rank those stages according to (a) the rate at which the current is changing, (b) the total flux through the inductor, and (c) the emf across the inductor, greatest first.

30. Suppose that in Question 29 and Fig. 31-19 the current has reached its equilibrium value when the switch is thrown to position *b*. Consider the inductor at the stages when the potential difference across the resistor is (1) 1 V, (2) 2 V, and (3) 3 V. Rank those stages according to (a) the rate at which the current is changing, (b) the total flux through the inductor, and (c) the emf across the inductor, greatest first.

31. The three *LR* circuits of Fig. 31-84 have identical resistors. In each, the switch *S* has been closed long enough for the currents to reach their steady-state value of 10 A through each resistor. (a) Rank the circuits according to the emf of their (ideal) batteries, greatest first. The switches are now reopened. Rank the circuits according to the currents just then through (b) resistor R_1 and (c) resistor R_2, greatest first.

32. *Organizing question:* The switch in Fig. 31-85 is closed at time $t = 0$. (a) Thereafter, what is the current through R_3 in terms of the currents through R_1 and R_2? Using that relation and given data, set up loop equations for (b) the left-hand loop and (c) the big loop. (d) To produce a single differential equation written in one unknown current, what step should now be taken and which current is then in the differential equation?

33. Two long, straight, parallel wires are shown in cross section in Fig. 31-86, with an *x* running through them. The wires carry currents of the same value. Which of the plots in the figure best represents the magnetic energy density u_B along the *x* axis if the currents are in (a) the same direction and (b) opposite directions?

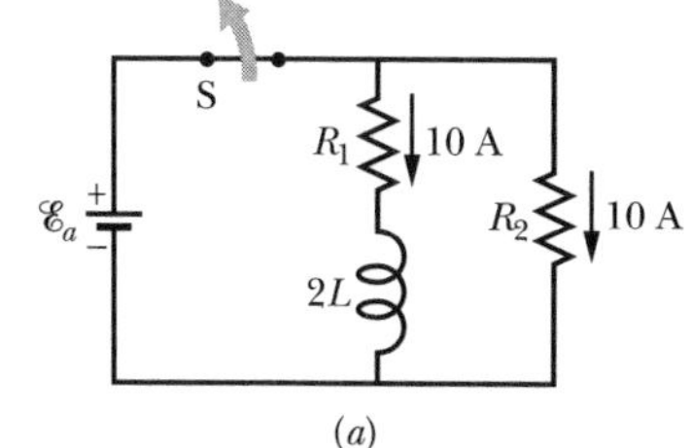

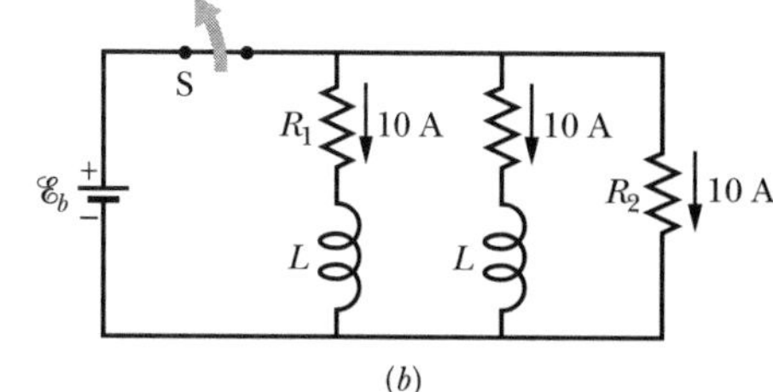

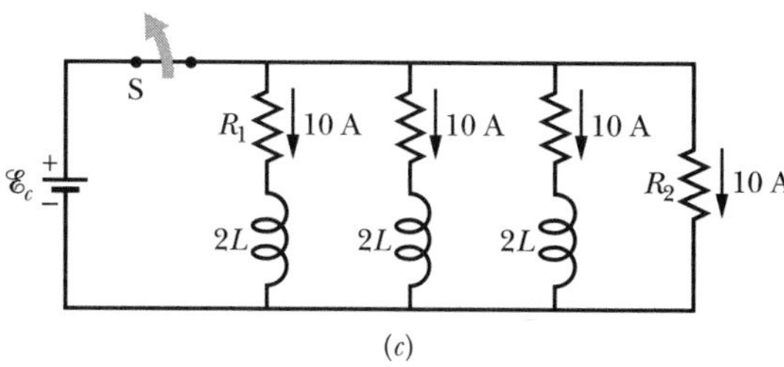

FIGURE 31-84 Question 31.

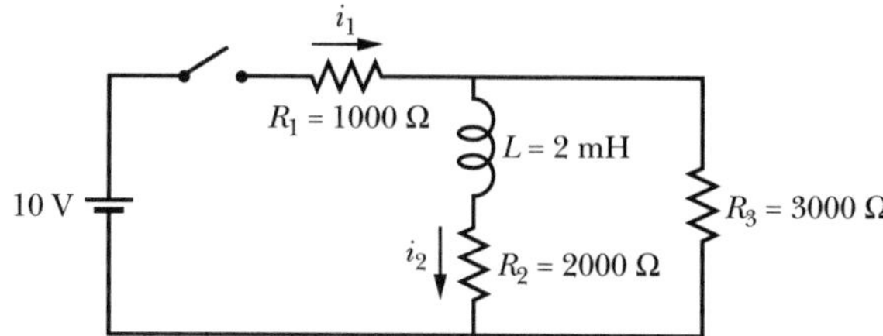

FIGURE 31-85 Question 32.

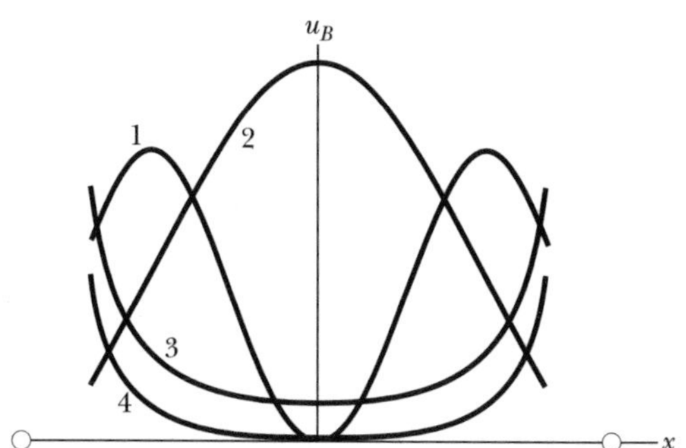

FIGURE 31-86 Question 33.

EXERCISES & PROBLEMS

101. A 50-turn circular coil (radius = 15 cm) with a total resistance of 4.0 Ω is placed in a uniform magnetic field directed perpendicularly to the plane of the coil. The magnitude of this field varies with time according to $B = A \sin(\omega t)$, where $A =$ 80 μT and $\omega = 50\pi$ rad/s. What is the magnitude of the current induced in the coil at 20 ms?

102. A circular loop (radius = 14 cm) of wire is placed in a magnetic field that makes an angle of 30° with the normal to the plane of the loop. The magnitude of this field increases at a constant rate from 30 mT to 60 mT in 15 ms. If the loop has a resistance of 5.0 Ω, what is the magnitude of the current induced in the loop when the field is 50 mT?

103. A square loop of wire is held fixed in a uniform magnetic field that is directed perpendicularly to the plane of the loop. The magnitude of the magnetic field is 0.24 T. The length of each side of the square is decreasing at a constant rate of 5.0 cm/s. What is the magnitude of the emf induced in the loop when the length of a side is 12 cm?

104. A rectangular loop (area = 0.15 m^2) turns in a uniform magnetic field with a magnitude of B = 0.20 T. At an instant when the angle between the magnetic field and the normal to the plane of the loop is $\pi/2$ rad and increasing at the rate of 0.60 rad/s, what is the magnitude of the emf induced in the loop?

105. In the circuit shown in Fig. 31-87, $\mathscr{E}$ = 12 V. The switch has been open for a long time before it is closed at t = 0. At what rate is the current in the inductor changing (a) immediately after the switch is closed and (b) when the current in the battery is 0.50 A? (c) What is the current in the battery when the circuit reaches its steady-state condition?

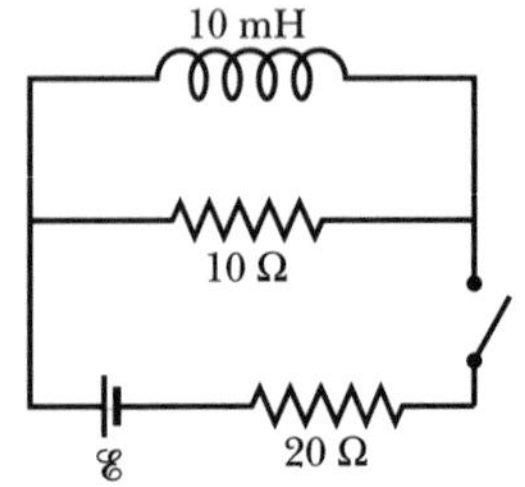

FIGURE 31-87 Problem 105.

106. In the circuit of Fig. 31-88, the switch has been open for a long time when it is closed at time t = 0. What are (a) the current through the battery and (b) the rate at which that current is changing immediately after the switch is closed? What are (c) the current and (d) the rate at time t = 3.0 μs? What are (e) the current and (f) the rate a long time later?

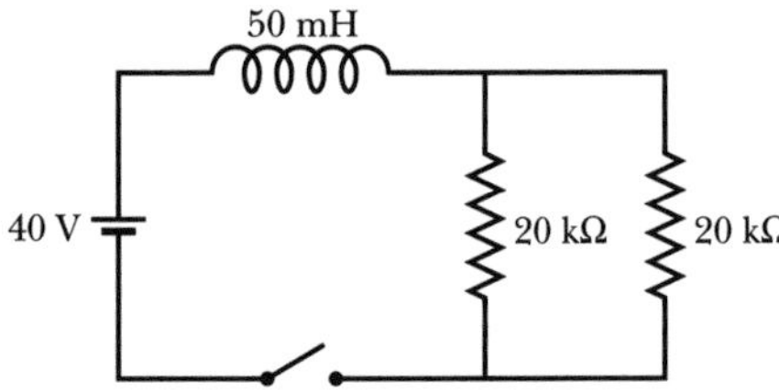

FIGURE 31-88 Problem 106.

107. In the circuit shown in Fig. 31-89, $\mathscr{E}$ = 6.0 V. (a) At what rate is the current in the 0.30 H inductor changing immediately after the switch is closed? (b) What is the current in the 0.30 H inductor when the circuit reaches its steady-state condition?

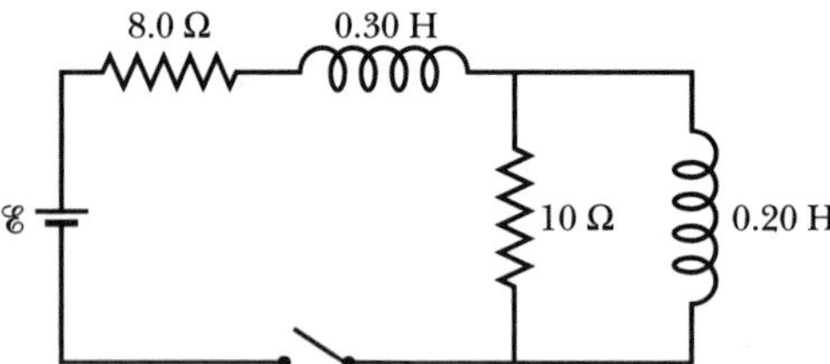

FIGURE 31-89 Problem 107.

108. A switch is closed at time t = 0 to connect a 10 V emf to a series combination of a 10 Ω resistor and a 10 mH inductor. What is the energy stored in the inductor at t = 2.0 ms?

109. In the circuit of Fig. 31-90, what are (a) the energy stored in the inductor and (b) the rate at which that energy is changing when the current through the inductor is increasing at a rate of 20 A/s? What are (c) the energy and (d) the rate when the current is decreasing at a rate of 20 A/s?

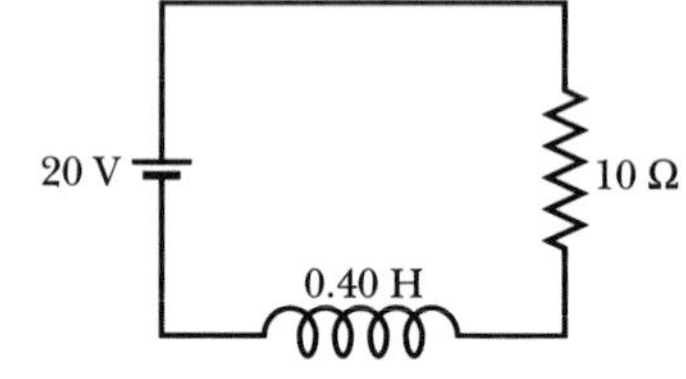

FIGURE 31-90 Problem 109.

110. Figure 31-91a shows a wire in the shape of a circle with an area of 3.0 m^2. The resistance of the wire is 9.0 Ω. The wire is immersed in a uniform magnetic field that is directed out of

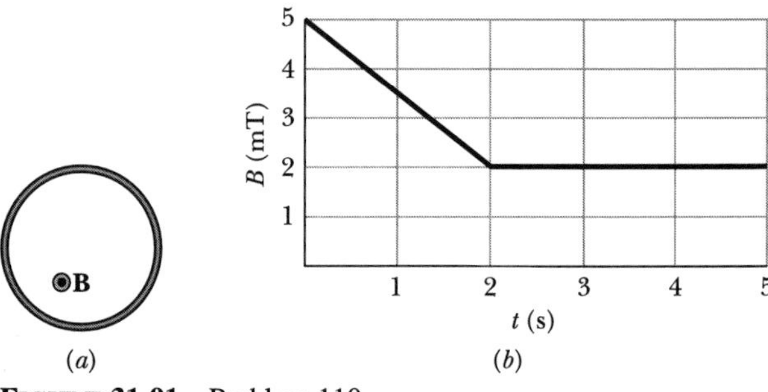

FIGURE 31-91 Problem 110.

the page; the magnitude of the field for $t \geq 0$ is shown in Fig. 31-91*b*. What are the sizes and directions of the currents induced in the wire at the times (a) 0.50 s, (b) 1.5 s, (c) 3.0 s, and (d) 4.0 s?

111. In Fig. 31-92, a wire forms a closed loop in the shape of a rectangle, with $L_1 = 20$ cm, $L_2 = 8.0$ cm, $L_3 = 6.0$ cm, and resistance $R = 40$ mΩ. A 50 mV battery is included in the loop. The top section (marked by a dashed line) is immersed in a magnetic field directed out of the page and having magnitude $B = (0.60\ \text{T/s}^2)t^2$. What is the current induced in this circuit when $t = 15$ s?

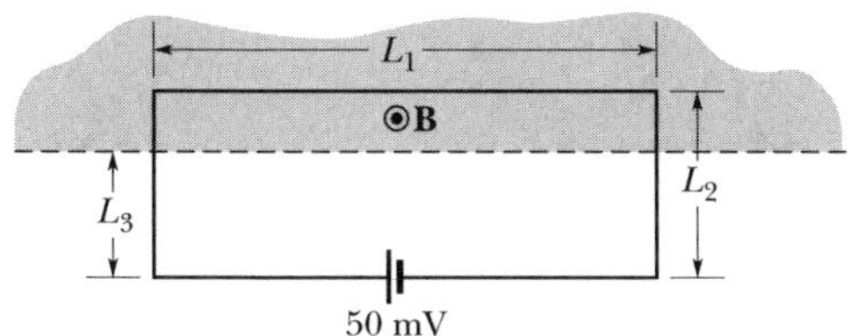

FIGURE 31-92 Problem 111.

112. In Fig. 31-93, a wire forms a closed loop in the shape of a circle, with radius $R = 2.0$ m and resistance 4.0 Ω. The circle is centered on a long straight wire; at time $t = 0$, the current in the wire is 5.0 A rightward. Thereafter, the current changes according to $i = 5.0\ \text{A} - (2.0\ \text{A/s}^2)t^2$. (The straight wire is insulated, so there is no electrical contact between it and the wire of the circle.) What are the size and direction of the current induced in the circle at time $t > 0$?

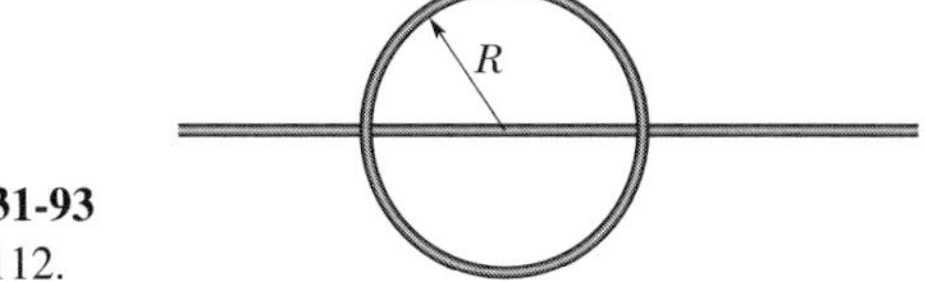

FIGURE 31-93 Problem 112.

113. In Fig. 31-94, a conducting bar is slid rightward along rails formed by a long wire in the shape of a U. The distance between the rails is $L = 2.0$ m. The distance between the bar and the left side of the wire is $x = (2.0\ \text{m/s}^3)t^3$. The apparatus is immersed in a uniform magnetic field of magnitude 30 mT, directed out of the page. What are the size and direction of the emf induced in the wire–bar loop at $t = 5.0$ ms?

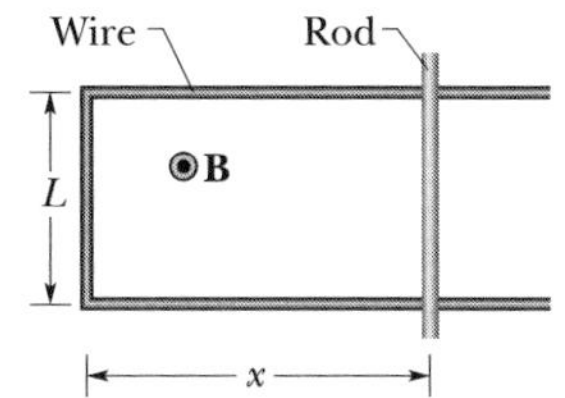

FIGURE 31-94 Problem 113.

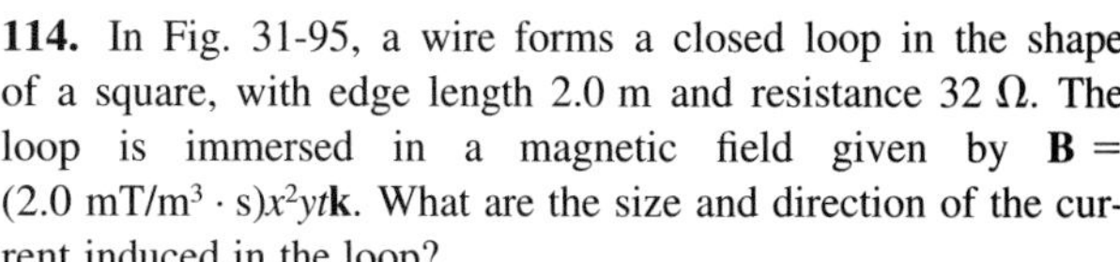

114. In Fig. 31-95, a wire forms a closed loop in the shape of a square, with edge length 2.0 m and resistance 32 Ω. The loop is immersed in a magnetic field given by $\mathbf{B} = (2.0\ \text{mT/m}^3 \cdot \text{s})x^2yt\mathbf{k}$. What are the size and direction of the current induced in the loop?

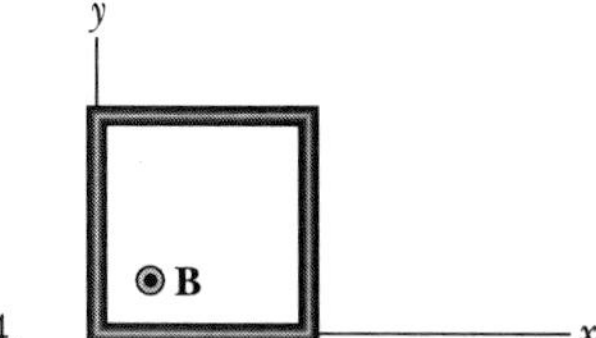

FIGURE 31-95 Problem 114.

115. The switch in Fig. 31-96 is closed at time $t = 0$. At $t = 5.0$ s, the current is 1.0 A. What is the value of the inductance L?

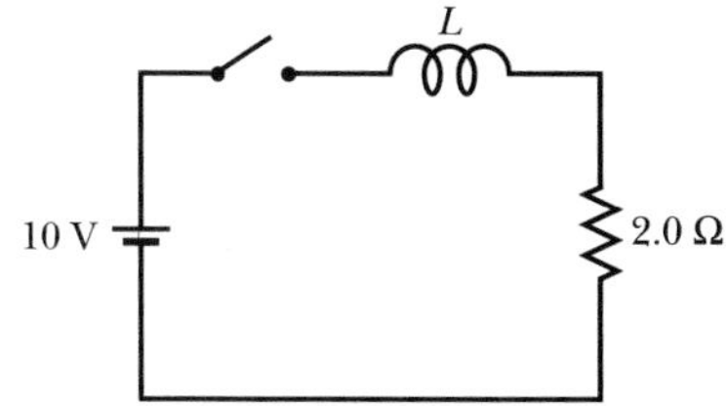

FIGURE 31-96 Problem 115.

116. In Fig. 31-97, what is the current through the battery (a) just as the switch is closed at time $t = 0$ and (b) long after the switch is closed?

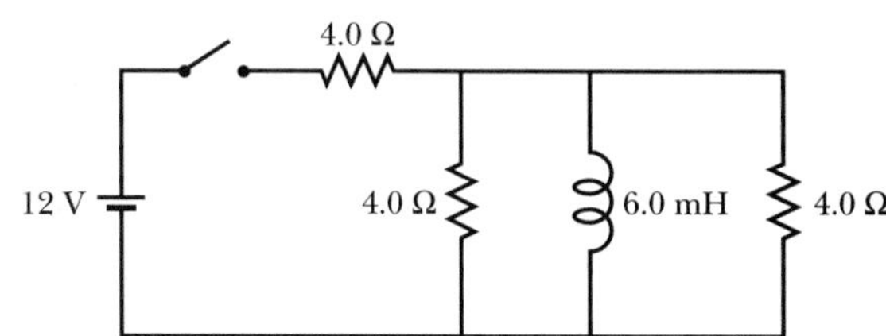

FIGURE 31-97 Problem 116.

117. Figure 31-98 shows a circuit consisting of a battery, a resistance, and an inductance. When the switch is closed, how long does the current take to build up to 2.0 mA?

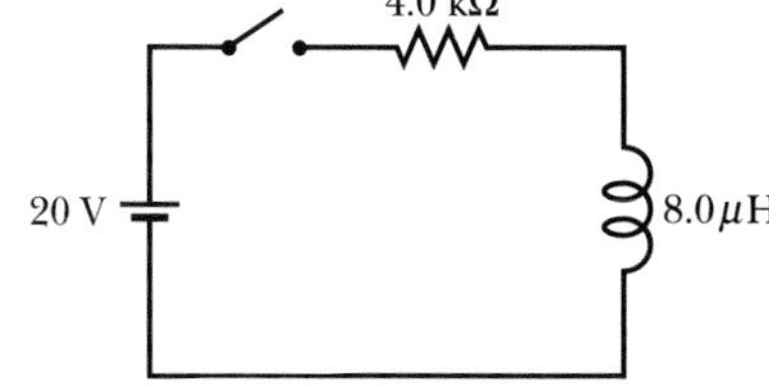

FIGURE 31-98 Problem 117.

118. At time $t = 0$, a 45 V potential difference is suddenly applied to the leads of a coil with inductance $L = 50$ mH and resistance $R = 180\ \Omega$. At what rate is the current through the coil increasing at $t = 1.2$ ms?

Clustered Problems

Cluster 1

The conducting loops described in these problems are fully or partially contained within a circular region of a magnetic field (see Fig. 31-99), that is directed into the page and decreasing in magnitude at a constant rate of 10.0 mT/s.

119. See this cluster's setup. In Fig. 31-99*a*, a circular conducting loop of radius 5.00 cm and resistance 4.00 Ω is in the region of the magnetic field. What are the magnitudes and directions of (a) the emf and (b) the current that are induced around the loop?

120. See this cluster's setup. In Fig. 31-99*b*, a conducting loop consists of an incomplete circle with a small gap, leads, and an 8.00 Ω resistor. The circle and the leads have a combined resistance of 4.00 Ω. The incomplete circle has a radius of 5.00 cm. What are the magnitudes and directions of (a) the emf and (b) the current that are induced around the loop?

121. See this cluster's setup. In Fig. 31-99*c*, a conducting loop consists of a semicircle with radius 5.00 cm and a straight section. The resistance of the loop is 8.00 Ω. What are the magnitudes and directions of (a) the emf and (b) the current that are induced around the loop?

122. See this cluster's setup. In Fig. 31-99*d*, a circular conducting loop of radius 5.00 cm and resistance 4.00 Ω is connected by leads of negligible resistance to a circular conducting loop of radius 2.50 cm and resistance 2.00 Ω. The loops are concentric and lie within the magnetic field. What are the magnitudes and directions of (a) the emf and (b) the current that are induced around the loop?

123. See this cluster's setup. In Fig. 31-99*e*, a circular conducting loop of radius 5.00 cm and resistance 4.00 Ω is connected by insulated leads of negligible resistance to a circular conducting loop of radius 2.50 cm and resistance 2.00 Ω. The loops are concentric and lie within the magnetic field; the leads are wrapped over each other once. What are the magnitudes and directions of (a) the emf and (b) the current that are induced around the loop?

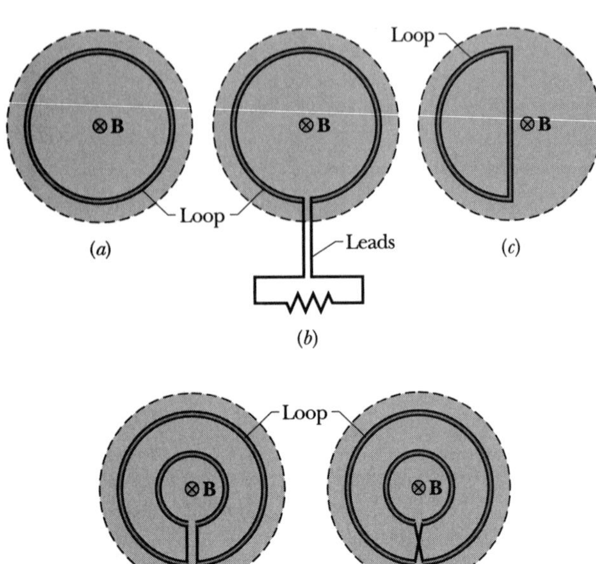

FIGURE 31-99 Problems 119 through 123.

Cluster 2

In the problems in this cluster, a wire (called a slider*) is forced to move at constant velocity* **v** *(of magnitude 12.0 m/s) along two long rails (see Fig. 31-100). A resistor of resistance $R =$ 5.00 Ω connects the rails at their other ends; the rails have negligible resistance. A uniform magnetic field* **B**, *directed into the page in the figures, exists throughout the region. The resistor, rails, and slider form a complete conducting loop.*

124. See this cluster's setup. In Fig. 31-100*a*, the magnitude of the magnetic field is a constant 200 mT. The rails are parallel, with a separation of 30.0 cm. What are the magnitude and direction of the current in the loop if (a) the slider has negligible resistance and (b) the slider has a resistance of 2.00 Ω?

125. See this cluster's setup. In Fig. 31-100*b*, the magnitude of the magnetic field is a constant 200 mT. The slider, which has negligible resistance, begins its motion at $t = 0$ at the resistor. There the rails are separated by 30 cm; their separation increases by 20.0 cm with every 1.00 m of distance from the resistor. (a) What are the magnitude and direction of the current in the loop when the slider reaches the point where the rail separation is 50 cm? (b) Write an expression for the current as a function of time t.

126. See this cluster's setup. In Fig. 31-100*a*, the slider begins at the resistor at time $t = 0$. The magnitude of the magnetic field is 200 mT just then but is increasing at a constant rate of 50.0 mT/s. The resistance of the slider is negligible; the rails are

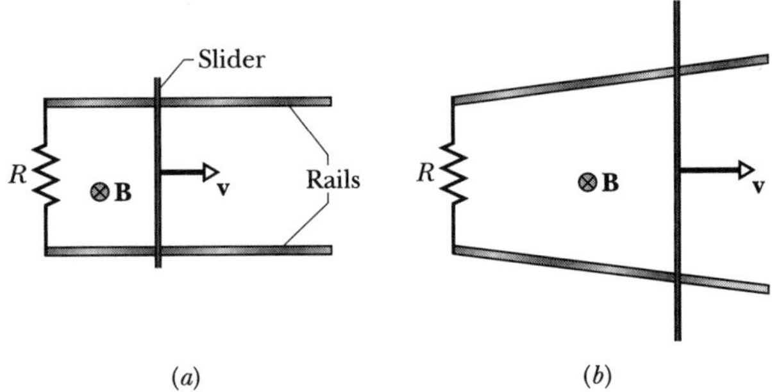

FIGURE 31-100 Problems 124 through 126.

parallel, with a separation of 30 cm. What are the magnitude and direction of the current in the loop at (a) $t = 0$ and (b) $t = 1.00$ s? (c) Write an expression for the current as a function of time.

Cluster 3

127. In the circuit of Fig. 31-101*a*, the switch has been open for a long time when it is closed at time $t = 0$. Just after it is closed, what are (a) the current i_L through the inductor, (b) the potential difference across the inductor, and (c) the rate di_L/dt at which current i_L is changing? (d) What is i_L long after the switch is closed? Find expressions for (e) current i_L and (f) rate di_L/dt as functions of t. (*Hint:* Write a loop equation for the right-hand loop and compare it with Eq. 31-48. Also compare the initial and final conditions here with those for Eq. 31-48. Then consider Eq. 31-49.)

128. In the circuit of Fig. 31-101*a*, the switch has been closed for a long time when it is opened at time $t = 0$. Just after it is opened, what are (a) the current i_L through the inductor, (b) the potential difference across the inductor, and (c) the rate di_L/dt at which current i_L is changing? (d) What is i_L long after the switch is opened? Find expressions for (e) current i_L and (f) rate di_L/dt as functions of t. (*Hint:* Write a loop equation for the circuit and compare it with Eq. 31-44. Also compare the initial and final conditions here with those for Eq. 31-44. Then consider Eq. 31-46.)

129. In the circuit of Fig. 31-101*b*, the switch has been open for a long time when it is closed at time $t = 0$. Just after it is closed, what are (a) the current i_L through the inductor, (b) the potential difference across either R_1 or R_2, (c) the potential difference across the inductor, and (d) the rate di_L/dt at which current i_L is changing? (e) What is i_L long after the switch is closed? Find expressions for (f) current i_L and (g) rate di_L/dt as functions of t. (*Hint:* Simplify the circuit and then write a loop equation for it. Put the resulting differential equation in the form of Eq. 28-38; here the variable is i_L. Then, starting with Eq. 28-39, go through the steps for a solution.)

130. In the circuit of Fig. 31-101*b*, the switch has been closed for a long time when it is opened at time $t = 0$. Just after it is opened, what are (a) the current i_L through the inductor, (b) the potential difference across R_1, (c) the potential difference across the inductor, and (d) the rate di_L/dt at which current i_L is changing? (e) What is i_L long after the switch is opened? Find expressions for (f) current i_L and (g) rate di_L/dt as functions of t. (*Hint:* Write a loop equation for the circuit. Put the resulting differential equation in the form of Eq. 28-38; here the variable is i_L. Then, starting with Eq. 28-39, go through the steps for a solution.)

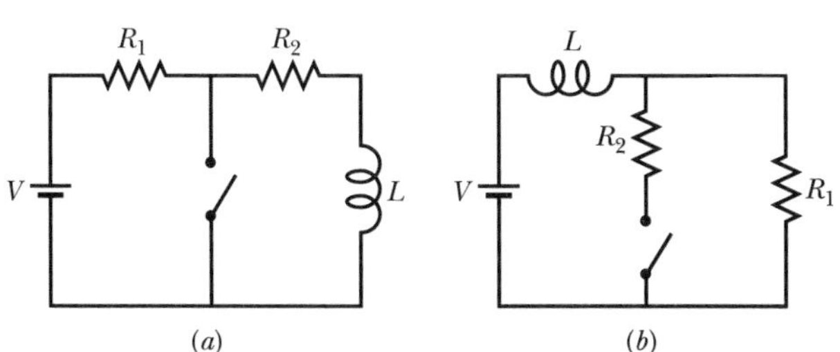

FIGURE 31-101 Problems 127 through 130.

Graphing Calculators

SAMPLE PROBLEM 31-13

Differential Equations for *RL* Circuits. In Fig. 31-102*a*, $\mathscr{E} = 12.0$ V, $R_1 = R_2 = R_3 = 1000\ \Omega$, and $L = 2.00$ mH. The circuit is initially in equilibrium (steady state) when switch S is closed at time $t = 0$ (Fig. 31-102*b*).

(a) Find the current $i_L(t)$ through the inductor for $t \geq 0$.

SOLUTION: We can find $i_L(t)$ by going through three steps that parallel those in Sample Problem 28-9 for *RC* circuits: (1) Find the initial current i_{Li} through the inductor when the switch is closed. (2) Apply the loop and junction rules to the circuit to generate an equation in which the only variable is i_L; this equation is a differential equation. (3) Solve the differential equation, either by hand or with a short program on a calculator.

STEP 1: find i_{Li}. The initial current i_{Li} through the inductor in Fig. 31-102*b* is equal to the current through that inductor in Fig. 31-102*a* as the switch is closed. Because the circuit in Fig. 31-102*a* is a single loop, the current i_{Li} through the inductor is also through R_2 and R_1. And because that circuit

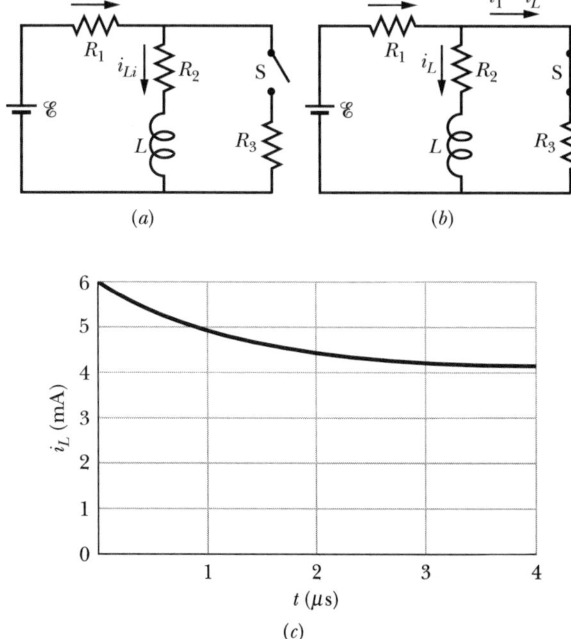

FIGURE 31-102 Sample Problem 31-13 and Problem 131.

is in equilibrium, i_{Li} is not changing; that is, $di_{Li}/dt = 0$. Writing a loop equation for the circuit with this information, we find

$$\mathscr{E} - i_{Li}R_1 - i_{Li}R_2 - L\frac{di_{Li}}{dt} = 0. \quad (31\text{-}65)$$

Substitution of given data then leads to

$$12\text{ V} - i_{Li}(1000\ \Omega + 1000\ \Omega) - (2.00 \times 10^{-3}\text{ H})(0) = 0$$

and $$i_{Li} = 6.00 \times 10^{-3}\text{ A} = 6.00\text{ mA}. \quad (31\text{-}66)$$

STEP 2: generate a differential equation in terms of i_L. We can solve the circuit of Fig. 31-102*b* to get an equation in terms of i_L in several ways. Here is one way. Writing a loop equation for the left-hand loop as we did in step 1 (but with variable i_L instead of value i_{Li}), we find

$$\mathscr{E} - i_1R_1 - i_LR_2 - L\frac{di_L}{dt} = 0. \quad (31\text{-}67)$$

Substituting known values, we get

$$12\text{ V} - i_1(1000\ \Omega) - i_L(1000\ \Omega) - (2.00 \times 10^{-3}\text{ H})\frac{di_L}{dt} = 0. \quad (31\text{-}68)$$

Although this is a differential equation involving i_L, it also contains the variable i_1. So, we need a second equation in those two variables. To produce a second equation we write a loop equation for the big loop:

$$\mathscr{E} - i_1R_1 - (i_1 - i_L)R_3 = 0. \quad (31\text{-}69)$$

Solving for i_1 and substituting known values, we find

$$i_1 = \frac{\mathscr{E} + i_LR_3}{R_2 + R_3} = \frac{12\text{ V} + (1000\ \Omega)i_L}{2000\ \Omega}. \quad (31\text{-}70)$$

Substituting this expression for i_1 into Eq. 31-68 and rearranging yield

$$\frac{di_L}{dt} + (7.50 \times 10^5\text{ s}^{-1})i_L = 3.00 \times 10^3\text{ A/s}, \quad (31\text{-}71)$$

which is a differential equation written in terms of known quantities and a single variable i_L, as we wanted.

STEP 3: solve the differential equation. Equation 31-71 is in the general form of a differential equation for *RL* circuits:

$$\frac{di_L}{dt} + Ai_L = B \qquad \text{(general } RL \text{ differential equation).} \quad (31\text{-}72)$$

Comparing Eqs. 31-71 and 31-72, we see that for the circuit of Fig. 31-102*b*

$$A = 7.50 \times 10^5\text{ s}^{-1} \quad (31\text{-}73)$$

and $$B = 3.00 \times 10^3\text{ A/s}. \quad (31\text{-}74)$$

The general solution to Eq. 31-72 is

$$i_L = i_{Lp} + Ke^{-At} = i_{Lp} + Ke^{-t/\tau} \qquad \text{(general solution).} \quad (31\text{-}75)$$

Here i_{Lp} is called a *particular solution, K* is a constant, and τ is the inductive time constant for the change in the current through the inductor. We see immediately that

$$\tau = \frac{1}{A}. \quad (31\text{-}76)$$

Substituting for A gives us

$$\tau = \frac{1}{7.50 \times 10^5\text{ s}^{-1}} = 1.333 \times 10^{-6}\text{ s} \approx 1.33\ \mu\text{s}. \quad (31\text{-}77)$$

To find i_{Lp} and K, we evaluate Eqs. 31-72 and 31-75 for the initial or final conditions of the inductor. First we get i_{Lp} from a final condition: When the circuit again reaches equilibrium, the current through the inductor will no longer change. That is, at $t = \infty$ (which means "a long time later"), $di_L/dt = 0$ and the current through the inductor has its final value $i_L = i_{Lf}$. Making these substitutions in Eq. 31-72, we have

$$0 + Ai_{Lf} = B,$$

or $$i_{Lf} = \frac{B}{A}. \quad (31\text{-}78)$$

But from Eq. 31-75 we also know that at $t = \infty$,

$$i_{Lf} = i_{Lp} + Ke^{-\infty/\tau} = i_{Lp} + 0 = i_{Lp}. \quad (31\text{-}79)$$

Comparing Eqs. 31-78 and 31-79, we see that the particular solution i_{Lp} is the final current i_{Lf} and that

$$i_{Lp} = \frac{B}{A}. \quad (31\text{-}80)$$

This is a general result, true for any inductor through which the current is changing in an *RL* circuit.

Next we get K from an initial condition: At time $t = 0$, the current through the inductor is $i_L = i_{Li}$. Making these substitutions in Eq. 31-75, we obtain

$$i_{Li} = i_{Lp} + Ke^{-0/\tau} = i_{Lp} + K. \quad (31\text{-}81)$$

Thus, $$K = i_{Li} - i_{Lp}. \quad (31\text{-}82)$$

This, too, is a general result, true for any inductor through which the current is changing in an *RL* circuit.

Substituting the values for A and B that we found earlier, Eq. 31-80 gives us the final current as

$$i_{Lp} = \frac{B}{A} = \frac{3.00 \times 10^3\text{ A/s}}{7.50 \times 10^5\text{ s}^{-1}} = 4.00 \times 10^{-3}\text{ A} \approx 4.00\text{ mA}. \quad (31\text{-}83)$$

Next, substituting this and the initial current $i_{Li} = 6.00$ mA we found in step 1, Eq. 31-82 yields

$$K = 6.00\text{ mA} - 4.00\text{ mA} = 2.00\text{ mA}. \quad (31\text{-}84)$$

Putting these results for i_{Lp}, K, and τ into Eq. 31-75, we find that when the switch is closed at $t = 0$, the current through the inductor varies as

$$i_L = 4.00\text{ mA} + (2.00\text{ mA})e^{-t/(1.33\ \mu\text{s})}. \quad \text{(Answer)}$$

(b) When the switch is closed at $t = 0$, does the current through the inductor increase or decrease as the circuit moves toward equilibrium? Graph i_L versus t.

SOLUTION: Comparing the final current $i_{Lp} = 4.00$ mA with the initial current $i_{Li} = 6.00$ mA, we see that the current decreases. Figure 31-102*c* is a graph for the first few time constants of the decrease.

(c) Write a program to solve the differential equation for i_L in an *RL* circuit where a current through the inductor is changing.

SOLUTION: Actually, we have already seen such a program, namely the one in Table 28-2. The only change we need to make is a mental one. Where you see a charge symbol, such as q, imagine it to be the corresponding current symbol, such as i_L.

131. Switch S in Fig. 31-102*b* has been closed a long time (the circuit is in equilibrium) when it is opened at $t = 0$. Find the current through the inductor for $t \geq 0$; graph the results.

132. In Fig. 31-103, $\mathscr{E} = 10.0$ V, $R_1 = 1000\ \Omega$, $R_2 = 2000\ \Omega$, and $L = 2.00$ mH. The circuit is initially in equilibrium (steady state) when switch S is closed at time $t = 0$. Find the current through the inductor for $t \geq 0$.

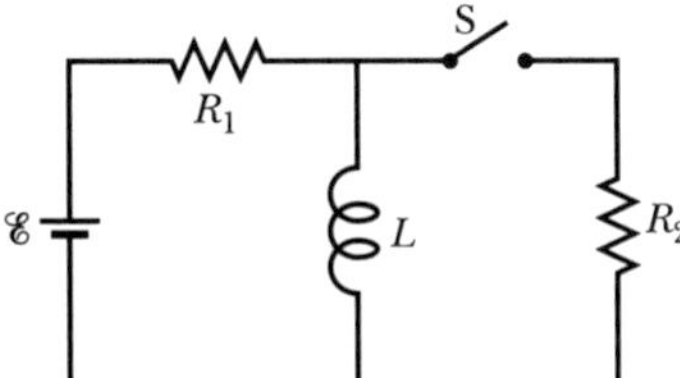

FIGURE 31-103 Problem 132.

133. In Fig. 31-104, $\mathscr{E} = 10.0$ V, $R_1 = R_2 = 1000\ \Omega$, and $L = 2.00$ mH. The circuit is initially in equilibrium (steady state) when switch S is closed at time $t = 0$. Find the current through (a) the inductor and (b) R_1 for $t \geq 0$; graph the results. When equilibrium is reached, the switch is opened. (c) Just as the switch is opened, how can the inductor maintain the current through itself?

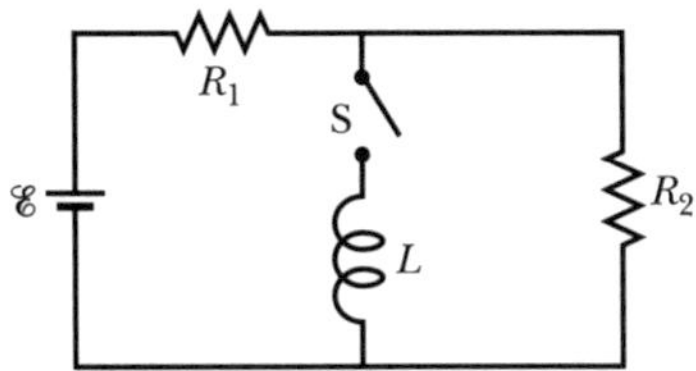

FIGURE 31-104 Problem 133.

134. In Fig. 31-105, $\mathscr{E} = 10.0$ V, $R_1 = 1000\ \Omega$, $R_2 = 2000\ \Omega$, $R_3 = 3000\ \Omega$, and $L = 4.00$ mH. The circuit is initially in equilibrium (steady state) when switch S is closed at time $t = 0$. Find the current through (a) the inductor and (b) R_3 for $t \geq 0$; graph the results. (c) What is the rate at which energy is being dissipated in R_1 at $t = \tau_L$, where τ_L is the inductive time constant of the circuit after the switch is closed? (d) Graph the rate at which energy is dissipated in R_1 for $t \geq 0$. When equilibrium is reached, the switch is reopened at $t = 0$ (the clock is reset then). (e) Find the current through the inductor for $t \geq 0$; graph the results.

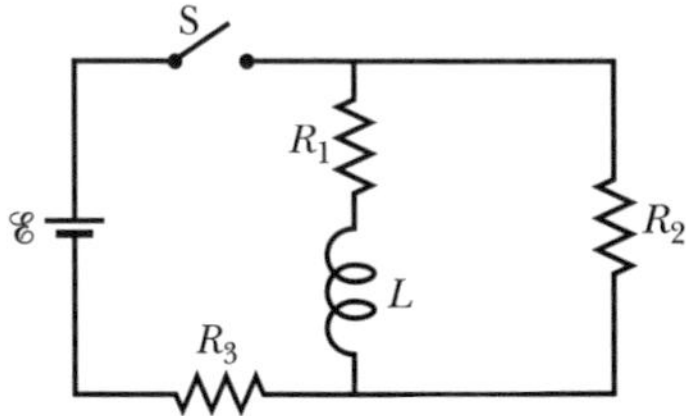

FIGURE 31-105 Problem 134.

135. In Fig. 31-106, $\mathscr{E} = 100$ V, $R_1 = 100\ \Omega$, $R_2 = 200\ \Omega$, $R_3 = 300\ \Omega$, and $L = \frac{4}{3}$ mH. The circuit is initially in equilibrium (steady state) when switch S is closed at time $t = 0$. (a) Find the current through the inductor for $t \geq 0$; graph the results. (b) Find the electric potential across R_1 for $t \geq 0$. When equilibrium is reached, the switch is reopened at $t = 0$ (the clock is reset then). (c) Find the current through the inductor for $t \geq 0$; graph the results.

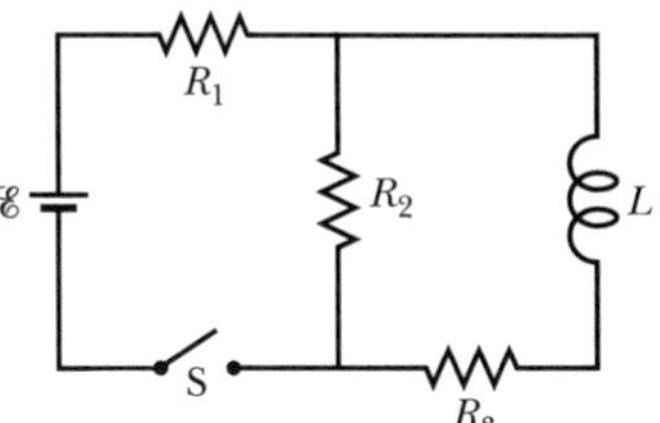

FIGURE 31-106 Problem 135.

Chapter Thirty-Two
Magnetism of Matter; Maxwell's Equations

QUESTIONS

17. The magnetic dipoles in a diamagnetic material are represented, for three situations, in Fig. 32-37. (For simplicity, the dipoles are assumed to point only up or down the page.) The three situations differ in the magnitude of a magnetic field applied to the material. (a) Is the applied field directed up or down the page? Rank the three situations according to (b) the magnitude of the applied field and (c) the magnetization of the material, greatest first.

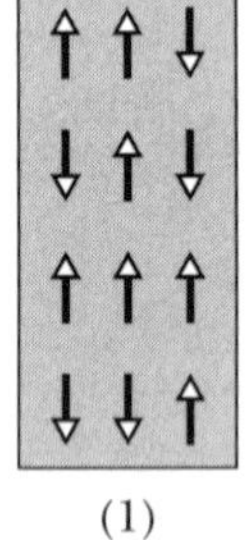
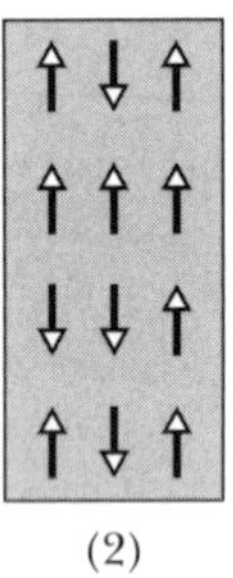
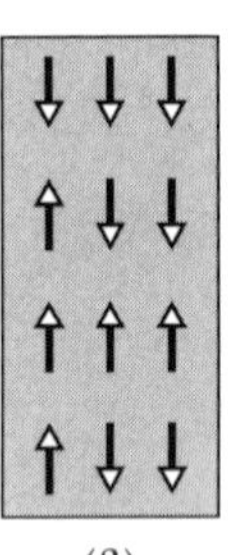

FIGURE 32-37 Question 17.

18. In Fig. 32-38, three orientations of a compass needle in a uniform magnetic field **B** are given, with the north and south ends of the needle labeled. The first two orientations involve the same angle θ_0. The needle can pivot about its mounting pin at its midpoint. (a) Rank the orientations according to the torque on the needle due to the magnetic field, greatest first. (b) For each orientation, the needle is released from rest. Rank the orientations according to the maximum kinetic energy the needle has during its rotation (assuming that the friction between needle and mounting pin is negligible), greatest first.

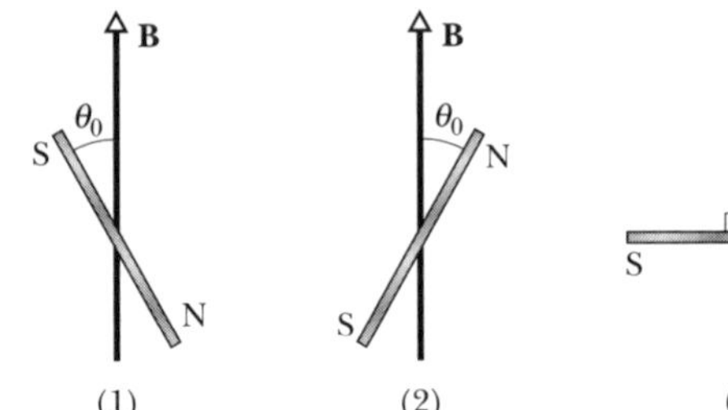

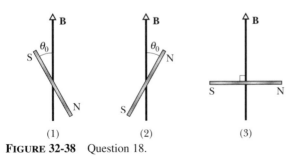

FIGURE 32-38 Question 18.

19. Figure 32-39 shows three situations in which identical compass needles are in uniform magnetic fields of magnitudes B_0 or $2B_0$. Each needle can rotate about its mounting pin at its midpoint. The needles are rotated from their usual alignment with the magnetic field by 1°, 2°, and 3°, respectively (the angles are exaggerated in the figure), and then released. Rank the three situations according to the resulting period of oscillation of the needle, greatest first.

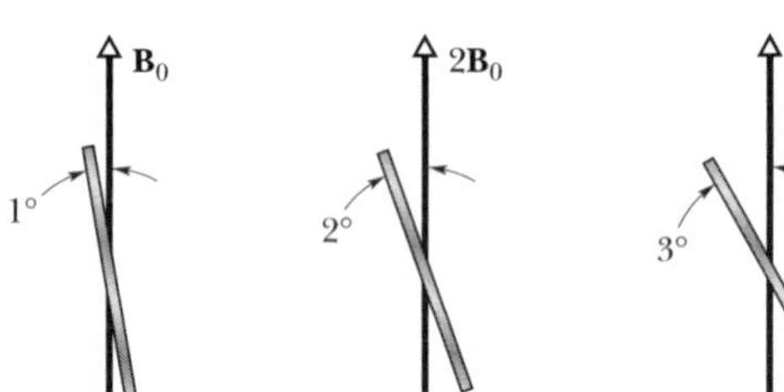

FIGURE 32-39 Question 19.

20. When two magnetic dipoles are near each other, they each experience the magnetic field of the other, and the two-dipole system has a certain potential energy. (a) If the magnetic dipoles are arranged side by side, does the system have greater potential energy when the dipole moments are antiparallel (as with the compass needles in Fig. 32-19*a*) or parallel (Fig. 32-19*b*)? (b) If, instead, the magnetic dipoles are arranged on the same axis, does the system have greater potential energy when the dipole moments are parallel (Fig. 32-19*c*) or antiparallel (Fig. 32-19*d*)?

21. When three magnetic dipoles are near each other, they each experience the magnetic field of the other two, and the three-dipole system has a certain potential energy. Figure 32-40 shows two arrangements in which three magnetic dipoles are side by side. Each dipole has the same magnitude of magnetic dipole moment, and the spacings between adjacent dipoles are the same.

In which arrangement is the potential energy of the three-dipole system greater?

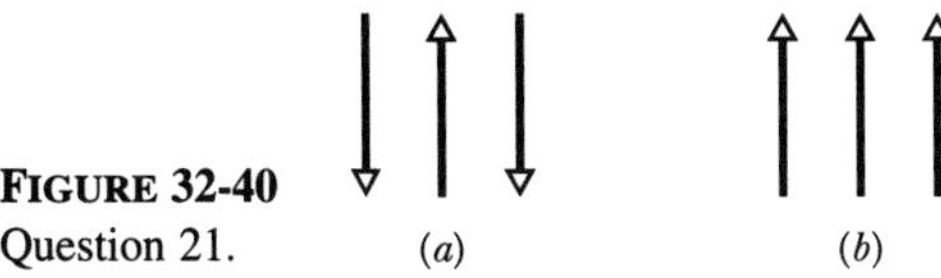

FIGURE 32-40 Question 21.

22. The hysteresis curves for three different materials are given in Fig. 32-41. To produce the curves, each material was placed in a coil (as in Fig. 32-11) and then the magnetic field B_0 of the coil was varied. (a) Are the materials paramagnetic or ferromagnetic? (b) Rank the materials according to the magnitude of their magnetic field when the current through the coil is zero, greatest first. (c) Rank the materials according to the magnitude of the applied magnetic field needed to cause the domains to begin changing in size, greatest first.

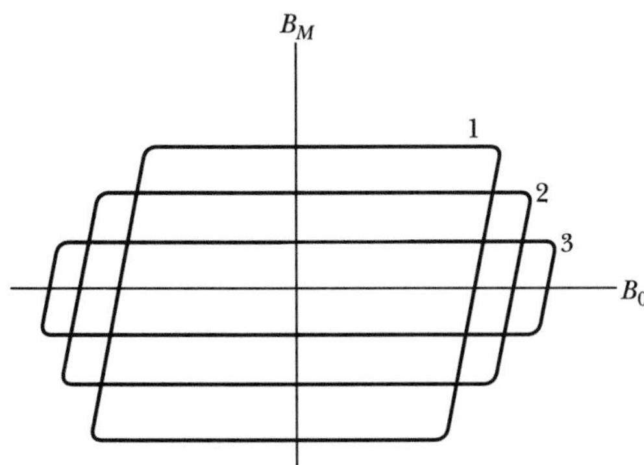

FIGURE 32-41 Question 22.

23. The Curie temperature of iron is 1043 K. If, instead, it were (a) 380 K or (b) 200 K, would VCRs and audiotape players work?

24. A sample of ferromagnetic material is thin enough to be considered planar; it is small enough to have only four domains. The sample, initially unmagnetized, is made magnetic by an applied field B_0. The dipoles of the domains, directed either up the page or down the page, are represented in Fig. 32-42 for three stages of the magnetizing process. Rank the stages according to (a) the magnitude of the applied field and (b) the magnetization of the sample, greatest first. (c) What is the direction of the applied field during the process?

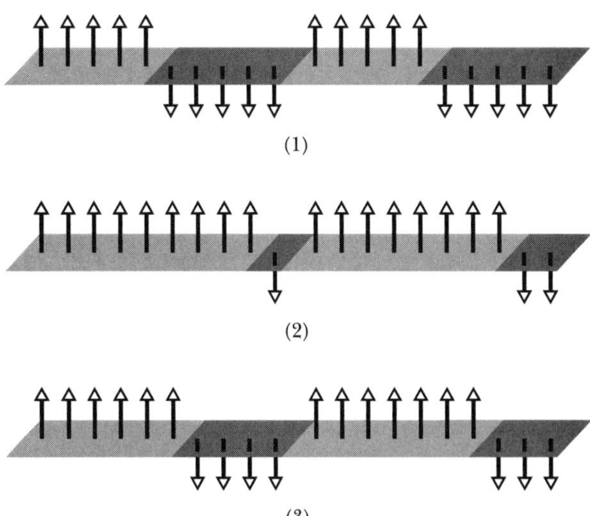

FIGURE 32-42 Question 24.

25. Figure 32-43 represents three rectangular samples of a ferromagnetic material in which the magnetic dipoles of the domains have been directed out of the page (encircled dot) by a very strong applied field B_0. In each sample, an island domain still has its magnetic field directed into the page (encircled ×). Sample 1 is one (pure) crystal. The other samples contain impurities collected along lines; domains cannot easily spread across such lines.

The applied field is now to be reversed and its magnitude kept moderate. The change causes the island domain to grow. (a) Rank the three samples according to the success of that growth, greatest growth first. Ferromagnetic materials in which the magnetic dipoles are easily changed are said to be *magnetically soft;* when the changes are difficult, requiring strong applied fields, the materials are said to be *magnetically hard.* (b) Of the three samples, which is the most magnetically hard?

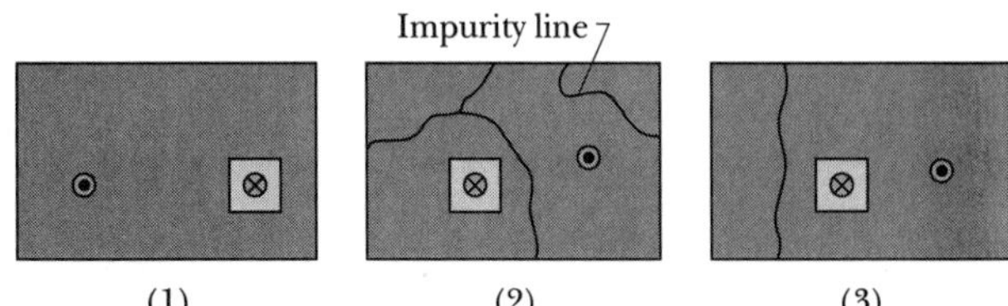

FIGURE 32-43 Question 25.

26. Figure 32-44 shows a section of a long straight wire through which there is an uniform current i in the positive direction of an x axis. Point a is just above the wire, in the xy plane. Assume that the current is constant. (a) What is the direction of the magnetic field **B** at point a due to current i? (b) What is the direction

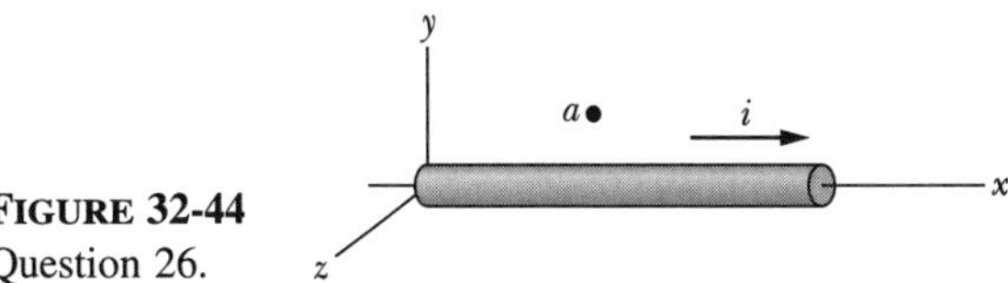

FIGURE 32-44 Question 26.

of the electric field **E** responsible for the current? (c) Is the electric field constant or is it varying with time?

Now assume that current i is increasing. (d) Is the magnitude of the electric field through the wire increasing, decreasing, or remaining the same? (e) What is the direction of $d\mathbf{E}/dt$? (f) What is the direction of the displacement current i_d associated with $d\mathbf{E}/dt$? (g) What is the direction of the magnetic field $\mathbf{B}_d$ at point a due to i_d; and is this direction the same as or opposite to that of the magnetic field **B** due to current i?

Finally, assume that current i is decreasing. (h) Are the magnetic fields at point a due to i and i_d in the same direction or in opposite directions?

27. *Organizing question:* Figure 32-45 shows a circular region of radius R in which an electric flux Φ_E is directed out of the page. It also shows two integration paths (of radii r_1 and r_2) that form circles concentric with the circular region. Set up the right side of the Ampere–Maxwell law (Eq. 32-30) for paths 1 and 2 for the following situations, where either total flux Φ_E, the flux $\Phi_{E,\text{enc}}$ encircled by a path of radius r, or the magnitude E of the associated electric field is given. Perform any differentiation that may be required, but not any integration (and be wary of pesky subscripts).

SITUATION	FLUX OR FIELD
a	$\Phi_E = (4\ \text{V}\cdot\text{m/s})t$, uniform
b	$E = (2\ \text{V/m}\cdot\text{s})t$, uniform
c	$E = (3\ \text{V/m})\left(\frac{r}{R}\right)$
d	$\Phi_{E,\text{enc}} = (5\ \text{V}\cdot\text{m/s})\left(\frac{r}{R}\right)t$
e	$E = (4\ \text{V/m}\cdot\text{s})\left(1-\frac{r}{R}\right)t$

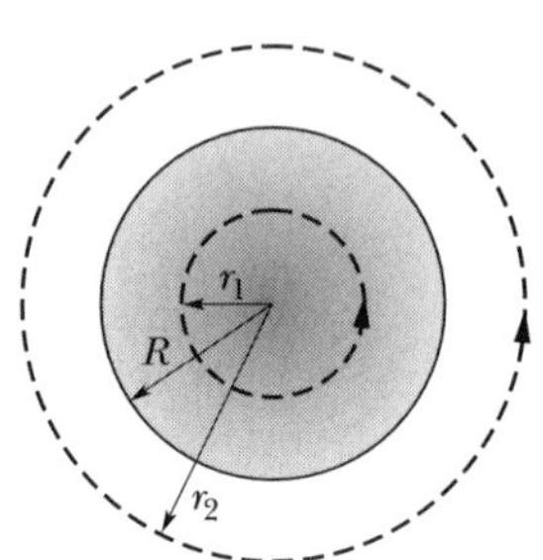

FIGURE 32-45
Questions 27 and 28.

28. *Organizing question:* Figure 32-45 shows a circular region of radius R in which a displacement current i_d is directed out of the page. It also shows two integration paths (of radii r_1 and r_2) that form circles concentric with the circular region. Set up the right side of the Ampere–Maxwell law (Eq. 32-30) for paths 1 and 2 for the following situations, where either the total displacement current i_d, the displacement current $i_{d,\text{enc}}$ encircled by a path of radius r, or the displacement current density J_d is given. (Be alert to irksome subscripts.)

SITUATION	CURRENT OR CURRENT DENSITY
a	$i_d = 2$ A, uniform
b	$J_d = 3\ \text{A/m}^2$, uniform
c	$i_{d,\text{enc}} = (5\ \text{A})\left(\frac{r}{R}\right)$
d	$J_d = (6\ \text{A/m}^2)\left(1-\frac{r}{R}\right)$

EXERCISES & PROBLEMS

52. A magnetic flux of 7.0 mW is directed outward through the bottom face of the closed surface shown in Fig. 32-46. Along the top face (which has a radius of 4.2 cm) there is a magnetic field **B** of magnitude 0.40 T directed perpendicular to the face. What is the magnetic flux (magnitude and direction) through the curved sides of the surface?

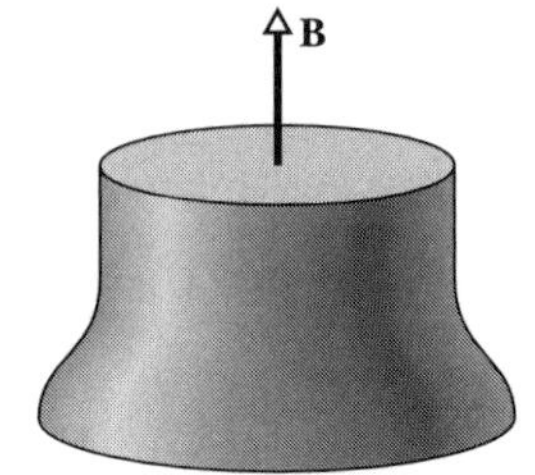

FIGURE 32-46
Problem 52.

53. The energy difference between parallel and antiparallel alignments of the z component of an electron's spin magnetic dipole moment with an external magnetic field **B** directed along the z component is 6.0×10^{-25} J. What is the magnitude of **B**?

54. What are the measured components of the orbital magnetic dipole moment of an electron with (a) $m_l = 3$ and (b) $m_l = -4$?

55. A parallel-plate capacitor with circular plates is being charged. Consider a circular loop centered on the central axis between the plates. If the loop's radius of 3.00 cm is larger than the plate radius, then what is the displacement current between the plates when the magnetic field along the loop has magnitude 2.00 μT?

56. A parallel-plate capacitor with circular plates is being charged. Consider a circular loop centered on the central axis between the plates. The loop radius is 0.20 m; the plate radius is

0.10 m; and the displacement current through the loop is 2.0 A. What is the rate at which the electric field between the plates is changing?

57. In Fig. 32-47, a parallel-plate capacitor is being discharged by a current of $i = 5.0$ A. The plates are square with edge length 8.0 mm. (a) What is the rate at which the electric field between the plates is changing? (b) What is the value of $\int \mathbf{B} \cdot d\mathbf{s}$ around the dashed path, where $H = 2.0$ mm and $W = 3.0$ mm?

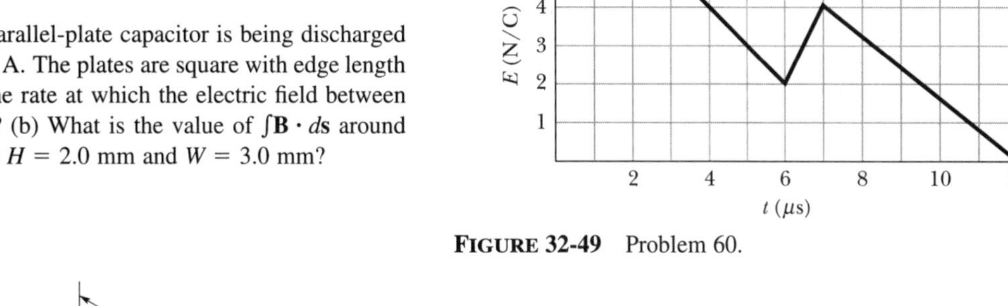

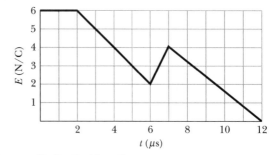

FIGURE 32-49 Problem 60.

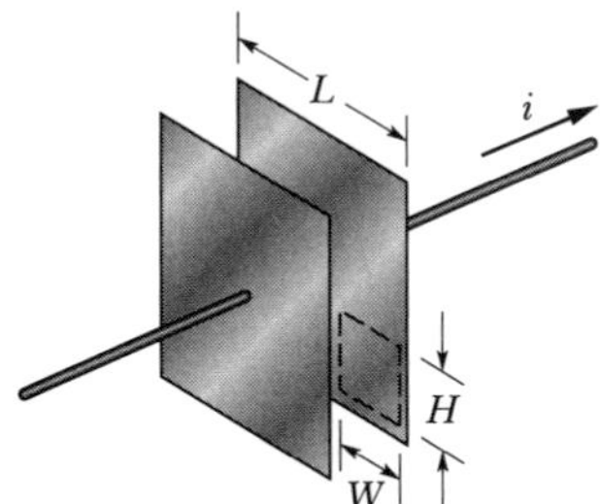

FIGURE 32-47 Problem 57.

58. A capacitor with parallel circular plates of radius R is discharging via a current of 12.0 A. Consider a loop of radius $R/3$ that is centered on the central axis between the plates. (a) How much displacement current is encircled by the loop? The maximum induced magnetic field has a magnitude of 12.0 mT. (b) At what radial distance from the central axis of the plate is the magnitude of the induced magnetic field 3.00 mT?

59. Figure 32-48 shows two charged plates with plate area 4.0×10^{-2} m^2. The electric field **E** between them is drawn for time $t = 0$, and the field magnitude is given by

$$E = 4.0 \times 10^5 \text{ V/m} - (6.0 \times 10^4 \text{ V/m} \cdot \text{s})t,$$

with positive values corresponding to a field up the page and negative values down the page. In the figure, is the induced magnetic field clockwise or counterclockwise at (a) $t = 0$ and (b) $t = 100$ s? What is the magnitude of the displacement current between the plates at (c) $t = 0$ and (d) $t = 100$ s?

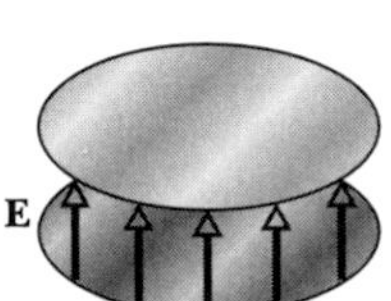

FIGURE 32-48 Problem 59.

60. Figure 32-49 gives the variation of an electric field that is perpendicular to a circular area of 2.0 m^2. During the variation, what is the greatest displacement current through the area?

Tutorial Problem

61. Maxwell's equations are written in the textbook in what is called their *integral form,* in which line and surface integrals of the electric and magnetic fields are related to various electric quantities associated with the enclosed surfaces and volumes, respectively. Maxwell's equations can also be written at single points in space in what is called their *differential form,* which involves partial derivatives. In a vacuum these equations are:

Gauss' law: $\nabla \cdot \mathbf{E} = 0$

Gauss' law for magnetism: $\nabla \cdot \mathbf{B} = 0$

Faraday's law: $\nabla \times \mathbf{E} = -\dfrac{\partial \mathrm{B}}{\partial t}$

Ampere–Maxwell law: $\nabla \times \mathbf{B} = \mu_0 \epsilon_0 \dfrac{\partial \mathrm{E}}{\partial t}$

In these expressions, ∇ (*del,* or *nabla*) represents the vector differential operator

$$\nabla \equiv \mathbf{i}\left(\frac{\partial}{\partial x}\right) + \mathbf{j}\left(\frac{\partial}{\partial y}\right) + \mathbf{k}\left(\frac{\partial}{\partial z}\right),$$

which is treated just like a vector but has to have some function on the right side on which it can operate. For example, its dot product with a vector is a scalar function called the *divergence* (abbreviated div):

$$\text{div } \mathbf{B} \equiv \nabla \cdot \mathbf{B} = \frac{\partial B_x}{\partial x} + \frac{\partial B_y}{\partial y} + \frac{\partial B_z}{\partial z}.$$

By Gauss' law for magnetism, this quantity must be zero for any real magnetic field, which greatly restricts the possible functional form of the magnetic field.

Similarly, the vector cross product of nabla with a vector gives another vector called the curl of that vector:

$$\text{curl } \mathbf{E} \equiv \nabla \times \mathbf{E} = \mathbf{i}\left(\frac{\partial E_z}{\partial y} - \frac{\partial E_y}{\partial z}\right) + \mathbf{j}\left(\frac{\partial E_x}{\partial z} - \frac{\partial E_z}{\partial x}\right) + \mathbf{k}\left(\frac{\partial E_y}{\partial x} - \frac{\partial E_x}{\partial y}\right).$$

Thus the differential form of Faraday's law says that the curl of the electric field vector must equal the negative of the partial time derivative of the magnetic field. This is equivalent to three separate equations for the three Cartesian coordinates; for example, the equation for the x components in Faraday's law gives

$$\left(\frac{\partial E_z}{\partial y} - \frac{\partial E_y}{\partial z}\right) = -\frac{\partial B_x}{\partial t}.$$

(a) Maxwell's treatise on electromagnetism was written before vectors were used in physics; so he had more than four equations. When written in terms of components, how many separate (scalar) differential equations are there in Maxwell's equations?

Now let's consider a plane electromagnetic wave moving in the positive x direction with speed of light c. That the wave is a plane wave means that its behavior (its electric and magnetic fields) is the same at all points in the yz plane. Or, in other words, the electric and magnetic fields have no dependence on y or z, only on x and t. Thus we expect the electric and magnetic fields to be expressible as $\mathbf{E}(x, t)$ and $\mathbf{B}(x, t)$. (b) Explain why the electric and magnetic fields of this plane electromagnetic wave must actually be functions of $x - ct$.

Now let's take the plane electromagnetic wave to be a sinusoidal wave with angular wave number k. Let's write its electric and magnetic fields in the form

$$\mathbf{E}(x - ct) = (E_{x0}\,\mathbf{i} + E_{y0}\,\mathbf{j} + E_{z0}\,\mathbf{k}) \sin k(x - ct)$$

and $$\mathbf{B}(x - ct) = (B_{x0}\,\mathbf{i} + B_{y0}\,\mathbf{j} + B_{z0}\,\mathbf{k}) \sin k(x - ct + \phi),$$

where we are allowing the possibilities that both fields have components along all three axes and the magnetic field has a different phase constant. (c) Explain how Gauss' law predicts that the electric field of the wave is transverse (in other words, E_{x0} is zero) and how Gauss' law for magnetism predicts that the magnetic field of the wave is transverse.

From here on, suppose that $E_{z0} = 0$. (d) Explain how our freedom to choose a coordinate system makes this possible, without loss of generality. (e) Apply Faraday's law to show that (1) the magnetic field must be perpendicular to the electric field (i.e., $B_{y0} = 0$), (2) there is no phase difference between the electric and magnetic fields (i.e., $\phi = 0$), and (3) the magnetic and electric field magnitudes are related by $B_{z0} = E_{y0}/c$.

If you have successfully proved the results of previous parts, you should now have

$$\mathbf{E}(x - ct) = [E_{y0} \sin k(x - ct)]\,\mathbf{j}$$

and

$$\mathbf{B}(x - ct) = \left[\left(\frac{E_{y0}}{c}\right) \sin k(x - ct)\right]\mathbf{k}.$$

(f) Substitute these expressions into the Ampere–Maxwell law. What new equation results from this substitution?

Which of Maxwell's equations can be used to prove the following characteristics of the plane electromagnetic wave: (g) the electric field is transverse; (h) the square of the speed of propagation of the waves is equal to $1/\mu_0\epsilon_0$; (i) there is no phase difference between the electric and magnetic fields; (j) the magnetic and electric field magnitudes are related by $cB = E$; (k) the magnetic field is transverse; (l) the electric and magnetic fields are perpendicular to each other? The grand finale! (m) Draw the coordinate axes for this wave, and sketch the appearance at $t = 0$ of the electric and magnetic field vectors along the x axis.

Answers

(a) There are eight: one each for Gauss' laws, and three each for the other two.

(b) For the wave to be a wave propagating with speed c in the positive x direction, the wave function must be a function of $x - ct$ if the values of the fields at some point x are to have shifted to a new value of x that is greater by ct.

(c) Gauss' law states that, in a vacuum, $\partial E_x/\partial x + \partial E_y/\partial y + \partial E_z/\partial z = 0$. Since there is no y or z dependence to the wave, this law reduces to $\partial E_x/\partial x = 0$, which can be true only if $E_{x0} = 0$ (not a nonzero constant because that wouldn't be a wave).

Gauss' law for magnetism states that, in a vacuum, $\partial B_x/\partial x + \partial B_y/\partial y + \partial B_z/\partial z = 0$. Since there is no y or z dependence to the wave, this law reduces to $\partial B_x/\partial x = 0$, which can be true only if $B_{x0} = 0$.

Thus the electric and magnetic fields can have components along only the y and z directions; in other words, they can have only components that are transverse to the direction of propagation of the wave, which is along x.

(d) Since we are free to choose any directions we want for two of our coordinate axes (the third is then fixed by our necessity to have a right-hand coordinate system), we can simply choose to have the y axis along the direction of the electric field. (We previously chose the x direction along the direction of propagation of the wave.) There is no loss of generality. This choice does not affect anything physical; it affects only the numerical values of the components. This choice obviously simplifies our expression for the electric field, since it has only one component. If we are lucky, it may not complicate the expression for the magnetic field.

(e) Faraday's law for an electric field in the y direction and a magnetic field in the y and z directions becomes (since there is no y or z dependence to the fields)

$$\mathbf{k}\left(\frac{\partial E_y}{\partial x}\right) = -\mathbf{j}\left(\frac{\partial B_y}{\partial t}\right) - \mathbf{k}\left(\frac{\partial B_z}{\partial t}\right).$$

(1) The y components of this vector equation imply that $0 = \partial B_y/\partial t$, which can be true only with $B_{y0} = 0$. So the magnetic field is only in the z direction; that is, it is perpendicular to the electric field (as well as to the direction of propagation). (2) The z components give $\partial E_y/\partial x = \partial B_z/\partial t$, or

$$kE_{y0} \cos k(x - ct) = -[-kcB_{z0} \cos k(x - ct - \phi)].$$

This equation can be satisfied at all times only if $\phi = 0$, which means that there is no phase difference between the electric and magnetic fields. (3) The equation then shows that

$$kE_{y0} = kcB_{z0},$$

which means that $E_{y0} = cB_{z0}$, or $B_{z0} = E_{y0}/c$.

(f) Substituting into the Ampere–Maxwell law, we find

$$\mathbf{j}\left(-\frac{\partial B_z}{\partial x}\right) = \mu_0\epsilon_0\left(\frac{\partial E_y}{\partial t}\right)\mathbf{j}.$$

Thus we have

$$-\frac{\partial B_z}{\partial x} = \mu_0\epsilon_0\left(\frac{\partial E_y}{\partial t}\right),$$

which leads to

$$-k\left(\frac{E_{y0}}{c}\right)\cos k(x-ct) = \mu_0\epsilon_0(-kcE_{y0})\cos k(x-ct)$$

in which most of the terms cancel out to give

$$\frac{1}{c} = \mu_0\epsilon_0 c$$

or

$$c = \sqrt{\frac{1}{\mu_0\epsilon_0}}.$$

This shows how the speed c of electromagnetic waves is related to the permittivity and permeability of the vacuum.

(g) Gauss' law (for electricity); **(h)** Ampere–Maxwell law; **(i)** Faraday's law of induction; **(j)** Faraday's law of induction; **(k)** Gauss' law for magnetism; **(l)** Faraday's law of induction. **(m)** See Fig. 34-5 on p. 845 in the textbook.

Chapter Thirty-Three Electromagnetic Oscillations and Alternating Current

QUESTIONS

17. For each of the curves of $q(t)$ in Fig. 33-37 for an LC circuit, determine the smallest positive phase constant ϕ required in Eq. 33-12 to produce the curve.

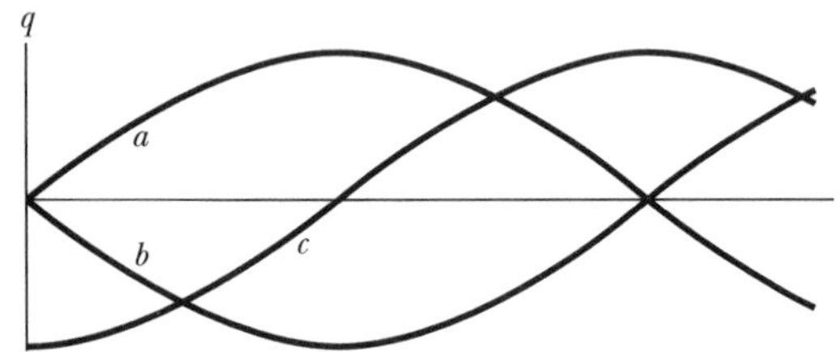

FIGURE 33-37 Question 17.

18. What is the smallest positive phase constant ϕ in Eq. 33-17 that will give the curve of Fig. 33-38 for the magnetic field energy U_B in an LC circuit?

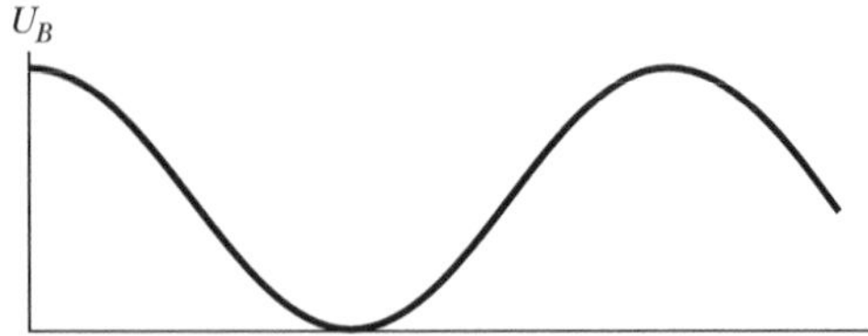

FIGURE 33-38 Question 18.

19. Figure 33-39 represents four stages in the rotation of a loop in a uniform magnetic field, as in Fig. 33-6 except that here **B** is directed up the page. One side of the loop is drawn with a broad line so that we can follow the rotation. In (*a*) the loop is tipped downward toward us (the near side is below the horizontal); in (*b*) it is tipped upward (the near side is above the horizontal); in (*c*) it is tipped downward; and in (*d*) it is tipped upward. In each stage, is the direction of the current induced in the side drawn with a broad line rightward or leftward?

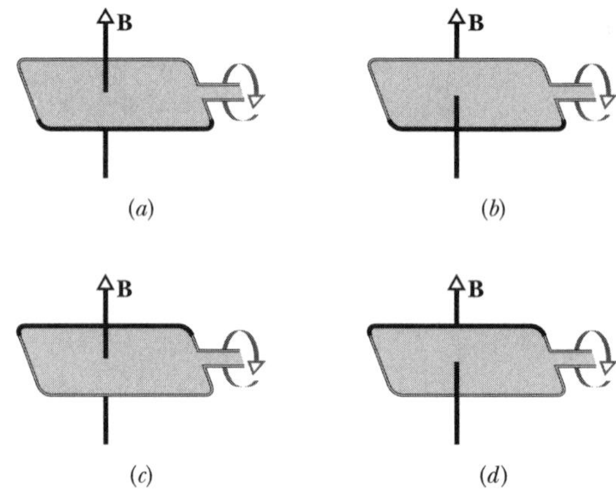

FIGURE 33-39 Question 19.

20. An alternating emf source with a certain emf amplitude is connected, in turn, to a resistor, a capacitor, and then an inductor. Once connected to one of the devices, the driving frequency f_d is varied and the amplitude I of the resulting current through the device is measured and plotted. Which of the three plots in Fig. 33-40 corresponds to which of the three devices?

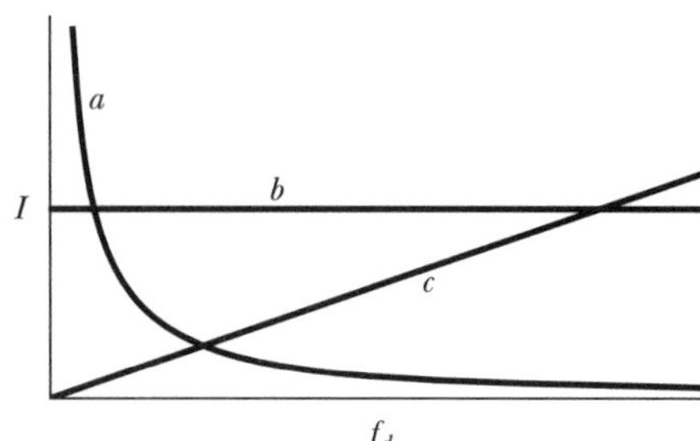

FIGURE 33-40 Question 20.

21. Figure 33-41*a* gives the potential v_L (t) across an inductor due to an alternating emf source, and Fig. 33-41*b* gives the phasor

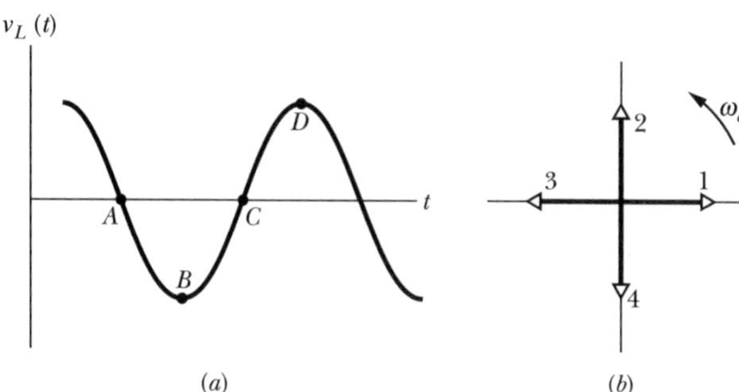

FIGURE 33-41 Question 21.

diagram for that potential at four times. (a) Which of the lettered points in Fig. 33-41*a* corresponds to which of the numbered phasor orientations in Fig. 33-41*b*? If the frequency of the emf source is increased without any change in the amplitude of the source's emf, do the following increase, decrease, or remain the same: (b) the amplitude of the potential across the inductor, (c) the rotation rate for the phasor of that potential, (d) the reactance of the inductor, and (e) the amplitude of the current through the inductor?

22. (a) Does the phasor diagram of Fig. 33-42 correspond to an alternating emf source connected to a resistor, a capacitor, or an inductor? (b) If the angular speed of the phasors is increased, does the length of the current phasor increase or decrease if the scale of the diagram is maintained?

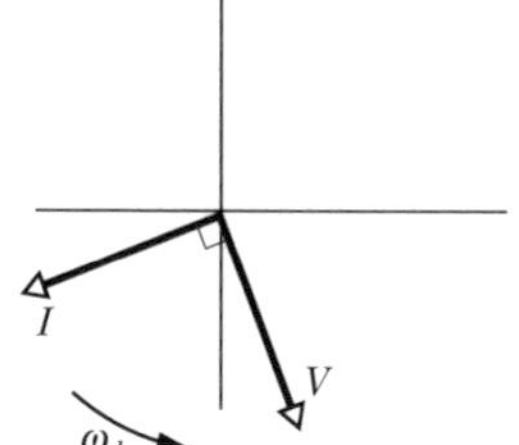

FIGURE 33-42 Question 22.

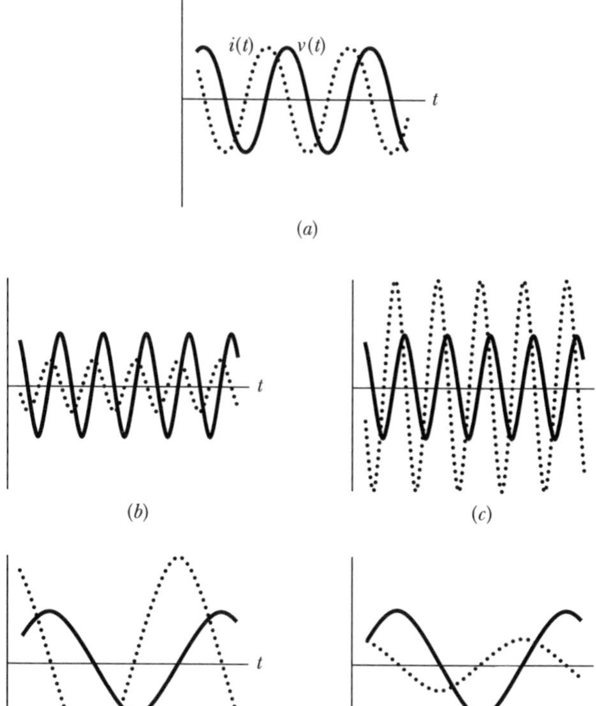

FIGURE 33-43 Question 23.

23. (a) Does the graph of Fig. 33-43*a* correspond to an alternating emf source connected to a resistor, a capacitor, or an inductor? (b) If the driving frequency of the emf source is increased, which one of the other four graphs in Fig. 33-43 (all to the same scale as Fig. 33-43*a*) then best gives $i(t)$? (c) Which one is best if the driving frequency is decreased?

24. *Organizing question (memory aids):* Fig. 33-44 is a sketch of the hunting grounds at Resonance Hill. Its features might help you remember the features of resonance curves (such as those in Fig. 33-13) and series *RLC* circuits.

Two "duc" hunters (with way-cool L. L. Bean caps) are below the hill when they spot three wild "ducs" flying beyond the hill. (a) Are the caps below or beyond Resonance Hill? (b) Are the ducs below or beyond the hill? (c) Thus, is the angular frequency of a mainly capacitive circuit below or beyond the resonance frequency of the circuit? (d) How about the angular frequency of a mainly inductive circuit? (e) Is the hunter with the higher cap (higher capacitance) to the right or left of the other hunter? (f) Is the highest duc (higher inductance) to the right or left of the other ducs? (g) Thus, if we increase the capacitance C or the inductance L in a driven series *RLC* circuit, does the circuit move to the right or left on a resonance curve such as those in Fig. 33-13?

Five pairs of $\mathscr{E}_m$ and I phasors are shown in Fig. 33-44, roughly angled, sized, and arranged (as in phasor diagrams) to correspond to their location relative to Resonance Hill. (h) Label the phasors with $\mathscr{E}_m$ or I. (*Hint:* In the vegetation of Resonance Hill, emf stalks thrust upward while they sprout "eye thorns" outward, away from the hill. Those thorns grow longer higher on the hill, which is why the hunters are not there.)

(i) Do positive ϕ values occur on the positive (rightward) or negative (leftward) side of the hill? (j) How about negative ϕ values? Label the angle between the $\mathscr{E}_m$ and I in each pair of phasors with the corresponding sign of the angle ϕ between them. (k) Do ϕ values increase or decrease with distance from the hill?

25. Suppose that a graph for a series *RLC* circuit is initially given by Fig. 33-12*b*. You wish to vary one of the following parameters of the circuit in order to change the graph first to that of Fig. 33-12*f* and then to that of Fig. 33-12*d*: (a) the driving frequency

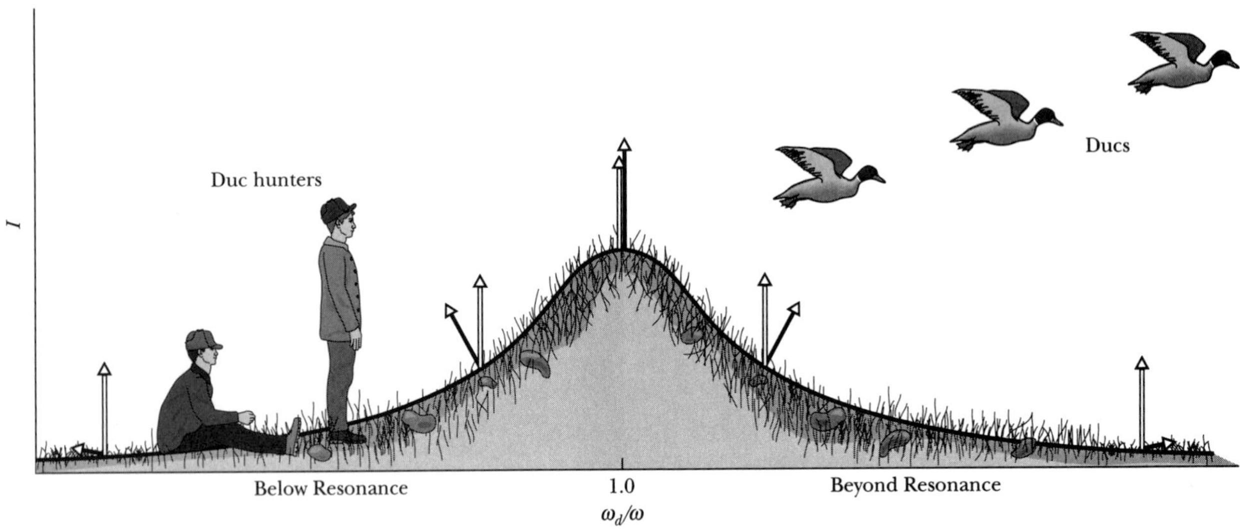

FIGURE 33-44 Question 24.

f_d; (b) the inductance L, (c) the capacitance C, or (d) the resistance R. For each parameter, determine if its variation can produce the desired changes and, if so, decide if the value of the parameter should be increased or decreased.

26. The current amplitude I for a particular series RLC circuit is given as a function of ω_d in two ways in Fig. 33-45. Does the peak in Fig. 33-45*a* shift leftward, rightward, upward, or downward or does it remain in place if we increase (a) the inductance, (b) the capacitance, and (c) the resistance? (d) Repeat the question for the peak in Fig. 33-45*b*.

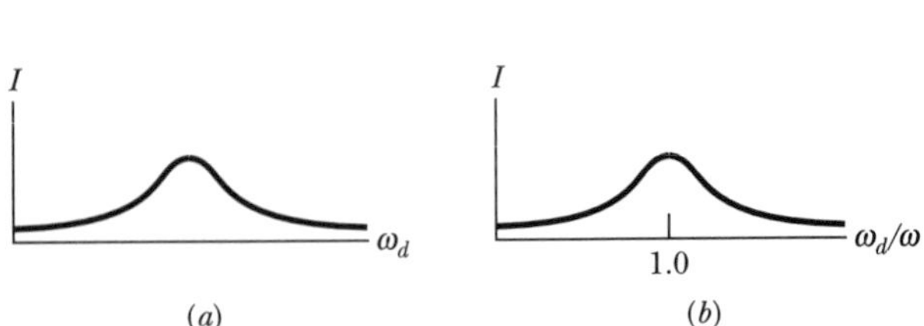

FIGURE 33-45 Question 26.

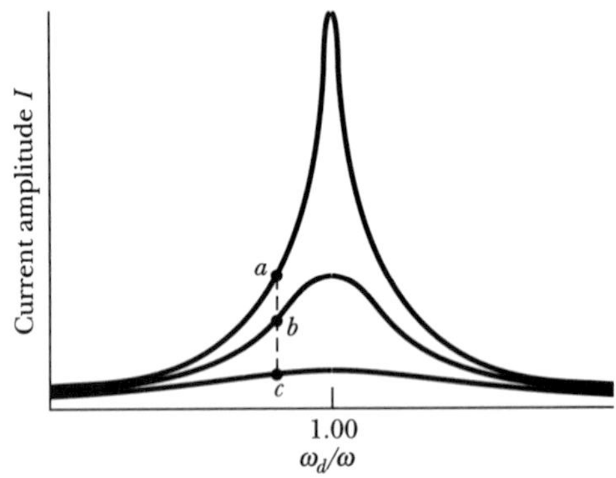

FIGURE 33-46 Question 27.

27. Figure 33-46 gives three resonance curves for the driven RLC circuit of Fig. 33-7, with the same values of L and C but different values of R. Three points are indicated at a particular value of ω_d/ω. Rank those points according to (a) the value of R and (b) the value of phase constant ϕ, greatest first.

28. Figure 33-47 shows an RLC circuit that is driven by an emf source of fixed amplitude $\mathscr{E}_m$. Initially the circuit consists of one resistor of resistance R, one inductor of inductance L, and one capacitor of capacitance C, and the driving frequency matches the natural frequency. Then switches S_1, S_2, S_3, and S_4 are closed, in

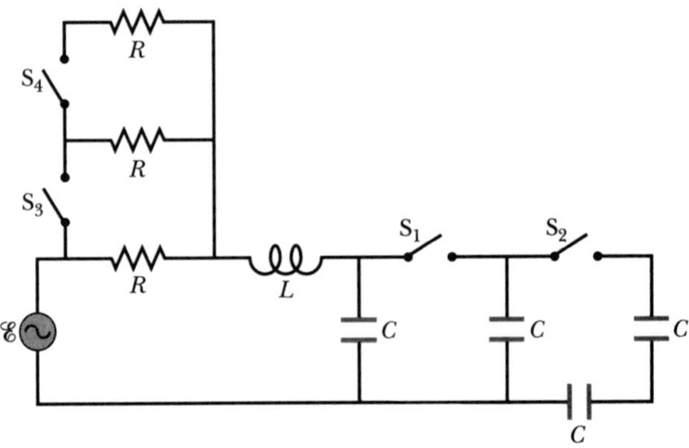

FIGURE 33-47 Question 28 and Problem 103.

that order. The closings bring in capacitors identical to the first one or resistors identical to the first one.

(a) Fill in the first blank column in the following table by indicating what happens to the equivalent capacitance C_{eq} of the circuit after each switch closing. Use inc for increase, dec for decrease, and un for unchanged. Similarly fill in (b) the second blank column for the natural (or resonant) frequency f of the circuit and (c) the equivalent resistance R_{eq}.

The fourth blank column is to indicate how a switch closing might shift a circuit left or right on a resonance curve (like one of the curves in Fig. 33-13) or, instead, how it might cause the circuit to jump up or down to a new, different resonance curve (like one of the other two curves in Fig. 33-13). (d) Fill in that column with either right, left, up, or down. In the next three columns indicate if the following increase, decrease, or go unchanged: (e) the impedance Z, (f) the current amplitude I, and (g) the magnitude (ϕ MAG) of the phase constant ϕ. (h) In the last column, give the sign of ϕ (+ or −) or indicate that it is zero.

CLOSING	C_{eq}	f	R_{eq}	SHIFT	Z	I	ϕ MAG	ϕ SIGN
S_1								
S_2								
S_3								
S_4								

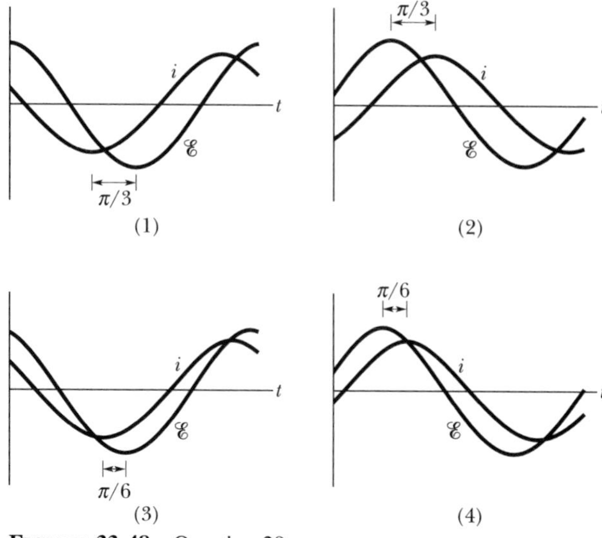

FIGURE 33-48 Question 29.

29. Figure 33-48 shows the current i and driving emf $\mathscr{E}$ for four series RLC circuits having the same I_{rms} and the same $\mathscr{E}_{rms}$. The phase (in radians) between the two curves for each circuit is indicated. (a) Rank the circuits according to their power factors, greatest first. Then rank them according to the rates at which energy is dissipated in (b) the resistor, (c) the inductor, and (d) the capacitor, greatest first.

30. The general configuration for electrical power transmission to your home is represented in Fig. 33-49. Transformers 1 and 2 are employed. For each transformer, determine (a) if it is a step-up or step-down transformer and (b) if the number of turns in the secondary is greater than or less than that in the primary.

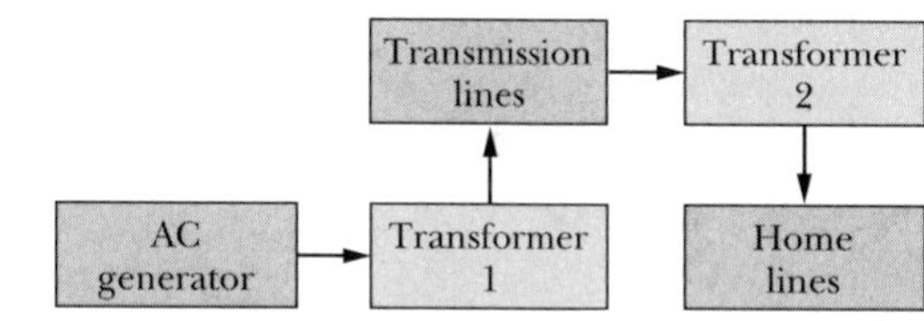

FIGURE 33-49 Question 30.

EXERCISES & PROBLEMS

96. A 7.00 μF capacitor has an initial potential of 12 V when it is connected across an inductor. The combination then oscillates at a frequency of 715 Hz. What is the inductance of the inductor?

97. An LC oscillator consists of a 2.00 nF capacitor and a 2.00 mH inductor. The maximum voltage is 4.00 V. What are (a) the frequency of the oscillations, (b) the maximum current, (c) the maximum energy stored in the inductor, and (d) the maximum rate di/dt at which the current changes?

98. An LC oscillator has a 1.00 nF capacitance, a 3.00 V maximum potential on the capacitor, and a 1.73 mA maximum current. (a) What is that oscillator's period of oscillation? What are the maximum energies stored in (b) the capacitor and (c) the inductor? (d) What is the maximum rate at which the current changes? (e) What is the maximum rate at which the inductor gains energy?

99. In an oscillating LC circuit, $L = 8.00$ mH and $C = 1.40$ μF. At time $t = 0$, the current is maximum at 12.0 mA. (a) What is the maximum charge on the capacitor during the oscillations? (b) At what times t is the rate of change of energy in the capacitor maximum? (c) What is that maximum rate of change?

100. A series RLC circuit is driven by a generator at frequency 1050 Hz. The inductance is 90 mH; the capacitance is 0.50 μF; and the phase constant has a magnitude of 60° (you should supply the appropriate sign for the angle). (a) What is the resistance? To increase the current amplitude in the circuit should we (b) increase or decrease the driving frequency, (c) increase or decrease the inductance, and (d) increase or decrease the capacitance?

101. A series RLC circuit is driven such that, during the oscillations, the maximum voltage across the inductor is 1.50 times the maximum voltage across the capacitor and 2.00 times the maximum voltage across the resistor. (a) What is the phase angle? (b) Is the circuit inductive, capacitive, or in resonance? The resistance is 49.9 Ω, and the current amplitude is 200 mA. (c) What is the amplitude of the driving emf?

102. A series *RLC* circuit is driven by a generator at a frequency of 2000 Hz and an emf amplitude of 170 V. The inductance is 60 mH, the capacitance is 0.40 μF, and the resistance is 200 Ω. (a) What is the phase constant in radians? (b) What is the current amplitude through the circuit?

103. In Question 28 and Fig. 33-47, $\mathscr{E}_m = 12.0$ V, $C = 2.00$ μF, $L = 2.00$ mH, and $R = 12.0$ Ω. (a) Fill in the table's first blank column with the values of the equivalent capacitance C_{eq} of the circuit after each switch closing. Similarly, fill in the other columns with values for (b) the natural (or resonance) frequency *f*, (c) the equivalent resistance R_{eq}, (d) the impedance *Z*, and (e) the current amplitude *I*. (Avoid rounding off the numbers until all calculations are finished.)

104. When under load, a certain electric motor draws an rms current of 3.00 A when operating under an rms voltage of 220 V. It has a resistance of 24.0 Ω and no capacitive reactance. What is its inductive reactance?

105. A series connection of resistor ($R = 35.0$ Ω), capacitor ($C = 8.65$ μF), and inductor ($L = 50.0$ mH) is driven at 1200 Hz by an alternating source operating at an rms voltage of 112 V. (a) What is the rms current? What are the rms voltages across (b) the resistor, (c) the capacitor, and (d) the inductor? What is the average rate at which energy is dissipated in (e) the resistor, (f) the capacitor, and (g) the inductor?

Tutorial Problems

106. Consider a series *RLC* circuit that has an emf of rms amplitude 120 V and frequency *f*. Suppose the inductance is 20.0 mH, the resistance is 30.0 Ω, and the capacitance is 40.0 μF. (a) Sketch the circuit. Make a list of the quantities whose numerical values are provided here, giving them the correct symbol, and also determine the resonant frequency of the circuit. (b) Several important physical quantities associated with an AC circuit have the unit of ohms. Determine all of them for the situation where $f_d = 60$ Hz.

(c) Determine the peak (maximum) current in the circuit and the phase constant. Write the complete numerical expressions for the emf $\mathscr{E}(t)$ and the current $i(t)$. (d) Determine the expression for the instantaneous power of the generator in this circuit. What is the average value of the power? Suppose the frequency of the emf were increased slightly. Determine qualitatively what would happen to each of the following quantities: (e) the inductive reactance, (f) the capacitive reactance, (g) the impedance, (h) the peak current, (i) the power factor, and (j) the average power. For what frequency of the generator will (k) the impedance be a minimum and (l) the average power be a maximum?

Answers

(a) See Fig. 33-7. $\mathscr{E}_{\text{rms}} = 120$ V, $L = 20.0$ mH, $R = 30.0$ Ω, and $C = 40.0$ μF. The resonant frequency of this circuit is

$$\begin{aligned} f &= \frac{1}{2\pi\sqrt{LC}} \\ &= \frac{1}{2\pi\sqrt{(20.0 \times 10^{-3}\text{ H})(40.0 \times 10^{-6}\text{ F})}} \\ &= 178 \text{ Hz.} \end{aligned}$$

(b) First, there is the resistance $R = 30$ Ω. Then there are the inductive reactance

$$X_L = \omega_d L = 2\pi f_d L = 2\pi(60\text{ Hz})(20.0 \times 10^{-3}\text{ H}) = 7.54\ \Omega,$$

the capacitive reactance

$$X_C = \frac{1}{\omega_d C} = \frac{1}{2\pi f_d C} = \frac{1}{2\pi(60\text{ Hz})(40.0 \times 10^{-6}\text{ F})} = 66.3\ \Omega,$$

and the impedance

$$\begin{aligned} Z &= \sqrt{R^2 + (X_L - X_C)^2} \\ &= \sqrt{(30.0\ \Omega)^2 + (7.54\ \Omega - 66.3\ \Omega)^2} \\ &= 66.0\ \Omega. \end{aligned}$$

(c) The peak current is

$$I = \frac{\mathscr{E}_m}{Z} = \frac{\sqrt{2}\mathscr{E}_{\text{rms}}}{Z} = \frac{\sqrt{2}(120\text{ V})}{66.0\ \Omega} = 2.57\text{ A}.$$

The phase constant is

$$\begin{aligned} \phi &= \tan^{-1}\left(\frac{X_L - X_C}{R}\right) \\ &= \tan^{-1}\left(\frac{7.54\ \Omega - 66.3\ \Omega}{30.0\ \Omega}\right) \\ &= -63.0^\circ = -1.10\text{ rad}. \end{aligned}$$

This negative sign means that the current leads the emf, which is consistent with the fact that the capacitive reactance has a greater magnitude than the inductive reactance. The emf is

$$\begin{aligned} \mathscr{E}(t) &= \mathscr{E}_m \sin \omega t \\ &= \sqrt{2}\mathscr{E}_{\text{rms}} \sin 2\pi f t \\ &= \sqrt{2}(120\text{ V}) \sin[2\pi(60\text{ Hz})t] \\ &= (170\text{ V}) \sin(377\text{ rad/s})t. \end{aligned}$$

The current is

$$\begin{aligned} i(t) &= I \sin(2\pi f t - \phi) \\ &= (2.57\text{ A}) \sin[2\pi(60\text{ Hz})t - (-1.10\text{ rad})] \\ &= (2.57\text{ A}) \sin[(377\text{ rad/s})t + 1.10\text{ rad}]. \end{aligned}$$

(d) The instantaneous power is equal to

$$\begin{aligned} i\mathscr{E} &= (2.57\text{ A}) \sin[(377\text{ rad/s})t + 1.10\text{ rad}] \\ &\quad \times (170\text{ V}) \sin(377\text{ rad/s})t \\ &= (437\text{ W}) \sin[(377\text{ rad/s})t + 1.10\text{ rad}] \\ &\quad \times \sin(377\text{ rad/s})t. \end{aligned}$$

The average value of the power is equal to

$$\begin{aligned} I_{\rm rms}\mathscr{E}_{\rm rms} \cos \phi &= \tfrac{1}{2} I\mathscr{E}_m \cos \phi \\ &= (0.5)(2.57 \text{ A})(170 \text{ V}) \cos(-1.10 \text{ rad}) \\ &= 99.1 \text{ W}. \end{aligned}$$

(e) Increasing the frequency always increases the inductive reactance of an AC circuit. **(f)** Increasing the frequency always decreases the capacitive reactance of an AC circuit. **(g)** Since the frequency is now closer to the resonant frequency, the impedance decreases. **(h)** Since the impedance decreases, the peak current increases. Since the frequency is now closer to the resonant frequency, **(i)** the power factor increases (it is now closer to its maximum value of 1) and **(j)** the average power increases. **(k)** The impedance is a minimum and **(l)** the average power is a maximum when $Z = R$, which occurs when $f_d = f = 178$ Hz.

107. In this problem we consider an *RLC* circuit with definite numerical values and calculate some of its important characteristics—average power, phase angle, power factor—as a function of the frequency of the AC generator providing the emf of the circuit. The numbers have been chosen to make the computation fairly easy to do with a calculator or computer if you rewrite the formulas in terms of ω_d/ω by replacing ω_d with $\omega(\omega_d/\omega)$. You might want to write a program that prints out and graphs these quantities.

Consider an *RLC* circuit that has an emf of rms amplitude 100 V and frequency f_d. Suppose the inductance is 1.00 mH, the resistance is 100 Ω, and the capacitance is such that the resonant frequency f of the system is 1.000 MHz. (a) Sketch the circuit. Make a list of the quantities whose numerical values are supplied here, giving them the correct symbol, and also determine the value of the capacitance.

(b) Let the frequency of the AC generator range from zero to twice the resonant frequency. Determine the numerical values of the following quantities for f_d/f = 0, 0.50, 1.0, 1.5, and 2.0.

f_d/f	0	0.50	1.0	1.5	2.0
Impedance (Ω)					
Phase angle (rad)					
Power factor					
Average power (W)					

(c) Plot the four quantities you determined in (b) versus ω_d/ω. Choose an appropriate scale. Under each graph explain in sentences the behavior of the graph and the reasons for that behavior; for example, explain why the quantity you are plotting has a minimum or maximum at a particular frequency or why it increases or decreases as the frequency changes.

Answers

(a) $\mathscr{E}_{\rm rms} = 100$ V; $L = 1.00$ mH; $R = 100$ Ω; and $f = 1.000$ MHz. Since $f = 1/(2\pi\sqrt{LC})$,

$$\begin{aligned} C &= \frac{1}{(2\pi f)^2 L} \\ &= \frac{1}{(2\pi)^2(1.000 \text{ MHz})^2(0.00100 \text{ H})} \\ &= 2.53 \times 10^{-11} \text{ F} = 25.3 \text{ pF}. \end{aligned}$$

(b)

f_d/f	0	0.50	1.0	1.5	2.0
Impedance (Ω)	∞	9426	100	5237	9425
Phase angle (rad)	$-\pi/2$	-1.56	0	1.55	1.56
Power factor	0	0.011	1.0	0.019	0.011
Average power (W)	0	0.011	100	0.036	0.011

The impedance

$$Z = \sqrt{R^2 + (X_L - X_C)^2} = \sqrt{R^2 + (\omega_d L - 1/\omega_d C)^2}.$$

The phase angle

$$\phi = \tan^{-1}\left(\frac{X_L - X_C}{R}\right) = \tan^{-1}\left(\frac{\omega_d L - 1/\omega_d C}{R}\right),$$

which is negative if $\omega_d < \omega$. The power factor is $\cos \phi$ and the average power is

$$P_{\rm av} = \frac{\mathscr{E}^2_{\rm rms} R}{Z^2}.$$

(c) See Fig. 33-50*a*. The impedance is a minimum at the resonant frequency. There the inductive and capacitive reactances cancel each other, so the impedance Z is just the resistance R. At lower frequencies the capacitive reactance is greater than the inductive reactance and the two reactances do not cancel. Thus the

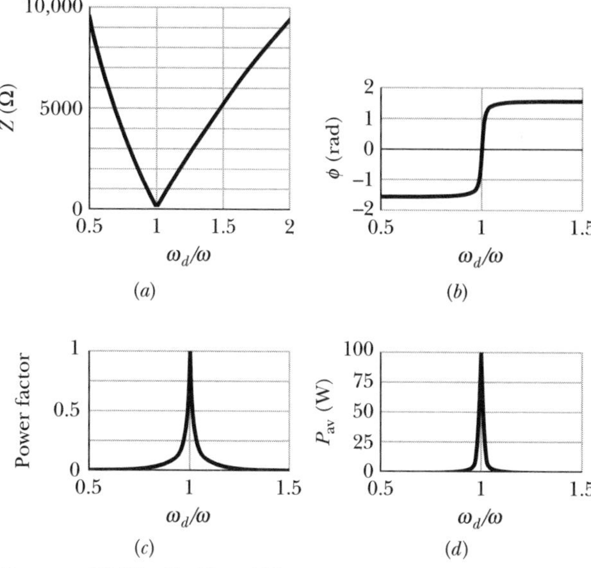

FIGURE 33-50 Problem 107.

impedance is greater than R. At higher frequencies the inductive reactance is greater than the capacitive reactance, so again the two reactances do not cancel and the impedance is greater than R.

See Fig. 33-50*b*. The phase angle is 0 at the resonant frequency because the inductive and capacitive reactances are equal in magnitude and thus cancel each other in the numerator of the expression for the tangent of the phase angle. At lower frequencies the capacitive reactance is greater than the inductive reactance, so the phase angle is negative. At higher frequencies the inductive reactance is greater than the capacitive reactance, so the phase angle is positive.

See Fig. 33-50*c*. The power factor, which is $\cos\phi$, is 1 at the resonance frequency, for which the phase angle is zero. At both lower and higher frequencies, where the phase angle is not zero, the power factor is less because the cosine of a nonzero angle is less than 1. The power factor approaches zero as the frequency approaches zero.

See Fig. 33-50*d*. The average power has its maximum at the resonance frequency, where the impedance is a minimum. It decreases at higher and lower frequencies because the impedance is greater (or, equivalently, the power factor is smaller). The average power goes to zero at zero frequency; this is the limit of the AC circuit's becoming a DC circuit, where the capacitor will block the current in the steady state.

108. In this problem we compare the electric currents and the energy dissipation when a resistor, a capacitor, and an inductor are, in turn, connected to an AC generator. Suppose that the AC generator produces an emf

$$\mathscr{E}(t) = (160\text{ V})\sin 2\pi(60\text{ Hz})t.$$

For each of the following parts, (1) draw a diagram for the circuit; (2) write the differential equation for the circuit; (3) determine the electric current as a function of time and find the maximum value of the current; (4) make a plot of the emf and the electric current on the same diagram; and (5) describe what's happening in the circuit, as a function of time, from an energy point of view and find the maximum power of the battery.

(a) Suppose the generator is first connected to a 40 Ω resistor. (b) Next, suppose the generator is connected to a 40 μF capacitor (assumed to have zero resistance). (c) Next, suppose the generator is connected to a 40 mH inductor (assumed to have no resistance).

Answers

(a) (1) See Fig. 33-8*a*. (2) Kirchhoff's law gives the differential equation $\mathscr{E} - iR = 0$. (3) The electric current is

$$i(t) = \frac{\mathscr{E}}{R} = \frac{160\text{ V}}{40\ \Omega}\sin 2\pi(60\text{ Hz})t = (4.0\text{ A})\sin 2\pi(60\text{ Hz})t.$$

The maximum current is 4.0 A. (4) See Fig. 33-8*b*, which shows that the emf $\mathscr{E}$ and the current i are exactly in phase. (5) From an energy point of view, the AC generator is supplying energy at a sinusoidally varying rate $i\mathscr{E} = i^2R$; that energy is being dissipated in the resistor. This power is always positive, but varies from 0 to (4.0 A)(160 V) = 640 W.

(b) (1) See Fig. 33-9*a*. (2) Kirchhoff's law gives the differential equation $\mathscr{E} - q/C = 0$. (3) The charge q on the capacitor is

$$\begin{aligned} q &= \mathscr{E}C \\ &= (160\text{ V})(40\ \mu\text{F})\sin 2\pi(60\text{ Hz})t \\ &= (0.0064\text{ A})\sin 2\pi(60\text{ Hz})t. \end{aligned}$$

The electric current can be determined by differentiating this result with respect to t:

$$\begin{aligned} i(t) &= \frac{dq}{dt} = 2\pi(60\text{ Hz})(0.0064\text{ A})\cos 2\pi(60\text{ Hz})t \\ &= (2.4\text{ A})\cos 2\pi(60\text{ Hz})t. \end{aligned}$$

The maximum current is 2.4 A. (4) See Fig. 33-9*b*, which shows that the electric current i leads the emf $\mathscr{E}$. (5) From an energy point of view, the AC generator is supplying energy at the varying rate $i\mathscr{E}$. The energy goes into or comes out of the electric field of the capacitor. Part of the time the power is positive, and part of the time it is negative.

(c) (1) See Fig. 33-10*a*. (2) Kirchhoff's law gives the differential equation $\mathscr{E} - L\,di/dt = 0$. (3) Integrating the expression $di/dt = \mathscr{E}/L$ gives the electric current as

$$\begin{aligned} i(t) &= -\frac{160\text{ V}}{2\pi(60\text{ Hz})(40\text{ mH})}\cos 2\pi(60\text{ Hz})t \\ &= -(10.6\text{ A})\cos 2\pi(60\text{ Hz})t. \end{aligned}$$

The maximum current is 10.6 A. (4) See Fig. 33-10*b*, which shows that the electric current i lags the emf $\mathscr{E}$. (5) From an energy point of view, the AC generator is supplying energy at the varying rate $i\mathscr{E}$. The energy goes into and comes out of the magnetic field of the inductor. This power is positive for part of the cycle and negative for part of the cycle.

Graphing Calculators

PROBLEM SOLVING TACTICS

Tactic 2: *Program for $X_L - X_C$*

Homework problems about driven *RLC* circuits often give the values of L, C, and either ω_d or f_d and then ask you to calculate $X_L - X_C$. To save time and keystroking, write a program XLC that prompts you for the given values and then displays the corresponding value of $X_L - X_C$.

In the standard homework problem, the resistance R and the driving emf $\mathscr{E}$ are also given and you are required to find the impedance Z via Eq. 33-52, the current amplitude I via Eq. 33-54, the phase constant ϕ via Eq. 33-56, and the resonance angular frequency ω via Eq. 33-58. If you anticipate needing to solve for these quantities many times with different values, you might write a program RLC that prompts you for the given values and then displays the results of these several equations.

Chapter Thirty-Four
Electromagnetic Waves

QUESTIONS

17. The point source in Fig. 34-8 emits electromagnetic waves isotropically. The electric component of a wave traveling along a radial axis extending from the source is shown in Fig. 34-58. In that figure, is the wave traveling toward the right or toward the left?

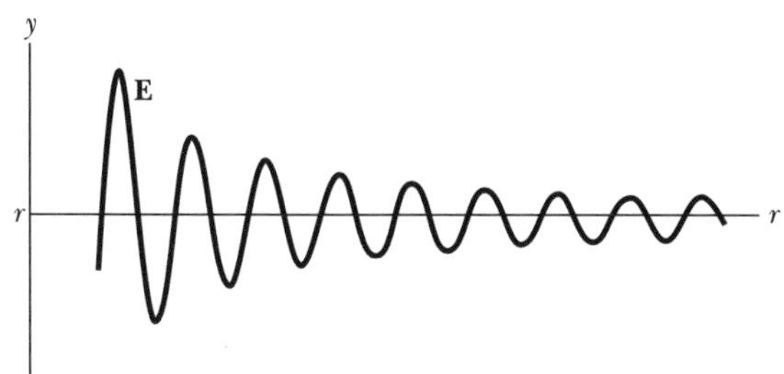

FIGURE 34-58 Question 17.

18. *Organizing question:* Which of the following are properly written functions to describe the electric and magnetic components of an electromagnetic wave?

Pair 1:

$$E = E_m \sin \frac{2\pi}{\lambda}(x - ct)$$

$$B = \frac{E_m}{c} \sin \frac{2\pi}{\lambda}(x - ct)$$

Pair 2:

$$E = cB_m \sin\left(\frac{2\pi x}{\lambda} - \frac{2\pi t}{T}\right)$$

$$B = B_m \sin\left(\frac{2\pi x}{\lambda} - \frac{2\pi t}{T}\right)$$

Pair 3:

$$E = E_m \sin \omega\left(\frac{x}{c} - t\right)$$

$$B = \sqrt{\mu_0 \epsilon_0}\, E_m \sin \omega\left(\frac{x}{c} - t\right)$$

Pair 4:

$$E = \frac{B_m}{\sqrt{\mu_0 \epsilon_0}} \sin 2\pi\left(\frac{x}{\lambda} - ft\right)$$

$$B = B_m \sin 2\pi\left(\frac{x}{\lambda} - ft\right)$$

19. A particular point source of light emits isotropically, with an intensity of $I = 8\ \text{W/m}^2$ at a radial distance of $r = 1$ m from the source. (a) Which of the curves in Fig. 34-59 best gives the intensity I of the light as a function of r? (b) If we place the source in sooty air, where the soot particles absorb the light they intercept, which of the curves then best gives $I(r)$?

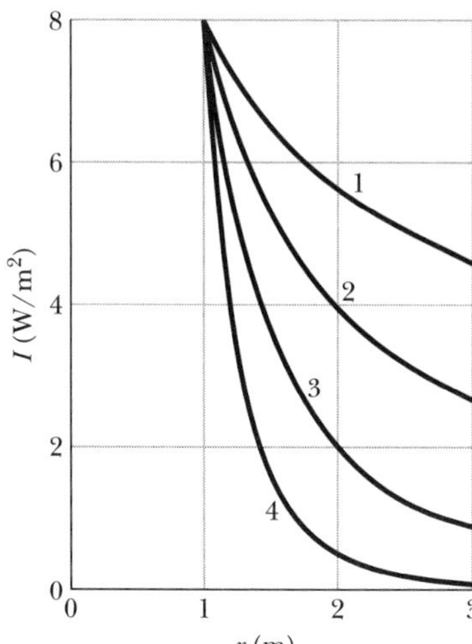

FIGURE 34-59 Question 19.

20. In Fig. 34-60, four objects are illuminated with light beams of the same intensity. The objects, shown in both side and front views, are either square plates of edge length L, circular plates of diameter L, or a sphere of diameter L. Objects 1, 2, and 3 totally absorb the light they intercept; object 4 totally reflects the light. Rank the objects according to (a) the radiation pressure on them and (b) the radiation force on them, greatest first.

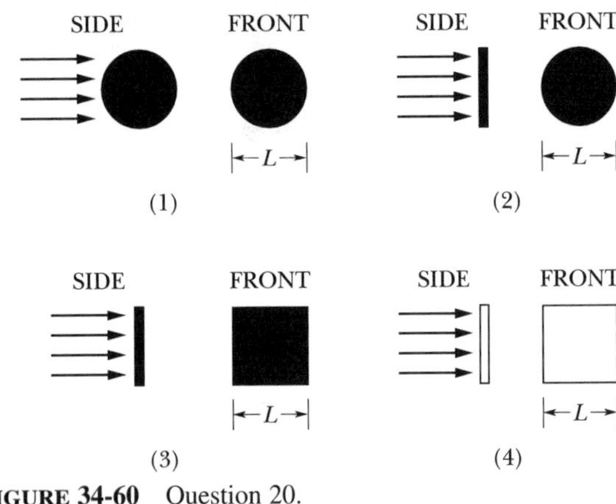

FIGURE 34-60 Question 20.

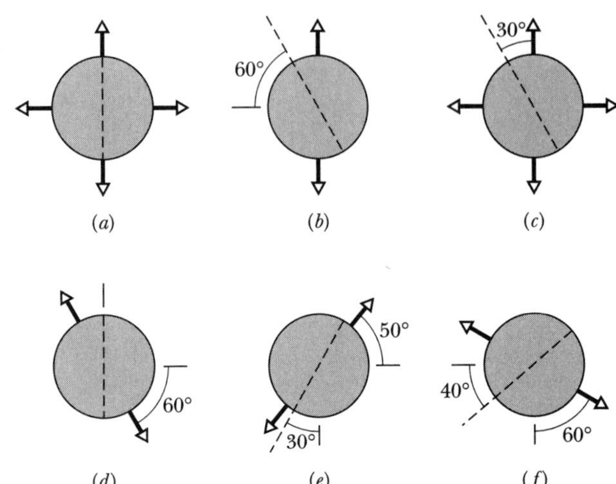

FIGURE 34-62 Question 23.

21. If the speed of light were significantly less than it is, would the force on your face from the flash of a camera then be greater or less than it is?

22. In Fig. 34-61, three spherical dust particles suddenly experience a radiation pressure when a beam of laser light is turned on. The radii of the particles are R, $2R$, and $3R$; the particles have the same density and totally absorb the light they intercept. Rank the three particles according to (a) the radiation pressure on them, (b) the radiation force on them, (c) their masses, (d) the accelerations they experience due to the radiation force, and (e) the distance they travel in a certain time after the laser is turned on, all greatest first.

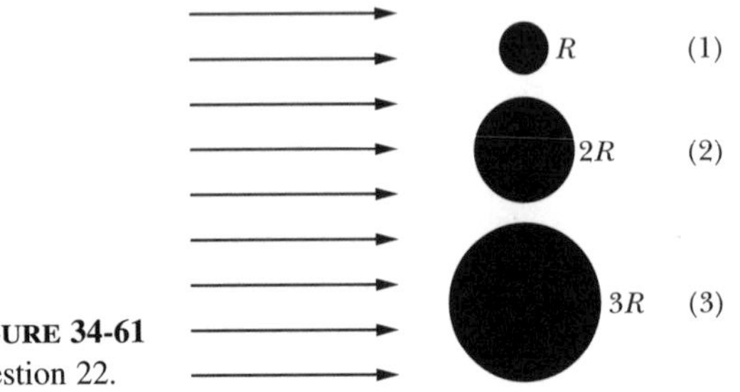

FIGURE 34-61 Question 22.

23. *Organizing question:* Figure 34-62 represents six situations in which light is sent toward you through a polarizing sheet. The general arrangement is like that of Fig. 34-12, except here the view of the sheet is head-on. The initial state of polarization of the light (before it reaches the sheet) is represented by the arrows. The polarizing direction of the sheet is represented by a dashed line. (a) For each situation, determine if the one-half rule or the cosine-squared rule applies. (b) For the latter, determine what angle should be used in the calculation of the intensity of the light transmitted by the sheet.

24. Figure 34-63a depicts total internal reflection for light inside a material with an index of refraction n_1 when air is outside the material. A light ray reaching point A from within the shaded region at the left (such as the ray shown) fully reflects at that point and ends up in the shaded region at the right. The other two parts of Fig. 34-63 show similar situations for two other materials. Rank the indices of refraction of the three materials, greatest first.

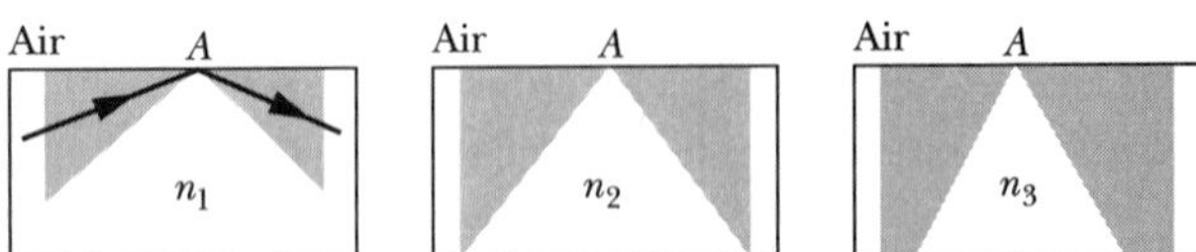

FIGURE 34-63 Question 24.

25. Figure 34-64 shows four situations in which a light ray is incident on an interface between two materials; the materials' indices of refraction are indicated. Determine in which situations total internal reflection is possible and then rank those situations according to their critical angle, greatest first.

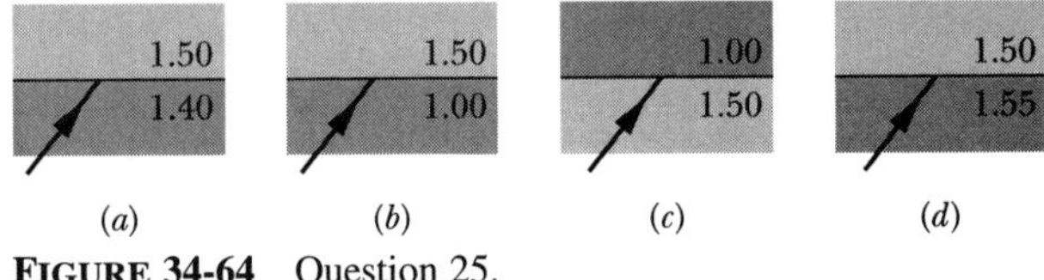

FIGURE 34-64 Question 25.

26. In Figure 34-65, a light ray refracts from air into a plastic rod and then partially reflects (at an angle of 60°) and partially refracts at a point along the side of the rod. What is the value of angle θ?

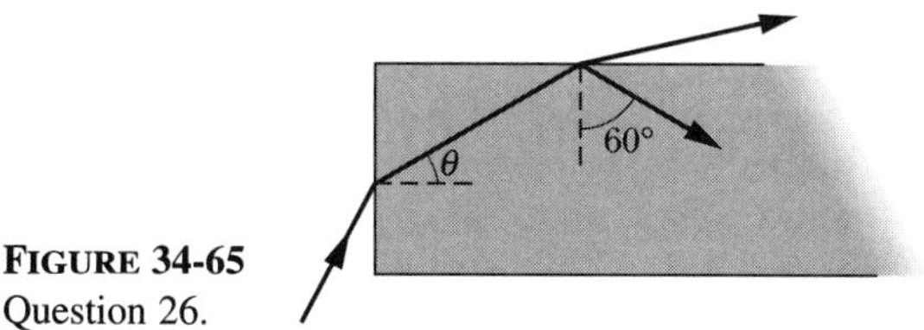

FIGURE 34-65 Question 26.

27. A beam of white light in plastic is incident on the interface between the plastic and the external air. If the incident angle is equal to the critical angle for yellow light, does any of (a) the red light and (b) the blue light in the beam refract from the plastic into the air?

28. Sound waves traveling through water can be trapped by total internal reflection somewhat as light can be trapped in an optical fiber. Figure 34-66*a* depicts a channel in the ocean where sound waves are trapped in a certain depth range as they travel rightward. (The paths taken by two waves are represented by the two curved rays.) As explained for Eq. 18-3, the speed of sound in a material depends on the ratio of the material's bulk modulus B to its density ρ. Figure 34-66*b* gives four choices for how that ratio might vary with depth for water. Which choice best corresponds to the entrapment shown in Fig. 34-66*a*?

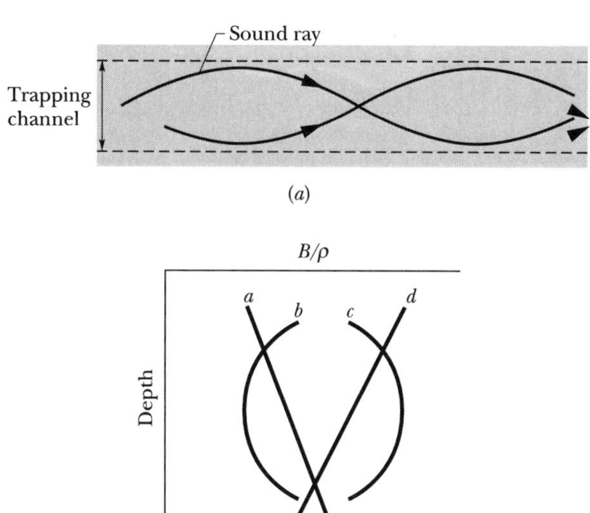

FIGURE 34-66 Question 28.

29. Materials with a distinct color usually have an index of refraction n that varies with λ as shown in Fig. 34-67*a*. Figure 34-67*b* represents the dispersion of light traveling from air into a solution that has this $n(\lambda)$ curve. Which wavelength in Fig. 34-67*a* corresponds to which refracted ray in Fig. 34-67*b*?

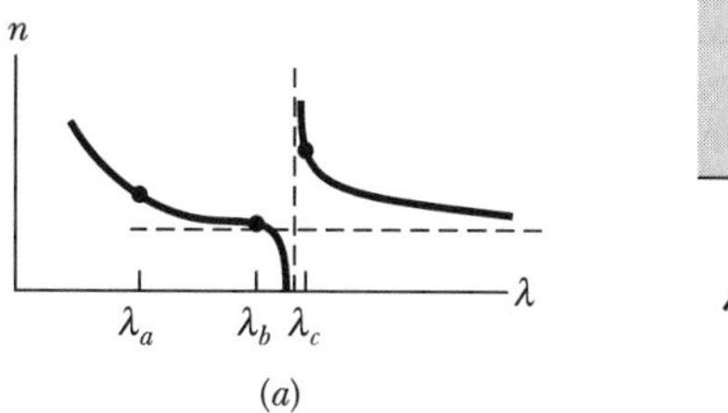

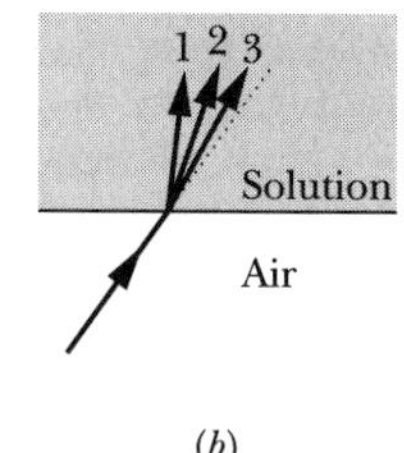

FIGURE 34-67 Question 29.

30. Light in air is incident on the surface of a certain glass. Is the Brewster angle for green light greater than, less than, or the same as that for yellow light?

31. In Fig. 34-68, a light ray in material 1 with index of refraction n_1 reflects at the Brewster angle from material 2 with index of refraction n_2. Suppose that n_2 is less than n_1 and that we could gradually increase n_2 until it is greater than n_1. (a) Would the Brewster angle increase or decrease? (b) If material 1 is air, what (approximately) is the smallest value of the Brewster angle possible?

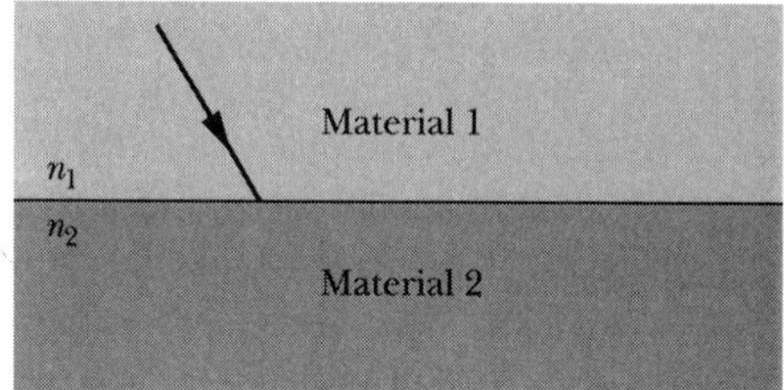

FIGURE 34-68 Question 31.

EXERCISES & PROBLEMS

89. The rms value of the electric field in a light wave is 0.200 V/m. What is the amplitude of the associated magnetic field?

90. A plane wave of light has an intensity of 1.00×10^4 W/m^2. What is the maximum strength of the magnetic field of the wave?

91. The magnetic component of an electromagnetic wave has an rms value of 56.0 nT. (a) What is the amplitude of the electric component of the wave? (b) What is the intensity of the light?

92. A point source of light emits isotropically with a power of 200 W. What is the force due to the light on a totally absorbing sphere of radius 2.0 cm at a distance of 20 m from the source?

93. At Earth's surface, what intensity of light is needed to suspend a totally absorbing spherical particle against its own weight if the mass of particle is 2.0×10^{-13} kg and its radius is 2.0 μm?

94. What is the radiation force on a totally reflecting sail at a distance of 3.0×10^{11} m from the Sun if the sail is square with edge length 2.0 m and has its surface perpendicular to the direction of the sunlight?

95. In a region of space where gravitational forces can be neglected, a sphere is accelerated by a uniform light beam of intensity 6.0 mW/m^2. The sphere is totally absorbing and has a radius of 2.0 μm and a uniform density of 5.0×10^3 kg/m^3. What is the acceleration of the sphere due to the light?

96. The magnetic component of a polarized wave of light is given by

$$B_x = (4.00\ \mu\text{T}) \sin[ky + (2.00 \times 10^{15}\ \text{s}^{-1})t].$$

(a) In which direction does the wave travel, (b) parallel to which axis is it polarized, and (c) what is its intensity? (d) Write an expression for the electric field of the wave, including a value for the angular wave number. (e) What is the wavelength? (f) In which region of the electromagnetic spectrum is this wave?

97. A light wave polarized parallel to a y axis is traveling in the negative direction of a z axis. The rms value of the electric field is 50.0 V/m and the wavelength is 250 nm. (a) Using this information, write an expression for the magnetic field component of the wave. Part of the wave is totally reflected by a chip in the shape of an equilateral triangle that is mounted on the z axis and perpendicular to that axis. The triangle has an edge length of 2.00 μm, and the chip is fully illuminated on the side facing the light source. (b) What is the force on the chip due to the light?

98. The electric component of a beam of polarized light is given by

$$E_y = (5.00\ \text{V/m}) \sin[(1.00 \times 10^6\ \text{m}^{-1})z + \omega t].$$

(a) Write an expression for the magnetic field component of the wave, including a value for ω. What are (b) the wavelength, (c) the period, and (d) the intensity? (e) Parallel to which axis does the magnetic field oscillate? (f) In which region of the electromagnetic spectrum is this wave?

99. In Fig. 34-69, initially unpolarized light is sent toward a system of three polarizing sheets. What fraction of the initial light intensity emerges from the system?

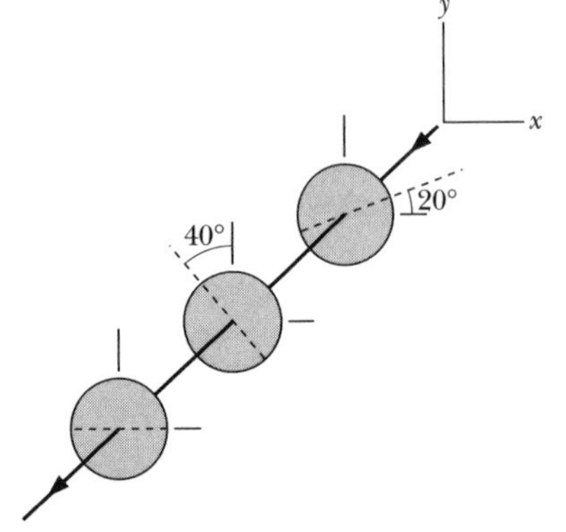

FIGURE 34-69
Problem 99.

100. In Fig. 34-70, light that is polarized parallel to the y axis is sent into a system of two polarizing sheets. The fraction of the initial light intensity that emerges from the system is 0.200. What is the angle θ shown for the second sheet?

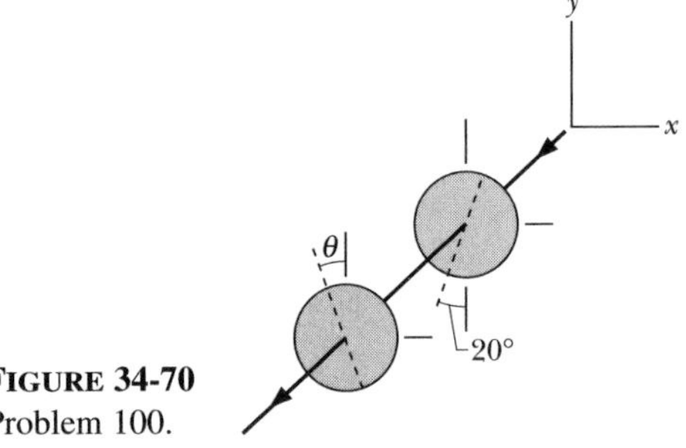

FIGURE 34-70
Problem 100.

101. In Fig. 34-71, light that is initially unpolarized is sent into a system of three polarizing sheets. What fraction of the initial light intensity is passed by the system?

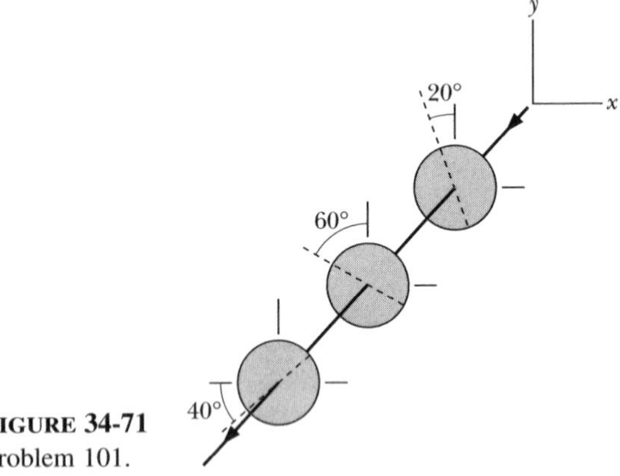

FIGURE 34-71
Problem 101.

102. In Fig. 34-72, light that is initially unpolarized is sent into a system of three polarizing sheets. What fraction of the initial light intensity emerges from the system?

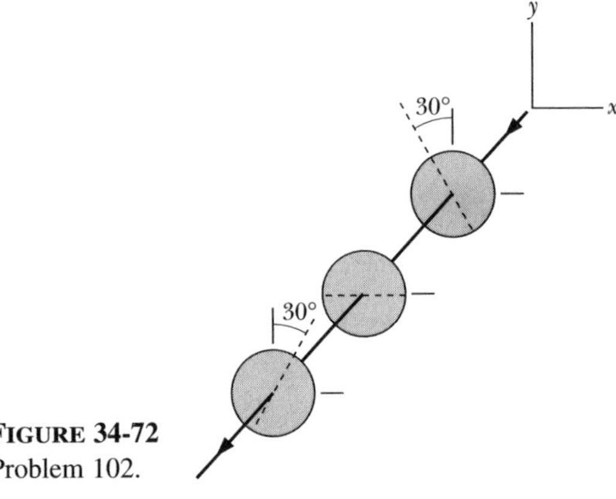

FIGURE 34-72
Problem 102.

103. A system of three polarizing sheets is shown in Fig. 34-73. When initially unpolarized light is sent into the system, the intensity of the transmitted light is 5.0×10^{-2} of the initial intensity. What is the value of θ?

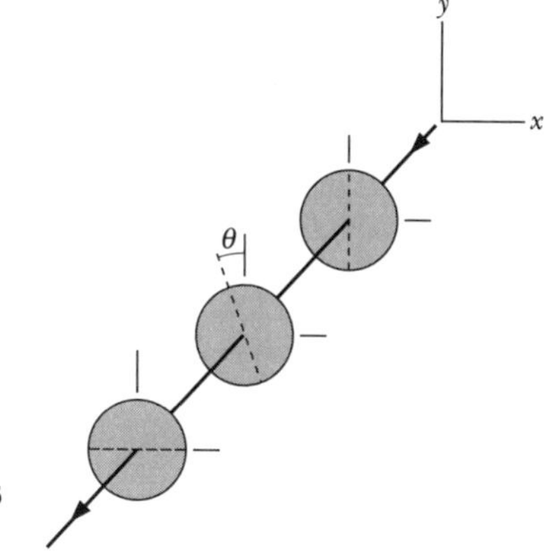

FIGURE 34-73 Problem 103.

104. In the ray diagram of Fig. 34-74, where the angles are not drawn to scale, the ray is incident at the critical angle on the interface between materials 2 and 3. Angle $\phi = 60.0°$. Find (a) index of refraction n_3 and (b) angle θ. (c) If θ is decreased, is there refraction of light into material 3?

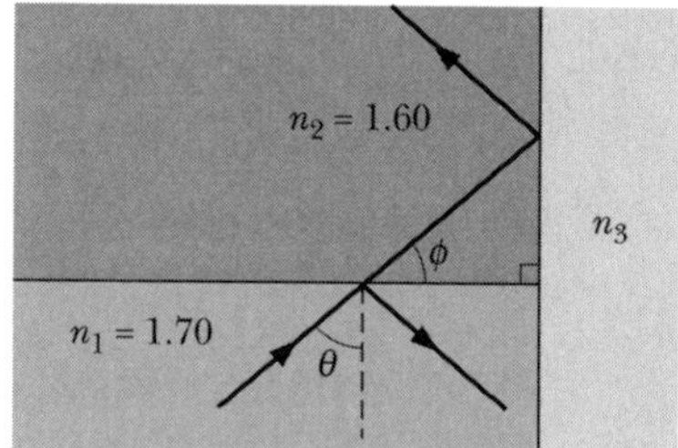

FIGURE 34-74 Problem 104.

105. In Fig. 34-75*a*, light refracts from material 1 into a thin layer of material 2, crosses that layer, and then is incident at the critical angle on the interface between materials 2 and 3. (a) What is angle θ? (b) If θ is decreased, is there refraction of light into material 3?

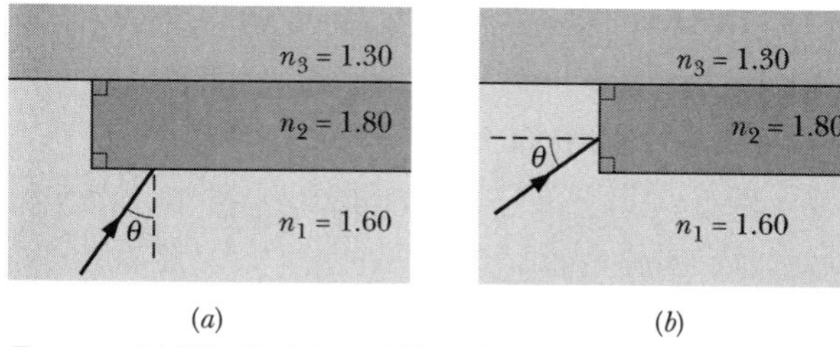

FIGURE 34-75 Problems 105 and 106.

106. In Fig. 34-75*b*, light refracts from material 1 into a thin layer of material 2, crosses that layer, and then is incident at the critical angle on the interface between materials 2 and 3. (a) What is angle θ? (b) If θ is decreased, is there refraction of light into material 3?

107. In Fig. 34-76, light refracts into material 2, crosses that material, and is then incident at the critical angle on the interface between materials 2 and 3. (a) What is angle θ? (b) If θ is increased, is there refraction of light into material 3?

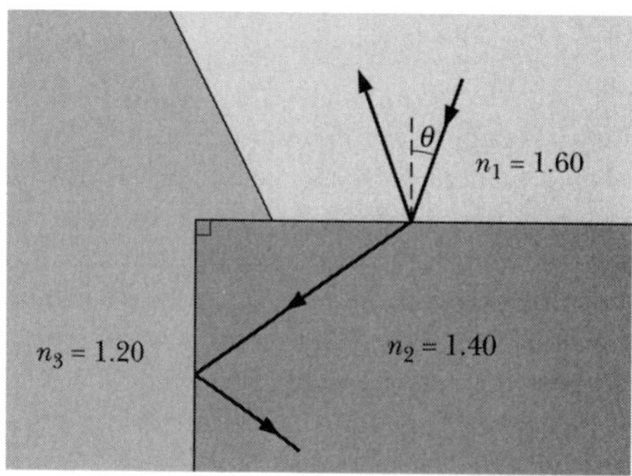

FIGURE 34-76 Problem 107.

108. In Fig. 34-77, light refracts from material 1 into material 2. If it is incident at point A at the critical angle for the interface between materials 2 and 3, what are (a) the angle of refraction at point B and (b) the initial angle θ? If, instead, it is incident at point B at the critical angle for the interface between materials 2 and 3, what are (c) the angle of refraction at point A and (d) the initial angle θ? If, instead of all that, it is incident at point A at Brewster's angle for the interface between materials 2 and 3, what are (e) the angle of refraction at point B and (f) the initial angle θ?

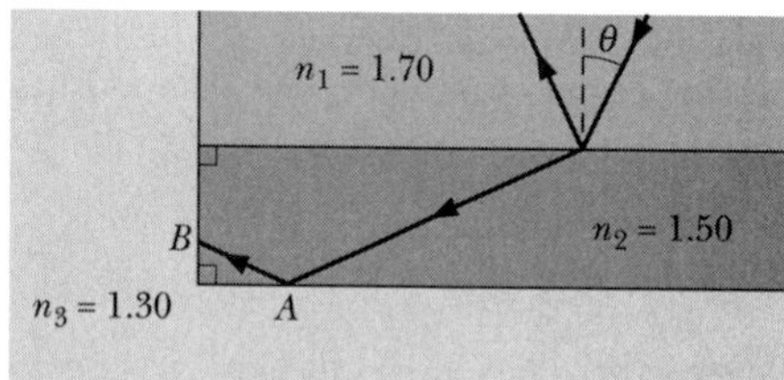

FIGURE 34-77 Problem 108.

109. In Fig. 34-78, light undergoes three refractions as it heads downward and a reflection and then a refraction to reach the air. The initial angle $\theta_1 = 40.1°$. What are the values of (a) θ_5 and (b) θ_4?

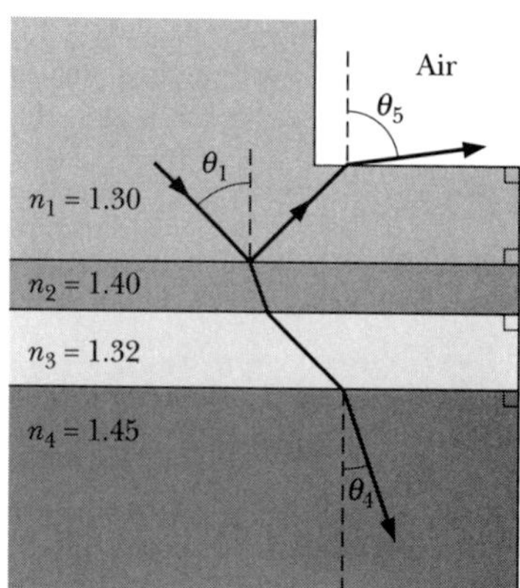

FIGURE 34-78 Problem 109.

Tutorial Problems

110. A plane electromagnetic wave with a wavelength of 200 nm is traveling in a vacuum in the positive x direction. Its magnetic field, whose maximum magnitude is 50 μT, is polarized parallel to the z axis. (a) Write the wave velocity as a vector. (b) Determine the following characteristics of the wave: its frequency f, its angular frequency ω, and its angular wave number k. What part of the electromagnetic spectrum is involved?

(c) Write the mathematical expression for the magnetic field of this wave. Take the phase constant to be zero, and remember that the magnetic field is a vector. (d) Similarly, write the expression for the electric field vector of this wave.

(e) Faraday's law applied to the electric and magnetic fields of a plane electromagnetic wave leads to the relation

$$\frac{\partial E_y}{\partial x} = -\frac{\partial B_z}{\partial t}$$

(Eq. 34-11, written in component notation). Verify that this equation is satisfied by the magnetic and electric fields derived in parts (c) and (d). (f) From his four electromagnetic equations, Maxwell derived the wave equations for electromagnetic waves as

$$\frac{\partial^2 \mathbf{E}}{\partial x^2} = \frac{1}{c^2}\frac{\partial^2 \mathbf{E}}{\partial t^2}$$

and

$$\frac{\partial^2 \mathbf{B}}{\partial x^2} = \frac{1}{c^2}\frac{\partial^2 \mathbf{B}}{\partial t^2}.$$

Verify that the electric field in this problem satisfies the wave equation for **E**.

Answers

(a) Since the wave is traveling in the positive x direction, which is along the unit vector **i**, the velocity of propagation of the wave is $\mathbf{v} = c\mathbf{i} = (3.00 \times 10^8 \text{ m/s})\mathbf{i}$.

(b) We have:

$$f = \frac{c}{\lambda} = \frac{3.00 \times 10^8 \text{ m/s}}{200 \times 10^{-9} \text{ m}} = 1.50 \times 10^{15} \text{ Hz}$$

$$\omega = 2\pi f = 2\pi(1.50 \times 10^{15} \text{ Hz}) = 9.42 \times 10^{15} \text{ rad/s}$$

$$k = \frac{2\pi}{\lambda} = \frac{2\pi}{200 \times 10^{-9} \text{ m}} = 3.14 \times 10^7 \text{ m}^{-1}$$

Since visible light has wavelengths in the range of about 400 to 700 nm, a wavelength of 200 nm is shorter than the shortest visible light by a factor of 2, so it falls in the ultraviolet part of the spectrum.

(c) The magnetic field is parallel to the z axis and depends only on x (and time), so it can be written in the form

$$\mathbf{B}(x, t) = B_m \sin(kx - \omega t)\mathbf{k}.$$

Since we were told that $B_{\max} = 50\ \mu\text{T}$ and $\lambda = 200$ nm, and we determined the values of k and ω in part (b), we must have

$$\mathbf{B}(x, t) = (50\ \mu\text{T}) \sin[(3.14 \times 10^7 \text{ m}^{-1})x - (9.42 \times 10^{15} \text{ s}^{-1})t]\mathbf{k}.$$

(d) The maximum magnitude of the magnetic field is

$$E_m = cB_m = (3.00 \times 10^8 \text{ m/s})(50\ \mu\text{T}) = 15 \text{ kV/m}.$$

Consequently, since the electric and magnetic fields are perpendicular to one another and are in phase,

$$\mathbf{E}(x, t) = (15 \text{ kV/m}) \sin[(3.14 \times 10^7 \text{ m}^{-1})x - (9.42 \times 10^{15} \text{ s}^{-1})t]\mathbf{j}.$$

Note: This is $+\mathbf{j}$, not $-\mathbf{j}$, as can be determined by checking that **E**, **B**, and **v** are mutually perpendicular and form a right-hand coordinate system (in that order).

(e) First,

$$\begin{aligned}\frac{\partial E_y}{\partial x} &= \frac{\partial}{\partial x}(15 \text{ kV/m}) \sin[(3.14 \times 10^7 \text{ m}^{-1})x - (9.42 \times 10^{15} \text{ s}^{-1})t] \\ &= (15 \text{ kV/m})(3.14 \times 10^7 \text{ m}^{-1}) \cos[(3.14 \times 10^7 \text{ m}^{-1})x - (9.42 \times 10^{15} \text{ s}^{-1})t] \\ &= (4.71 \times 10^{11} \text{ V/m}^2) \cos[(3.14 \times 10^7 \text{ m}^{-1})x - (9.42 \times 10^{15} \text{ s}^{-1})t].\end{aligned}$$

Next,

$$\begin{aligned}-\frac{\partial B_z}{\partial t} &= -\frac{\partial}{\partial t}(50\ \mu\text{T}) \sin[(3.14 \times 10^7 \text{ m}^{-1})x - (9.42 \times 10^{15} \text{ s}^{-1})t] \\ &= -(50\ \mu\text{T})(-9.42 \times 10^{15} \text{ s}^{-1})\cos[(3.14 \times 10^7 \text{ m}^{-1})x - (9.42 \times 10^{15} \text{ s}^{-1})t] \\ &= (4.71 \times 10^{11} \text{ V/m}^2) \cos[(3.14 \times 10^7 \text{ m}^{-1})x - (9.42 \times 10^{15} \text{ s}^{-1})t].\end{aligned}$$

Yes, these are equal.

(f) There is only one component to check, namely, E_y.

$$\begin{aligned}\frac{\partial E_y}{\partial x} &= \frac{\partial}{\partial x}(15 \text{ kV/m}) \sin[(3.14 \times 10^7 \text{ m}^{-1})x - (9.42 \times 10^{15} \text{ s}^{-1})t] \\ &= (15 \text{ kV/m})(3.14 \times 10^7 \text{ m}^{-1}) \cos[(3.14 \times 10^7 \text{ m}^{-1})x - (9.42 \times 10^{15} \text{ s}^{-1})t] \\ &= (4.71 \times 10^{11} \text{ V/m}^2) \cos[(3.14 \times 10^7 \text{ m}^{-1})x - (9.42 \times 10^{15} \text{ s}^{-1})t];\end{aligned}$$

$$\begin{aligned}\frac{\partial^2 E_y}{\partial x^2} &= \frac{\partial^2}{\partial x^2}(15 \text{ kV/m}) \sin[(3.14 \times 10^7 \text{ m}^{-1})x - (9.42 \times 10^{15} \text{ s}^{-1})t] \\ &= -(15 \text{ kV/m})(3.14 \times 10^7 \text{ m}^{-1})^2 \sin[(3.14 \times 10^7 \text{ m}^{-1})x - (9.42 \times 10^{15} \text{ s}^{-1})t] \\ &= -(1.48 \times 10^{19} \text{ V/m}^3) \sin[(3.14 \times 10^7 \text{ m}^{-1})x - (9.42 \times 10^{15} \text{ s}^{-1})t];\end{aligned}$$

$$\frac{\partial E_y}{\partial t} = \frac{\partial}{\partial t}(15\ \text{kV/m})\sin[(3.14 \times 10^7\ \text{m}^{-1})x - (9.42 \times 10^{15}\ \text{s}^{-1})t]$$
$$= -(15\ \text{kV/m})(9.42 \times 10^{15}\ \text{s}^{-1})\cos[(3.14 \times 10^7\ \text{m}^{-1})x - (9.42 \times 10^{15}\ \text{s}^{-1})t]$$
$$= (1.41 \times 10^{20}\ \text{V/m}\cdot\text{s})\cos[(3.14 \times 10^7\ \text{m}^{-1})x - (9.42 \times 10^{15}\ \text{s}^{-1})t];$$

$$\frac{\partial^2 E_y}{\partial t^2} = \frac{\partial^2}{\partial t^2}(15\ \text{kV/m})\sin[(3.14 \times 10^7\ \text{m}^{-1})x - (9.42 \times 10^{15}\ \text{s}^{-1})t]$$
$$= -(15\ \text{kV/m})(9.42 \times 10^{15}\ \text{s}^{-1})^2\sin[(3.14 \times 10^7\ \text{m}^{-1})x - (9.42 \times 10^{15}\ \text{s}^{-1})t]$$
$$= -(1.33 \times 10^{36}\ \text{V/m}\cdot\text{s}^2)\sin[(3.14 \times 10^7\ \text{m}^{-1})x - (9.42 \times 10^{15}\ \text{s}^{-1})t].$$

Comparing, we see that

$$\frac{\partial^2 E_y}{\partial x^2} = (1.11 \times 10^{-17}\ \text{s}^2/\text{m}^2)\left(\frac{\partial^2 E_y}{\partial t^2}\right) = \frac{1}{c^2}\left(\frac{\partial^2 E_y}{\partial t^2}\right),$$

which is the wave equation.

111. In this problem we begin with the numerical value of the average intensity of the electromagnetic radiation received by Earth from the Sun. From this we calculate all sorts of interesting quantities, including the rate at which the Sun is losing mass, the rate at which photons from sunlight are absorbed by Earth, the rate of transfer of momentum to Earth from the Sun, and the solar radiation pressure on Earth.

The *solar constant* of 1370 W/m^2 is the average intensity of the solar radiation at Earth's mean distance from the Sun (149×10^6 km). This is the intensity of the solar radiation on a surface perpendicular to the solar rays just above the atmosphere. (a) What is the textbook's name and notation for this solar constant? (b) Assuming the solar radiation is isotropic (equal in all directions) what is the total rate at which energy is emitted by the Sun (the Sun's power)? Explain your reasoning carefully; don't just produce a formula from out of nowhere. (c) What is the total rate at which energy strikes Earth? Explain your reasoning.

The radiation of energy by the Sun corresponds to a loss of mass by the Sun, the energy E being related to the mass loss Δm by $E = (\Delta m)c^2$. (d) Using the result of part (b), estimate the rate, in kilograms per second, at which the Sun is losing mass. At this rate, what fraction of the total mass of the Sun is lost each year?

(e) Estimate as best you can the radiation pressure of the Sun on Earth, making whatever additional assumptions you need. (f) Use that estimate to estimate the force exerted on Earth by solar radiation. Compare the force magnitude with that of the gravitational attraction exerted on Earth by the Sun.

Electromagnetic radiation is emitted by a source in discrete and tiny amounts of energy and momentum called photons. Similarly, it is absorbed as photons. The energy E of each discrete emission or absorption is

$$E = hf = \frac{hc}{\lambda},$$

where h is Planck's constant (6.625×10^{-34} J · s), and f and λ are the frequency and wavelength of the associated electromagnetic radiation.

The spectrum of electromagnetic radiation emitted by the Sun extends over a large range of wavelengths, but most is concentrated in the range 300 to 1000 nm. We can take the average wavelength (or typical wavelength) to be 550 nm. (g) Using this wavelength, determine the rate of photons involved in the emission of electromagnetic radiation by the Sun and in the absorption of that radiation by Earth.

Answers

(a) The textbook refers to this as the average intensity and denotes it $\bar{S}$; it is the average magnitude of the Poynting vector.

(b) If the solar radiation is isotropic, the same intensity of solar radiation will be found in all directions at the same distance from the Sun. We can then determine the total solar power by determining the surface area of the giant sphere whose radius R equals the mean distance from Earth to the Sun and multiplying it by the solar constant:

$$P = (4\pi R^2)(1370\ \text{W/m}^2) = 4\pi(1.49 \times 10^{11}\ \text{m})^2(1370\ \text{W/m}^2) = 3.82 \times 10^{26}\ \text{W}.$$

(c) The total power striking Earth is just the power striking a disk whose radius equals the radius of Earth. That radius R_E = 6370 km, so the total power is

$$(\pi R_E^2)(1370\ \text{W/m}^2) = \pi(6370 \times 10^3\ \text{m})^2(1370\ \text{W/m}^2) = 1.75 \times 10^{17}\ \text{W}.$$

(d) The rate at which the Sun is losing mass is thus the negative of the power of the solar radiation divided by c^2:

$$\frac{dm}{dt} = \frac{dE/dt}{c^2} = -\frac{P}{c^2} = -\frac{3.82 \times 10^{26}\ \text{W}}{(3.00 \times 10^8\ \text{m/s})^2} = -4.24 \times 10^9\ \text{kg/s}.$$

The Sun is losing 4.24×10^9 kg/s, or

$$(4.24 \times 10^9\ \text{kg/s})(365\ \text{days/y})(24\ \text{h/day})(3600\ \text{s/h}) = 1.34 \times 10^{17}\ \text{kg/y}.$$

Since the mass of the Sun is 1.99×10^{30} kg, the fraction of its mass lost each year as solar radiation is

$$(1.34 \times 10^{17}\ \text{kg/y})/(1.99 \times 10^{30}\ \text{kg}) = 6.7 \times 10^{-14}/\text{y}.$$

This is a very small fraction (luckily!).

(e) The radiation pressure p_r should be somewhere between I/c and $2I/c$, depending on whether there is total absorption or total reflection back along the path. These quantities are

$$\frac{I}{c} = \frac{1370\ \text{W/m}^2}{3.00 \times 10^8\ \text{m/s}} = 4.6 \times 10^{-6}\ \text{N/m}^2$$

and $$\frac{2I}{c} = 9.2 \times 10^{-6}\ \text{N/m}^2.$$

Let's take an intermediate value of 7×10^{-6} N/m^2 as an estimate, since solar radiation is partially absorbed and partially reflected back into space.

(f) The force exerted on Earth can be estimated by multiplying the radiation pressure by the cross-sectional area of Earth, similar to what was done in part (c):

$$(\pi R_E^2)(7 \times 10^{-6}\ \text{N/m}^2) = \pi(6370 \times 10^3\ \text{m})^2(7 \times 10^{-6}\ \text{N/m}^2)$$
$$= 8.9 \times 10^8\ \text{N}.$$

By comparison, the gravitational force exerted on Earth by the Sun has the magnitude

$$\frac{GmM}{d^2} = \frac{(6.67 \times 10^{-11}\ \text{N}\cdot\text{m}^2/\text{kg}^2)(5.98 \times 10^{24}\ \text{kg})(1.99 \times 10^{30}\ \text{kg})}{(1.549 \times 10^{11}\ \text{m})^2}$$
$$\approx 3.6 \times 10^{22}\ \text{N},$$

which is much greater, by a factor of more than 10^{13}.

(g) A photon with an associated wavelength of 550 nm has an energy

$$E = \frac{hc}{\lambda} = \frac{(6.625 \times 10^{-34}\ \text{J}\cdot\text{s})\,(3.00 \times 10^8\ \text{m/s})}{550 \times 10^{-9}\ \text{m}}$$
$$= 3.61 \times 10^{-19}\ \text{J}.$$

As calculated in part (b), the Sun emits energy at the rate of 3.82×10^{26} W $= 3.82 \times 10^{26}$ J/s, which corresponds to

$$\frac{3.82 \times 10^{26}\ \text{J/s}}{3.61 \times 10^{-19}\ \text{J/photon}} = 1.06 \times 10^{45}\ \text{photons/s}.$$

Earth is receiving 1.75×10^{17} W of solar power, or

$$\frac{1.75 \times 10^{17}\ \text{J/s}}{3.61 \times 10^{-19}\ \text{J/photon}} = 4.85 \times 10^{35}\ \text{photons/s}.$$

Graphing Calculators

112. *Rainbow.* Figure 34-79 shows a light ray entering and then leaving a falling, spherical raindrop after one internal reflection (compare with Fig. 34-22). The ray is deviated (turned) from its original direction of travel by angular deviation θ_{dev}. (a) Show that θ_{dev} is given by

$$\theta_{dev} = 180° + 2\theta_i - 4\theta_r,$$

where θ_i is the angle of incidence of the ray on the drop and θ_r is the angle of refraction of the ray within the drop. (b) Using Snell's law, substitute for θ_r in terms of θ_i and the index of refraction n of the water. Then graph θ_{dev} versus θ_i for the range of possible θ_i values, but in doing so, replace n with a list: 1.331 for red light and 1.343 for blue light.

The red-light curve and the blue-light curve that appear on the calculator have different minima, which means that there is a different *angle of least deviation* for each color. The light of any given color that leaves the drop at that color's angle of least deviation is especially bright because rays bunch up at that angle. So the bright red light leaves the drop at one angle and the bright blue light leaves at another angle. The story is the same for other drops in your view (Fig. 34-22).

Determine the angles of least deviation from the curves of θ_{dev} for (c) red light and (d) blue light. Let us assume that these colors form the inner and outer edges of a rainbow. (e) What then is the angular width of the rainbow?

113. The *first-order* (or *primary*) *rainbow* described in Problem 112 is the type commonly seen in regions where rainbows appear. It is produced by light reflecting once inside the drops. More rare is the *second-order* (or *secondary*) *rainbow* that is produced by light reflecting twice inside the drops (Fig. 34-80*a*). (a) Show that the angular deviation of such light is given by

$$\theta_{dev} = (180°)k + 2\theta_i - 2(k + 1)\theta_r,$$

where k is the number of internal reflections. Using the procedure of Problem 112, find the angles of (b) red light and (c) blue light in the second-order rainbow. (d) What is the angular width of that rainbow?

The *third-order rainbow* depends on three internal reflections (Fig. 34-80*b*). It probably occurs but cannot be seen because it is faint and lies in the bright sky surrounding the direction of the Sun. What are the angles of (e) the red light and (f) the blue light in this rainbow and (g) what is the rainbow's angular width?

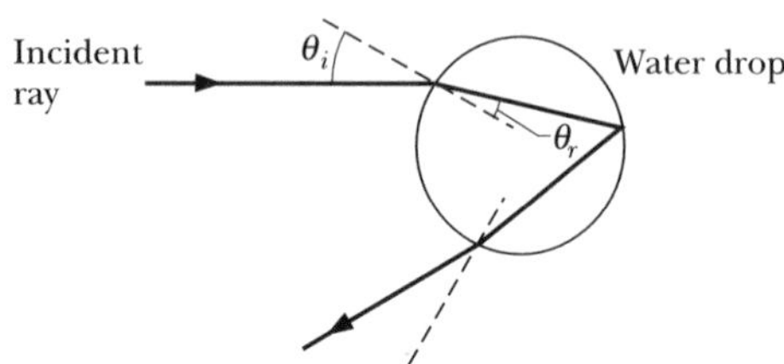

FIGURE 34-79 Problem 112.

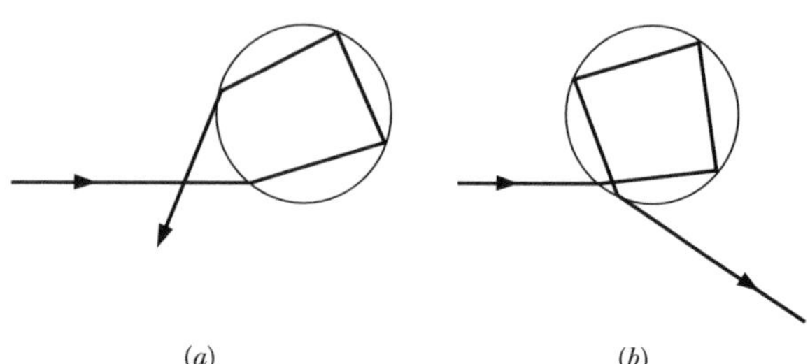

FIGURE 34-80 Problem 113.

Chapter Thirty-Five
Images

QUESTIONS

16. (a) When the Sun is low over a perfectly calm body of water that stretches away from you to the horizon, you can see a single image of the Sun in the water via reflections of the sunlight to you. If the Sun is 15° above the horizon, how far below the horizon is its image? (b) The scene differs when waves sweep over the water because the tilted surfaces produced by the waves shift the image of the Sun upward toward the horizon or downward from it. The composite of all the images from the various changing tilts of the water surfaces results in a luminous *glitter path* of sunlight over the water. The shape of the luminous region might be similar to that shown in Fig. 35-47. Is the nearest point due to a surface that is tilted up toward you or up away from you? (*Hint:* Experiment with a flat mirror. Lay the mirror on a table so that in it you see the image of an object on the other side of the table. Then tilt the mirror's front edge or its back edge upward. Does the image move toward or away from you?)

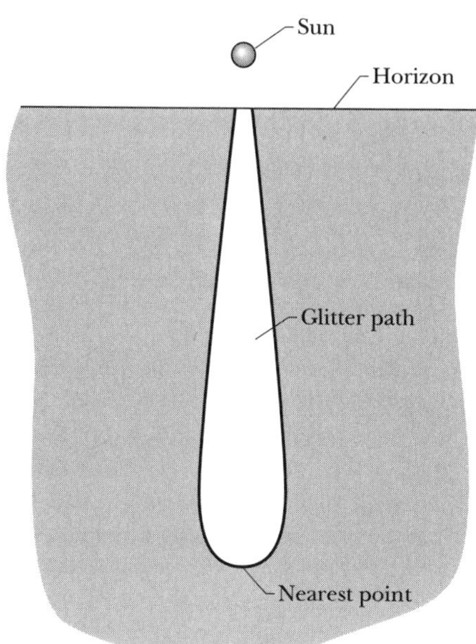

FIGURE 35-47 Question 16.

17. Figure 35-48*a* shows, from an overhead view, a person standing between two vertical mirrors, looking into one of them. The other parts of the figure are suggestions for what the person might see in the mirror. Which of the suggestions correspond to the situations where (a) the mirrors are parallel and (b) the mirror separation is somewhat greater at the left end than at the right end? (*Hint:* Experiment with two flat mirrors, one fairly large and mounted on a wall. Stand in front of the mounted mirror; hold the second mirror at eye level and parallel to the mounted mirror to mimic Fig. 35-48*a*. Then look into the second mirror as you tilt a vertical edge toward you or away from you. What do you see in the reflections?)

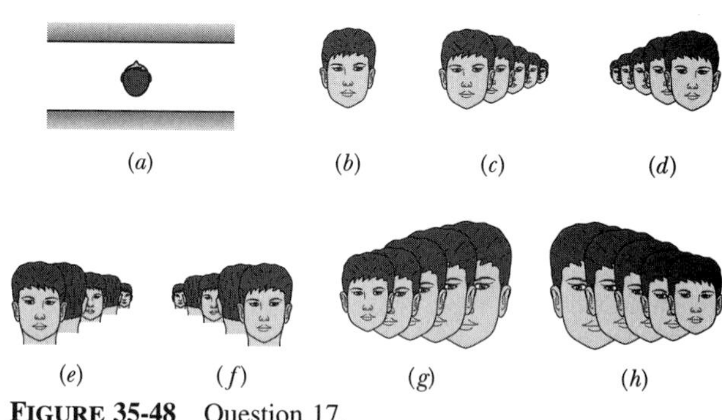

FIGURE 35-48 Question 17.

18. A toy soldier is placed, in turn, in front of four mirrors, *A, B, C,* and *D.* The following table gives the corresponding object distances *p* and image distances *i,* all in centimeters. (a) Rank the mirrors according to the size of the (lateral) height of the image, greatest first. (b) Which mirrors produce an image of the toy soldier that could appear on a sheet of paper?

	A	*B*	*C*	*D*
p	2	4	2	6
i	4	−8	−6	2

19. You hastily erect a meter stick in front of a carnival "funny mirror" before being driven away by the manager (who resents physics students because he did not pass his physics course). Your hasty sketch of the image of the meter stick in the mirror is given in Fig. 35-49*a*. Two possible side views of the mirror are given in the other two parts of the figure. Which one best corresponds to the carnival's mirror?

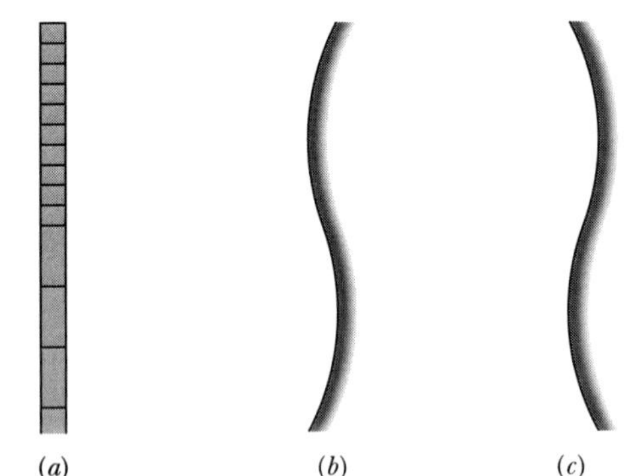

FIGURE 35-49 Question 19.

20. *Organizing question:* In Fig. 35-50 stick figure O sits in front of a spherical mirror that is mounted within the boxed region; the central axis through the mirror is shown. Some of the four stick figures I suggest the general locations and orientations of the images that might be produced by the mirror. (The figures are only sketched in; neither their heights nor their distances from the mirror are drawn to scale.) (a) Which of the stick figures could not possibly represent images? Of the possible images, (b) which would be due to a concave mirror, (c) which would be due to a convex mirror, (d) which would be virtual, and (e) which would involve negative magnification?

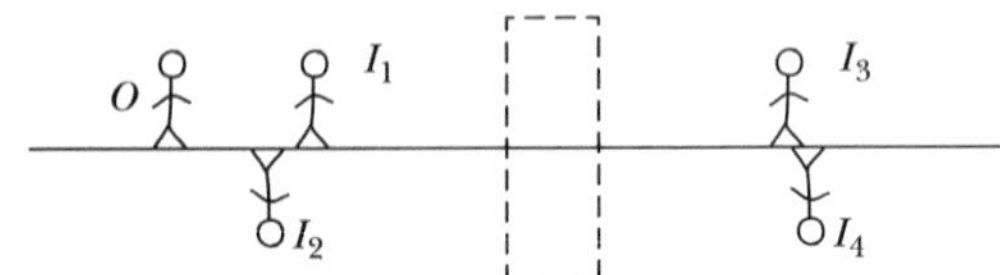

FIGURE 35-50 Questions 20 and 22.

21. *Organizing question:* Figure 35-51 shows three situations in which a stick figure stands in front of a spherical mirror; the focal points F are indicated. For each situation determine if (a) the focal length, (b) the image distance, and (c) the lateral magnification are positive or negative. (*Hint:* If you have filled out Table 35-1, this should be a snap, rather than a crackle or a pop.)

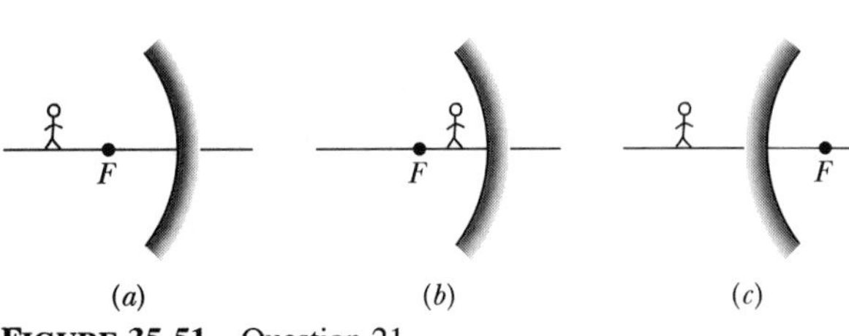

FIGURE 35-51 Question 21.

22. *Organizing question:* In Fig. 35-50 stick figure O sits in front of a thin, symmetric lens that is mounted within the boxed region; the central axis through the lens is shown. Some of the four stick figures I suggest the general locations and orientations of the images that might be produced by the lens. (The figures are only sketched in; neither their height nor their distance from the lens is drawn to scale.) (a) Which of the stick figures could not possibly represent images? Of the possible images, (b) which would be due to a converging lens, (c) which would be due to a diverging lens, (d) which would be virtual, and (e) which would involve negative magnification?

23. *Organizing question:* Figure 35-52 shows three situations in which a stick figure stands in front of a symmetric thin lens; the focal points F are indicated. For each situation determine if (a) the focal length, (b) the image distance, and (c) the lateral magnification are positive or negative. (*Hint:* If you have filled out Table 35-2, this should be a breeze, rather than a twister or a willy-nilly.)

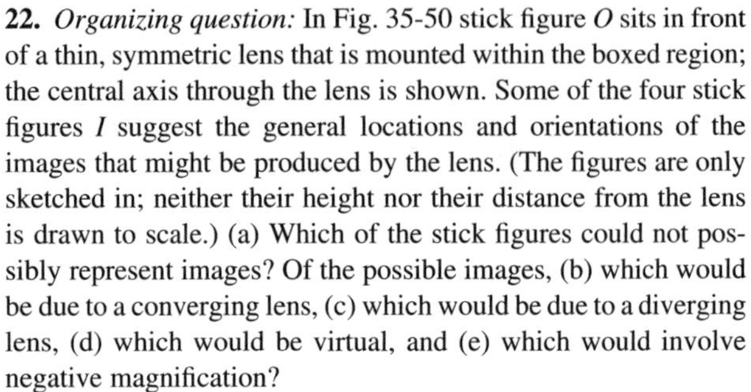

(a) (b) (c)

FIGURE 35-52 Question 23.

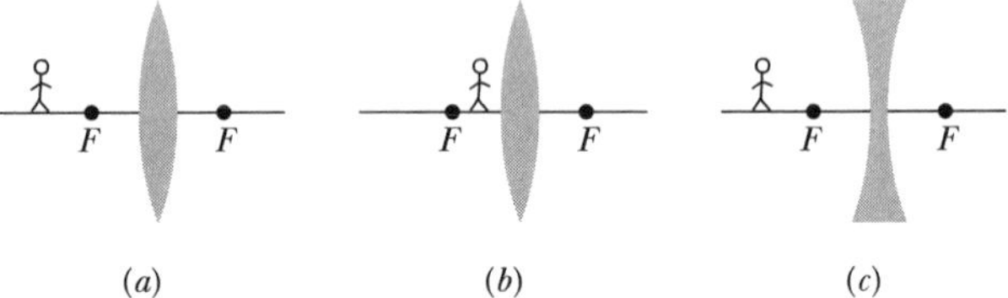

24. (a) If an object is outside the focal point of a converging lens, does its image height increase, decrease, or stay the same when the object moves away from the lens? (*Hint:* Consider how the rays in Fig. 35-15*a* would change.) (b) If an object is in front of a diverging lens, does its image height increase, decrease, or stay the same when the object moves away from the lens? (*Hint:* Consider how the rays in Fig. 35-15*c* would change.)

25. As a penguin waddles toward a large converging lens, the image of it formed by the lens moves. (a) If the penguin is outside the focal point, is that motion toward the lens or away from the lens? (b) Repeat the question for the penguin inside the focal point, waddling toward the mirror.

26. A turnip sits before a thin converging lens, outside the focal point of the lens. The lens is filled with a transparent gel so that it is flexible; by squeezing its ends toward its center (as indicated in Fig. 35-53*a*), you can increase the curvature of its front and rear sides. (a) When you squeeze the lens, does the image of the

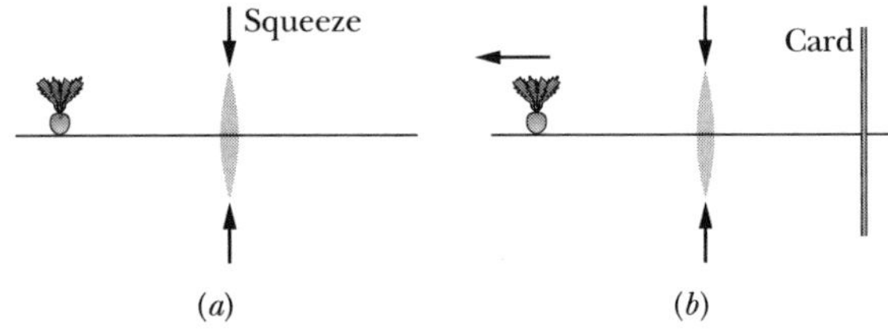

FIGURE 35-53 Question 26.

turnip move toward or away from the lens? (b) Does the size of the lateral height of the image increase, decrease, or stay the same? (c) Suppose that you must keep the image on a card at a certain distance behind the lens (Fig. 35-53*b*) while you move the turnip away from the lens. Must you increase or decrease your squeeze on the lens during the move?

27. (a) A beam of white light is sent into a thin converging lens along the central axis of the lens. Is the focal length of the red light in the beam greater than, less than, or the same as that of the green light? (b) Repeat the question if, instead, the light is sent into a concave spherical mirror along the central axis.

28. An air cavity in the shape of a normal converging lens has been formed in plastic of index n_1. If rays that are parallel to the central axis through the cavity are sent through the cavity, (a) do they converge or diverge and (b) do they form a real or virtual focal point? We can fill the cavity with one of two liquids: liquid *A* has an index that is less than n_1; liquid *B* has an index that is more than n_1. If we use liquid *A*, (c) will rays parallel to the central axis converge or diverge and (d) will they form a real or a virtual focal point? (e) What are the answers if, instead, we use liquid *B*?

EXERCISES & PROBLEMS

55. Two plane mirrors are placed parallel to each other and 40 cm apart. An object is placed 10 cm from one mirror. What is the distance from the object to the image for each of the five images that are closest to the object?

56. You are standing 1.0 m in front of a large shiny sphere that is 0.70 m in diameter. (a) How far from the surface of the sphere closest to you and on which side of the surface does your image appear to be? (b) If you are 2.0 m tall, how tall is your image? (c) Is the image inverted or erect with respect to you?

57. A concave mirror has a radius of curvature of 24 cm. How far is an object from the mirror if an image is formed that is (a) virtual and 3.0 times the size of the object, (b) real and 3.0 times the size of the object, and (c) real and 1/3 the size of the object?

58. An object is 30.0 cm from a spherical mirror, along the central axis. The absolute value of lateral magnification is 0.500. The image produced is inverted. What is the focal length of the mirror?

59. A grasshopper hops to a point on the central axis of a spherical mirror. The absolute magnitude of the mirror's focal length is 40.0 cm, and the lateral magnification of the grasshopper image produced by the mirror is +0.200. (a) Is the mirror convex or concave? (b) How far from the mirror is the grasshopper?

60. A small cup of green tea is positioned on the central axis of a spherical mirror. The magnification of the cup is +0.250, and the distance between the mirror and its focal point is 2.00 cm. (a) What is the distance between the mirror and the image it produces? (b) Is the focal length positive or negative? (c) Is the image real or virtual?

61. A glass sphere has a radius of 5.0 cm and a refractive index of 1.6. A paperweight is constructed by slicing through the sphere on a plane that is 2.0 cm from the center of the sphere and perpendicular to a radius of the sphere that passes through the center of the circle formed by the intersection of the plane and the sphere. The paperweight is placed on a table and viewed from directly above by an observer who is 8.0 cm from the tabletop, as shown in Fig. 35-54. When viewed through the paperweight, how far away does the tabletop appear to be to the observer?

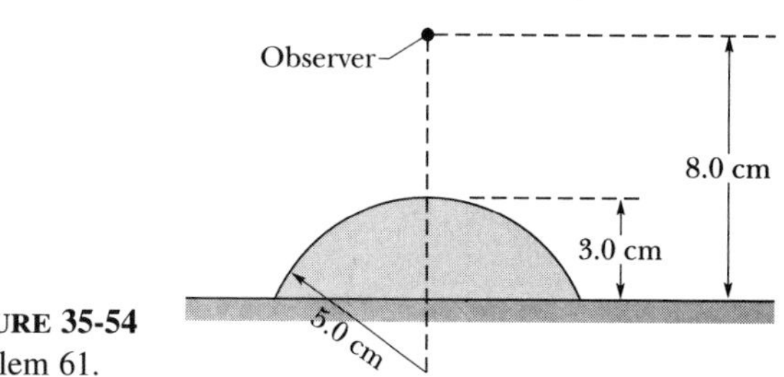

FIGURE 35-54 Problem 61.

62. One end of a long glass rod (n = 1.5) is formed into the shape of a convex surface of radius 6.0 cm. An object is located in air along the axis of the rod, at a distance of 10 cm from the end of the rod. (a) How far apart are the object and the image formed by the glass rod? (b) For what range of distances from the end of the rod must the object be located in order to produce a virtual image?

63. In Fig. 35-55, a fish watcher watches a fish through a 3.0-cm-thick glass wall of a fish tank. The watcher is level with the fish; the index of refraction of the glass is 8/5 and that of the water is 4/3. (a) To the fish, how far away does the watcher appear to be? (*Hint:* The watcher is the object. Light from that object passes through the wall's outside surface, which acts as a refracting surface. Find the image produced by that surface. Then treat that image as an object whose light passes through the wall's inside surface, which acts as another refracting surface. Find the image produced by that surface, and there is the answer.) (b) To the watcher, how far away does the fish appear to be?

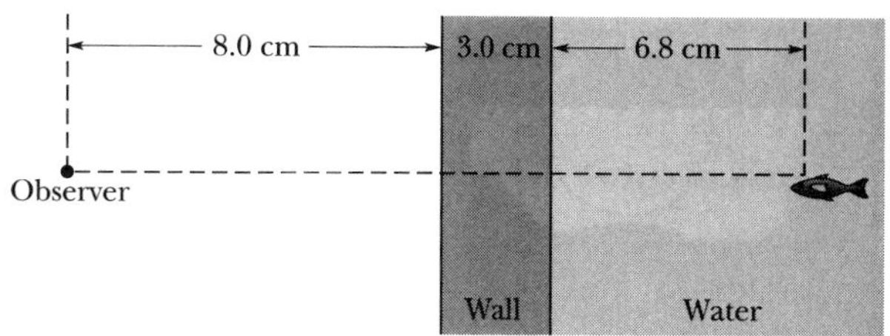

FIGURE 35-55 Problem 63.

64. A cheese enchilada is 4.00 cm in front of a converging lens. The magnification of the enchilada is -2.00. What is the focal length of the lens?

65. A pepper seed is placed in front of a lens. The lateral magnification of the seed is $+0.300$. The absolute value of the focal length is 40.0 cm. How far from the lens is the image?

66. An object is located to the left of a thin lens, on the central axis of the lens. The lens forms an image of the object with the same orientation as the object, and the image–object distance is 20 cm. What are the focal length of the lens and the lens–object distance if the image has (a) 2.0 times the height of the object and (b) 0.50 times the height of the object?

67. In Fig. 35-56*a,* a pea sits at a focal point of the first (nearer) thin diverging lens, 4.00 cm from that lens. The lenses are identical and separated by 10.0 cm, with a common central axis. (a) Where is the image of the pea produced by the second lens? (b) Is that image inverted from the pea or does it have the same orientation? (c) Is it real or virtual?

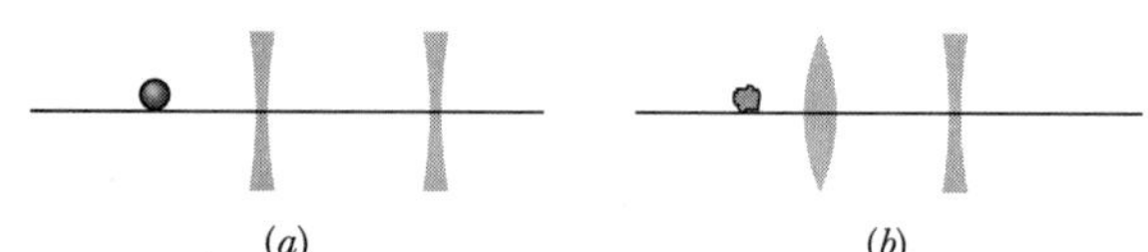

FIGURE 35-56 Problems 67 and 68.

68. In Fig. 35-56*b,* a sand grain is 3.00 cm from the first (nearer) thin lens, on the central axis through the two symmetric lenses. The distance between focal point and lens is 4.00 cm for both lenses; the lenses are separated by 8.00 cm. (a) What is the distance between the second lens and the image that it produces of the sand grain? (b) Is the image to the left or right of the second lens? (c) Is it real or virtual? (d) Is it inverted from the sand grain?

69. In Fig. 35-57, a box is somewhere at the left, on the central axis of the thin converging lens. The image of the box produced by the plane mirror is 4.00 cm "inside" the mirror. The lens–mirror separation is 10.0 cm, and the focal length of the lens is 2.00 cm. (a) What is the distance between the box and the lens? Light reflected by the mirror travels back through the lens, which produces a final image of the box. (b) What is the distance between the lens and that final image?

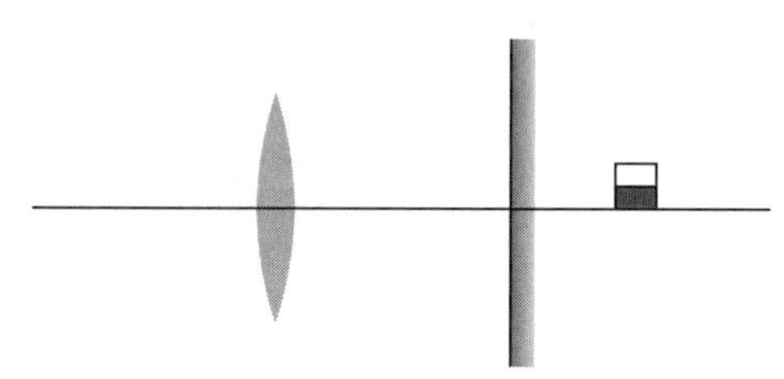

FIGURE 35-57 Problem 69.

70. A diverging lens with a focal length of -15 cm and a converging lens with a focal length of 12 cm have a common central axis. Their separation is 12 cm. An object of height 1.0 cm is 10 cm in front of the diverging lens, on the common central axis. (a) Where does the lens combination produce the final image of the object (the one produced by the second, converging lens)? (b) What is the height of that image? (c) Is the image real or virtual? (d) Does it have the same orientation as the object or is it inverted?

Tutorial Problems

71. In this problem we look at the images formed on either side of a two-sided spherical mirror. Use a ruler, marked in metric units, to make straight light rays and to measure all distances as accurately as possible. The image location and size will be determined in two ways, using ray tracing and using the mirror equation.

(a) A two-sided mirror (convex on one side and concave on the other) has a radius of curvature of 5.00 cm. Make a sketch of the mirror and its center of curvature, placing the center to the right of the mirror. Make an accurate 1:1 sketch. Locate and mark the focal point or points. (b) Suppose an object is located on the left side of the mirror. Is the mirror then acting as a convex mirror or a concave mirror? What are the magnitude and sign of the focal length of this mirror? On which side would a real image be found? On which side would a virtual image be found?

(c) Suppose a small object 1.00 cm high is located on the left side a distance of 5.00 cm from the mirror. Draw the object at its correct size at its correct location. Draw at least two light rays and use them to locate the image. Approximately where is the image? What is its approximate image distance (including the correct sign)? Is the image real or virtual? Is the image erect or inverted? Is the image larger than or smaller than the object? Estimate its approximate height. Use ray construction and your ruler only in answering this part; a quantitative check will be carried out later. (d) Using the mirror equation, determine numerically the image location, magnification *m,* and image height. (e) If the object were brought closer to the mirror, how would the image position and height change?

(f) Repeat everything in parts (a) through (d) for an object 1.00 cm high located 6.00 cm from the mirror on its right side. Think about where you should place the mirror in order to have the most room for your diagram.

Answers

(a) The center of curvature is 5.00 cm to the right of the mirror, and the focal point is halfway between the center and the mirror, 2.50 cm from each.

(b) With the object on the left, the mirror acts as a convex mirror. The focal length is negative (since the focal point is on the opposite side of the mirror from the object), so it is

$$f = -\frac{r}{2} = -\frac{5.00\text{ cm}}{2} = -2.50\text{ cm}.$$

Real images are found on the side to which the light rays reflect, which, in this case, is the same side as the object, the left side. A virtual image would be found on the other side, the right side.

(c) See Figs. 35-10*b* and *c* for the general features of the diagram. The diagram should show that the image is on the right side, so it is a virtual image. Also, the image is apparently erect and smaller than the object. It is located about 1.5 to 2 cm on the right side, so the image distance is about -2 cm. Its height should be about 1/3 that of the object, or about 0.3 cm, but this is difficult to measure accurately.

(d) The mirror equation

$$\frac{1}{f} = \frac{1}{p} + \frac{1}{i}$$

becomes

$$\frac{1}{-2.50 \text{ cm}} = \frac{1}{5.00 \text{ cm}} + \frac{1}{i},$$

so
$$\frac{1}{i} = -\frac{1}{2.50 \text{ cm}} - \frac{1}{5.00 \text{ cm}} = -0.600 \text{ cm}^{-1}$$

and
$$i = -1.67 \text{ cm}.$$

This is the image location. The magnification is

$$m = -\frac{i}{p} = -\frac{-1.67 \text{ cm}}{5.00 \text{ cm}} = +0.33.$$

The image size is then

$$|m|(\text{object size}) = (0.33)(1.00 \text{ cm}) = 0.33 \text{ mm}.$$

(e) Bringing the object closer to the mirror would reduce p and increase $1/p$, so it would cause $1/i$ to decrease and i to increase. The image distance i would then become a smaller negative number, corresponding to an image closer to the mirror than before. The effect on the magnification is more difficult to determine, because its magnitude is i/p and both i and p are smaller in magnitude than before. A close analysis shows that the magnification increases as the object is brought closer to the mirror. When the object is very close to the mirror, the mirror acts nearly as a plane mirror, with magnification 1.

(f) See Fig. 35-10*a* for the general features of the diagram (left-right reversed). Since the object is on the right, the mirror acts as a concave mirror. Its focal length is positive:

$$f = \frac{r}{2} = \frac{5.00 \text{ cm}}{2} = 2.50 \text{ cm}.$$

A real image is found on the same side as the object, the right side, because that's the side on which the rays really are. A virtual image appears to be on the other side, the left side.

The ray diagram construction shows that the image is on the right side, so it is a real image. It is apparently inverted and larger than the object. It is located a little inside the center of curvature, about 4 cm to the right of the mirror, so the image distance is approximately $+4$ cm. Its height appears to be about the same as that of the object, about 1.0 cm.

To determine the image location and magnification numerically, we start with the mirror equation:

$$\frac{1}{f} = \frac{1}{p} + \frac{1}{i}$$

becomes

$$\frac{1}{2.50 \text{ cm}} = \frac{1}{6.00 \text{ cm}} + \frac{1}{i},$$

so
$$\frac{1}{i} = \frac{1}{2.50 \text{ cm}} - \frac{1}{6.00 \text{ cm}} = 0.233 \text{ cm}^{-1}$$

and
$$i = 4.29 \text{ cm}.$$

This is the image location. The magnification is

$$m = -\frac{i}{p} = -\frac{4.29 \text{ cm}}{5.00 \text{ cm}} = -0.86.$$

The image size is then

$$|m|(\text{object size}) = (0.86)(1.00 \text{ cm}) = 0.86 \text{ cm}.$$

72. In this problem we consider the path of light rays from air into glass and vice versa. The glass has a flat, horizontal surface, and air is above it (Fig. 35-58*a*). We want to determine the angles for the reflected and refracted rays. We also consider the case of multiple reflections in the glass.

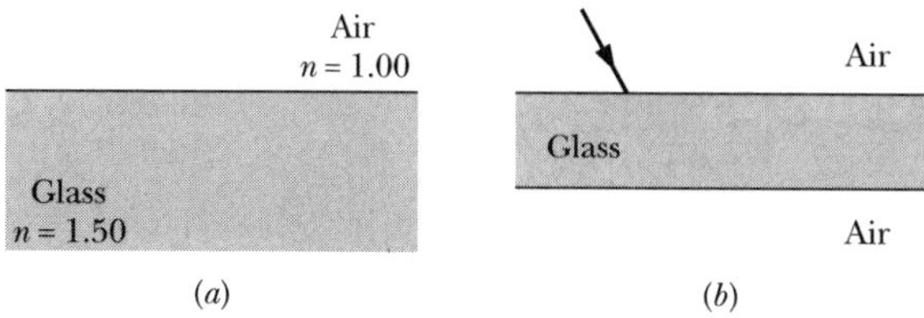

FIGURE 35-58 Problem 72.

(a) Suppose a light ray is incident from the air at an angle of 35°. Determine the angles of the reflected and refracted rays and show the rays and angles on a diagram. (b) Suppose, instead, that the light ray is incident from the glass at an angle of 35°. Determine the angles of the reflected and refracted rays and show the rays and angles on a diagram. (c) For light rays incident from the air into the glass, determine the angles of reflection and of refraction for angles of incidence at 10° intervals from 0° to 90°. Do the same for light rays incident from the glass into the air. Enter the results in a table with the following headings:

FROM AIR INTO GLASS, ANGLES OF INCIDENCE REFLECTION REFRACTION FROM GLASS INTO AIR, ANGLES OF INCIDENCE REFLECTION REFRACTION

(d) In part (c) there is a particular angle of incidence for which something interesting occurs. Determine that angle. Name the phenomenon involved and describe the physical difference between light rays with a smaller angle of incidence and light rays with a greater angle of incidence than this particular angle. (e) Suppose the glass has a flat lower side (Fig. 35-58*b*). For a light ray incident from the left at an angle of 60°, as shown, draw as many reflected and refracted light rays as possible. Remember that reflections and refractions occur every time a light ray strikes an interface. Label the angles in this diagram.

Answers

(a) See Fig. 35-59*a*. The reflected ray is directed back into the air at an angle of 35° on the opposite side of the normal to the surface. The angle of the refracted ray can be found by the use of Snell's law:

$$n_1 \sin \theta_1 = n_2 \sin \theta_2$$

$$(1.00)\sin 35° = (1.50)\sin \theta_2$$

$$\theta_2 = \sin^{-1}\left(\frac{1.00}{1.50} \sin 35°\right) = 22.5°.$$

(b) See Fig. 35-59*b*. In this case the reflected ray is directed at an angle of 35°, and the refracted ray has an angle of refraction of

$$\theta_2 = \sin^{-1}\left(\frac{1.50}{1.00} \sin 35°\right) = 59.4°.$$

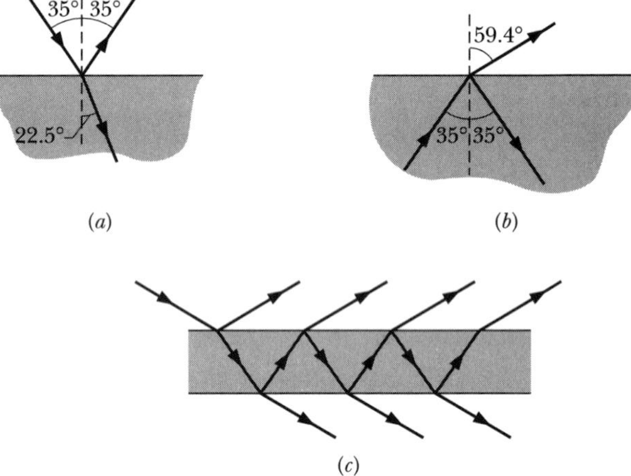

FIGURE 35-59 Problem 72.

(c)

FROM AIR INTO GLASS, ANGLES OF			FROM GLASS INTO AIR, ANGLES OF		
INCIDENCE	REFLECTION	REFRACTION	INCIDENCE	REFLECTION	REFRACTION
0°	0°	0°	0°	0°	0°
10°	10°	6.6°	10°	10°	15.1°
20°	20°	13.2°	20°	20°	30.9°
30°	30°	19.5°	30°	30°	48.6°
40°	40°	25.4°	40°	40°	74.6°
50°	50°	30.7°	50°	50°	none
60°	60°	35.3°	60°	60°	none
70°	70°	38.8°	70°	70°	none
80°	80°	41.0°	80°	80°	none
90°	90°	41.8°	90°	90°	none

(d) At the angle of incidence for light rays from glass into air for which the angle of refraction is 90°, the angle of incidence must be

$$\theta_1 = \sin^{-1}\left(\frac{1.00}{1.50} \sin 90°\right) = 41.8°.$$

This is the critical angle for total internal reflection. At smaller angles of incidence, the ray from the glass can be refracted out into the air (as well as reflected), but at larger angles it can only be reflected, as there is then no solution to Snell's law.

(e) The original ray has angles of incidence and reflection of 60° and an angle of refraction of $\theta = \sin^{-1}[(1.00/1.50)\sin 60°] \approx 35°$. The reflected ray is gone forever, but the refracted ray will reflect off the bottom, and the ray reflected off the bottom can reflect and refract at the air–glass interface, and so on. Thus, there can be many internal reflections and many points at which light refracts from the glass (Fig. 35-59*c*).

73. In this problem we consider the images of several lens systems. You should make a scale sketch of the physical situation described. Use both ray tracing and the lens equation to characterize the image qualitatively and numerically, that is, to determine properties of the image such as its location, orientation, size, and magnification. Use a ruler to draw straight lines and make accurate measurements. Use a solid line to show a real ray, and a dashed line to show a *virtual ray,* that is, a backward extrapolation of a real ray.

(a) Consider first a thin convergent lens with a focal length of 4.00 cm. An object 1.00 cm high is placed on the central axis of the lens, 2.50 cm to the left of the lens. (b) Next consider a thin convergent lens with a focal length of 4.00 cm. A 1.00 cm high object is located 6.00 cm to the left of the lens. (c) Finally, use a thin divergent lens with a focal length of magnitude 4.00 cm. The object is 1.00 cm high is located 8.00 cm to the left of the lens.

Answers

(a) See Fig. 35-15*b* for the general features of the diagram. The diagram should show that the image is on the left; so it must be a virtual image because only the virtual rays intersect at the image location. In other words, there isn't a real image there; the image appears only to an observer on the right side looking back through the lens. The image is erect, and it is more than twice as large as the object.

To find the image distance, we can write

$$\frac{1}{f} = \frac{1}{p} + \frac{1}{i},$$

so $$\frac{1}{i} = \frac{1}{f} - \frac{1}{p} = \frac{1}{4.00 \text{ cm}} - \frac{1}{2.50 \text{ cm}} = -0.150 \text{ cm}^{-1}$$

and $$i = -6.67 \text{ cm}.$$

The image magnification is given by

$$m = -\frac{i}{p} = -\frac{-6.67 \text{ cm}}{2.50 \text{ cm}} = +2.67.$$

The image size is

$$|m|(\text{object size}) = (2.67)(1.00 \text{ cm}) = 2.67 \text{ cm}.$$

(b) See Fig. 35-15*a* for the general features of the diagram. The diagram should show that the image is on the right, where real light rays intersect, so the image must be a real image. It is inverted and it is about the same height as the object. To find image distance, we can write

$$\frac{1}{f} = \frac{1}{p} + \frac{1}{i},$$

so $$\frac{1}{i} = \frac{1}{f} - \frac{1}{p} = \frac{1}{4.00 \text{ cm}} - \frac{1}{6.00 \text{ cm}} = 0.083 \text{ cm}^{-1}$$

and $$i = 12.0 \text{ cm}.$$

The image magnification is

$$m = -\frac{\text{i}}{\text{p}} = -\frac{12.00 \text{ cm}}{6.00 \text{ cm}} = -2.00,$$

and the image size is

$$|m|(\text{object size}) = (2.00)(1.00 \text{ cm}) = 2.00 \text{ cm}.$$

(c) See Fig. 35-15*c* for the general features of the diagram. The diagram should show that the image is on the left, where virtual rays rather than real rays intersect, so this image must be a virtual image. It is erect and only about 1/3 the size of the object. To find image distance, we can write

$$\frac{1}{f} = \frac{1}{p} + \frac{1}{i},$$

so $$\frac{1}{i} = \frac{1}{f} - \frac{1}{p} = \frac{1}{-4.00 \text{ cm}} - \frac{1}{8.00 \text{ cm}} = -0.375 \text{ cm}^{-1}$$

and $$i = -2.67 \text{ cm}.$$

The image magnification is

$$m = -\frac{i}{p} = -\frac{-2.67 \text{ cm}}{8.00 \text{ cm}} = +0.33,$$

and the image size is

$$|m|(\text{object size}) = (0.33)(1.00 \text{ cm}) = 0.33 \text{ cm}.$$

Chapter Thirty-Six
Interference

QUESTIONS

20. In Fig. 36-46, the waves along rays 1 and 2 are initially in phase while they travel through plastic material. Along the way, ray 2 goes through an air cavity of length L. Is the number of wavelengths of ray 2 in that length L in the air greater than, less than, or equal to the number of wavelengths of ray 1 in that same length in the plastic?

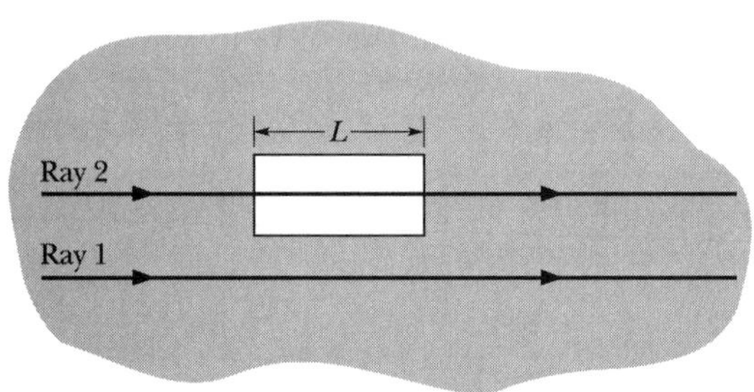

FIGURE 36-46 Question 20.

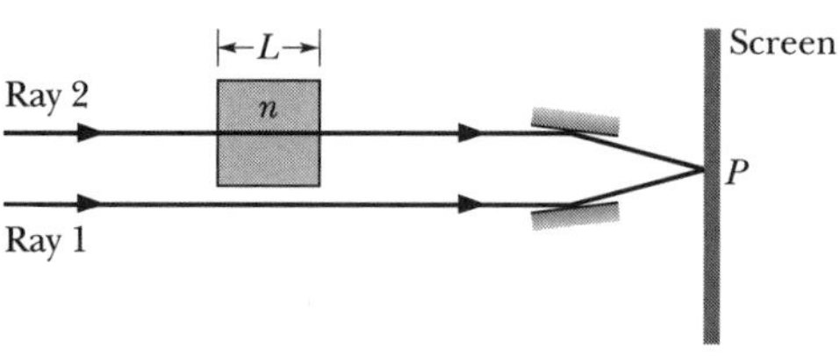

(*a*)

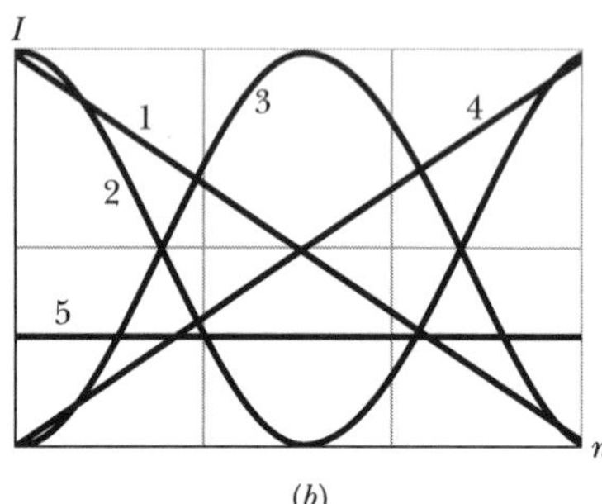

(*b*)

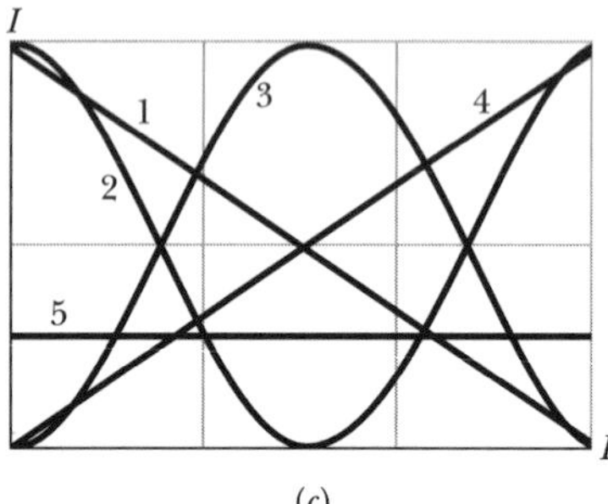

(*c*)

FIGURE 36-47 Question 21.

21. In Fig. 36-47*a*, the waves along rays 1 and 2 are initially in phase. Ray 2 goes through a material with length L and index of refraction n. The rays are then reflected by mirrors to a common point P on a screen. (a) Suppose that we can vary n from an initial value matching that of air to larger values. Which of the curves in Fig. 36-47*b* would then best give the intensity I of the light at point P versus n? (b) Suppose, instead, that we can vary L from an initial value of zero to larger values. Then which of the curves in Fig. 36-47*c* would best give the intensity I of the light at point P versus L?

22. Monochromatic yellow light is used in a Young's interference experiment; then monochromatic green light is used. Graphs of the intensity I versus position x on the screen in the two experiments are superimposed in Fig. 36-48. In that figure, is the central bright fringe to the left or to the right?

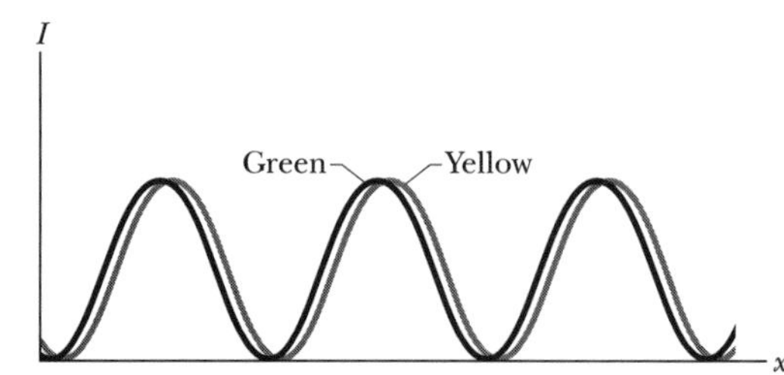

FIGURE 36-48 Question 22.

23. Figure 36-49*a* gives the intensity I versus position x on the screen for the central portion of a two-slit interference pattern. The other parts of the figure give phasor diagrams for the electric field components of the waves arriving at the screen from the two slits (as in Fig. 36-10*a*). Which points on the screen best correspond to which phasor diagram?

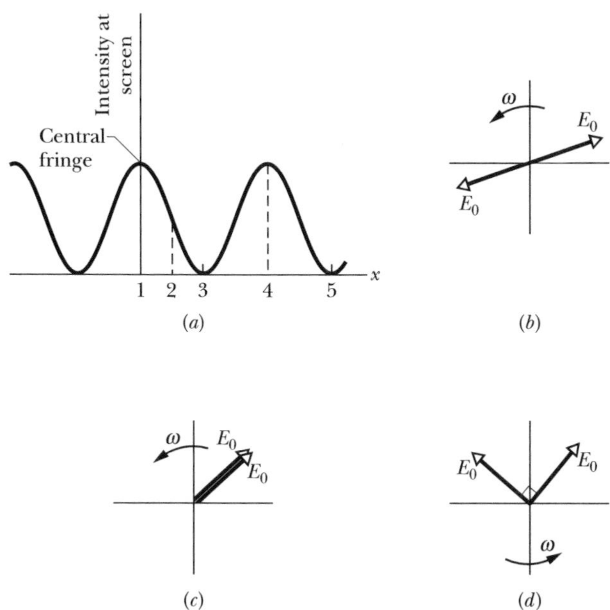

FIGURE 36-49 Question 23.

24. Figure 36-50 shows two sources A and B that emit radio waves of the same wavelength λ. The waves are detected at point P, which is along a perpendicular bisector to the straight line between A and B. (a) In terms of λ, what is the phase difference between the waves arriving at P from A and B if A and B emit their waves in phase? (b) What is that phase difference if, instead, the phase of the emission from A is ahead of that from B by 90°? (Here, *ahead* means that the "crest" of the wave from A, say, its peak electric field, emerges before the "crest" of the wave from B emerges.)

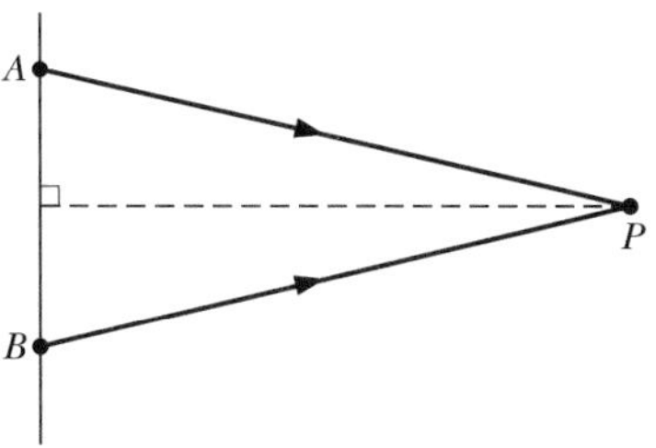

FIGURE 36-50 Question 24.

25. Figure 36-51*a* shows an arrangement similar to that given in Sample Problem 36-1: waves with identical wavelengths and amplitudes and that are initially in phase travel through different media, ray 2 through a plastic layer of thickness L and ray 1 through only air. The number of wavelengths in length L is N_2 for ray 2 and N_1 for ray 1. The following table gives values for N_2 and N_1 for four situations. For each situation, the rays reach a common point on a screen. Rank the situations according to the intensity of the light at that common point, greatest first.

SITUATION	1	2	3	4
N_2	2.75	2.80	3.25	4.00
N_1	2.25	1.80	3.00	3.25

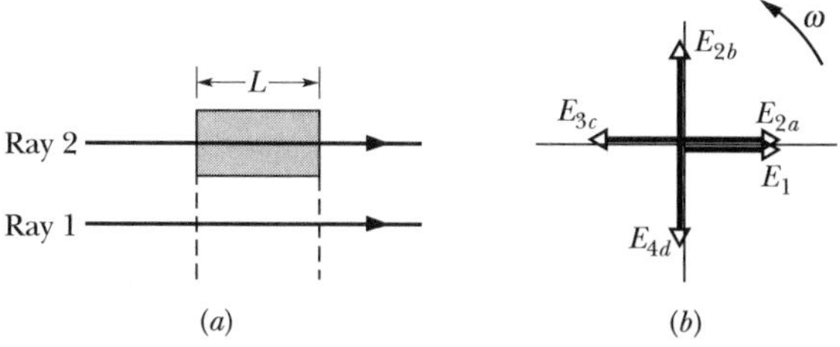

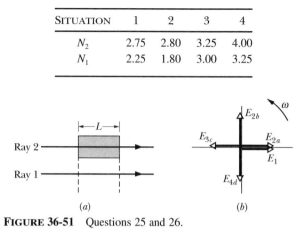

FIGURE 36-51 Questions 25 and 26.

26. For the rays of Question 25, let phasor E_1 represent the electric field component of ray 1 and phasor E_2 represent that of ray 2. In the phasor diagram of Fig. 36-51*b*, which choice of E_2 best corresponds to which situation in Question 25?

27. *Organizing question:* Figure 36-52 shows four situations in which light reflects perpendicularly from a thin film of thickness L (as in Fig. 36-12), with the indices of refraction as given. In which situations does Eq. 36-34 correspond to the reflections yielding maxima (that is, a bright film)?

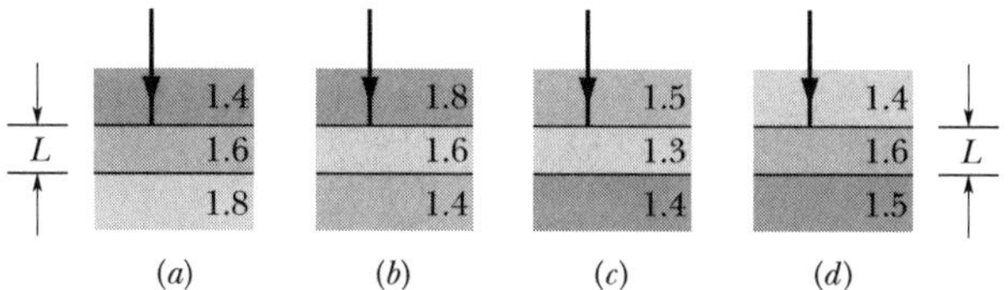

FIGURE 36-52 Question 27.

28. Figure 36-53 shows light that passes through a thin film of water on glass, either directly (ray r_1) or via reflections (ray r_2). The indices of refraction are indicated. The incident ray is actually perpendicular to the film but is drawn slanted only for clarity. (a)

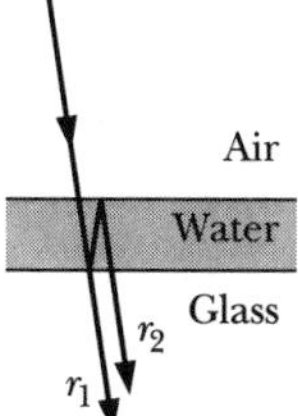

FIGURE 36-53 Question 28.

How many reflections does the light that emerges as r_2 undergo? (b) What is the reflection phase shift of that light at each of those reflections?

29. Suppose that the incident light rays in Fig. 36-12 are initially perpendicular to the thin film ($\theta = 0$). If we increase their angle ($\theta > 0$) somewhat, do the following increase, decrease, or stay the same: (a) the path length of ray r_2 within the film and (b) the wavelength producing a bright film?

EXERCISES & PROBLEMS

79. S_1 and S_2 are two point sources of radiation that are radiating waves in phase with each other of equal amplitude R and of wavelength 400 nm. The sources are located on an x axis at $x = 6.5\ \mu m$ and $x = -6.0\ \mu m$, respectively. (a) Determine the phase difference (in radians) at the origin between the radiation from S_1 and the radiation from S_2. (b) Suppose a slab of transparent material with thickness 1.5 μm and index of refraction $n = 1.5$ is placed between $x = 0$ and $x = 1.5\ \mu m$. What then is the phase difference (in radians) at the origin between the radiation from S_1 and the radiation from S_2?

80. A light beam of wavelength 600 nm in air passes through film 1 ($n_1 = 1.2$) of thickness 1.0 μm, then through film 2 (air) of thickness 1.5 μm, and finally through film 3 ($n_3 = 1.8$) of thickness 1.0 μm. (a) Which film does the light cross in the least time, and what is that least time? (b) What are the total number of wavelengths (at any instant) across all three films together?

81. Figure 36-54 shows the design of a Texas arcade game. Four laser pistols are pointed toward the center of an array of plastic layers where a clay armadillo is the target. The refractive indices of the various plastic layers are indicated. The layers have a thickness of either 2.00 mm or 4.00 mm, as drawn. If the pistols are fired simultaneously, (a) which laser burst hits the target first and (b) what are the times of flight for each burst?

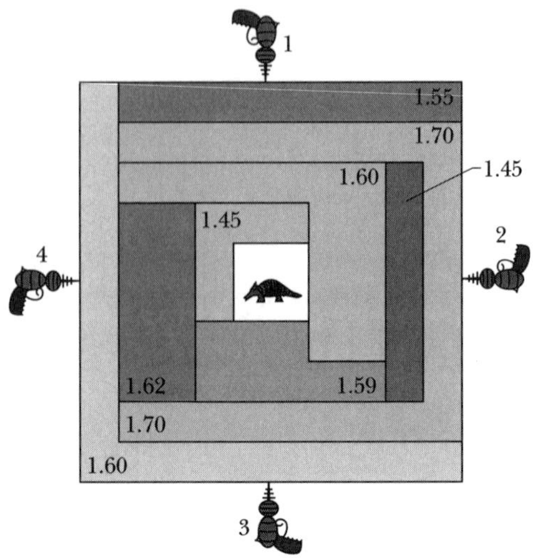

FIGURE 36-54 Problem 81.

82. Light of wavelength 700.0 nm is sent along a route of length 2000 nm. The route is then filled with a medium having an index of refraction of 1.400. In degrees, by how much does the medium phase shift the light? Give the full shift and then the equivalent shift of less than 360°.

83. Two light rays, initially in phase and having wavelength 6.00×10^{-7} m, go through different plastic layers of the same thickness, 7.00×10^{-6} m. The indices of refraction are 1.65 for one layer and 1.49 for the other. (a) What is the equivalent phase difference, in terms of wavelength, between the rays when they emerge? (b) If those two rays then reach a common point, does the interference result in complete darkness, maximum brightness, intermediate illumination but closer to complete darkness, or intermediate illumination but closer to maximum brightness? (c) If the two rays are, instead, initially exactly out of phase, what are the answers to (a) and (b)?

84. Two light rays, initially in phase and with a wavelength of 500 nm, go through different paths by reflecting from the various mirrors shown in Fig. 36-55. (Such a reflection does not itself produce a phase shift.) (a) What least value of distance d will put the rays exactly out of phase when they emerge from the region? (Ignore the slight tilt of the path for ray 2.) (b) Repeat the question assuming that the entire apparatus is immersed in a protein solution with an index of refraction of 1.38.

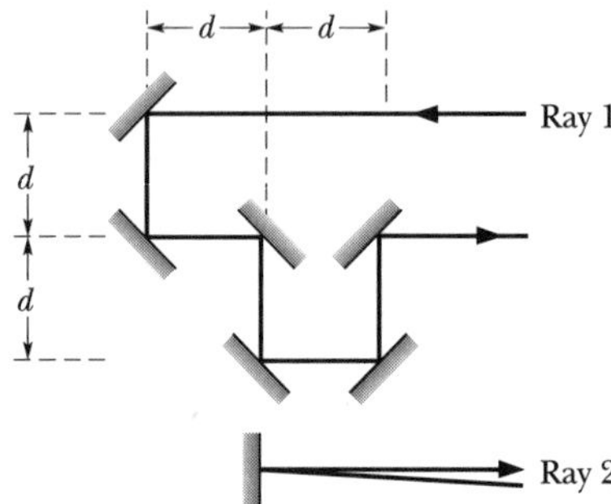

FIGURE 36-55 Problem 84.

85. Two coherent radio point sources that are separated by 2.0 m are radiating in phase with a wavelength of 0.25 m. If a detector moves in a large circle around their midpoint, at how many points will the detector show a maximum signal?

86. The second dark band in a double-slit interference pattern is 1.2 cm from the central maximum of the pattern. The separation between the two slits is equal to 800 wavelengths of the monochromatic light incident (perpendicularly) onto the slits. What is the distance between the plane of the slits and the viewing screen?

87. In a double-slit interference pattern, the slit separation is 2.00 μm, the wavelength is 500 nm, and the separation between the slits and the screen is 4.00 m. (a) What is the angle to the third side bright band? (b) If we decrease the frequency of light by 10.0% of its initial value, how far along the screen and in what direction does that bright band shift?

88. In Fig. 36-8, let the angle θ of the two rays be 20°, slit separation d be 58.00 μm, and wavelength λ be 500.9 nm. (a) In terms of wavelengths, what is the phase difference of the two rays when they reach a common point on a distant screen? (b) Does their interference result in complete darkness, maximum brightness, intermediate illumination but closer to complete darkness, or intermediate illumination but closer to maximum brightness?

89. A double-slit arrangement produces interference fringes for sodium light (wavelength = 589 nm) that are angularly separated by 0.30° near the center of the pattern. What is the angular fringe separation if the entire arrangement is immersed in water, which has an index of refraction of 1.33?

90. Two parallel slits are illuminated with monochromatic light of wavelength = 500 nm. An interference pattern is formed on a screen some distance from the slits, and the fourth dark band is located 1.68 cm from the central bright band on the screen. (a) What is the path length difference corresponding to the fourth dark band? (b) What is the distance on the screen between the central bright band and the first bright band on either side of the central band? (*Hint:* The angles to the fourth dark band and the first bright band are small enough that $\tan\theta \approx \sin\theta$.)

91. Three electromagnetic waves travel through a certain point along an x axis. They are polarized parallel to a y axis, with the following variations in their amplitudes. Find the resultant.

$$E_1 = (10.0\ \mu\text{V/m})\sin[(2.0 \times 10^{14}\ \text{rad/s})t]$$

$$E_2 = (5.00\ \mu\text{V/m})\sin[(2.0 \times 10^{14}\ \text{rad/s})t + 45°]$$

$$E_3 = (5.00\ \mu\text{V/m})\sin[(2.0 \times 10^{14}\ \text{rad/s})t - 45°]$$

92. A 600 nm thick soap (n = 1.40) film in air is illuminated with white light in a direction perpendicular to the film. For how many different wavelengths in the 300 to 700 nm range is there (a) fully constructive interference and (b) fully destructive interference in the reflected light?

93. A thin film (n = 1.25) is deposited on a glass plate (n = 1.40). What are the minimum (nonzero) thicknesses for the film that will (a) maximally transmit light with a wavelength of 550 nm and (b) maximally reflect light with a wavelength of 550 nm?

94. Two microscope slides are lying horizontally one on top of the other when a piece of tissue paper is inserted between them at one end, thereby forming a wedge of air between them. When light of 500 nm wavelength shines vertically down onto the top plate, dark interference fringes are observed with a 1.2 mm separation. What is the angle between the two slides?

95. Two rectangular, optically flat, glass plates (n = 1.60) are in contact along one edge and are separated along the opposite edge by a thin foil of unknown thickness. Light with a wavelength of 600 nm is incident perpendicularly onto the top plate. Nine dark fringes and eight bright fringes are observed across the top plate. If the distance between the two plates along the separated edge is increased by 600 nm, how many dark fringes will there then be across the top plate?

96. White light is sent downward onto a horizontal thin film that has been formed between two materials. The indices of refraction are 1.80 for the top material, 1.70 for the thin film, and 1.50 for the bottom material. The film thickness is 5.00×10^{-7} m. (a) Which visible wavelengths (400 to 700 nm) result in fully constructive interference at an observer above the film? The materials and film are then heated so that the film thickness increases. (b) Does the light resulting in fully constructive interference shift toward longer or shorter wavelengths?

97. In Fig. 36-56, two glass plates (n = 1.60) form a wedge, and a fluid (n = 1.50) fills the interior. At the left end the plates touch; at the right, they are separated by 580 nm. Light with a wavelength (in air) of 580 nm shines downward on the assembly, and an observer intercepts light sent back upward. Do dark or bright bands lie at (a) the left end and (b) the right end? (c) How many dark bands are along the plates?

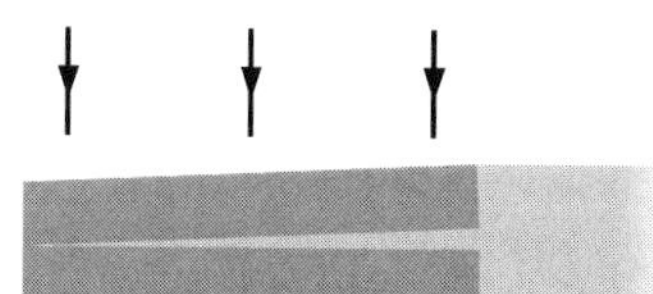

FIGURE 36-56 Problem 97.

Tutorial Problem

98. Coherent monochromatic light of wavelength 600 nm is incident on a pair of slits with a separation of 0.500 mm and then falls on a screen 1.50 m past the slits. (a) Sketch the system, showing the location of some of the expected bright and dark fringes. (Locate the fringes qualitatively only, since you have not yet calculated their separations on the screen.)

(b) What is the color of this light? What is the ratio of the wavelength of the light to the slit separation? (c) What is the phase condition (the condition on the phase angles of interfering waves) if constructive interference is to occur at the angle θ? Convert this phase condition into an equation for $\sin\theta$ by using the path length difference. (d) Determine the angle (in both radians and degrees) of several bright fringes near the center of the interference pattern. (e) Determine the distance on the screen between adjacent bright fringes.

(f) Using the answer for (e), make a rough sketch of the intensity of the interference pattern versus position on the screen, from 6 mm on one side of the pattern's center to 6 mm on the other side. Mark the center with 0. Label the bright fringes B and the dark fringes D. Include a horizontal dashed line to mark the average intensity. (g) Repeat part (f), but assume radiation of wavelength 900 nm instead of 600 nm. Where is 900 nm light in the electromagnetic spectrum? *Suggestion:* You can just determine the separation of the fringes and adjust the scale of the pattern for 600 nm light. (h) The intensity of the light pattern on the screen is related to the distribution of the light energy. Explain how the law of conservation of energy applies to the two-slit interference pattern.

Answers

(b) Light of wavelength 600 nm is yellow-green. The ratio is

$$\frac{\lambda}{d} = \frac{600\text{ nm}}{0.500\text{ mm}} = 0.00120 = 1.20 \times 10^{-3}.$$

(c) Constructive interference occurs between the two slits when their phase difference is a multiple of 2π rad. Their phases involve $kx = (2\pi/\lambda)x$, where x is distance. The path length difference is $\Delta L = d \sin \theta$, where d is the distance between the slits and θ is the angle at which the waves go off toward the screen. So the phase condition for constructive interference is

$$\frac{2\pi}{\lambda}\Delta L = \frac{2\pi}{\lambda} d \sin \theta = 2\pi m,$$

or, equivalently, $\sin \theta = m\lambda/d$.

(d) The bright fringes are at angles θ given by $\sin \theta = m\lambda/d$, where $m = 0, \pm 1, \pm 2, \ldots$. So the angles corresponding to the bright fringes are

$m = 0$ $\quad \theta = 0$

$m = \pm 1$ $\quad \theta = \pm 0.00120\text{ rad} = \pm 0.069°$

$m = \pm 2$ $\quad \theta = \pm 0.00240\text{ rad} = \pm 0.138°$

(e) From the result of part (d), the angular distance between adjacent bright fringes is $\lambda/d = 0.00120$ rad. For small θ this ratio is also equal to $\Delta y/D$, where Δy is the distance between the bright fringes and D is the distance from the slits to the screen. Thus

$$\Delta y = (\lambda/d)D = (0.00120)(1.50\text{ m}) = 0.00180\text{ m} = 1.80\text{ mm}.$$

(f) See Fig. 36-57*a*.

(g) The fringes are 1.5 times as far apart as they were in part (f). See Fig. 36-57*b*. Such 900 nm light is in the near infrared, that is, in the infrared but near the visible spectrum.

(h) With one slit open, a certain amount of light energy passes through the slit and lands on the screen. At some point on the screen the light intensity is, say, I_0. With two slits open, knowing nothing about interference, we'd naively expect intensity $2I_0$ at that point. Actually, the intensity ranges from 0 to $4I_0$ on the screen. However, the average intensity is half as much, $2I_0$, so the law of conservation of energy applies. Interference redistributes the intensity (and thus the energy) but does not change the total energy.

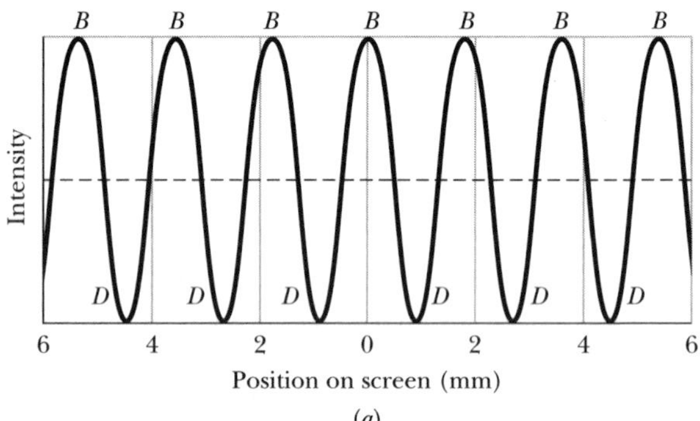

(*a*)

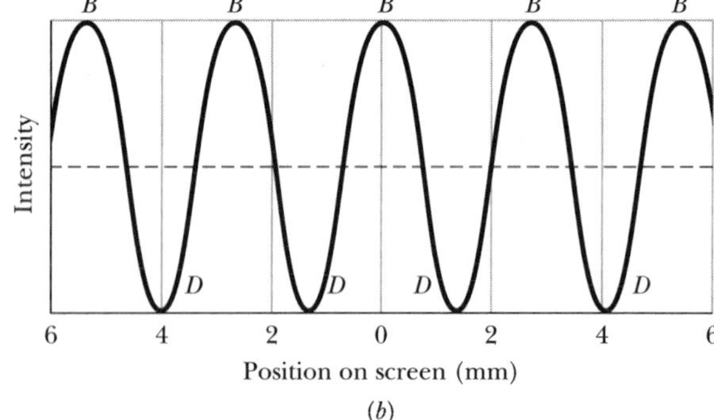

(*b*)

FIGURE 36-57 Problem 98.

Chapter Thirty-Seven
Diffraction

QUESTIONS

13. In Fig. 37-45*a,* three light waves of the same amplitude E_0 and wavelength λ are initially in phase, travel along different paths (because of reflections by mirrors for at least two of the rays), and then end up at a common point P on a distant screen. (The reflections from the mirrors do not introduce reflection phase shifts; neglect the tilt of the two tilted rays.) The other parts of the figure show three other arrangements of mirrors using the same light waves. In each arrangement distance $d = \lambda/2$. Rank the four arrangements according to the intensity of the light at the point P on the screen, greatest first.

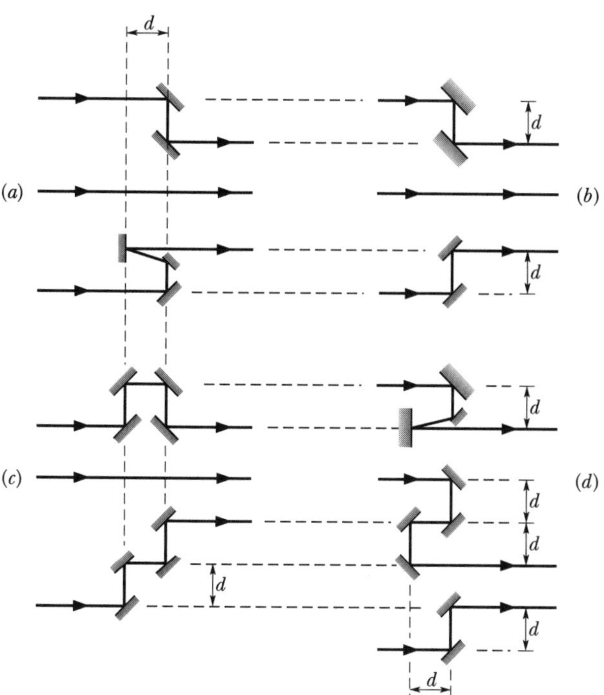

FIGURE 37-45 Question 13.

14. *Organizing question:* A slit of width a is illuminated with light of wavelength λ as in Figs. 37-4 and 37-5. However, here we mentally split the slit into three zones of width $a/3$, as in Fig. 37-46 (compare this with Fig. 37-5*b*). The parallel rays that are shown make an angle θ with the horizontal such that the path length difference between rays r_1 and r_2 is $\lambda/2$ and that between rays r_2 and r_3 is also $\lambda/2$. At that angle θ in the interference pattern is there fully constructive interference, fully destructive interference, or intermediate interference?

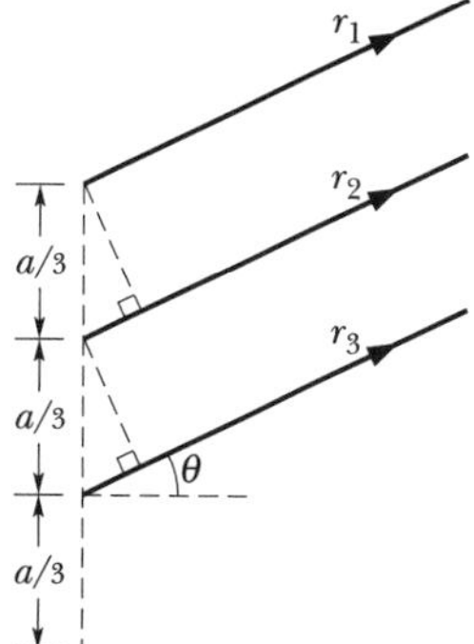

FIGURE 37-46 Question 14.

15. *Organizing question:* Figure 37-47*a* shows rays from the top and bottom of a slit in a single-slit diffraction experiment (as in Figs. 37-4 and 37-5). The rays, which are at angle θ to the horizontal, reach a distant viewing screen at the first minimum in the diffraction pattern there. (a) What is their path length difference

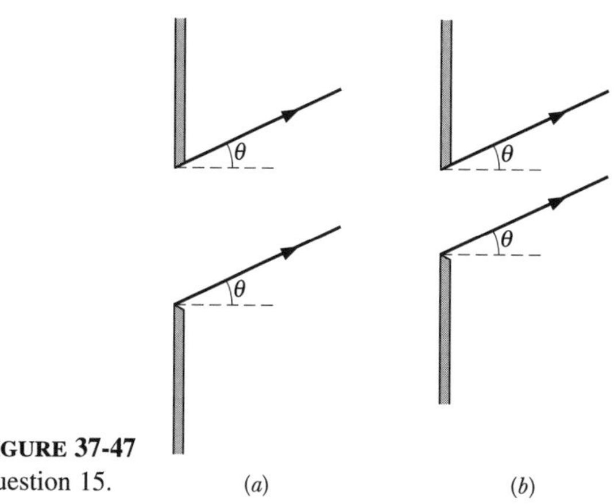

FIGURE 37-47 Question 15.

in terms of the wavelength λ of the light? The arrangement in Fig. 37-47*b* is identical (in particular, the top and bottom rays are still at angle θ), but now the slit is narrower. (b) Is the path length difference between the top and bottom rays greater than, less than, or equal to that in Fig. 37-47*a*? (c) What part of the diffraction pattern do these rays reach on the viewing screen: part of the central maximum, the first minimum, or some part of the pattern beyond the first minimum?

16. Figure 37-48 is an overhead view of ocean waves (assumed to be plane waves) approaching an opening in a breakwater. The waves diffract through the opening and then hit a beach; the erosion rate on the beach depends on the amplitude of the waves. As the waves also erode the sides of the opening in the breakwater, increasing the width of the opening, does the erosion rate near point P on the beach just opposite the opening increase, decrease, or stay the same?

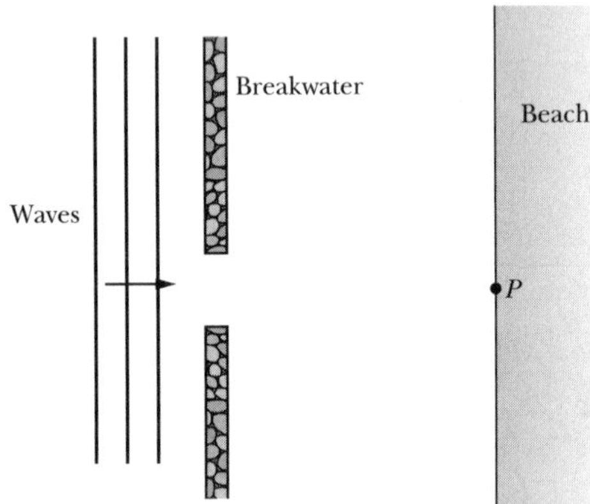

FIGURE 37-48 Question 16.

17. In a single-slit diffraction experiment, the ratio λ/a of wavelength to slit width is 1/3.5. Without written calculation or use of a calculator, determine which of the first five minima appear in the diffraction pattern.

18. In Fig. 37-49*a*, a plane wave of light is sent through a small rectangular opening in an otherwise opaque screen and the resulting diffraction pattern is studied on a distant viewing screen. The coordinate system shown has its origin at the center of that pattern. Which drawing, Fig. 37-49*b* or *c*, better represents the pattern?

19. In Eq. 37-5, intensity minima occur for $\alpha = m\pi$, as noted for Eq. 37-7. Without using a calculator, determine if an intensity maxima occurs for α equal to $\pi/2$, somewhat more than $\pi/2$, or somewhat less than $\pi/2$.

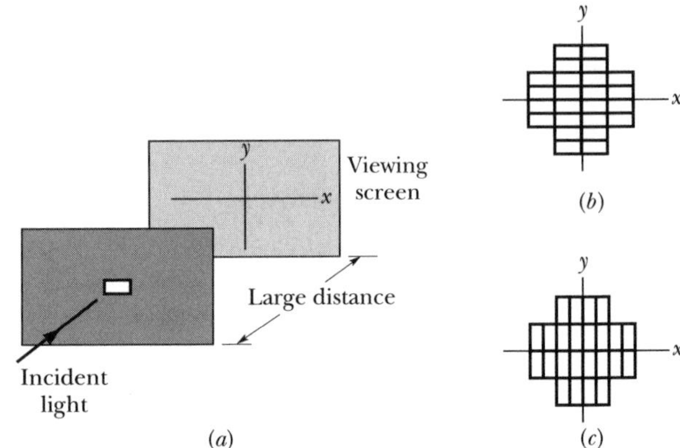

FIGURE 37-49 Question 18.

20. In three arrangements you view two closely spaced small objects that are the same large distance from you. The angles that the objects occupy in your field of view and their distances from you are the following: (1) 2ϕ and R; (2) ϕ and $2R$; (3) $\phi/2$ and $R/2$. (a) Rank the arrangements according to the separation between the objects, greatest first. If you can just barely resolve the two objects in arrangement 2, can you resolve them in (b) arrangement 1 and (c) arrangement 3?

21. The first diffraction minima in a double-slit diffraction pattern happens to coincide with the fourth side bright fringes. (a) How many bright fringes are in the central diffraction envelope? (b) To shift the coincidence to the fifth side bright fringes, should the distance between the slits be increased or decreased? (c) If that shift is, instead, made by changing the slit widths, should they be increased or decreased?

22. For a certain diffraction grating, the ratio λ/a of wavelength to ruling spacing is 1/3.5. Without written calculation or use of a calculator, determine which of the orders beyond the zeroth order appear in the diffraction pattern.

23. Figure 37-50 gives the intensity versus diffraction angle for the diffraction of a monochromatic x-ray beam by a particular family of reflecting planes in a crystal. Rank the three intensity peaks according to the associated path length differences of the x rays, greatest first.

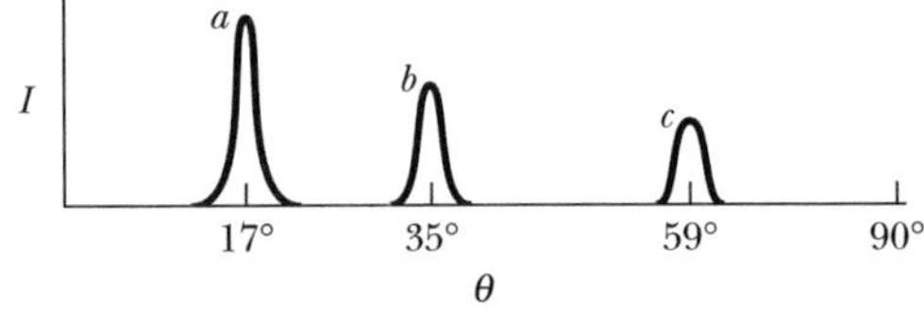

FIGURE 37-50 Question 23.

EXERCISES & PROBLEMS

85. Monochromatic light (wavelength = 450 nm) is incident perpendicularly on a single slit (width = 0.40 mm). A screen is placed parallel to the slit plane, and the distance between the two minima on either side of the central maximum is 1.8 mm. (a)

What is the distance from the slit to the screen? (*Hint:* The angle to either minimum is small enough that $\sin\theta \approx \tan\theta$.) (b) What is the distance on the screen between the first minimum and the third minimum on the same side of the central maximum?

86. In a single-slit diffraction experiment, there is a minimum of intensity for orange light ($\lambda = 600$ nm) and a minimum of intensity for blue-green light ($\lambda = 500$ nm) at the same angle of 1.00 mrad. What is the minimum width of the slit for which this is possible?

87. Light of wavelength 500 nm diffracts through a slit of width 2.00 μm and onto a screen that is 2.00 m away. On the screen, what is the distance between the center of the diffraction pattern and the third diffraction minimum?

88. Suppose that two points are separated by 2.0 cm. If they are viewed by an eye with a pupil opening of 5.0 mm, what distance from the viewer puts them at the Rayleigh limit of resolution? Assume a wavelength of light that is 500 nm.

89. If you look at something 40 m from you, what is the smallest length (perpendicular to your line of sight) that you can resolve, according to Rayleigh's criterion? Assume your pupil (opening) has a diameter of 4.00 mm, and use 500 nm as the wavelength of light.

90. In two-slit interference, if the slit separation is 14 μm and the slit widths are each 2.0 μm, then (a) how many two-slit maxima (bright bands of the two-slit pattern) are in the central diffraction envelope and (b) how many are in the first diffraction envelope to either side?

91. If there are 17 bright fringes within the central diffraction envelope in a two-slit interference pattern, then what is the ratio of slit separation to the slit width?

92. A double-slit system with individual slit widths of 0.030 mm and a slit separation of 0.18 mm is illuminated with 500 nm light directed perpendicular to the plane of the slits. What is the total number of complete bright fringes appearing between the two first-order minima of the diffraction pattern? (Do not count the fringes that coincide with the minima of the diffraction pattern.)

93. A diffraction grating having 180 lines/mm is illuminated with a light signal containing only two wavelengths, $\lambda_1 = 400$ nm and $\lambda_2 = 500$ nm. The signal is incident perpendicularly on the grating. (a) What is the angular separation for the second-order maxima of these two wavelengths? (b) What is the smallest angle at which two of the resulting maxima are superimposed? (c) What is the highest order for which maxima for both wavelengths are present?

94. How many orders to one side of the central fringe can be produced of the entire visible spectrum (400–700 nm) by a grating of 500 lines/mm?

95. A diffraction grating has 8900 slits across 1.20 cm. If light with a wavelength of 500 nm is sent through it, how many orders (maxima) lie to one side of the central maximum?

96. A diffraction grating has 200 lines/mm. Light consisting of a continuous range of wavelengths between 550 nm and 700 nm is incident perpendicular on the grating. (a) What lowest order spectrum is overlapped by another spectrum? (b) What is the highest order for which the complete spectrum is present?

97. A diffraction grating with a width of 2.0 cm contains 1000 lines/cm across that width. For a wavelength of 600 nm, what is the smallest wavelength difference this grating can resolve in the second order?

98. A beam of x rays with all wavelengths between 0.12 nm and 0.070 nm scatters from a family of planes whose separation is 0.25 nm. It is observed that scattered beams are produced for 0.10 nm and 0.075 nm. What is the angle between the incident and scattered beams?

Tutorial Problems

99. Coherent monochromatic light of wavelength 600 nm is incident on a pair of slits of width 0.100 mm and center-to-center separation 0.500 mm. The light then falls on a screen 1.50 m on the other side of the slits.

(a) Make a rough sketch of this system. (b) Determine the angle (in radians and degrees) of the first few diffraction minima, taking $\theta = 0$ at the center of the pattern. (c) Determine the distance on the screen, from the central diffraction maximum, of the first two diffraction minima on either side. (d) There is no simple expression for the locations of the diffraction maxima, but you should be able to estimate them. Where, approximately, is the first noncentral diffraction maximum for this system?

(e) The intensity of the diffraction pattern is given by

$$I(\theta) = I_0\left(\frac{\sin^2[(\pi a \sin\theta)/\lambda]}{[(\pi a \sin\theta)/\lambda]^2}\right),$$

where I_0 is the intensity at the center of the pattern. Use this to determine the approximate intensity ratios $I(\theta)/I_0$ near the first and second diffraction maxima.

(f) Using the results of (d) and (e), sketch the intensity of the diffraction pattern versus position on the screen, from 12 mm on one side of the pattern's center to 12 mm on the other side. Mark the center with 0. Put in the label of I_0. (g) Now make a similar graph for the intensity of the double-slit pattern for 600 nm light (it should show the effects of both interference between the slits and diffraction through each slit). (h) What would change in the graph of (f) if you changed the separation of the slits but not their widths? What would change if you changed the widths but not the separation?

Answers

(b) The diffraction minima occur at angles θ given by

$$\sin\theta = \frac{m\lambda}{a} = m\frac{600 \text{ nm}}{0.100 \text{ mm}} = m(0.00600 \text{ m}) = m(6.0 \text{ mm}),$$

where $m = \pm 1, \pm 2, \ldots$, (but not 0). The angles corresponding to these minima are

$$m = \pm 1 \quad \theta = \pm 0.0060 \text{ rad} = \pm 0.344°$$
$$m = \pm 2 \quad \theta = \pm 0.0120 \text{ rad} = \pm 0.688°$$

(c) From the result of part (b), the angular distance from the central diffraction maximum to the first two diffraction minima on either side are $\pm\lambda/a = 0.0060$ rad and $\pm 2\lambda/a = 0.0120$ rad. For small θ these equal $\pm\Delta y/D$, where Δy is the distance on the screen from the central diffraction maximum and D is the distance from the slits to the screen. Thus

$$\Delta y = \left(\frac{\lambda}{a}\right)D = (0.0060)(1.50\text{ m}) = 0.0090\text{ m} = 9.00\text{ mm}$$

on either side of the center for the first diffraction minimum, and

$$2\Delta y = 2\left(\frac{\lambda}{a}\right)D = (0.0120)(1.50\text{ m}) = 0.0180\text{ m} = 18.0\text{ mm}$$

on either side of the center for the second diffraction minimum.

(d) The angle θ should be approximately (but not exactly) halfway between the diffraction minima:

$$\frac{0.0060\text{ rad} + 0.0120\text{ rad}}{2} = 0.0090\text{ rad}.$$

In terms of distance, this would be about

$$\frac{9.0\text{ mm} + 18.0\text{ mm}}{2} = 13.5\text{ mm}$$

on either side of the central maximum.

(e) The argument $(\pi a \sin\theta)/\lambda$ is approximately 1.5π for the first diffraction maximum and approximately 2.5π for the second, leading to the following ratios:

First diffraction maximum:

$$I(\theta)/I_0 \approx [\sin^2(1.5\pi)]/[1.5\pi]^2 = 1/[1.5\pi]^2 = 0.045$$

Second diffraction maximum:

$$I(\theta)/I_0 \approx [\sin^2(2.5\pi)]/[2.5\pi]^2 = 1/[2.5\pi]^2 = 0.016$$

(f) See Fig. 37-51*a*.

(g) See Fig. 37-51*b*.

(h) Changing the separation of the slits but not their widths would affect the location of the bright and dark interference fringes but not the diffraction envelope. A smaller separation would spread out the interference fringes, and a larger separation would bring them closer together. Changing the widths of the slits but not their separation would change the diffraction envelope but not the locations of the bright and dark interference fringes. Smaller slits would spread out the diffraction pattern, and wider slits would narrow it.

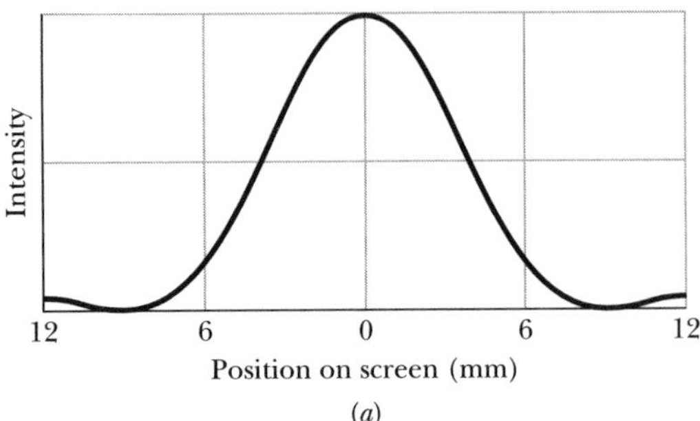

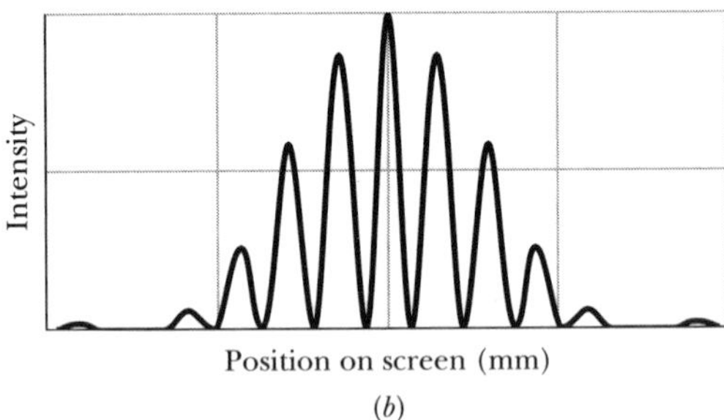

FIGURE 37-51 Problem 99.

100. In this problem we look at the expression for the intensity of light in a single-slit diffraction pattern and show how it can be used to determine the positions of the central diffraction maximum (trivial), the diffraction minima (easy), and the secondary diffraction maxima (not so easy). We then calculate the positions of the first few maxima and minima for a specific example.

The intensity of the single-slit diffraction pattern is given by

$$I = I_m\left(\frac{\sin\alpha}{\alpha}\right)^2$$

where

$$\alpha = \frac{\pi a \sin\theta}{\lambda}.$$

(a) What is the physical meaning of the quantity α? (b) Use the first two terms of the Taylor series expansion for $\sin\alpha$ (see Appendix E of the textbook) to determine the intensity as a function of α for small values of α. What is the limit of the intensity as α approaches 0? Does the intensity have an extremum (minimum or maximum) at $\alpha = 0$? (c) Locate the minima of the diffraction pattern in terms of α and θ.

(d) Between each pair of diffraction minima on either side of the central diffraction maximum there is a secondary maximum. It is approximately (but not exactly) halfway between the diffraction minima. From your study of differential calculus, you should know that the locations of these secondary maxima can be found by differentiating the expression for $I(\alpha)$ with respect to α and setting the derivative equal to 0. Differentiate the expression for $I(\alpha)$ and show that setting it equal to 0 enables you to locate the central maximum, the diffraction minima, and the secondary maxima. Show that the secondary maxima occur at the values of α for which $\tan\alpha = \alpha$ (where α is in radians).

(e) Plot the functions $\tan\alpha$ and α on the same graph and mark their intersections. Choose a scale showing at least two intersections. Explain why the values of α at the intersections are the values of α at the secondary maxima. Explain how the graph shows that the secondary maxima are a little less than an odd multiple of $\pi/2$. (f) Consider a slit of width 0.400 mm illuminated by monochromatic light of wavelength 550 nm. List the values of θ (in radians) for the first few diffraction maxima and minima, including the central maximum. On a screen 80.0 cm away, how far apart are the diffraction minima?

Answers

(a) Quantity α is the phase difference of the waves between the middle and either end of the slit. The α appears because it determines whether the Huygens wavelets over one-half of the slit interfere constructively or destructively with those on the other half.

(b) The general expansion is

$$\sin x = x - \frac{1}{6}x^3 + \cdots,$$

so

$$\frac{\sin \alpha}{\alpha} = 1 - \frac{1}{6}\alpha^2 + \cdots$$

and

$$I = I_m\left(\frac{\sin \alpha}{\alpha}\right)^2 = I_m(1 - \frac{1}{6}\alpha^2 + \cdots)^2 = I_m(1 - \frac{1}{3}\alpha^2 + \cdots).$$

The limit as $\alpha \to 0$ is $I = I_m$. That point is a maximum, since the next term is negative.

(c) The minima of the intensity occur where $(\sin \alpha/\alpha) = 0$. These coincide with the points where $\sin \alpha = 0$, with the exception of $\alpha = 0$ (which we found in part (a) to be a maximum). Thus, $\sin \alpha = 0$ (with $\alpha \neq 0$) when $\alpha = m\pi$ and $m = \pm 1, \pm 2, \pm 3, \ldots$. The corresponding values of the angle θ are $\theta = \sin^{-1}(\alpha\lambda/\pi a) = \sin^{-1}(m\lambda/a)$.

(d) First, let's calculate the derivative of $I(\alpha) = I_m(\sin \alpha/\alpha)^2$:

$$\frac{d}{d\alpha}I(\alpha) = I_m\frac{d}{d\alpha}\left(\frac{\sin \alpha}{\alpha}\right)^2 = 2I_m\left(\frac{\sin \alpha}{\alpha}\right)\frac{d}{d\alpha}\left(\frac{\sin \alpha}{\alpha}\right)$$

$$= 2I_m\left(\frac{\sin \alpha}{\alpha}\right)\left(\frac{\cos \alpha}{\alpha} - \frac{\sin \alpha}{\alpha^2}\right)$$

$$= \frac{2I_m}{\alpha^3}(\sin \alpha)(\alpha \cos \alpha - \sin \alpha).$$

This derivative is zero if either

(1) $\sin \alpha = 0$, which is the condition for the diffraction minima (and, in the case of $\alpha = 0$, for the central diffraction maximum), or

(2) $\alpha \cos \alpha - \sin \alpha = 0$, which is the condition for the secondary diffraction maxima.

The condition for the secondary maxima is then $\alpha \cos \alpha - \sin \alpha = 0$, which is equivalent to $\tan \alpha = \alpha$.

(e) In Figs. 37-52*a* and *b*, the plot of the function α is just a straight line. The function $\tan \alpha$ has branches since the function has singularities at the values $\alpha = \pi/2, 3\pi/2, 5\pi/2, \ldots = 1.57, 4.71, 7.85$ rad, ... The intersections of the two functions are the values of α for which $\alpha = \tan \alpha$, so they are the values of α meeting the condition (derived in part (c)) for the locations of the secondary maxima. These intersections can occur only just before odd multiples of $\pi/2$ because that is where $\tan \alpha$ is a large positive quantity. From Fig. 37-52*a* we see that the first of the secondary maxima occurs for $\alpha \approx 4.5$ rad. Thus for that maximum,

$$\alpha = \frac{\pi a \sin \theta}{\lambda} \approx 4.5 \text{ rad.}$$

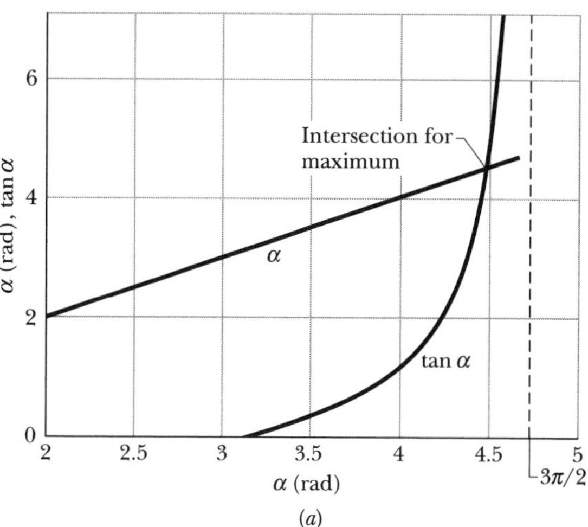

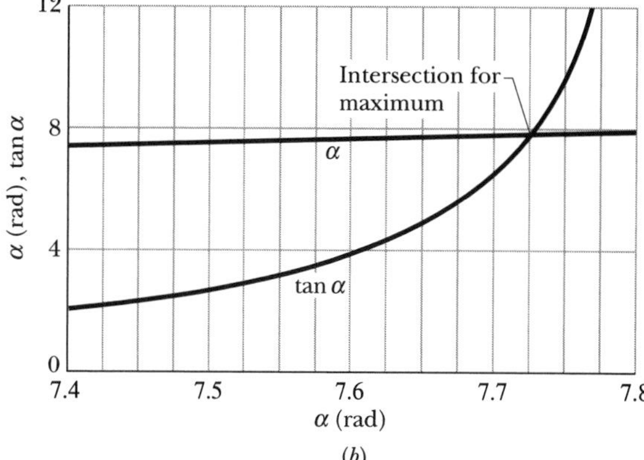

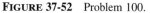

FIGURE 37-52 Problem 100.

Similarly, from Fig. 37-52*b*, we see that the next secondary maximum occurs for

$$\alpha = \frac{\pi a \sin \theta}{\lambda} \approx 7.25 \text{ rad.}$$

(f) Central maximum: $\theta = 0$;

First minimum:

$$\theta = \sin^{-1}(\lambda/a) = \sin^{-1}(550 \text{ nm}/0.400 \text{ mm}) = 0.0138 \text{ rad;}$$

First secondary maximum:

$$\theta = \sin^{-1}(1.43\lambda/a) = 0.00197 \text{ rad;}$$

Second minimum:

$$\theta = \sin^{-1}(2\lambda/a) = 0.00275 \text{ rad;}$$

Second secondary maximum:

$$\theta = \sin^{-1}(2.31\lambda/a) = 0.00318 \text{ rad;}$$

Third minimum:

$$\theta = \sin^{-1}(3\lambda/a) = 0.00413 \text{ rad.}$$

The diffraction minima are separated by 0.001375 rad, which on a screen 80.0 cm away would mean that they are separated by (0.001375)(80.0 cm) = 0.11 cm = 1.1 mm.

Chapter Thirty-Eight
Special Theory of Relativity

QUESTIONS

15. The plane of clocks and measuring rods in Fig. 38-27 is like that in Fig. 38-3. The clocks along the x axis are separated (center to center) by 1 light-second, as are the clocks along the y axis, and all the clocks are synchronized via the procedure described in Section 38-3. When the initial synchronizing signal of $t = 0$ from the origin reaches (a) clock A, (b) clock B, and (c) clock C, what initial time is then set on those clocks? An event occurs at clock A when it reads 10 s. (d) How long does the signal of that event take to travel to an observer stationed at the origin? (e) What time does that observer assign to the event?

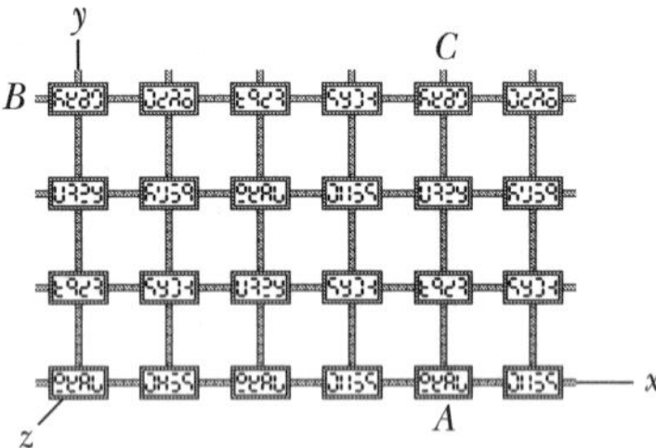

FIGURE 38-27 Question 15.

16. In the situation of Fig. 38-4, if, instead, the light from event Red and the light from event Blue had reached Sally simultaneously, which of those events would have occurred first according to Sam?

17. In Checkpoint 1, suppose that the laser pulse returns to the hobo via a reflection from a mirror at the left end of the boxcar. (a) Is the hobo's measurement of the flight time of the pulse (from its emission to its return to him) a proper time? (b) Are his measurement and our measurement of that flight time related by Eq. 38-8?

18. (a) In the muon experiment described on p. 965, if the muon speed were, instead, $0.9995c$ relative to the laboratory clocks, would the average muon lifetime measured in the laboratory frame then be greater than or less than the stated 63.5 μs? (b) Would the distance it could travel in the laboratory frame then be greater than, less than, or the same as in the actual experiment?

19. *Organizing question:* An observer in frame S' of Fig. 38-9 measures two events as occurring at the following locations and times:

event Yellow	5.0 m and 20 ns
event Green	−2.0 m and 45 ns

The velocity of S' relative to frame S is $0.90c$. Set up equations, complete with known data, to find (a) the displacement of event Yellow from event Green and (b) the corresponding temporal separation between the events, both according to an observer in frame S.

20. *Organizing question:* Figure 38-28 shows four situations in which a reference frame S' moves with speed $0.60c$ either leftward or rightward (as indicated by vector **v**) relative to reference frame S. In each situation a particle moves either leftward or rightward (as indicated by vector **u**$'$) with speed $0.70c$ relative to S'. For each situation, set up equations, complete with known data, to find the velocity **u** of the particle relative to frame S.

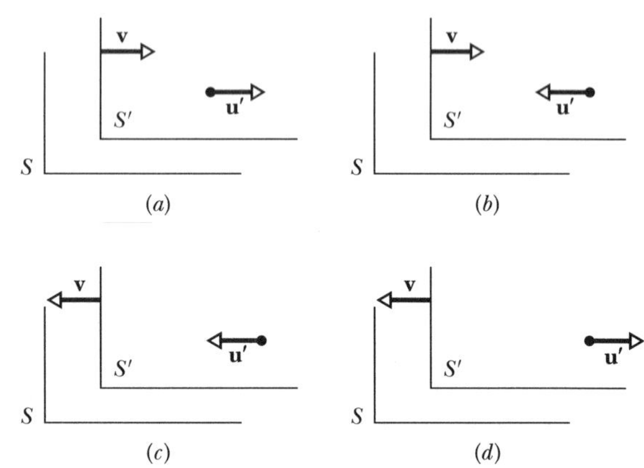

FIGURE 38-28 Question 20.

21. The elementary particle K-zero (K^0) can spontaneously decay to (suddenly transform into) pions π via the reaction $K^0 \rightarrow \pi^+ + \pi^-$. Assume that the K^0 is stationary. (a) Is the mass of the K^0 greater than, less than, or equal to the sum of the masses of the π^+ and π^-? Can the angle between the paths of the pions be (b) 10°, (c) 90°, and (d) 180°? (e) Can the magnitude of the linear momentum of the π^+ be greater than that of the π^-, or must the two linear momenta be equal in magnitude?

22. Three electrons have the following initial speeds: (1) $0.20c$, (2) $0.40c$, and (3) $0.90c$. Work is to be done on them to increase their total energy by 2.0 keV. Rank them according to (a) the required work and (b) the resulting change in the magnitude of the linear momentum, greatest first.

23. The following chart gives the initial speed and the desired decrease in speed for five electrons. Rank the electrons according to the energy they must lose to achieve the desired decrease in speed, greatest loss first.

ELECTRON	1	2	3	4	5
SPEED	$0.25c$	$0.25c$	$0.32c$	$0.90c$	$0.90c$
SPEED DECREASE	$0.02c$	$0.01c$	$0.02c$	$0.02c$	$0.03c$

24. Figure 38-29 shows the triangle of Fig. 38-14 for six particles; the slanted lines 2 and 4 have the same length. Rank the particles according to (a) their masses, (b) the magnitudes of their momenta, and (c) their Lorentz factors, greatest first. (d) Identify which two particles have the same total energy. (e) Rank the particles of the lower mass according to their kinetic energy, greatest first.

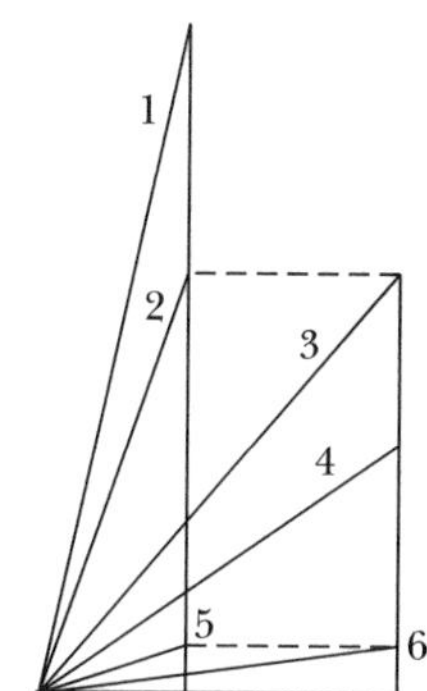

FIGURE 38-29 Question 24.

EXERCISES & PROBLEMS

68. An elementary particle produced in an experiment travels 0.230 mm through the lab at a relative speed of $0.960c$ before it decays (becomes another particle). (a) What is the proper lifetime of the particle? (b) What was the distance as measured from the rest frame of the particle?

69. The premise of the *Planet of the Apes* movies and book is that hibernating astronauts travel into Earth's future far enough that the human civilization has been replaced with an ape civilization. Considering just special relativity, determine how far into Earth's future the astronauts travel if they sleep for 120 y while traveling relative to Earth with a speed of $0.9990c$, first outward from Earth and then back again.

70. Bullwinkle in coordinate system S' of Fig. 38-9 passes you in coordinate system S with a relative speed of $0.9990c$. (a) If Bullwinkle carries a stick that is 2.50 m long according to him and aligned parallel to his direction of motion, what is the length of the stick according to you?

He measures that two events occur at the following locations and times:

event alpha	4.0 m and 40 ns
event beta	− 4.0 m and 80 ns

According to your measurements, (b) what is the distance between the events, (c) what is their temporal separation, and (d) which event occurred first?

71. In Question 7 and Fig. 38-19, the ship passes us with speed $0.950c$, the proton has speed $0.980c$ relative to the ship, and the proper length of the ship is 760 m. What are the temporal separations between the two events (the firing of the proton and the impact of the proton) according to (a) a passenger in the ship and (b) us? Suppose that, instead, the proton is fired from the rear toward the front. What then are the temporal separations according to (c) the passenger and (d) us?

72. A Foron cruiser moves directly toward a Reptulian scout ship when it fires a decoy toward the scout ship. Relative to the scout ship, the speed of the decoy is $0.980c$ and the speed of the Foron cruiser is $0.900c$. What is the speed of the decoy relative to the cruiser?

73. Bullwinkle chases Rocky along an x axis in our reference frame S. Relative to us, Bullwinkle has velocity $0.800c\mathbf{i}$ and Rocky has velocity $0.990c\mathbf{i}$. What is the velocity of Rocky relative to Bullwinkle?

74. A light source travels directly away from you at speed $v = 0.500c$, emitting light at frequency 2.00×10^{14} Hz as measured in its own rest frame. What is the wavelength (in meters) of the light as measured by you?

75. A proton passes us with speed v and a total energy of 14.242 nJ. What is v? (*Hint:* Use the proton mass given in Appendix B under "Best Value" and not the commonly remembered rounded number.)

76. What is the momentum in MeV/c of an electron with a kinetic energy of 2.00 MeV?

77. An alpha particle is a helium nucleus (with two protons and two neutrons) and has a mass of 3727 MeV/c^2. What is the particle's linear momentum (in mega-electron-volts per speed of light) if the particle is accelerated from rest by an electric potential difference of 300 MV?

78. How much work is needed to accelerate a proton from a speed of $0.9850c$ to a speed of $0.9860c$?

79. When spacecraft were lifted from Florida to the Moon in the 1960s and 1970s, each was propelled by engines providing about

10^{14} J. Suppose that the technology could be improved enough to provide 1000 times as much energy. If that much energy were used to propel a small spacecraft of 6.60×10^4 kg, about how much time would be required for the spacecraft to make a trip from Earth to Bernard's Star, which is 5.9 ly distant, according to (a) an Earth observer and (b) someone on the spacecraft?

80. If we intercept an electron with total energy of 1533 MeV that came from Vega, which is 26 ly from us, how far in light-years was the trip according to the rest frame of the electron?

81. A particle with a total energy of 2000 MeV and a rest energy of 40 MeV travels along a diameter of a galaxy that has a proper length of 4.5×10^5 ly. (a) What is the length of the path traveled as measured in the frame of the particle? How long did the passage take as measured in the rest frame of (b) the particle and (c) the galaxy?

82. As you read this book, a cosmic ray proton passes along the left-right width of the book with relative speed v and a total energy of 14.24 nJ. According to your measurements, that left-right width 21.0 cm. (a) What is the width according to the proton's reference frame? How long did the passage take according to (b) your frame and (c) the proton's frame?

Tutorial Problems

83. Consider two inertial reference frames (IRF) S and S' whose origins coincide at time $t = 0.00$ s and whose Cartesian coordinate axes are parallel ($x \| x'$, $y \| y'$, and $z \| z'$). Suppose that the origin of the IRF S' is moving with a constant velocity $\mathbf{v} = \frac{1}{2}c\mathbf{i}$ relative to the origin of the IRF S. We are interested in how we can transform data from S to S' (or vice versa) using Galilean transformations and Lorentz transformations.

(a) Write the Galilean transformations that relate (x', y', z', t') to (x, y, z, t) and vice versa. (b) Write the Lorentz transformation that relates (x', y', z', t') to (x, y, z, t). (c) Suppose an event E_1 occurs at $(x_1, y_1, z_1, t_1) = (2.00 \text{ m}, 3.00 \text{ m}, 4.00 \text{ m}, 0.00 \text{ s})$ and another event E_2 occurs at $(x_2, y_2, z_2, t_2) = (3.00 \text{ m}, 4.00 \text{ m}, 5.00 \text{ m}, 1.00 \text{ s})$. Using both Galilean and Lorentz transformations, determine the coordinates of these two events in the IRF S'. (d) How far apart (both spatially and temporally) were two events as measured by an observer stationary in the IRF S? (e) Using both Galilean and Lorentz transformations, determine how far apart (both spatially and temporally) these two events were as measured by an observer stationary in the IRF S'.

(f) Comment on the differences (if any) between your answers to parts (d) and (e). (g) If the speed v had been much smaller, say, only a few meters per second, how different would the results have been between the Galilean transformation separations and the Lorentz transformation separations? Which, if either, can be used all the time without error?

Answers

(a)

$$x' = x - \tfrac{1}{2}ct \qquad y' = y \qquad z' = z \qquad t' = t$$
$$x = x' + \tfrac{1}{2}ct' \qquad y = y' \qquad z = z' \qquad t = t'$$

(b) First, let's compute

$$\gamma = \frac{1}{\sqrt{1-(v/c)^2}} = \frac{1}{\sqrt{1-0.25}} = 1.15.$$

Then the Lorentz transformations relating S' to S are

$$x' = \gamma(x - v_0 t) = \gamma(x - \tfrac{1}{2}ct) = 1.15x - 0.58ct$$
$$y' = y$$
$$z' = z$$
$$t' = \gamma(t - xv_0/c^2) = 1.15(t - x/2c).$$

(c) The coordinates in S' of event E_1 are, with the Galilean transformations,

$$x_1' = 2.00 \text{ m} - 0.5(3.00 \times 10^8 \text{ m/s})(0.00 \text{ s}) = 2.00 \text{ m}$$
$$y_1' = y_1 = 3.00 \text{ m}$$
$$z_1' = z_1 = 4.00 \text{ m}$$
$$t_1' = t_1 = 0.00 \text{ s},$$

and, with the Lorentz transformations,

$$x_1' = 1.15[2.00 \text{ m} - 0.5(3.00 \times 10^8 \text{ m/s})(0.00 \text{ s})] = 2.30 \text{ m}$$
$$y_1' = y_1 = 3.00 \text{ m}$$
$$z_1' = z_1 = 4.00 \text{ m}$$
$$t_1' = 1.15[0.00 \text{ s} - (2.00 \text{ m})/2(3.00 \times 10^8 \text{ m/s})] \approx 0.00 \text{ s}.$$

The coordinates in S' of event E_2 are, with the Galilean transformations,

$$x_2' = 3.00 \text{ m} - 0.5(3.00 \times 10^8 \text{ m/s})(1.00 \text{ s}) \approx -1.50 \times 10^8 \text{ m}$$
$$y_2' = y_2 = 4.00 \text{ m}$$
$$z_2' = z_2 = 5.00 \text{ m}$$
$$t_2' = t_2 = 1.00 \text{ s},$$

and, with the Lorentz transformations,

$$x_2' = 1.15[3.00 \text{ m} - 0.5(3.00 \times 10^8 \text{ m/s})(1.00 \text{ s})] \approx -1.72 \times 10^8 \text{ m}$$
$$y_2' = y_2 = 4.00 \text{ m}$$
$$z_2' = z_2 = 5.00 \text{ m}$$
$$t_2' = 1.15[1.00 \text{ s} - (3.00 \text{ m})/2(3.00 \times 10^8 \text{ m/s})] \approx 1.15 \text{ s}.$$

(d) To the observer in S, the spatial distance between these two events was

$$\sqrt{(x_2 - x_1)^2 + (y_2 - y_1)^2 + (z_2 - z_1)^2} = \sqrt{(1.00 \text{ m})^2 + (1.00 \text{ m})^2 + (1.00 \text{ m})^2} = \sqrt{3.00 \text{ m}^2} = 1.73 \text{ m},$$

and the temporal separation was $t_2 - t_1 = 1.00 \text{ s} - 0.00 \text{ s} = 1.00 \text{ s}$.

(e) To the observer in S', the spatial distance between these two events was, with the Galilean transformation,

$$\sqrt{(x_2' - x_1')^2 + (y_2' - y_1')^2 + (z_2' - z_1')^2} = \sqrt{(-1.5 \times 10^8 \text{ m} - 2.00 \text{ m})^2 + (1.00 \text{ m})^2 + (1.00 \text{ m})^2} \approx 1.50 \times 10^8 \text{ m},$$

and with the Lorentz transformation,

$$\sqrt{(x_2' - x_1')^2 + (y_2' - y_1')^2 + (z_2' - z_1')^2}$$
$$= \sqrt{(-1.72 \times 10^8 \text{ m} - 2.30 \text{ m})^2 + (1.00 \text{ m})^2 + (1.00 \text{ m})^2}$$
$$\approx 1.72 \times 10^8 \text{ m}.$$

To the observer in S', the temporal separation between these two events was, with the Galilean transformation,

$$t_2' - t_1' = 1.00 \text{ s} - 0.00 \text{ s} = 1.00 \text{ s},$$

and with the Lorentz transformation,

$$t_2' - t_1' = 1.15 \text{ s} - 0.00 \text{ s} = 1.15 \text{ s}.$$

(f) The difference in spatial separations is enormous, regardless of whether the Galilean or Lorentz transformation is used, but that is intuitively what we would expect because of the rapid relative motion of the two IRFs. The spatial separation observed in S' is slightly larger with the Lorentz transformation.

With the Galilean transformation, there is no difference in temporal separations as measured by the two observers. With the Lorentz transformation there is a small but significant difference, the temporal separation being slightly larger in S'. This is not an intuitive result, but is correct according to the laws of physics.

(g) As the speed between the IRFs decreases, there is less and less difference between the separations in the two transformations. The Lorentz transformation is always correct, but for small values of v the difference becomes negligible.

84. In elementary particle physics it is convenient to measure energies in electron-volts (eV) rather than joules. One electron-volt is the work done by an electric force acting on a charge e accelerated through a potential of one volt:

$$1 \text{ eV} = 1.602 \times 10^{-19} \text{ J}.$$

SI prefixes are used with eV; for example, 1 keV = 10^3 eV = 1.602×10^{-16} J. A "1 keV electron" is an electron that has been given a kinetic energy of 1 keV.

Consider three electrons: a 1 keV electron, a 1 MeV electron, and a 1 GeV electron (remember the SI prefixes: M = 10^6 and G = 10^9). (a) Determine each electron's kinetic energy in both electron-volts and joules. (b) Determine each electron's mass energy in both electron-volts and joules. The mass of an electron is 9.11×10^{-31} kg. (c) Determine the total energy of each electron in both electron-volts and joules. (d) Determine the magnitude of the linear momentum of each electron in SI units for the energies given. (*Hint:* Start with the relativistic expression relating energy, mass, and momentum, and rewrite it to give the magnitude of the momentum in terms of the other quantities.)

(e) Prove that the speed v of a particle is related to its energy E and the magnitude p of its linear momentum by the expression $v = pc^2/E$. (f) Use the expression given in part (e) to determine the speed (in meters per second) of each electron. (g) Determine whether each electron is *nonrelativistic, relativistic,* or *extremely relativistic.* A particle is considered nonrelativistic if $mc^2 \gg K$, extremely relativistic if $K \gg mc^2$, and relativistic in intermediate situations.

Answers

(a) For the 1 keV electron,

$$K = 10^3 \text{ eV} = 1.60 \times 10^{-16} \text{ J}.$$

For the 1 MeV electron,

$$K = 10^6 \text{ eV} = 1.60 \times 10^{-13} \text{ J}.$$

For the 1 GeV electron,

$$K = 10^9 \text{ eV} = 1.60 \times 10^{-10} \text{ J}.$$

(b) The mass energy is mc^2, independent of the velocity and kinetic energy of the electron, so in every case

$$mc^2 = (9.11 \times 10^{-31} \text{ kg})(3.00 \times 10^8 \text{ m/s})^2 = 8.20 \times 10^{-14} \text{ J}.$$

Dividing this by 1.602×10^{-19} J/eV gives 0.512×10^6 eV, or 0.512 MeV.

(c) For the 1 keV electron,

$$E = mc^2 + K$$
$$= (0.512 + 0.001) \text{ MeV} = 0.513 \text{ MeV} = 8.22 \times 10^{-14} \text{ J}.$$

For the 1 MeV electron,

$$E = mc^2 + K$$
$$= (0.512 + 1.000) \text{ MeV} = 1.512 \text{ MeV} = 2.42 \times 10^{-13} \text{ J}.$$

For the 1 GeV electron,

$$E = mc^2 + K$$
$$= (0.512 + 1000) \text{ MeV} \approx 1000 \text{ MeV} = 1.60 \times 10^{-10} \text{ J}.$$

(d) From $E^2 = (mc^2)^2 + (pc)^2$ we can write the magnitude of the linear momentum as

$$p = \sqrt{(E/c)^2 - (mc)^2}.$$

However, it is more useful to rewrite this in terms of the kinetic energy $K = E - mc^2$, which leads to

$$p = \sqrt{2mK + (K/c)^2}.$$

For the 1 keV electron,

$$p = [2(9.11 \times 10^{-31} \text{ kg})(1.60 \times 10^{-16} \text{ J}) + (1.60 \times 10^{-16} \text{ J})^2/(3.00 \times 10^8 \text{ m/s})^2]^{0.5}$$
$$= 1.71 \times 10^{-23} \text{ kg} \cdot \text{m/s}.$$

For the 1 MeV electron,

$$p = 7.59 \times 10^{-22} \text{ kg} \cdot \text{m/s}.$$

For the 1 GeV electron,

$$p = 5.34 \times 10^{-19} \text{ kg} \cdot \text{m/s} \approx \frac{K}{c} \approx \frac{E}{c}.$$

(e) We can write

$$\frac{pc}{E} = \frac{(\gamma m v)c}{\gamma m c^2} = \frac{v}{c},$$

so

$$v = \frac{pc^2}{E}.$$

(f) For the 1 keV electron,

$$v = 1.87 \times 10^7 \text{ m/s}.$$

For the 1 MeV electron,

$$v = 2.82 \times 10^8 \text{ m/s}.$$

For the 1 GeV electron,

$$v = 3.00 \times 10^8 \text{ m/s (to three significant figures)}.$$

(g) By comparing the results in parts (a) and (b) we can see that a 1 keV electron is nonrelativistic, a 1 MeV electron is relativistic, and a 1 GeV electron is extremely relativistic. This classification is consistent with part (c), where $E \approx mc^2$ for the 1 keV electron and $E \approx K$ for the 1 GeV electron. It is also consistent with part (f), where $v < c/10$ for the 1 keV electron and $v \approx c$ for the 1 GeV electron.

Graphing Calculators

PROBLEM SOLVING TACTICS

Tactic 1: *Programs for γ and v*

In the homework you are often given the relative velocity v between two reference frames and asked to find the Lorentz factor γ, or vice versa. To save time and keystroking, write a program VGA (for Velocity and GAmma) that prompts you for a value of v in terms of c and then displays the corresponding value of γ. Similarly, write a program GAV that prompts you for a value of γ and then displays the corresponding value of v in terms of c or meters per second.

TACTIC 2: *Mass–Energy Conversions*

The conversion menu on a TI-85/86 allows you to switch quickly between joules and electron-volts. You might like to store the following additional conversions between mass and energy units as constants under the USER portion of the CONS menu:

$$1 \text{ u} = 931.4943 \text{ MeV}$$

which you can store as 931.4943 under the name UMEV, and

$$1 \text{ kg} = 5.609586 \times 10^{29} \text{ MeV},$$

which you can store as 5.609586E29 under the name KMEV

You can then use these constants in the home screen. For example, you can determine the rest energy E_0 of an electron from its mass in kilograms (which is a built-in constant called Me) by pressing in Me*KMEV and then pressing ENTER. You'll find that $E_0 = 0.511$ MeV.